D1200371

Document and Image Compression

Signal Processing and Communications

1. Digital Signal Processing for Multimedia Systems, *edited by Keshab K. Parhi and Takao Nishitani*
2. Multimedia Systems, Standards, and Networks, *edited by Atul Puri and Tsuhan Chen*
3. Embedded Multiprocessors: Scheduling and Synchronization, *Sundararajan Sriram and Shuvra S. Bhattacharyya*
4. Signal Processing for Intelligent Sensor Systems, *David C. Swanson*
5. Compressed Video over Networks, *edited by Ming-Ting Sun and Amy R. Reibman*
6. Modulated Coding for Intersymbol Interference Channels, *Xiang-Gen Xia*
7. Digital Speech Processing, Synthesis, and Recognition: Second Edition, Revised and Expanded, *Sadaoki Furui*
8. Modern Digital Halftoning, *Daniel L. Lau and Gonzalo R. Arce*
9. Blind Equalization and Identification, *Zhi Ding and Ye (Geoffrey) Li*
10. Video Coding for Wireless Communication Systems, *King N. Ngan, Chi W. Yap, and Keng T. Tan*
11. Adaptive Digital Filters: Second Edition, Revised and Expanded, *Maurice G. Bellanger*
12. Design of Digital Video Coding Systems, *Jie Chen, Ut-Va Koc, and K. J. Ray Liu*
13. Programmable Digital Signal Processors: Architecture, Programming, and Applications, *edited by Yu Hen Hu*
14. Pattern Recognition and Image Preprocessing: Second Edition, Revised and Expanded, *Sing-Tze Bow*
15. Signal Processing for Magnetic Resonance Imaging and Spectroscopy, *edited by Hong Yan*
16. Satellite Communication Engineering, *Michael O. Kolawole*
17. Speech Processing: A Dynamic and Optimization-Oriented Approach, *Li Deng*
18. Multidimensional Discrete Unitary Transforms: Representation: Partitioning and Algorithms, *Artyom M. Grigoryan, Sos S. Agaian, S.S. Agaian*

19. High-Resolution and Robust Signal Processing, *Yingbo Hua, Alex B. Gershman and Qi Cheng*
20. Domain-Specific Processors: Systems, Architectures, Modeling, and Simulation, *Shuvra Bhattacharyya; Ed Deprettere; Jurgen Teich*
21. Watermarking Systems Engineering: Enabling Digital Assets Security and Other Applications, *Mauro Barni, Franco Bartolini*
22. Biosignal and Biomedical Image Processing: MATLAB-Based Applications, *John L. Semmlow*
23. Broadband Last Mile Technologies: Access Technologies for Multimedia Communications, *edited by Nikil Jayant*
24. Image Processing Technologies: Algorithms, Sensors, and Applications, *edited by Kiyoharu Aizawa, Katsuhiko Sakaue and Yasuhito Suenaga*
25. Medical Image Processing, Reconstruction and Restoration: Concepts and Methods, *Jiri Jan*
26. Multi-Sensor Image Fusion and Its Applications, *edited by Rick Blum and Zheng Liu*
27. Advanced Image Processing in Magnetic Resonance Imaging, *edited by Luigi Landini, Vincenzo Positano and Maria Santarelli*
28. Digital Video Image Quality and Perceptual Coding, *edited by H.R. Wu and K.R. Rao*
29. Document and Image Compression, *edited by Mauro Barni*

Document and Image Compression

edited by
Mauro Barni

Taylor & Francis
Taylor & Francis Group
Boca Raton   London   New York

CRC is an imprint of the Taylor & Francis Group,
an informa business

Published in 2006 by
CRC Press
Taylor & Francis Group
6000 Broken Sound Parkway NW, Suite 300
Boca Raton, FL 33487-2742

International Standard Book Number-10: 0-8493-3556-6 (Hardcover)
International Standard Book Number-13: 978-0-8493-3556-3 (Hardcover)

Taylor & Francis Group
is the Academic Division of Informa plc.

**Visit the Taylor & Francis Web site at
http://www.taylorandfrancis.com**

**and the CRC Press Web site at
http://www.crcpress.com**

Dedication

For
Franco

Preface

Still image coding is one of the oldest research topics in the image processing field. Activity in this area has spanned the last four decades, producing a huge volume of scientific literature and an innumerable amount of algorithms and techniques, each with their own merits and drawbacks. Starting from the mid-1980s, an intense standardization activity has been carried out, leading to the release of the worldwide popular JPEG standard in 1992. Since then, the JPEG format has reached an ubiquitous diffusion, becoming the privileged means to exchange images stored in digital format. Along with JPEG, other coding standards have been developed, like the JBIG standard, just to mention one, that meet the requirements set by particular applications. Spurred by new advances in image processing and the availability of increasing computational power, a new standardization activity began in the mid-1990s that resulted in the release of two new coding standards: the JPEG-LS and JPEG 2000. Image coding is still a mature research field, leading one to assume that little room exists for new significant advances. In this framework, a question naturally arises: Is image coding research still worth it? As strange as it may seem, the answer is definitely yes. The reasons for such an answer are rather simple and are basically the same as those that spurred the research in the last decades: The computing power made available by technological advancements is continuously increasing, hence making it possible for the development of new approaches that were unfeasible until a few years ago. At the same time, new image representation tools are being developed whose exploitation for image coding looks particularly promising. Finally, new application areas are appearing that call for the development of ad hoc image coding algorithms.

Having answered the above basic question about the usefulness of image coding research, it is also possible to answer a second question: Is a new book on still image coding worth it? Once again the answer is yes. One reason is that the process that led to the abandonment of the old block-DCT approach that is typical of the JPEG standard is not properly covered by most image coding textbooks. Second, new promising research directions are still appearing, making image coding research as exciting as it has been during the last decades. Finally, while many books deal with the application of image coding to a particular application scenario, it is not easy to find a reference in which several such scenarios are gathered together. It is the aim of this book to provide a snapshot of the latest advancements in the image coding field, providing an up-to-date reference of the best performing schemes proposed so far, while at the same time highlighting some new research directions that are likely to assume great importance in the years to come. In this spirit, this book is targeted to scientists and practitioners who, while possessing an elementary knowledge of conventional image coding theory including basic notions of image processing and information theory, want to keep abreast of the latest results in the field.

The book is organized into three parts. Part I describes the current state of the art of image coding. The first three chapters of this part are of a general nature. Chapter 1 ("Multiresolution Analysis for Image Compression," by Alparone, Argenti, and Bianchi) provides an introduction to multiresolution image representation and its application to image coding, with multiresolution analysis being the common denominator of most of the post-JPEG image coding algorithms. Chapter 2 ("Advanced Modeling and Coding Techniques for Image Compression" by Taubman) presents advanced coding and modeling techniques that are applied in some of the newest image compression algorithms. Chapter 3 ("Perceptual Aspects of Image Coding," by Neri, Carli, and Mitra) deals with perceptual issues in image coding, a topic that has received renewed interest in recent years. The next two chapters focus on two classes of coding schemes; the former ("The JPEG Family of Coding Standards," by Magli) presents the JPEG family of coding standards with particular attention to JPEG 2000 and the latter ("Lossless Image Coding," by Forchhammer and Memon) discusses lossless image coding, a topic that has also received increased interest in recent years due to its importance in applications like remote sensing and biomedical imaging. The first part of the book ends with a chapter on fractal image coding ("Fractal Image Compression," by Hamzaoui and Saupe), a completely different approach to the coding problem that was developed during the 1990s.

The second part of the book introduces the reader to four new research fields: image coding by means of image representation paradigms that go beyond the wavelet-based framework ("Beyond Wavelets: New Image Representation Paradigms," by Führ, Demaret, and Friedrich); image coding by means of redundant dictionaries, an approach whose interest is going to increase with the availability of increasing computing power ("Image Coding Using Redundant Dictionaries," by Vandergheynst and Frossard); application of distributed source coding paradigms to image compression ("Distributed Compression of Field Snapshots in Sensor Networks," by Servetto); and the exploitation of novel data-hiding techniques to improve the effectiveness of current codecs both in terms of coding efficiency and robustness ("Data Hiding for Image and Video Coding," by Campisi and Piva).

The third and last part of the book is devoted to the description of coding techniques expressly developed to compress particular classes of images, to which the general-purpose algorithms could not be applied successfully. In particular, Chapter 11 ("Binary Image Compression," by Boncelet) discusses the compression of binary images, Chapter 12 ("Two-Dimensional Shape Coding," by Ostermann and Vetro) deals with the coding of binary shapes, and Chapter 13 ("Compressing Compound Documents," by de Queiroz) presents coding schemes available to efficiently compress compound documents, i.e., images depicting both natural scenes and text. In Chapter 14 ("Trends in Model-Based Coding of Multidimensional Medical Data," by Menegaz) the compression of biomedical images is addressed, whereas in Chapter 15 ("Remote-Sensing Image Coding," by Aiazzi, Baronti, and Lastri) coding of remote sensing imagery is considered. The book ends with a chapter ("Lossless Compression of VLSI Layout Image Data," by Dai and Zakhor) devoted to the compression of VLSI image data.

As is evident from the outline given above, this book is the result of the efforts of several researchers spread all over the world. I am grateful to all of them for agreeing to share their expertise and time. Without them, this book would have never been possible. I also thank my wife, Francesca, and my sons, Giacomo, Margherita, and Simone; even if they did not actively participate in the preparation of the book, their closeness has been an invaluable help to me. Finally, I would like to thank my colleague and friend, Franco Bartolini, who suddenly passed away on January 1, 2004 at a young age of 39. As with many other initiatives, we conceived this book together, though in the end Franco was not able to accompany me in this project. There is still much of his view in it.

Mauro Barni
Siena

About the Editor

Mauro Barni has carried out his research activities for over 14 years, first at the Department of Electronics and Telecommunication of the University of Florence, then at the Department of Information Engineering of the University of Siena, where he presently works as Associate Professor. During the last decade, his research activity has focused on digital image processing. More recently, he has been studying the application of image processing techniques to copyright protection and authentication of multimedia (digital watermarking), and to the transmission of images and video signals in error-prone, wireless environments. He is the author/co-author of more than 160 papers published in international journals and conference proceedings. He is the co-author of the book *Watermarking Systems Engineering: Enabling Digital Assets Security and Other Applications*, published by Dekker Inc. in February 2004.

He serves as the associate editor of *IEEE Signal Processing Letters*, the *IEEE Signal Processing Magazine* (column and forum section), the *IEE Proceedings on Information Security*, the Springer *LNCS Transactions on Data Hiding and Multimedia Security*, and is on the editorial board of the *Eurasip Journal of Applied Signal Processing*. He has served as an associate editor of the *IEEE Transactions on Multimedia* and was part of the conference board of the IEEE Signal Processing Society. Professor Barni is a member of the IEEE Multimedia Signal Processing Technical Committee (MMSP-TC) and the Technical Committee on Information Forensics and Security (IFS-TC) of the IEEE Signal Processing Society.

Contributors

Bruno Aiazzi
IFAC-CNR: Institute of Applied Physics
  "N. Carrara"
National Research Council
Rome, Italy

Luciano Alparone
Department of Electronics and
  Telecommunications
University of Florence
Rome, Italy

Fabrizio Argenti
Department of Electronics and
  Telecommunications
University of Florence
Rome, Italy

Stefano Baronti
IFAC-CNR: Institute of Applied Physics
  "N. Carrara"
National Research Council
Rome, Italy

Tiziano Bianchi
Department of Electronics and
  Telecommunications
University of Florence
Rome, Italy

Charles Boncelet
Department of Electrical and Computer
  Engineering
University of Delaware
Newark, Delaware, USA

Patrizio Campisi
Department of Applied Electronics
University of Rome TRE
Rome, Italy

Marco Carli
Department of Applied Electronics
University of Rome TRE
Rome, Italy

Vito Dai
Department of Electrical Engineering and
  Computer Science
University of California
Berkeley, California, USA

Laurent Demaret
GSF National Research Center for Environment
  and Health
IBB Institute of Biomathematics and Biometry
Neuherberg, Germany

Søren Forchhammer
Research Center COM Department of
  Communications, Optics and Materials
Technical University of Denmark
Copenhagen, Denmark

Felix Friedrich
GSF National Research Center for Environment
  and Health
IBB Institute of Biomathematics and Biometry
Neuherberg, Germany

Pascal Frossard
Signal Processing Institute
Swiss Federal Institute of Technology
EPFL Ecole Polytechnique Fédérale de
  Lausanne
Zurich, Switzerland

Hartmut Führ
GSF National Research Center for Environment
  and Health
IBB Institute of Biomathematics and Biometry
Neuherberg, Germany

Raouf Hamzaoui
Department of Computer and Information
  Science
University of Konstanz
Konstanz, Germany

Cinzia Lastri
IFAC-CNR: Institute of Applied Physics
  "N. Carrara"
National Research Council
Rome, Italy

Enrico Magli
Department of Electronics
Politecnico di Torino
Torino, Italy

Nasir Memon
Department of Computer Science
Polytechnic University
Brooklyn, New York, USA

Gloria Menegaz
Department of Information Engineering
University of Siena
Siena, Italy

Sanjit K. Mitra
Department of Electrical and Computer
    Engineering
University of California
Santa Barbara, California, USA

Alessandro Neri
Department of Applied Electronics
University of Rome TRE
Rome, Italy

Joern Ostermann
Institute of Theoretical Communications
    Engineering and Informations, Verarbeitung
University of Hannover
Hannover, Germany

Alessandro Piva
Department of Electronics and
    Telecommunications
University of Florence
Florence, Italy

Ricardo L. de Queiroz
University of Brasilia
Brasilia, Brasil

Dietmar Saupe
Department of Computer and Information
    Science
University of Konstanz
Konstanz, Germany

Sergio D. Servetto
School of Electrical and Computer Engineering
Cornell University
Ithaca, New York, USA

David Taubman
School of Electrical Engineering and
    Telecommunications
University of New South Wales
Sydney, Australia

Pierre Vandergheynst
Signal Processing Institute
Swiss Federal Institute of Technology
EPFL Ecole Polytechnique Fédérale de
    Lausanne
Zurich, Switzerland

Anthony Vetro
Mitsubishi Electric Research Labs
Murray Hill
New Jersey, USA

Avideh Zakhor
Department of Electrical Engineering and
    Computer Science
University of California
Berkeley, California, USA

Table of Contents

Part I State of the Art

Chapter 1
Multiresolution Analysis for Image Compression 3
Luciano Alparone, Fabrizio Argenti, and Tiziano Bianchi

Chapter 2
Advanced Modeling and Coding Techniques for Image Compression 35
David Taubman

Chapter 3
Perceptual Aspects of Image Coding 69
Alessandro Neri, Marco Carli, and Sanjit K. Mitra

Chapter 4
The JPEG Family of Coding Standards 87
Enrico Magli

Chapter 5
Lossless Image Coding 113
Søren Forchhammer and Nasir Memon

Chapter 6
Fractal Image Compression 145
Raouf Hamzaoui and Dietmar Saupe

Part II New Directions

Chapter 7
Beyond Wavelets: New Image Representation Paradigms 179
Hartmut Führ, Laurent Demaret, and Felix Friedrich

Chapter 8
Image Coding Using Redundant Dictionaries 207
Pierre Vandergheynst and Pascal Frossard

Chapter 9
Distributed Compression of Field Snapshots in Sensor Networks 235
Sergio D. Servetto

Chapter 10
Data Hiding for Image and Video Coding 255
Patrizio Campisi and Alessandro Piva

Part III Domain-Specific Coding

Chapter 11
Binary Image Compression 285
Charles Boncelet

Chapter 12
Two-Dimensional Shape Coding 299
Joern Ostermann and Anthony Vetro

Chapter 13
Compressing Compound Documents 323
Ricardo L. de Queiroz

Chapter 14
Trends in Model-Based Coding of Multidimensional Medical Data 351
Gloria Menegaz

Chapter 15
Remote-Sensing Image Coding 389
Bruno Aiazzi, Stefano Baronti, and Cinzia Lastri

Chapter 16
Lossless Compression of VLSI Layout Image Data 413
Vito Dai and Avideh Zakhor

Index 427

Part I

State of the Art

1 Multiresolution Analysis for Image Compression

Luciano Alparone, Fabrizio Argenti, and Tiziano Bianchi

CONTENTS

1.1   Introduction 3
1.2   Wavelet Analysis and Filter Banks 4
    1.2.1   Continuous Wavelet Transform and Frames 5
    1.2.2   Multiresolution Spaces 5
    1.2.3   Multiresolution Analysis and Filter Banks 6
    1.2.4   Orthogonal and Biorthogonal Filter Banks 8
    1.2.5   Reconstruction at Boundaries 9
    1.2.6   Lifting Scheme 11
        1.2.6.1   Implementing Wavelet Transforms via Lifting 12
    1.2.7   Wavelet Decomposition of Images 14
    1.2.8   Quantization of Mallat's DWT 15
1.3   Enhanced Laplacian Pyramid 16
    1.3.1   Quantization of ELP with Noise Feedback 18
1.4   Coding Schemes 20
    1.4.1   Embedded Zero-Tree Wavelet Coder 20
    1.4.2   Set Partitioning in Hierarchical Trees Coder 22
    1.4.3   Embedded Block Coding with Optimized Truncation 26
    1.4.4   Content-Driven ELP Coder 27
        1.4.4.1   Synchronization Tree 29
    1.4.5   Results 30
1.5   Conclusions 32
References 32

1.1   INTRODUCTION

With the advent of the multimedia era and the growth of digital packet networks, the total amount of image data accessed and exchanged by users daily has reached the huge value of several petabytes. Therefore, the compression of continuous-tone still images, either grayscale or color, has grown tremendously in importance.

Compression algorithms are said to be reversible, or *lossless*, when the images that are reconstructed from the coded bit stream are identical to the originals, or *lossy*. The difference in performance expressed by the compression ratio between lossy and lossless algorithms can be of one order of magnitude without a significant perceptual degradation. For this reason, lossy algorithms are extremely interesting and are used in all those applications in which a certain distortion may be tolerated. Whenever reversibility is recommended, compression ratios higher than 2 can hardly be obtained because the attainable bit rate is lower bounded by the entropy of the imaging sensor noise [34].

The classical image compression scheme consists of a decorrelator, followed by a quantizer and an entropy coding stage. The purpose of the decorrelator is to remove spatial redundancy; hence it must be tailored to the specific characteristics of the data to be compressed. Examples are orthogonal transforms, e.g., discrete cosine transforms (DCTs) and discrete wavelet transforms (DWTs) [25], and differential pulse code modulations (DPCMs), either *causal*, i.e., prediction-based, or *noncausal*, i.e., interpolation-based, or *hierarchical* [4]. The quantizer introduces a distortion to allow a decrement in the entropy rate to be achieved. Once a signal has been decorrelated, it is necessary to find a compact representation of its coefficients, which may be sparse data. Eventually, an entropy coding algorithm is used to map such coefficients into codewords in such a way that the average codeword length is minimized.

After the introduction of DWTs, it was noticed that a full-frame DWT allows long-range correlation to be effectively removed, unlike DCTs in which full-frame processing leads to a spread of energy in the transformed plane due to the fact that DCTs are not suitable for the analysis of nonstationary signals. In fact, DCTs are not usually applied to the full frame, but only to small blocks in which the assumption of stationarity approximately holds [32]. Hence, they fail in exploiting long-range correlations and can effectively remove only short-range correlations. The DC component of each block that is coded stand-alone (e.g., by spatial DPCMs) is a typical drawback of first-generation transform coders, e.g., JPEGs [31]. The emerging standard JPEG 2000 [19] was devised to overcome such limitations and to cope with scalability issues dictated by modern multimedia applications, thereby leading to substantial benefits over JPEGs.

Another purpose of this chapter is to compare "early" wavelet coding with "second-generation" wavelet coding. Image coding was one of the first applications of the newly discovered wavelet theory [5]. The reason for this was that wavelet analysis was very similar to the well-established subband analysis, which meant that the techniques of subband coding [16] could be directly applied to wavelet coding. More recent wavelet coders use techniques that are significantly different from those of subband coding and are based on ideas originating with embedded zero-tree wavelets (EZWs) [36]. The main innovation is that coding of sparse data (nonzero coefficients after quantization) is expedited by exploiting the self-similarity of details across scales [33,43] or, equivalently, the absence of details across scales [35,36].

The remainder of the chapter is organized as follows. Section 1.2 reviews the theoretical fundamentals of wavelet analysis and discusses the basic properties of the wavelet transform that are pertinent to image compression. The material in this section builds on the background material in generic transform coding. This section shows that boundary effects motivate the use of biorthogonal wavelets and introduces the lifting paradigm as a promising means to implement wavelet transforms. Optimal, in the minimum mean-squared error (MMSE) sense, quantization strategies of wavelet coefficients are reported as well. Section 1.3 presents the enhanced Laplacian pyramid (ELP), a redundant multiresolution decomposition suitable for lossless/near-lossless compression through quantization noise feedback loops at intermediate resolution levels. Section 1.4 describes the second-generation wavelet coders that extend the ideas found in the last decade as well as a lesser-known encoder that exploits inter- and intrascale statistical dependencies of ELP coefficients. Wavelet and ELP coders are assessed together with two state-of-the-art spatial DPCM schemes on the test image *Lenna*. Concluding remarks are drawn in Section 1.5.

1.2 WAVELET ANALYSIS AND FILTER BANKS

The theoretical fundamentals of multiresolution analyses will be briefly reviewed in this section. Following Mallat's approach, a basis for the space of square summable functions satisfying certain multiresolution properties is constructed based on *dilations* and *translations* of a lowpass *scaling function* $\phi(t)$ and a bandpass *wavelet function* $\psi(t)$. This multiresolution representation is strictly related to the continuous wavelet transform (CWT) used for analyzing a continuous function $f(t)$

and is defined starting only from the function $\psi(t)$. In the following, first the CWT is introduced and then a rapid tour of the principal properties of multiresolution representations and filter bank theory is given.

1.2.1 Continuous Wavelet Transform and Frames

Let $L^2(\mathbb{R})$ denote the Hilbert space of real, square summable functions, with a scalar product $\langle f, g \rangle = \int f(x)g(x)\,dx$. The CWT represents the projection of $f(t)$ onto the function $\psi_{a,b}(t) = 1/\sqrt{a}\psi((t-b)/a)$, that is,

$$W_f(a,b) = \langle f, \psi_{a,b} \rangle = \int_{-\infty}^{\infty} f(t)\frac{1}{\sqrt{a}}\,\psi^*\left(\frac{t-b}{a}\right)dt \tag{1.1}$$

where a and b are the *scale* and *shift* parameters, respectively. The transform is invertible and $f(t)$ can be reconstructed from $W_f(a,b)$. According to the uncertainty principle, the analyzing functions $\psi_{a,b}(t)$ are characterized by a variable time and frequency support and, hence, by a variable time and frequency resolution capability. This trade-off is ruled by the choice of the scale parameter a. Thanks to this property, the CWT is able to capture fast-decaying transients and resolve (long-duration) tones very close in frequency.

As can be seen, however, a two variate transform is associated to a one-dimension function. From a compression point of view, this is not an appealing property. It is apparent that such a representation is redundant, and, if appropriate conditions are met, a sampling of the parameters a and b does not affect the reconstruction property. The set of coefficients obtained by discretizing the scale and shift parameters, a and b, is called a *frame*. If we let $a = a_0^j$, $b = kb_0a_0^j$, and $\psi_{j,k}(t) = a_0^{-j/2}\psi(ta_0^{-j} - kb_0)$, $j, k \in \mathbb{Z}$, it is possible to demonstrate that, with reasonable choice of $\psi(t)$, a_0, and b_0, there exist dual wavelet functions $\tilde{\psi}_{j,k}(t)$ such that

$$f(t) = \sum_{j,k} \langle f, \psi_{j,k} \rangle \tilde{\psi}_{j,k}(t) \tag{1.2}$$

Under suitable hypotheses, the sets of functions $\{\psi_{j,k}(t), j, k \in \mathbb{Z}\}$ and $\{\tilde{\psi}_{j,k}(t), j, k \in \mathbb{Z}\}$ constitute a basis for $L^2(\mathbb{R})$, so that they yield a minimum redundancy representation for $f(t)$. The basis can be *orthogonal*, and in this case $\tilde{\psi}_{j,k}(t)$ coincide with $\psi_{j,k}(t)$. The bases are *biorthogonal* if they satisfy $\langle \psi_{j,k}, \tilde{\psi}_{m,n} \rangle = \delta(j-m)\delta(n-k)$. The existence of two different sets of functions, one for the analysis and one for the synthesis of $f(t)$, allows the biorthogonal bases to present more degrees of freedom than the orthogonal ones, so that special features can be added during their design. The most valuable of such features is the symmetry of the basis functions, which cannot be achieved in the orthogonal case. This property is of particular interest for image processing applications.

In the following, multiresolution spaces with specific reference to the dyadic case, i.e., an analysis whose scales vary as powers of 2, are considered. This is obtained with the choice of $a_0 = 2$ and $b_0 = 1$.

1.2.2 Multiresolution Spaces

Multiresolution analysis with J levels of a continuous signal f having finite energy is a projection of f onto a basis $\{\phi_{J,k}, \{\psi_{j,k}\}_{j \leq J}\}_{k \in \mathbb{Z}}$ [14]. First, the case of orthogonal bases is considered; then, the modifications that must be introduced in the case of biorthogonal wavelets are summarized.

The approximation of f at the scale 2^j is given by the projection of f onto the set of functions $\phi_{j,k}(t) = 2^{-j/2}\phi(2^{-j}t - k)$, obtained from the translations and dilations of the same lowpass function $\phi(t)$, called the scaling function, verifying $\int \phi(t)\,dt = 1$. The family $\{\phi_{j,k}\}_{k \in \mathbb{Z}}$ spans a subspace

$V_j \subset L^2(\mathbb{R})$ so that the projection of f onto V_j is described by the coefficients

$$\{a_{j,k} = \langle f, \phi_{j,k} \rangle\}_{k \in \mathbb{Z}} \tag{1.3}$$

Analogously, the basis functions $\psi_{j,k}(t) = 2^{-j/2} \psi(2^{-j}t - k)$ are the results of dilations and translations of the same bandpass functions $\psi(t)$, called the wavelet functions, which fulfills $\int \psi(t)\, dt = 0$. The family $\{\psi_{j,k}\}_{k \in \mathbb{Z}}$ spans a subspace $W_j \subset L^2(\mathbb{R})$. The projection of f onto W_j yields the wavelet coefficients of f, defined as

$$\{w_{j,k} = \langle f, \psi_{j,k} \rangle\}_{k \in \mathbb{Z}} \tag{1.4}$$

representing the details between two successive approximations, i.e., the data to be added to V_{j+1} to obtain V_j. Hence, W_{j+1} is the orthogonal complement of V_{j+1} in V_j, i.e., $V_j = V_{j+1} \oplus W_{j+1}$.

The subspaces V_j realize the multiresolution analysis [25]. They present several properties, among which we have

$$\cdots \subset V_1 \subset V_0 \subset V_{-1} \subset \cdots \tag{1.5}$$

and

$$f(t) \in V_{j+1} \quad \Leftrightarrow \quad f(2t) \in V_j \tag{1.6}$$

Eventually, multiresolution analysis with J levels yields the following decomposition of $L^2(\mathbb{R})$:

$$L^2(\mathbb{R}) = \left(\bigoplus_{j \leq J} W_j \right) \oplus V_J \tag{1.7}$$

As a consequence of this partition, all functions $f \in L^2(\mathbb{R})$ can be decomposed as follows:

$$f(t) = \sum_k a_{J,k}\, \phi_{J,k}(t) + \sum_{j \leq J} \sum_k w_{j,k}\, \psi_{j,k}(t) \tag{1.8}$$

In the biorthogonal case, the spaces V_j and W_j are no longer orthogonal to each other. In this case, two new functions, $\tilde{\phi}(t)$ and $\tilde{\psi}(t)$, dual to the scaling and wavelet functions $\phi(t)$ and $\psi(t)$, can be defined, and a new basis for $L^2(\mathbb{R})$ can be constructed from their dilations and translations. Similar to what was previously exposed, dual spaces $\tilde{V}_j$ and $\tilde{W}_j$ can be constructed from $\tilde{\phi}(t)$ and $\tilde{\psi}(t)$. The new basis is called *biorthogonal* if $\phi(t - k)$ is orthogonal to $\tilde{\psi}(t - k)$ and $\tilde{\phi}(t - k)$ is orthogonal to $\psi(t - k)$, $\forall k \in \mathbb{Z}$. This choice leads to $V_j \perp \tilde{W}_j$ and $W_j \perp \tilde{V}_j$.

As for CWTs, the basis $\{\phi_{J,k}, \{\psi_{j,k}\}_{j \leq J}\}_{k \in \mathbb{Z}}$ is used for analysis, whereas $\{\tilde{\phi}_{J,k}, \{\tilde{\psi}_{j,k}\}_{j \leq J}\}_{k \in \mathbb{Z}}$ is used for synthesis; in the biorthogonal case, (1.8) becomes

$$f(t) = \sum_k a_{J,k}\, \tilde{\phi}_{J,k}(t) + \sum_{j \leq J} \sum_k w_{j,k}\, \tilde{\psi}_{j,k}(t) \tag{1.9}$$

where $a_{J,k}$ and $w_{j,k}$ are defined in (1.3) and (1.4), respectively.

1.2.3 MULTIRESOLUTION ANALYSIS AND FILTER BANKS

The connection between filter banks and wavelets stems from dilation equations that allow us to pass from a finer scale to a coarser one [13,14,25]. Since $V_1 \subset V_0$, the functions $\phi(x/2)$ and $\psi(x/2)$

can be decomposed as

$$\phi(t/2) = \sqrt{2} \sum_i h_i \, \phi(t - i)$$
$$\psi(t/2) = \sqrt{2} \sum_i g_i \, \phi(t - i)$$

(1.10)

where $h_i = \langle \phi_{1,0}, \phi_{0,i} \rangle$ and $g_i = \langle \psi_{1,0}, \phi_{0,i} \rangle$. Normalization of the scaling function implies $\sum_i h_i = \sqrt{2}$. Analogously, $\int \psi(t)\,dt = 0$ implies $\sum_i g_i = 0$.

It can be shown [25] that the coefficients of the multiresolution analysis of a signal f at a resolution j can be computed from those relative to the immediately higher resolution by means of a filter bank composed of the lowpass analysis filter $\{h_i\}$ and the highpass analysis filter $\{g_i\}$, that is,

$$a_{j+1,k} = \langle f, \phi_{j+1,k} \rangle = \sum_i h_{i-2k} \, a_{j,i} = (a_{j,i} * h_{-i}) \!\downarrow\! 2$$
$$w_{j+1,k} = \langle f, \psi_{j+1,k} \rangle = \sum_i g_{i-2k} \, a_{j,i} = (a_{j,i} * g_{-i}) \!\downarrow\! 2$$

(1.11)

In other words, a coarser approximation of f at scale 2^{j+1} is provided by lowpass filtering the approximation coefficients at the scale 2^j with the filter h_{-i} and by downsampling the result by a factor 2. Wavelet coefficients at scale 2^{j+1} are obtained by highpass filtering the same sequence with the filter g_{-i} followed by downsampling by a factor 2.

The reconstruction of the approximation coefficients are derived from

$$a_{j,k} = \langle f, \phi_{j,k} \rangle = \sum_i h_{k-2i} \, a_{j+1,i} + \sum_i g_{k-2i} \, w_{j+1,i}$$
$$= (a_{j+1,i}\!\uparrow\!2) * h_i + (w_{j+1,i}\!\uparrow\!2) * g_i$$

(1.12)

In words, to reconstruct the approximation of a signal at the resolution 2^j, the approximation and detail coefficients at the immediately coarser resolution are upsampled by a factor 2 and filtered by the synthesis filters $\{h_i\}$ and $\{g_i\}$. The scheme of wavelet decomposition and reconstruction is shown in Figure 1.1. For orthogonal bases, the set of relationships between analysis and synthesis filters is completed by [25]

$$g_n = (-1)^n h_{-n+1}$$

(1.13)

The wavelet analysis can be directly applied to a discrete sequence. Consider a sampled signal $f(n) = \{f(t)|_{t=nT}\}$, where T is the sampling period. The samples can be regarded as the coefficients of the projection of a continuous function $f(t)$ onto V_0, that is, $f(n) = a_{0,n}$. The output sequences of the analysis stage represent the approximation and the detail of the input sequence $\{f(n)\}$, i.e., a smoothed version and the rapid changes occurring within the signal, respectively.

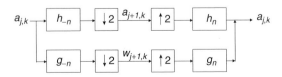

FIGURE 1.1 Dyadic wavelet analysis and synthesis.

1.2.4 ORTHOGONAL AND BIORTHOGONAL FILTER BANKS

The wavelet representation is closely related to a subband decomposition scheme [46], as can be seen from Figure 1.1. Thanks to this relationship, multirate filter banks theory and wavelet theory benefit from each other. Several results that were known in the former field have suggested further developments in the latter, and vice versa. For example, the rigorous mathematical formalism and the explanation of some properties of filter banks related to the regularity of the filter frequency responses can be borrowed from the wavelet theory. On the other side, scaling and wavelet functions can be derived from the frequency responses of the filters h_n and g_n [25], so that filter bank design methods can also be used for wavelet construction. Considering the importance of two-channel filter banks for wavelet theory, some facts about their properties are now summarized. The two-channel filter bank scheme shown in Figure 1.2(a), using z domain representation, will be taken into consideration in this section. The analysis and synthesis filters have been renamed and the notation has been slightly changed. The output sequence $\hat{x}(n)$ is the reconstruction of the input sequence $x(n)$.

Useful results can be obtained from the polyphase representation of the analysis and synthesis filters, given by

$$\begin{cases} H_0(z) = H_{0,0}(z^2) + zH_{0,1}(z^2) \\ G_0(z) = G_{0,0}(z^2) + zG_{0,1}(z^2) \end{cases} \tag{1.14}$$

$$\begin{cases} H_1(z) = H_{1,0}(z^2) + z^{-1}H_{1,1}(z^2) \\ G_1(z) = G_{1,0}(z^2) + z^{-1}G_{1,1}(z^2) \end{cases} \tag{1.15}$$

Let the analysis and synthesis polyphase matrices be defined as

$$\mathbf{E}(z) = \begin{pmatrix} H_{0,0}(z) & H_{0,1}(z) \\ G_{0,0}(z) & G_{0,1}(z) \end{pmatrix}, \quad \mathbf{R}(z) = \begin{pmatrix} H_{1,0}(z) & H_{1,1}(z) \\ G_{1,0}(z) & G_{1,1}(z) \end{pmatrix} \tag{1.16}$$

By using these definitions and the noble identities [46], which allow the order of transfer functions and upsamplers/downsamplers to be changed, the two-channel filter bank can be redrawn as shown in Figure 1.2(b).

Applying upsampler and downsamplers input/output relationships, the reconstructed signal $\hat{x}(n)$ can be written as [46,47]

$$\begin{aligned} \hat{X}(z) = {} & \tfrac{1}{2}[H_0(z)H_1(z) + G_0(z)G_1(z)]X(z) \\ & + \tfrac{1}{2}[H_0(-z)H_1(z) + G_0(-z)G_1(z)]X(-z) \end{aligned} \tag{1.17}$$

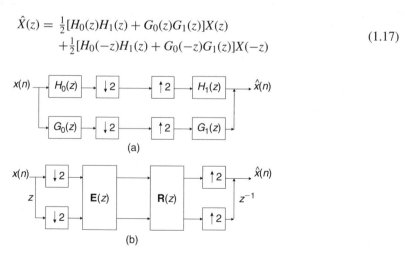

FIGURE 1.2 (a) Two-channel analysis/synthesis filter banks; (b) filter bank polyphase representation.

The second term on the right-hand side of (1.17) is called *aliasing component*. In a *perfect reconstruction* (PR) system, the aliasing component is null and the overall transfer function is equal to a constant times a delay, that is,

$$H_0(z)H_1(z) + G_0(z)G_1(z) = cz^{-k}$$

$$H_0(-z)H_1(z) + G_0(-z)G_1(z) = 0$$

(1.18)

As discussed in the previous section, an orthogonal multiresolution representation is characterized by *orthogonal filter banks*, which satisfy the following relationships:

$$H_1(z) = H_0(z^{-1})$$

$$G_1(z) = G_0(z^{-1})$$

$$G_0(z) = -z^{-1}H_0(-z^{-1})$$

(1.19)

As can be seen, all the filters of the bank can be derived from a single lowpass prototype, namely $H_0(z)$. The filter banks are PR if the following *power complementary* property is verified:

$$|H_0(\omega)|^2 + |H_0(\omega + \pi)|^2 = 1$$

(1.20)

which is obtained by substituting (1.19) into the first equation appearing in (1.18). This class of PR filter banks and methods for its design, based on (1.20), were known in the multirate signal processing community as conjugate mirror filter (CMF) banks [29,38,39].

However, using orthogonal filter banks is not the only way to obtain the PR property [26,48]. Actually, less restrictive conditions than those shown in (1.19) can be used to achieve PR. If the synthesis filter bank is derived from the analysis filter bank by

$$\begin{pmatrix} H_1(z) \\ G_1(z) \end{pmatrix} = \begin{pmatrix} cz^{2k+1}G_0(-z) \\ -cz^{2k+1}H_0(-z) \end{pmatrix}$$

(1.21)

where c is a constant and k an integer, then cancellation of the aliasing term is still verified. Let $P(z) = H_0(z)H_1(z)$. It can be shown that $G_0(z)G_1(z) = P(-z)$; hence, from the first of (1.18) we have PR if the condition $P(z) + P(-z) = 2$ holds. This constraint defines a *valid* polynomial, characterized by all even-indexed coefficients equal to zero, apart from $p(0) = 2$. The filter bank design problem may be stated as searching for a valid polynomial $P(z)$ with lowpass characteristics and, from its factorization, the filters of the bank are derived. This bank is denoted as *biorthogonal filter bank*.

The constraints imposed on orthogonal banks do not allow for the design of PR banks with filters having simultaneously the properties of linear-phase frequency response and finite impulse response (FIR). Orthogonal, linear-phase, FIR filter banks can be designed if the PR constraint is relaxed and substituted by quasi-PR, i.e., $\hat{x}(n) \approx x(n)$. Quadrature mirror filter (QMF) banks [12,21] are an example of filter banks belonging to this class, in which the prototype $H_0(z)$ satisfies only approximately the power complementary property. On the contrary, thanks to the larger number of degrees of freedom, symmetric filters are compatible with biorthogonal conditions. This fact makes this type of filter bank particularly appealing for image processing.

1.2.5 RECONSTRUCTION AT BOUNDARIES

An important problem that must be tackled when the DWT is applied to a finite length sequence is representing the signal without an expansion of the number of coefficients in the wavelet domain. Consider, for example, a finite sequence composed of L samples and a DWT implemented by a lowpass and a highpass filter, both having an N-tap impulse response. Before subsampling, the lowpass and highpass subbands are sequences composed of $L + N - 1$ samples and, after subsampling, each subband comprises $\lceil (L + N - 1)/2 \rceil$ samples. Hence, in the wavelet domain, the signal is represented by about $L + N$ samples, that is, the number of coefficients needed to represent the signal is expanded.

An appealing feature of a wavelet transform applied to finite length signals would be the use of a number of coefficients equal to the length of the signal. Different methods have been devised to solve this problem. *Boundary filters* for two-channel [18] and M-channel [17] filter banks can be constructed to satisfy the PR property at the beginning and at the end of the signal. Other methods consider a *periodic* extension of the input signal and exploit the fact that the resulting subbands are also periodic sequences. This method can be applied when both linear and nonlinear phase filters, either FIR or IIR [6,40], are used. A major problem in using a periodic extension is that the last sample of the signal becomes adjacent to the first sample, so that a discontinuity may be created. If the jump of the discontinuity is quite large, high-amplitude samples — thus, highly costly to be coded — are generated in the highpass subband at the beginning and at the end of the signal.

One way to overcome this problem is to use a *periodic symmetric* extension of the signal. Thanks to the symmetric extension, discontinuities do not appear in the periodic repetition of the signal and smoother subband samples are produced at the borders. A symmetric extension, e.g., around the origin, can be obtained by reflecting the signal around $n = 0$. If the sample in $n = 0$ is not repeated, the sequence becomes symmetric around this sample; if the sample is repeated, the symmetry axis is halfway between two samples. These two types of extensions are called *whole-sample symmetric* (WS) and *half-sample symmetric* (HS), respectively. Antisymmetric extensions can be defined in a similar way, yielding *whole-sample antisymmetric* (WA) and *half-sample antisymmetric* (HA) sequences.[1]

The period length of the extended input sequence is about $2L$ samples: different period lengths must be considered according to the different types of symmetric extensions used at both borders. The important fact is that if also the impulse responses of the wavelet filters are symmetric or antisymmetric,[2] then the lowpass and highpass filtered signals are, before subsampling, still (about) $2L$-periodic sequences. Their period is a symmetric or antisymmetric sequence, so that only (about) L samples are necessary to represent each subband. If the subsampled versions of such signals are still symmetric or antisymmetric, then they can be represented with (about) $L/2$ samples and the coefficient expansion problem does not occur.

Several studies have investigated the necessary conditions that allow the decimated subband samples to be periodic-symmetric or -antisymmetric sequences. The case of a subsampling factor $M = 2$ was first dealt with in [40] and extensions to generic values of M can be found, e.g., in [7,9,22,27]. The choice of the correct type of symmetric extensions is illustrated with an example.

Example. Consider a finite length signal composed of $L = 9$ samples. Consider a biorthogonal wavelet decomposition [14] with lowpass and highpass filters transfer functions given by[3]

$$H_0(z) = -0.354 + 1.061z^{-1} + 1.061z^{-2} - 0.354z^{-3}$$

$$G_0(z) = -0.177 + 0.530z^{-1} - 0.530z^{-2} + 0.177z^{-3}$$

(1.22)

The impulse responses are HS and HA, respectively, and have a group delay of $\tau_{H_0} = \tau_{G_0} = 3/2$ samples. The signal has been HS-extended at both ends, obtaining the 18-periodic signal shown in Figure 1.3(a). The lowpass and highpass subbands before decimation are shown in Figure 1.3(b) and Figure 1.3(d), respectively. They are 18-periodic sequences with WS and WA extensions, respectively,

---

[1] In WA extensions, the sample coinciding with the center of symmetry has a null amplitude.

[2] A symmetric or antisymmetric filter impulse response can also be classified as WS, HS, WA, or HA by observing if the center of symmetry or antisymmetry coincides with a sample or it is halfway between two samples.

[3] The filters coefficients used in this example have been obtained with the Matlab® call wfilters('bior3.1').

at both ends. This is the result of filtering the original HS sequence by means of either an HS or an HA sequence. This fact can be seen also in terms of the delay introduced by the filters: the half-sample symmetry of the left extreme of the input can be seen as a $-1/2$ delay of the symmetry axis; after the filter delay, the final position of the symmetry axis is at $n = 1$, as can be verified from Figure 1.3(b). Figure 1.3(c) and Figure 1.3(e) show the subbands obtained after decimation and upsampling by a factor 2 (upsampling has been included to improve the readability of the plots): these are the signals that will be filtered by the synthesis stage. The number of significant samples in the lowpass and highpass subbands is five and four, respectively, so that a nonexpansive transform is implemented.

1.2.6 Lifting Scheme

Construction of wavelet filters usually relies on certain polynomial factorizations in the frequency domain. The lifting scheme is an alternative approach that allows a generic biorthogonal wavelet to be defined directly in the spatial domain [41,42]. The main reason behind the introduction of the lifting scheme is its ability to define wavelets in the case of complex geometries and irregular sampling, i.e., in settings where the translation and dilation approach cannot be applied. Usually, these kinds of wavelets are referred to as *second-generation wavelets*, to distinguish them from classical wavelets, also called *first-generation wavelets*.

The typical lifting stage is composed of three steps: *split*, *predict* ($\mathcal{P}$), and *update* ($\mathcal{U}$), as shown in Figure 1.4. Consider a set of data $\mathbf{x}$, whose spatial correlation properties are known. The aim is to exploit this correlation to find a more compact representation of $\mathbf{x}$.

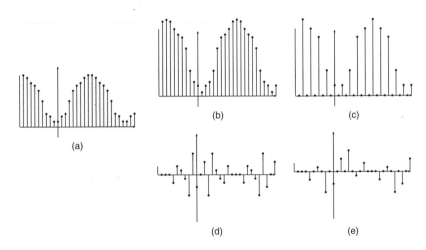

FIGURE 1.3 Signal composed of $L = 9$ samples and HS-extended at both sides. (a) Periodic symmetric extension; (b) lowpass subband before subsampling; (c) lowpass subband after subsampling and upsampling; (d) highpass subband before subsampling; and (e) highpass subband after subsampling and upsampling.

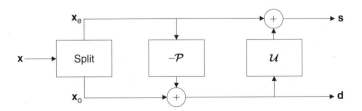

FIGURE 1.4 Canonical lifting stage.

First, the original sequence $\mathbf{x}$ is split into two nonoverlapping sets, $\mathbf{x}_e$ and $\mathbf{x}_o$ comprising the even and odd samples of $\mathbf{x}$, respectively. Owing to the structure of the data in the initial set, we can suppose that the data in $\mathbf{x}_e$ and $\mathbf{x}_o$ are highly correlated and that the values of $\mathbf{x}_o$ can be predicted from the data in $\mathbf{x}_e$.

The second step of the lifting scheme consists of a predictor $\mathcal{P}$ that processes the even part of the input data set in order to obtain a prediction for the odd part. The data in $\mathbf{x}_o$ are then replaced by the prediction residual given by

$$\mathbf{d} = \mathbf{x}_o - \mathcal{P}(\mathbf{x}_e) \tag{1.23}$$

The form of the predictor depends on some model that takes into account the correlation structure of the data. Obviously, the predictor $\mathcal{P}$ cannot depend directly on the data $\mathbf{x}$, otherwise a part of the information in $\mathbf{x}$ would be hidden in the predictor. Given $\mathbf{x}_e$, $\mathbf{d}$, and $\mathcal{P}$, we can reconstruct exactly the initial data set. Hence, the predict step preserves all information contained in $\mathbf{x}$. Moreover, if the predictor fits well the original data set, we expect the residual $\mathbf{d}$ to contain much less information than the odd data set. The new representation obtained after the predict step is indeed more compact than the initial one. However, it does not preserve some global properties of the original data. In particular, the data set $\mathbf{x}_e$ may not have the same average value as the original data set.

The lifting scheme uses a third step that updates the values of $\mathbf{x}_e$, based on the residual $\mathbf{d}$. The updated version is given by

$$\mathbf{s} = \mathbf{x}_e + \mathcal{U}(\mathbf{d}) \tag{1.24}$$

where $\mathcal{U}$ is referred to as the update operator. Usually, the update operator is so designed that $\mathbf{s}$ maintains some desired properties of the original data set.

Once the predictor $\mathcal{P}$ and the update operator $\mathcal{U}$ have been chosen, the lifting stage can be immediately inverted by reversing its operations, as shown in Figure 1.5. Given $\mathbf{d}$ and $\mathbf{s}$, the even and odd parts of the original data set can be obtained as

$$\mathbf{x}_e = \mathbf{s} - \mathcal{U}(\mathbf{d}) \tag{1.25}$$

$$\mathbf{x}_o = \mathbf{d} + \mathcal{P}(\mathbf{x}_e) \tag{1.26}$$

Therefore, these two data sets can be merged to recompose the original data set. An important property is that the inverse lifting scheme holds irrespective of the choice of predictor and update operator, i.e., the lifting scheme is implicitly biorthogonal, thereby yielding PR.

1.2.6.1 Implementing Wavelet Transforms via Lifting

The lifting scheme is a generic tool that can be applied to arbitrary data sets. Examples include the extension of the wavelet transform to nonuniformly spaced sampling and to complex geometries as the sphere [42] as well as the definition of wavelets in arbitrary dimensions [23]. However, it can also be used to implement any classical DWT or two-channel filter bank [15]. The main advantages of implementing a wavelet transform by means of lifting are the reduced complexity as well as the possibility of a fully in-place calculation of the wavelet coefficients. Furthermore, lifting can be used to define wavelet transforms that map integers to integers [11].

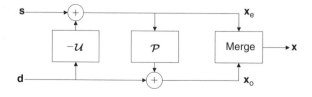

FIGURE 1.5 Inverse lifting stage.

Consider a signal $x(n)$. The even and odd polyphase components can be defined as $x_e(n) = x(2n)$ and $x_o(n) = x(2n+1)$, respectively. If the predictor and the update operator are modeled as linear time-invariant (LTI) filters with transfer functions $P(z)$ and $U(z)$, then the lifting stage is equivalent to a critically sampled two-channel perfect reconstruction filter bank defined by the pair of polyphase matrices

$$\mathbf{E}(z) = \begin{bmatrix} 1 & U(z) \\ 0 & 1 \end{bmatrix} \begin{bmatrix} 1 & 0 \\ -P(z) & 1 \end{bmatrix} \tag{1.27}$$

and

$$\mathbf{R}(z) = \begin{bmatrix} 1 & 0 \\ P(z) & 1 \end{bmatrix} \begin{bmatrix} 1 & -U(z) \\ 0 & 1 \end{bmatrix} \tag{1.28}$$

The interconnections between the lifting scheme and a two-channel filter bank are highlighted in Figure 1.6.

It is demonstrated in [15] that any two-channel perfect reconstruction filter bank with FIR filters can be factorized into a finite sequence of simple filtering steps implemented via lifting. Given an arbitrary biorthogonal wavelet transform, the synthesis polyphase matrix associated with it can always be expressed as

$$\mathbf{R}(z) = \prod_{i=1}^{m} \begin{bmatrix} 1 & 0 \\ P_i(z) & 1 \end{bmatrix} \begin{bmatrix} 1 & -U_i(z) \\ 0 & 1 \end{bmatrix} \begin{bmatrix} \gamma & 0 \\ 0 & 1/\gamma \end{bmatrix} \tag{1.29}$$

where γ is a nonzero constant. The above factorization means that any FIR filter wavelet transform can be obtained by starting with a twofold polyphase decomposition of the input signal followed by m lifting steps and a final scaling. We can start from the simplest wavelet decomposition, i.e., a polyphase decomposition, and then "lift" it through a finite number of steps until we obtain a wavelet having the desired properties.

The relationships between wavelet transforms and lifting can be explained more clearly by means of an example.

Consider the Le Gall's wavelet transform, implemented with 5- and 3-tap filters whose coefficients are shown in Table 1.1 [24]. In this case, the synthesis polyphase matrix can be factorized as

$$\begin{aligned} \mathbf{R}(z) &= \begin{bmatrix} 1 & (-2 - 2z^{-1})/8 \\ (z+1)/2 & (-z + 6 - z^{-1})/8 \end{bmatrix} \\ &= \begin{bmatrix} 1 & 0 \\ (z+1)/2 & 1 \end{bmatrix} \begin{bmatrix} 1 & (-1 - z^{-1})/4 \\ 0 & 1 \end{bmatrix} \end{aligned} \tag{1.30}$$

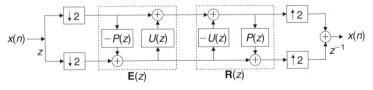

FIGURE 1.6 Analogies between lifting and filter banks.

TABLE 1.1
Le Gall's 5/3 Wavelet Filter Coefficients

	Analysis Filters		Synthesis Filters	
n	$f_L(n)$	$f_H(n)$	$g_L(n)$	$g_H(n)$
0	6/8	1	1	6/8
±1	2/8	−1/2	1/2	−2/8
±2	−1/8			−1/8

yielding the expressions $P(z) = (z + 1)/2$ and $U(z) = (1 + z^{-1})/4$. Therefore, the lifting stage that implements Le Gall's 5/3 wavelet transform is given by

$$d[n] = x[2n + 1] - \tfrac{1}{2}(x[2n + 2] + x[2n]) \tag{1.31}$$

$$s[n] = x[2n] + \tfrac{1}{4}(d[n] + d[n - 1]) \tag{1.32}$$

As can be seen, the expressions in (1.31) and (1.32) are very attractive from a computational point of view, since they can be implemented relying only on additions and divisions by powers of 2. Moreover, the lifting scheme allows an integer-to-integer version of Le Gall's transform to be derived. This transform is obtained by simply rounding the output of $P(z)$ and $U(z)$ and can be expressed as [11]

$$d[n] = x[2n + 1] - \left\lfloor \tfrac{1}{2}(x[2n + 2] + x[2n]) \right\rfloor \tag{1.33}$$

$$s[n] = x[2n] + \left\lfloor \tfrac{1}{4}(d[n] + d[n - 1]) + \tfrac{1}{2} \right\rfloor \tag{1.34}$$

where the constant term in (1.34) is needed to avoid bias effects on rounding. Note that the transform defined by (1.33) and (1.34) is no longer the equivalent of the two-channel filter bank based on Le Gall's filters because the new predictor and update operator are nonlinear. However, due to the implicit biorthogonality, and hence PR, of the lifting scheme, it can be perfectly inverted relying on the general equations in (1.25) and (1.26). This version of the Le Gall's transform has a particular relevance in practical applications, since it is used as lossless transform in the JPEG 2000 standard [37].

1.2.7 WAVELET DECOMPOSITION OF IMAGES

Image multiresolution analysis [25] can be implemented by using 2-D separable filters. Filtering and downsampling, as in the 1-D case, is applied to image rows and, successively, to image columns. Since a lowpass and a highpass subband are generated in both row and column processing, the lower resolution representation consists of four subbands, denoted with the superscripts LL, HL, LH, and HH. Let $A_j^{LL}(m, n)$ be the approximation coefficients at the jth level of decomposition, in which (m, n) is the pixel position. Relationships to obtain the $(j + 1)$ level coefficients from the jth level are

$$A_{j+1}^{LL}(m, n) = \sum_l \sum_k h_l h_k A_j^{LL}(2m + k, 2n + l)$$

$$W_{j+1}^{LH}(m, n) = \sum_l \sum_k h_l g_k A_j^{LL}(2m + k, 2n + l)$$

FIGURE 1.7 Two-level wavelet decomposition of test image *Lenna*.

$$W_{j+1}^{HL}(m,n) = \sum_l \sum_k g_l h_k A_j^{LL}(2m+k, 2n+l)$$

$$W_{j+1}^{HH}(m,n) = \sum_l \sum_k g_l g_k A_j^{LL}(2m+k, 2n+l) \tag{1.35}$$

In Figure 1.7, the wavelet decomposition applied to the image *Lenna* is shown.

1.2.8   QUANTIZATION OF MALLAT'S DWT

The output of the transform stage of a K-level octave decomposition on image data is $M = 3K + 1$ separate subbands of coefficients. By the design of an orthogonal filter bank, the coefficients in each subband are uncorrelated from coefficients in other subbands. As a result, the coefficients in each subband can be quantized independently of coefficients in other subbands without loss in performance. The variance of the coefficients in each of the subbands is typically different, and thus each subband requires a different amount of bit resources to obtain best coding performance. The result is that each subband will have a different quantizer, with each quantizer having its own separate rate (bits/sample). The only issue to be resolved is that of bit allocation, or the number of bits to be assigned to each individual subband to give the best performance.

Let us derive the solution to the bit allocation problem for the case of uniform scalar quantization in each of the subbands. The goal is to assign each subband, containing a fraction α_k of the total number of coefficients ($\sum_{k=1}^{M} \alpha_k = 1$), a bit rate, denoted as R_k bits/coefficient, such that the average rate is

$$R = \sum_{k=1}^{M} \alpha_k R_k \tag{1.36}$$

and the reconstruction distortion is minimized. If uniform scalar quantization is used, the quadratic distortion, or error energy, introduced by the quantizer in each subband can be modeled by

$$D_k = c_k \sigma_k^2 2^{-2R_k} \tag{1.37}$$

where σ_k^2 is the variance of coefficients in the kth subband, R_k the subband bit rate, and c_k a parameter which depends on the probability distribution in the subbands, e.g., $\pi e/6$ and $e^2/6$ for Gaussian and Laplacian PDF, respectively [20]. Equation (1.37) makes intuitive sense since the more bits/sample allocated to the subband (R_k), the lower the resulting distortion from that subband. Since all subbands are orthogonal to each other, the total quadratic distortion D, or reconstruction error of the wavelet coefficients, assuming the same $c_k = c$ in each subband, will be given by

$$D = c \sum_{k=1}^{M} \sigma_k^2 2^{-2R_k} \qquad (1.38)$$

Equations (1.36) and (1.38) can be combined to form a constrained minimization problem that can be solved using the Lagrange multipliers method, where the Lagrangian to be minimized is

$$J = c \sum_{k=1}^{M} \sigma_k^2 2^{-2R_k} - \lambda \left(R - \sum_{k=1}^{M} \alpha_k R_k \right) \qquad (1.39)$$

Minimization of this function results in the best bit allocation

$$R_k = R + \frac{1}{2} \log_2 \frac{\alpha_k \sigma_k^2}{\prod_{k=1}^{M} (\alpha_k \sigma_k^2)^{\alpha_k}} \qquad (1.40)$$

According to (1.40), distortion is allocated among subbands proportionally to their size, i.e., to the α_ks, to minimize the total distortion D, given the rate R. If $\alpha_k = 1/M$, $\forall k$, (1.40) reduces to

$$R_k = R + \frac{1}{2} \log_2 \frac{\sigma_k^2}{\left(\prod_{k=1}^{M} \sigma_k^2 \right)^{1/M}} \qquad (1.41)$$

For the case of a biorthogonal wavelet decomposition, it holds that $D = \sum_{k=1}^{M} w_k D_k$. In fact, depending on the synthesis filters, the distortion coming from each subband is differently weighted. Therefore the set coefficients w_k must be introduced in (1.38) and (1.39) as well. In order to expedite the distortion allocation task, biorthogonal filters yielding $w_k \approx 1$ are desirable. The widely used 9/7 biorthogonal filter set deviates by only a few percent from the orthogonal filter weighting ($w_k = 1$, $\forall k$) [45].

1.3 ENHANCED LAPLACIAN PYRAMID

In this section, another multiresolution decomposition suitable for image compression is reviewed. The main advantage with respect to Mallat's orthogonal pyramid is that the intrinsic redundancy of this structure is exploited through the design of a suitable quantizer to obtain an L_∞-bounded, or near-lossless, coding scheme, analogous to spatial DPCM, but with the extra feature of spatial scalability.

The Laplacian pyramid, originally introduced by Burt and Adelson [10] several years before wavelet analysis, is a bandpass image decomposition derived from the Gaussian pyramid (GP), which is a multiresolution image representation obtained through a recursive *reduction* (lowpass filtering and decimation) of the image data set.

A modified version of Burt's LP, known as ELP [8], can be regarded as a redundant wavelet transform in which the image is lowpass-filtered and downsampled to generate a lowpass subband, which is re-expanded and subtracted pixel by pixel from the original image to yield the 2-D detail signal having zero mean. Thus, the output of a separable 2-D filter is recursively downsampled along rows and columns to yield the next level of approximation. Again, the detail is given as the difference between the lowpass approximation at a certain scale and an expanded version of the lowpass approximation at the previous (finer) scale. Unlike the baseband approximation, the 2-D detail signal cannot be decimated if PR is desired.

The attribute enhanced depends on the zero-phase expansion filter having cutoff at exactly one half of the bandwidth, and chosen independently of the reduction filter, which may have *half-band* cutoff as well or not. The ELP outperforms the former Burt's LP [10] when image compression is concerned [4], thanks to its layers being almost completely uncorrelated with one another.

Figure 1.8 shows the GP and ELP applied of the test image *Lenna*. Notice the *lowpass* octave structure of GP layers as well as the *bandpass* octave structure of ELP layers. An octave LP is oversampled by a factor $4/3$ at most (when the baseband is 1 pixel wide). Overhead is kept moderate, thanks to decimation of lowpass components.

Burt's LP [10] achieves PR only in the case of infinite precision. When dealing with finite arithmetics, roundoff is introduced in order to yield integer-valued pyramids, denoted with superscript stars.

Let $\mathbf{G}_0^* = \{G_0^*(m,n), m = 0, \dots, M-1; n = 0, \dots, N-1\}$, $M = p \times 2^K$ and $N = q \times 2^K$ be an integer-valued image with p, q, and K being positive integers. The set of images $\{\mathbf{G}_k^*, k = 0, 1, \dots, K\}$, with $\mathbf{G}_k^* = \{G_k^*(m,n), m = 0, \dots, M/2^k - 1, n = 0, \dots, N/2^k - 1\}$, constitutes a *rounded* GP in which, for $k > 0$, $\mathbf{G}_k^*$ is a *reduced* version of $\mathbf{G}_{k-1}^*$:

$$\mathbf{G}_k^* = round[reduce(\mathbf{G}_{k-1}^*)] \tag{1.42}$$

in which roundoff to integer is applied component-wise, and reduction consists of linear lowpass filtering followed by downsampling by 2:

$$\left[reduce(\mathbf{G}_{k-1}^*)\right](m,n) \triangleq \sum_{i=-I_R}^{I_R} \sum_{j=-I_R}^{I_R} r(i) \times r(j) \times G_{k-1}^*(2m+i, 2n+j) \tag{1.43}$$

(a)   (b)

FIGURE 1.8 (a) GP and (b) ELP of *Lenna*.

where k identifies the pyramid level (K is the top or *root*). Burt's parametric kernel $\{r(i), i = -2, \ldots, 2\}$ ($I_R = 2$) has been widely used [8,10,30]:

$$R(z) \triangleq \sum_{i=-2}^{2} r(i)z^{-i} = a + 0.25(z + z^{-1}) + (0.25 - 0.5a)(z^2 + z^{-2}) \tag{1.44}$$

An integer-valued ELP [8] $\{\mathbf{L}_k^*, k = 0, 1, \ldots, K\}$, with $\mathbf{L}_k^* = \{L_k^*(m,n), m = 0, \ldots, M/2^k - 1,$ $n = 0, \ldots, N/2^k - 1\}$ may be defined from $\mathbf{G}_k^*$: for $k = K$, $\mathbf{L}_K^* \equiv \mathbf{G}_K^*$, while for $k = 0, \ldots, K - 1$,

$$L_k^*(m,n) = G_k^*(m,n) - round\{[expand(\mathbf{G}_{k+1}^*)](m,n)\} \tag{1.45}$$

in which *expansion* signifies upsampling by 2, followed by linear lowpass filtering:

$$[expand(\mathbf{G}_{k+1}^*)](m,n) \triangleq \sum_{\substack{i=-I_E}}^{I_E} \sum_{\substack{j=-I_E \\ (j+n) \bmod 2 = 0 \\ (i+m) \bmod 2 = 0}}^{I_E} e(i) \times e(j) \times G_{k+1}^*\left(\frac{i+m}{2}, \frac{j+n}{2}\right) \tag{1.46}$$

The following 7-taps half-band kernel has been employed for expansion:

$$E(z) \triangleq \sum_{i=-3}^{3} e(i)z^{-i} = 1 + b(z + z^{-1}) + (0.5 - b)(z^3 + z^{-3}) \tag{1.47}$$

Even though $I_E = 3$, the filter has the same computational cost as for Burt's kernel, thanks to the null coefficients. Analogous to the a appearing in the reduction filter (1.44), b is an adjustable parameter, which determines the shape of the frequency response (see [1,4,8]) but not the extent of the passband, which is variable for Burt's kernel [8,10]. Burt's LP uses *the same* kernel (1.44) for both reduction (1.43) and expansion (1.46), apart from a multiplicative factor to adjust the DC gain in (1.46), and does not feature roundoff to integer values in (1.42) and (1.45).

1.3.1 QUANTIZATION OF ELP WITH NOISE FEEDBACK

When quantizers at each level are not independent of each other, the analysis leading to optimal distortion allocation is more complex, because the rate distortion (RD) plot at the kth level depends on the working point at level $k + 1$, containing the lower frequency components. To explain the meaning of dependent quantizers, let us consider the Block diagram of an ELP coder with quantization noise feedback, shown in Figure 1.9. Let $Q_k(\cdot)$ indicate quantization with a step size Δ_k, $Q_k(t) = round[t/\Delta_k]$, and $Q_k^{-1}(\cdot)$ the inverse operation, $Q_k^{-1}(l) = l\Delta_k$. We immediately note that, for perfect reconstruction, an arbitrary step size Δ_k can be used for quantizing L_k^*, provided that $\Delta_0 = 1$. With nonunity step sizes, $\hat{G}_k^*$, the integer-valued GP reconstructed at the receiving end for $k < K$ will be recursively given by the sum of the expanded $\hat{G}_{k+1}^*$ and of an approximate version of $\hat{L}_k^*$ due to quantization errors, in which

$$\hat{L}_k^*(m,n) = G_k^*(m,n) - round\{[expand\{\hat{G}_{k+1}^*\}](m,n)\} \tag{1.48}$$

Since the term $\hat{G}_{k+1}^*$ recursively accounts for previous quantization errors starting from the root level K down to level $k + 1$ inclusive, setting $\Delta_0 = 1$ causes all errors previously introduced and delivered to the pyramid base to be compensated irrespective of the other step sizes. Absolute errors are upper-bounded as well, if $\Delta_0 > 1$.

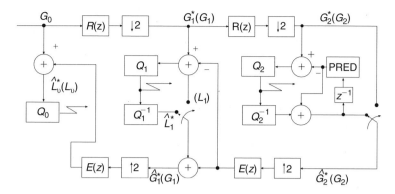

FIGURE 1.9 Block diagram of a hybrid (ELP + DPCM) encoder with switchable quantization noise feedback. The output of reduction/expansion filtering blocks, $R(z)$ and $E(z)$, is rounded to integer values.

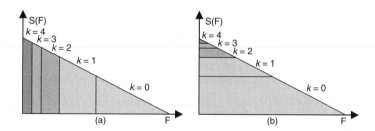

FIGURE 1.10 Spectral energy subdivision of Laplacian pyramids with ideal filters: (a) independent quantizers (no feedback); (b) quantization noise feedback with suitable step sizes.

Figure 1.10 highlights the meaning of *independent* and *dependent* quantizers for the ELP. In the case without feedback (a), analogous to Mallat's wavelet pyramid, the power spectral density is vertically sliced in disjoint frequency octaves (*bandpass* representation) that are quantized independently of one another. In the case of feedback (b), the lower frequency components are coarsely quantized, and their errors, having flat spectra, are added to the higher frequency layers before these are quantized as well. The resulting pyramid levels embrace all the octaves starting from the baseband, and thus constitute a *lowpass* representation. It is also evident that quantization errors coming from upper levels affect the quantizers at lower levels. The RD curve of a quantizer at a given pyramid level depends on the distortion value introduced by the quantizer at the upper level, i.e., on the work point on the previous RD plot. If the set of quantization step sizes is chosen as in Figure 1.10(b), the spectra will be vertically sliced, resulting in a maximum flatness of the coefficients to be coded at each level. From the above considerations, such a set of step sizes is expected to be data-dependent and, thus, to vary from one image to another.

Upon these premises, let us derive the optimum set of dependent quantizers for the ELP coder in Figure 1.9. Let D_k and R_k, respectively, denote the MSE distortion produced at the kth level of the ELP with quantization noise feedback (1.48), and the mean binary rate needed to transmit one ELP coefficient at the same level. Assuming that R_k is large, and therefore D_k is small, the relationship between D_k and R_k may be written as [20]

$$D_k = c_k \hat{\sigma}_k^{*2} 2^{-2R_k} \qquad (1.49)$$

where $\hat{\sigma}_k^{*2}$ is the variance of $\hat{\mathbf{L}}_k^*$ and c_k is a coefficient that depends on the shape of the PDF of the data and is independent of D_k. Inverting (1.49) yields

$$R_k = \frac{1}{2} \log_2 \frac{c_k \hat{\sigma}_k^{*2}}{D_k} \tag{1.50}$$

An equivalent entropy rate of ELP may be defined as

$$R_{eq} = \sum_{k=0}^{K} \frac{R_k}{4^k} = \frac{1}{2} \sum_{k=0}^{K} \frac{1}{4^k} \log_2 \frac{c_k \hat{\sigma}_k^{*2}}{D_k} \tag{1.51}$$

Due to quantization noise feedback, a dependence exists between $\hat{\sigma}_k^{*2}$ and D_{k+1}, which can be modeled as

$$\hat{\sigma}_k^{*2} = \sigma_k^{*2} + D_{k+1} P_E^2 \tag{1.52}$$

where P_E denotes the power gain of the 1-D expansion filter $E(z)$ (1.47). $\{\sigma_k^*, k = 0, 1, \ldots, K-1\}$ represent the square roots of the variances of $\mathbf{L}_k^*$ (1.45). The latter equals $\hat{\mathbf{L}}_k^*$ (1.48) calculated with unit quantization step sizes. Denoting by D the overall distortion of the reconstructed image, $D \equiv D_0$.

The problem may be stated as: given D, i.e., D_0, find $D_k, k = 1, \ldots, K$, that minimize the overall rate R_{eq}. It can be solved by replacing (1.52) into (1.51), taking its partial derivatives with respect to $D_k, k = 0, \ldots, K$, and imposing them to be null with the constraint $D_0 = D$. This procedure yields $\hat{D}_k, k = 1, \ldots, K$, the distortion values that minimize (1.51),

$$\hat{D}_k = \frac{\sigma_{k-1}^{*2}}{3 P_E^2} \tag{1.53}$$

If $\{\hat{\Delta}_k, k = 1, \ldots, K\}$ is the set of step sizes of the optimum uniform quantizer, and a piecewise linear PDF approximation is taken within each quantization interval, in the middle of which the dequantized value lies, the relationship between step size and distortion is $\hat{D}_k = \hat{\Delta}_k^2 / 12$. Replacing (1.53) yields the optimum step sizes

$$\hat{\Delta}_k = \frac{2 \sigma_{k-1}^*}{P_E} \tag{1.54}$$

The step sizes $\hat{\Delta}_k$ may be approximated with odd integers, so as to minimize the PE of the lower resolution image versions, without introducing a significant performance penalty, as shown in [4]. The step size Δ_0 is set up based on quality requirements, since the optimal quantizer is independent of its value.

1.4   CODING SCHEMES

1.4.1   EMBEDDED ZERO-TREE WAVELET CODER

The wavelet representation offers a set of features that are not fully exploited in classical subband or transform coding schemes. First, the wavelet transform is multiresolution, i.e., it offers different representations of the same image with increasing levels of detail. Moreover, the wavelet transform has a superior ability in decorrelating the data, i.e., the energy of the signal is compacted into a very small subset of wavelet coefficients. This permits to code only a relatively small amount of data, thereby achieving very good compression performance. The main drawback of this approach is that the information relative to the position of the significant wavelet coefficients also needs to be coded.

A possible way to overcome this problem could be to exploit the multiresolution properties of the wavelet representation for coding this information in a more efficient way.

The EZW coding algorithm proposed by Shapiro [36] was the first coding scheme that exploited the multiresolution of the wavelet transform to yield a very compact representation of the significant coefficients. From this point of view, it can be considered as the ancestor of the subsequent coding schemes that have led to the JPEG 2000 standard.

The two key concepts of EZW are the coding of the significance map by means of zero-trees and successive approximation quantization. The significance map is defined as the indication of whether a particular coefficient is zero or nonzero with respect to a given quantization level. The main idea of the EZW algorithm consists of the fact that the zeros of the significance map can be coded in a very efficient way if we rely on the self-similarity of the wavelet representation. Given a coefficient at a coarse scale, the coefficients having the same spatial location at finer scales are called *descendants* and can be represented by means of a tree, as shown in Figure 1.11. Experimental evidence shows that if a zero is present in the significance map at a coarse scale, then its descendants are likely to be zero. Hence, all the coefficients belonging to the tree can be coded with a single symbol. Usually, this set of insignificant coefficients is referred to as zero-tree. Relying on zero-trees, a very compact description of a significance map can be obtained. Moreover, this representation is completely embedded in the bit stream, without the need of sending any side information.

A significance map can be thought as a partial bit plane that codes only the most significant bit of each coefficient. In order to obtain a finer representation of the wavelet coefficients, the EZW algorithm uses a successive approximation strategy. This consists of a refinement step in which, for each coefficient that was found to be significant at the previous levels, a new bit is produced according to the current quantization level. In practice, given the bit plane relative to a particular quantization threshold, this step codes the bits that cannot be deduced from the significance map. Thanks to this coding strategy, the EZW algorithm is also SNR-scalable, i.e., the output bit stream can be stopped at different positions yielding reconstructed images of different quality.

The details of the EZW algorithm can be better explained by using a simple example. Consider the three-level wavelet decomposition of an 8×8 image tile shown in Figure 1.12. We will denote the four subbands at the ith decomposition level as LL_i, HL_i, LH_i, and HH_i, using the same order in which they are visited by the EZW algorithm. Smaller subscript values correspond to finer resolution levels. Coarser resolution levels are visited first. Obviously, subband LL is present only at the coarsest level and it is the first subband that is coded.

The EZW algorithm considers two lists referred to as the *dominant list* and the *subordinate list*. The dominant list contains all the coefficients that are not already found to be significant. When a

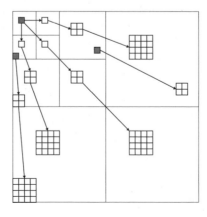

FIGURE 1.11 Example of descendant trees in a three-level wavelet decomposition.

127	26	69	11	0	3	10	1
−81	33	7	−12	1	8	2	0
22	17	−12	11	5	2	−4	0
30	13	7	−6	3	−2	−5	−3
0	5	13	0	0	−2	1	0
3	13	5	0	−1	1	1	0
13	6	1	−3	−1	0	1	0
11	0	0	4	0	2	0	−4

FIGURE 1.12 An example of three-level wavelet decomposition.

coefficient is found to be significant with respect to a given threshold, that coefficient is moved to the subordinate list. In the initialization pass, all wavelet coefficients are inserted in the dominant list and the quantization threshold is set to $T = 2^{\lfloor \log_2 x_{\max} \rfloor}$. Referring to Figure 1.12, we have $T_0 = 64$.

The algorithm is composed of two main passes. In the first pass, referred to as *dominant pass*, all the coefficients in the dominant list are compared with the threshold and coded in the significance map by using one of four possible symbols:

- *ps* (*positive significant*): the coefficient is positive and its magnitude is greater than the threshold.
- *ns* (*negative significant*): the coefficient is negative and its magnitude is greater than the threshold.
- *iz* (*isolated zero*): the coefficient magnitude is less than the threshold, but some of the descendants are significant.
- *zt* (*zero-tree*): the magnitude of the coefficient and of all the descendants is less than the threshold.

Considering Figure 1.12, the output of the first dominant pass is given in Table 1.2. In particular, when a zero-tree is found, the descendants are not coded. Referring to Table 1.2, we find that subband HH_2 as well as most coefficients in the first decomposition level do not need to be coded.

In the second pass, referred to as the *subordinate pass*, the coefficients in the subordinate list are coded with a 1 or a 0 depending on whether the coefficient magnitude is in the upper or lower half of the quantization interval. The first subordinate pass relative to the proposed example is shown in Table 1.3. Note that this pass considers the threshold $T = 32$. This is due to the fact that since the subordinate list contains only the coefficients that are found to be significant at previous passes, for $T = 64$, this list is empty.

The above passes are repeated in a loop, halving the threshold at each step. The coding process ends when the desired bit rate is achieved. Usually, the string of symbol produced by the EZW algorithm is entropy coded by means of an arithmetic encoder to meet the required compression rate.

1.4.2 SET PARTITIONING IN HIERARCHICAL TREES CODER

The set partitioning in hierarchical trees (SPIHT) algorithm proposed by Said and Pearlman [35] can be thought of as an improved version of the EZW scheme. It is based on the same concepts that are

TABLE 1.2
Output of the First Dominant Pass in EZW ($T = 64$)

Subband	Coefficient	Symbol	Reconstructed Value
LL_3	127	ps	96
HL_3	26	iz	0
LH_3	−81	ns	−96
HH_3	33	zt	0
HL_2	69	ps	96
HL_2	11	zt	0
HL_2	7	zt	0
HL_2	−12	zt	0
LH_2	22	zt	0
LH_2	17	zt	0
LH_2	30	zt	0
LH_2	13	zt	0
HL_1	0	iz	0
HL_1	3	iz	0
HL_1	1	iz	0
HL_1	8	iz	0

TABLE 1.3
Output of the First Subordinate Pass in EZW ($T = 32$)

Coefficient	Symbol	Reconstructed Value
127	1	112
−81	0	80
69	0	80

involved in EZW, i.e., the coding of the significance of the wavelet coefficients relative to a given threshold and the successive refinement of the significant coefficients. However, the significance is coded in a completely different way than in the EZW technique. In particular, the representation offered by SPIHT is so compact that it achieves performance comparable or superior to the EZW scheme even without using an entropy coder.

The main feature of SPIHT is a set partitioning rule that is used to divide the set of wavelet coefficients into significant and insignificant subsets. The rule is based on the well-known self-similarity properties of the wavelet decomposition and aims at obtaining insignificant subsets with a large number of elements, so that they can be coded at very little expense.

The set partitioning algorithm relies on the same tree structure, often referred to as *spatial orientation tree*, as shown in Figure 1.11. Relying on the spatial orientation tree, the set partitioning algorithm defines some particular sets of wavelet coefficients that are used for sorting the coefficients according to their significance. Given a wavelet decomposition, the set of four coefficient at the coarsest resolution is called $\mathcal{H}$ and is used to initialize the algorithm. Then, with respect to a generic coefficient denoted by the index k, the following sets are defined:

- $\mathcal{O}(k)$: set of the *offspring*. With respect to a particular coefficient, the offsprings are the four direct descendants in the next finer level of the wavelet decomposition.

- $\mathcal{D}(k)$: set of all the descendants of the coefficient k.
- $\mathcal{L}(k)$: defined as $\mathcal{D}(k) - \mathcal{O}(k)$.

The above-described sets are evaluated by means of a significance function that decides whether a set is to be partitioned. Given the generic set $\mathcal{T}$, the significance of $\mathcal{T}$ with respect to the nth bit plane is defined as

$$S_n(\mathcal{T}) = \begin{cases} 1 & \text{if there is a } k \in \mathcal{T} \text{ so that } c(k) \geq 2^n \\ 0 & \text{elsewhere} \end{cases} \tag{1.55}$$

where $c(k)$ denotes the magnitude of the coefficient with index k.

The rationale of the set partitioning algorithm is as follows. First, the set of wavelet coefficients is partitioned into the four single-element sets $\{k\}$, for $k \in \mathcal{H}$, plus the three sets $\mathcal{D}(k)$ corresponding to the coefficients in $\mathcal{H}$ that have descendants. Therefore, each significant $\mathcal{D}(k)$ is divided into the four single-element sets $\{l\}$, for $l \in \mathcal{O}(k)$, plus the set $\mathcal{L}(k)$. Finally, for each set $\mathcal{L}(k)$ that is found to be significant, a new partition is obtained considering the four sets $\mathcal{D}(l)$, for $l \in \mathcal{O}(k)$. The above partitioning rules are applied iteratively until all the sets $\mathcal{L}(k)$ prove to be empty, i.e., there are no descendants other than the four offsprings.

In the SPIHT algorithm, the order of the splitting pass is stored by means of three ordered lists, namely, the list of significant pixel (LSP), the list of insignificant pixel (LIP), and the list of insignificant set (LIS). Here we use the same terminology as in [35], where "pixel" denotes a wavelet coefficient. The LIS contains either set $\mathcal{D}(k)$ or $\mathcal{L}(k)$. If an entry refers to a set $\mathcal{D}(k)$ it is called of type A, otherwise type B. The LIP contains sets of single insignificant pixels. The LSP contains sets of single pixels that are found to be significant. Both the encoder and the decoder use the same control list, so that the decoder can recover the splitting ordering from the execution path.

In the following, the entire SPIHT encoding algorithm is presented as given in [35]. As can be seen, in the *sorting pass* the LIP and the LIS are processed, whereas in the *refinement pass* the LSP is processed.

1. **Initialization**: output $n = \lfloor \log_2 (\max_k \{c(k)\}) \rfloor$; set the LSP as an empty list, add all $k \in \mathcal{H}$ to the LIP and only the three with descendants also to the LIS, as type A entries;
2. **Sorting Pass**:
   2.1. for each k in the LIP do
        2.1.1. output $S_n(\{k\})$;
        2.1.2. if $S_n(\{k\}) = 1$ then move k to the LSP; output the sign of $c(k)$;
   2.2. for each k in the LIS do
        2.2.1. if the entry is of type A then
               - output $S_n(\mathcal{D}(k))$;
               - if $S_n(\mathcal{D}(k)) = 1$ then
                 − for each $l \in \mathcal{O}(k)$ do
                    * output $S_n(\{l\})$;
                    * if $S_n(\{l\}) = 1$ then add l to the LSP; output the sign of $c(l)$;
                    * if $S_n(\{l\}) = 1$ then add l to the LIP;
                 − if $\mathcal{L}(k) \neq \emptyset$ then move k to the LIS, as an entry of type B; go to 2.2.2;
                 − if $\mathcal{L}(k) = \emptyset$ then remove k from LIS;
        2.2.2. if the entry is of type B then
               - output $S_n(\mathcal{L}(k))$;
               - if $S_n(\mathcal{L}(k)) = 1$ then

     – add each $l \in \mathcal{O}(k)$ to the LIS, as an entry of type A;
     – remove k from the LIS;

3. **Refinement Pass**: for each k in the LSP, except those included in the last sorting pass, output the nth bit of $c(k)$;
4. **Quantization Step Update**: $n = n - 1$; go to 2.

The above algorithm can be better understood by means of a simple example considering again the wavelet decomposition in Figure 1.12. At the beginning of the algorithm we have $n = 6$ and the control list are set to LSP $\rightarrow$ {}, LIP $\rightarrow$ $\{127, 26, -81, 33\}$, and LIS $\rightarrow$ $\{26(A), -81(A), 33(A)\}$. Step 2.1 in the first sorting pass along with its output is shown in Table 1.4. For simplicity, in the example, each coefficient k is denoted by its value $c(k)$. As can be seen, two significant coefficients are found in the LIP and they are indicated in the output stream as 1 followed by a sign symbol (p and n to denote the positive and negative sign, respectively). At the end of this pass the control list are LSP $\rightarrow$ $\{127, -81\}$ and LIP $\rightarrow$ $\{26, 33\}$. The LIS is kept unchanged.

TABLE 1.4
Output of the First Sorting Pass in SPIHT ($n = 6$): Step 2.1

LIP	$S_n(\{k\})$	Output
127	1	1
		p
26	0	0
-81	1	1
		n
33	0	0

TABLE 1.5
Output of the First Sorting Pass in SPIHT ($n = 6$): Step 2.2

LIS	Type	$S_n(\mathcal{D}(k))$	$S_n(\mathcal{L}(k))$	$l \in \mathcal{O}(k)$	$S_n(\mathcal{O}(\{l\}))$	Output
26	A	1	—	—	—	1
				69	1	1
						p
				11	0	0
				7	0	0
				-12	0	0
-81	A	0	—	—	—	0
33	A	0	—	—	—	0
26	B	—	0	—	—	0

Step 2.2 in the first sorting pass is shown in Table 1.5. In this case, we note that one significant descendant set is found. For this set, one of the four offsprings is found to be significant and is coded in the output stream as in the previous step. Note that the root of this set ($c(k) = 26$) is moved to the end of the LIS as an entry of type B and evaluated in the same sorting pass. This means that if either a significant $\mathcal{D}(k)$ or $\mathcal{L}(k)$ is found in the LIS, then this set is recursively partitioned until either an insignificant set or a single element set is found. After step 2.2, the control lists become LSP → {128, −81, 69}, LIP → {26, 33, 11, 7, −12}, and LIS → {−81(A), 33(A), 26(B)}.

As to the refinement pass, this is identical to that in the EZW algorithm and gives exactly the same output. It is worth noting that the SPIHT algorithm yields an output stream of binary symbols. This stream can be either directly mapped onto a bit stream or entropy-coded, according to the desired trade-off between complexity and compression performance.

1.4.3 EMBEDDED BLOCK CODING WITH OPTIMIZED TRUNCATION

The embedded block coding with optimized truncation (EBCOT) algorithm [43] shares with its predecessors, namely the EZW and the SPIHT coders, the use of the wavelet transform to decorrelate the data of the input image. The coding engine that follows the transform, however, is completely different. The principal aim of the coder is "scalable compression," in its various forms. *Resolution scalability* refers to the fact that the compressed information contains a subset of data that can be decoded independently of the remainder of the bit stream, which allows a lower spatial resolution image to be reconstructed. *SNR scalability* refers to the possibility for the decoder to reconstruct different quality images decompressing only a subset of the bit stream. The EBCOT algorithm achieves these forms of scalability in an extremely simple way. For this reason it has been chosen as the coding algorithm for the standard JPEG 2000 [44]. Thanks to the scalability feature, an image can be compressed once and decoded by many decoders with different needs of resolution and quality.

In EBCOT, each subband is partitioned in smaller blocks of samples, which are termed *code-blocks*. The bit planes of each code-block are coded by starting from the most significant bits of its samples. This resembles what is done in EZW and SPIHT, but unlike these coders, in EBCOT, no dependence among samples belonging to different subbands is created, so that code-blocks are coded independently of each other. Resolution scalability is obtained by decoding all the code-blocks belonging to the subbands that form a certain level of resolution of the wavelet decomposition. SNR scalability is achieved by inserting into the bit stream *truncation points*, which allows samples to be reconstructed with different quantization step sizes. "Natural" truncation points are placed after a bit plane has been coded. Fractional bit plane coding is introduced to create further truncation points within the compressed bit stream. Some details about each pass of the coding procedure are now given.

Code-blocks are rectangular partitions of the subbands obtained from the wavelet transform. Typical code-block dimensions are 32×32 or 64×64. The 9/7-taps Daubechies' and 5/3-taps Le Gall's biorthogonal wavelet transforms are used for irreversible (or lossy) and reversible (or lossless) coding, respectively. These transforms are implemented by using a multistage lifting scheme as shown in Section 1.2.6.

Each sample $s_i(k_1, k_2)$, belonging to code-block B_i of the bth subband, is quantized with a deadzone quantizer having quantization step size equal to Δ_b and deadzone width equal to $2\Delta_b$. The information to be coded is the sign of $s_i(k_1, k_2)$ and the quantized magnitude $v_i = \lfloor s_i(k_1, k_2)/\Delta_b \rfloor$. Let $v_i^p(k_1, k_2)$ be the pth bit of the binary representation of the positive integer-valued quantity $v_i(k_1, k_2)$. Bit plane coding means transmitting first the information related to the bits $v_i^{p_i^{\max}}(k_1, k_2)$, where $p_i^{\max}$ is the most significant bit of the samples in the code-block B_i, then the information related to $v_i^{p_i^{\max}-1}(k_1, k_2)$, and so forth. Truncating the bit stream after decoding j bit planes is equivalent to a coarser representation of the samples, i.e., with a quantization step size equal to $2^{(p_i^{\max}-j)}\Delta_b$.

Bit planes are coded by means of the *MQ* conditional arithmetic coder, whose probability models are derived from the context of the bit that must be encoded and is based on the neighboring samples. The context is identified by means of the state of some variables associated to each sample in B_i. The significance $\sigma_i(k_1, k_2)$ is a variable that is initialized to zero and is set to one as soon as its most significant bit is coded. Four primitives are used to encode a bit or a set of bits in each bit plane. The *zero coding* (ZC) primitive is used to code a bit given that it did not become significant in previous bit planes. The significance of $s_i(k_1, k_2)$ is strongly influenced by the state of the significance of its immediate eight neighbors. Based on these values, nine ZC contexts have been identified and are used by the arithmetic coder. The *run-length coding* (RLC) primitive is invoked when a run of four insignificant samples (with insignificant neighbors) is encountered. A bit of the group, however, can become significant in the current bit plane and its position must be coded. The use of this primitive is motivated more by decreasing the number of calls to the arithmetic coder than by a compression performance improvement, which is quite limited. The *sign coding* (SC) primitive is used when a code-block sample becomes significant during a ZC or a RLC primitive. It is used once for each sample. Since sample signs are correlated, five SC contexts have been identified based on the significance and the sign of four (horizontal and vertical) neighbor samples. The *magnitude refinement* (MR) primitive is used to code a bit of a sample that became significant in a previous bit plane. It uses three MR contexts, which are derived both by the significance of its eight immediate neighbors and by a new state variable, which is initialized to 0 and is set to 1 after the MR primitive is invoked.

As already mentioned, bit plane coding indicates the positions where truncation points are to be placed. The rate R and the distortion D obtained after coding a bit plane identifies a point (R, D) that lies on the RD curve characteristic of the EBCOT coder, which is obtained by modulating the quantization step sizes and decoding all bit planes. All the points belonging to this curve identify "optimum coders," i.e., characterized by minimum distortion for a given rate or minimum rate for a given distortion. Truncating the bit stream between the pth and the $(p + 1)$th bit plane truncation point allows the decoder to reconstruct the image using the pth bit plane and a part of the $(p + 1)$th bit plane information. The resulting (R, D) point usually does not lie on the optimum coder R–D curve. The reason for this fact is that the most effective samples in reducing the distortion not necessarily occupy the first positions scanned during the block coding process. To avoid loss of performance, fractional bit plane coding is used in EBCOT. It consists of three coding passes: in each pass, the samples of the code-block are scanned in a determined order and the bits of the samples that are estimated to yield major distortion reduction are selected and coded. The first pass selects the bits of those samples that are insignificant but at least one of its immediate neighbors is significant; the second pass selects the bits that became significant in a previous bit plane; and the third pass selects the remainder bits. Every bit of each bit plane is coded in only one of the three coding passes. Truncation points are placed after each coding pass is completed and the bits of the selected samples have been coded. Truncation of the bit stream at one of these points again yields reconstructed images characterized by (R, D) points on the optimum R–D curve.

The main achievement of the fractional bit plane coding is a refinement of the truncation points distribution. The more dense the positions of the truncation points, the larger the number of optimum coders, characterized by a given bit rate or distortion, embedded into the bit stream. This fact makes the EBCOT algorithm extremely flexible and suitable for a wide range of applications.

1.4.4 Content-Driven ELP Coder

The minimum-distortion quantizers described in Section 1.3.1 would lead to optimum encoding if the pyramid were a space-invariant memoryless source. However, such hypotheses are far from being valid when true images are being dealt with. Therefore, the ELP scheme has been further specialized in order to take advantage of both the residual local correlation and the nonstationarity by selecting only areas comprising significant features at low resolution and by demanding full reconstruction at the bottom level through the feedback mechanism.

To this end, ELP levels are partitioned into adjacent 2×2 blocks that may not be split further, and are considered as information *atoms*. A quartet of nodes at level k is assumed to have a parent node at level $k + 1$; also, each node from a quartet at level $k > 0$ is regarded as the parent node of a 2×2 underlying block at level $k - 1$. A *quad-tree* hierarchy is thus settled, to enable the introduction of a split decision rule based on image content.

Content-driven progressive transmission, introduced for quad-tree structured gray-scale images, has been slightly modified and extended within the framework of LP [30]. It originally featured multiple breadth-first tree-scan steps, each driven by a different set of thresholds $\{T_k, k = 1, \ldots, K\}$ related to pyramid levels. An activity function of the parent node at level k, A_k, was computed on each underlying block at level $k - 1$ of the LP, possibly within a neighborhood of the 2×2 block. If $A_k > T_k$, the four interpolation errors were encoded; otherwise, they were taken to be null. In this way, the information was prioritized: the most important pyramid coefficients were considered from the early stages. However, only nodes underlying a split node were retained for content-driven transmission, while all the nodes underlying an unsplit node were automatically disregarded, same as a zero-tree [36]. Therefore, the content-driven feature precluded error-free reconstruction in a single-step scanning manner, since skipped nodes were no longer expanded [8]. In the present scheme, still featuring a single-step pyramid scanning, all the quartets at any level are to be checked for their activity, to permit reversibility.

A crucial point of the above outline is the choice of a suitable activity measure. Several functions have been reviewed in [30]. In subsequent works [1,4,8] as well in the present work, $A_k(m,n)$, the activity function of node (m,n) at level k, is taken as the L_∞ norm, or *maximum absolute value* (MAV) of its four offspring, i.e., four underlying interpolation errors,

$$A_k(m, n) \triangleq \max_{i,j=0,1} \left\{ |\hat{L}_{k-1}^*(2m + i, 2n + j)| \right\} \tag{1.56}$$

for $k = K, K - 1, \ldots, 1$; $m = 0, \ldots, M/2^k - 1$; $n = 0, \ldots, N/2^k - 1$. In terms of both visual quality and objective errors, this choice represents a very simple, yet efficient, selection criterion. In fact, due to the residual spatial correlation of ELP, four neighboring values are likely to be of similar magnitude; therefore, thresholding their MAV guarantees a selection that is efficient for the whole quartet of values.

The content-driven decision rule with uniform quantization of the retained quartets may be regarded as a data-dependent threshold quantization in which the quantization levels of the quartet of coefficients to be discarded are all zero, and quantization errors equal the values themselves. All the errors introduced by the decision rule can therefore be recovered at higher-resolution pyramid layers, thus extending quantization feedback also to content-driven pyramid schemes. In addition, when the split decision is embedded in the noise feedback loop, T_k, the decision threshold at level $k = 1, \ldots, K - 1$, is not critical for the lossless, or near-lossless, performance, since the error is thoroughly determined by the quantizer at the bottom level. In practice, the set of thresholds can be related to the step sizes of the uniform quantizer (1.54), e.g., by taking $\lfloor \hat{\Delta}_{k-1}/2 \rfloor < T_k \leq \hat{\Delta}_{k-1}$, analogously to a *dead-zone* quantizer [16].

The Kth level of the GP (*root* image or baseband) is encoded followed by the quantized and coded ELP at levels from $K - 1$ to 0. The feedback loops enclosing the pairs of quantizer/dequantizer are evident in the flow-chart of Figure 1.9, in which the block Q_i, $i = 0, 1$, stands for content-driven quantizer. A causal spatial prediction is employed on the root set at level $K = 2$.

It is better to exploit the local correlation of the root by DPCM encoding than to overly reduce the size of pyramid layers. Quantization errors of the root are interpolated as well and delivered to subsequent coding stages. The optimum step size $\hat{\Delta}_K$ is no longer optimum for minimizing entropy. Since the quantizer at level K is embedded in the DPCM loop (see Figure 1.9), too coarse a quantization would reduce the prediction efficiency. The best value of Δ_K is chosen empirically.

A simple prediction scheme employs a linear regression of four neighboring pixels, three in the previous row and one in the current row. The predicted value at (m, n), $\tilde{G}_K^*(m, n)$, is given by

$$
\begin{aligned}
\tilde{G}_K^*(m, n) = & \ round[\phi_1 G_K^*(m - 1, n - 1) + \phi_2 G_K^*(m - 1, n) \\
& + \phi_3 G_K^*(m - 1, n + 1) + \phi_4 G_K^*(m, n - 1)]
\end{aligned}
\tag{1.57}
$$

where the optimal coefficients $\{\phi_i, i = 1, \ldots, 4\}$ are found by *least squares* minimizing the difference between $G_K^*(m, n)$ and $\tilde{G}_K^*(m, n)$. For an *auto-regressive* (AR) model of the field [20], the inversion of a correlation matrix of a size equal to the prediction support is required. Nonzero prediction errors are quantized and entropy-coded. Positions of zero/nonzero quantization levels are run-length-encoded. Quantization errors of the root are interpolated as well and fed back to subsequent pyramid layers.

1.4.4.1 Synchronization Tree

The introduction of a selective choice of the nodes to be split is effective in the coding scheme, but requires a synchronization overhead. Flag bits may be arranged to form a binary tree whose root corresponds to the Kth layer and whose bottom corresponds to the first layer of the ELP. Each one-bit marks the split of the related node into the quartet of its offspring. Conversely, each zero-bit indicates that the underlying quartet of interpolation errors has not been considered. A straightforward encoding of the *split-tree* is not efficient because of the survived correlation, which reflects the features of the ELP: contours and active regions are marked by clustered one-bits, while homogeneous areas are characterized by gatherings of zero-bits.

An easy way to exploit such a correlation is to run-length encode the sequences of zeros and ones. Each level of the quad-tree is partitioned into square blocks. Each block is alternately scanned to yield runs of zeros and ones, possibly continuing inside one of the adjacent blocks. Experiments on x-ray images showed that the average run-length is maximized when 8×8 blocks are considered. Figure 1.13 depicts the quad-tree of synchronization bits and outlines how the run-length scan mode takes advantage of the spatial correlation within each tree level, in the case of 4×4 blocks at level $k - 1$. Notice that the scheme does not feature zero-trees: due to activity introduced by error feedback, also an unsplit node at level k may become father of any split nodes at level $k - 1$.

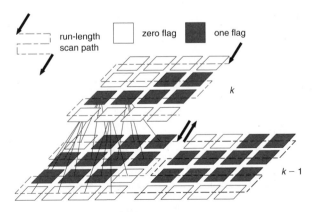

FIGURE 1.13 Detail of a sample synchronization flag tree associated to content-driven ELP coding, with inter-block run-length of its levels (4×4 blocksize scan mode).

1.4.5 Results

In this section, coding results are briefly reported with the intent of highlighting the difference in performance among a new-generation wavelet coder (namely EBCOT, standardized as JPEG 2000 [JPEG2K]), the ELP coder, an advanced DPCM coder (RLPE [3]), relying on a crisp adaptive spatial prediction and a context modeling for entropy coding suitable for near-lossless coding [2], and the DPCM standard JPEG-LS [49].

Let $\{g(i,j)\}$, $0 \le g(i,j) \le g_{fs}$, denote an N-pixel digital image and $\{\tilde{g}(i,j)\}$ its distorted version achieved by compression. Widely used distortion measurements are, MSE or L_2^2,

$$\text{MSE} = \frac{1}{N} \sum_i \sum_j [g(i,j) - \tilde{g}(i,j)]^2 \tag{1.58}$$

maximum absolute distortion (MAD), or *peak error*, or L_∞

$$\text{MAD} = \max_{i,j}\{|g(i,j) - \tilde{g}(i,j)|\} \tag{1.59}$$

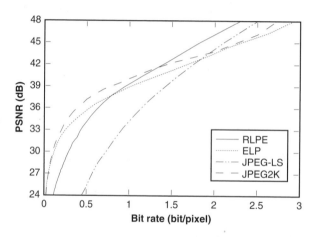

FIGURE 1.14 Lossy compression performances of RLPE, ELP, JPEG-LS, and JPEG 2000.

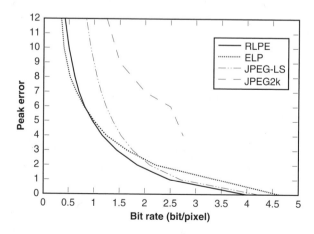

FIGURE 1.15 L_∞ performance of RLPE, ELP, JPEG-LS, and JPEG 2000.

peak signal-to-noise ratio (PSNR)

$$\text{PSNR}_{(dB)} = 10 \log_{10} \frac{g_{fs}^2}{\text{MSE} + \frac{1}{12}} \tag{1.60}$$

in which the MSE is incremented by the variance of the integer roundoff error, to handle the limit lossless case, when MSE $= 0$. Thus, PSNR will be upper-bounded by $10 \log_{10}(12g_{fs}^2)$, in the lossless case.

Rate distortion plots are shown in Figure 1.14 for RLPE, ELP, and JPEG-LS, all working in near-lossless, that are L_∞ error-constrained as well as for the lossy JPEG 2000.

FIGURE 1.16 *Lenna* coded at 0.8 bit/pel by (a) JPEG 2000 and (b) ELP; at 0.4 bit/pel by (c) JPEG 2000 and (d) ELP; at 0.2 bit/pel by (e) JPEG 2000 and (f) ELP.

The wavelet coder gives the best performance for a wide interval of bit rates. RLPE is always superior to JPEG-LS, especially for low rates, and gains over JPEG 2000 only for very high rates (>1.25 bit/pel). The rapidly decaying curve for decreasing rates shown by RLPE and by JPEG-LS is typical of all the causal DPCM schemes and is an effect of quantization noise feedback in the prediction loop. The "noisy" data that are reconstructed at the decoder are utilized for prediction, thus making it increasingly poorer as the quality, and hence the bit rate, decreases. This effect is also exhibited, though at rates even lower, by the DCT-based JPEG, and is due to DPCM coding of the DC component [31]. The noncausal prediction of ELP performs better for low rates, with an RD curve similar to that of JPEG 2000, which, however, is not L_∞-constrained. The near-lossless plots, i.e., L_∞ vs. bit rate, shown in Figure 1.15, demonstrate that the error-bounded encoders are obviously far superior to JPEG 2000. RLPE and ELP share the best results, the former for rates higher than 0.75 bit/pel, the latter otherwise.

The visual quality of decoded images at high, medium, and low bit rates is shown in Figure 1.16 for the test image *Lenna*, the original of which is shown in Figure 1.8(a). Only JPEG 2000 and ELP are compared, since the causal DPCM schemes, RLPE, and especially JPEG-LS, perform rather poorly at low bit rates. Although the difference in PSNR of coded images between JPEG 2000 and ELP is about 1 dB, the peak error of JPEG 2000 is more than three times greater than that of ELP. The thin features of the hat in Figure 1.16(e) and Figure 1.16(f) reveal that the JPEG 2000 coded image is more pleasant, but the ELP coded one more accurate, at low bit rates.

1.5 CONCLUSIONS

This survey has pointed out the potentialities of multiresolution analysis for lossy compression of still images, especially for low/medium bit rates. Concepts native of scale-space analysis, developed by a new generation of wavelet coders during the last decade, have given rise to the new JPEG 2000 standards, which trades off coding efficiency with requirements of scalability and layered coding dictated by multimedia applications. Although the scientific quality of image data is better preserved by predictive near-lossless coders, like JPEG-LS and RLPE, we wish to remind that a new coder [28], in which a three-level DWT pyramid is split into as many trees as pixels in the lowpass approximation (see Figure 1.11), which are arranged into 8×8 blocks and coded analogously to DCT-JPEG, is being recommended by CCSDS (Consultative Committee for Space Data Systems) [50] for high-quality lossy compression of images acquired from space.

REFERENCES

1. B. Aiazzi, L. Alparone, and S. Baronti, A reduced Laplacian pyramid for lossless and progressive image communication, *IEEE Trans. Commun.*, 44 (1996) 18–22.
2. B. Aiazzi, L. Alparone, and S. Baronti, Context modeling for near-lossless image coding, *IEEE Signal Process. Lett.*, 9 (2002) 77–80.
3. B. Aiazzi, L. Alparone, and S. Baronti, Near-lossless image compression by relaxation-labelled prediction, *Signal Process.*, 82 (2002) 1619–1631.
4. B. Aiazzi, L. Alparone, S. Baronti, and F. Lotti, Lossless image compression by quantization feedback in a content-driven enhanced Laplacian pyramid, *IEEE Trans. Image Process.*, 6 (1997) 831–843.
5. M. Antonini, M. Barlaud, P. Mathieu, and I. Daubechies, Image coding using wavelet transform, *IEEE Trans. Image Process.*, 1 (1992) 205–220.
6. F. Argenti, V. Cappellini, A. Sciorpes, and A.N. Venetsanopoulos, Design of IIR linear-phase QMF banks based on complex allpass sections, *IEEE Trans. Signal Process.*, 44 (1996) 1262–1267.
7. R.H. Bamberger, S.L. Eddins, and V. Nuri, Generalized symmetric extension for size-limited multirate filter banks, *IEEE Trans. Image Process.*, 3 (1994) 82–87.
8. S. Baronti, A. Casini, F. Lotti, and L. Alparone, Content-driven differential encoding of an enhanced image pyramid, *Signal Process.: Image Commun.*, 6 (1994) 463–469.

9. C.M. Brislawn, Preservation of subband symmetry in multirate signal coding, *IEEE Trans. Signal Process.*, 43 (1995) 3046–3050.

10. P.J. Burt and E.H. Adelson, The Laplacian pyramid as a compact image code, *IEEE Trans. Commun.*, 31 (1983) 532–540.

11. A.R. Calderbank, I. Daubechies, W. Sweldens, and B.-L. Yeo, Wavelet transforms that map integers to integers, *Appl. Comput. Harmon. Anal.*, 5 (1998) 332–369.

12. A. Croisier, D. Esteban, and C. Galand, Perfect channel splitting by use of interpolation/decimation/tree decomposition techniques, *International Conference on Information Sciences and Systems*, 1976, pp. 443–446.

13. I. Daubechies, Orthonormal bases of compactly supported wavelets, *Commun. Pure Appl. Math.*, 41 (1988) 909–996.

14. I. Daubechies, *Ten Lectures on Wavelets*, CBMS-NSF Reg. Conf. Series Appl. Math., vol. 61, SIAM, Philadelphia, PA, 1992.

15. I. Daubechies and W. Sweldens, Factoring wavelet transforms into lifting steps, *J. Fourier Anal. Appl.*, 4 (1998) 247–269.

16. H. Gharavi and A. Tabatabai, Subband coding of monochrome and color images, *IEEE Trans. Circuits Syst.*, 35 (1988) 207–214.

17. C. Herley, Boundary filters for finite-length signals and time-varying filter banks, *IEEE Trans. Circuits Syst. II: Analog Digital Signal Process.*, 42 (1995) 102–114.

18. C. Herley and M. Vetterli, Orthogonal time-varying filter banks and wavelet packets, *IEEE Trans. Signal Process.*, 42 (1994) 2650–2663.

19. ISO/IEC JTC 1/SC 29/WG1, *ISO/IEC FCD 15444-1: Information technology–JPEG 2000 image coding system: Core coding system [WG 1 N 1646]*, March 2000.

20. N.S. Jayant and P. Noll, *Digital Coding of Waveforms: Principles and Applications to Speech and Video*, Prentice Hall, Englewood Cliffs, NJ, 1984.

21. J. Johnston, A filter family designed for use in quadrature mirror filter banks, *Proceedings of IEEE the International Conference on Acoustics, Speech, and Signal Processing*, 1980, pp. 291–294.

22. H. Kiya, K. Nishikawa, and M. Iwahashi, A development of symmetric extension method for subband image coding, *IEEE Trans. Image Process.*, 3 (1994) 78–81.

23. J. Kovačević and W. Sweldens, Wavelet families of increasing order in arbitrary dimensions, *IEEE Trans. Image Process.*, 9 (2000) 480–496.

24. D. Le Gall and A. Tabatabai, Sub-band coding of digital images using symmetric short kernel filters and arithmetic coding techniques, *Proceedings of the IEEE International Conference on Acoustics Speech, Signal Processing*, vol. 2, Apr. 1988, pp. 761–764.

25. S. Mallat, A theory for multiresolution signal decomposition: the wavelet representation, *IEEE Trans. Pattern Anal. Machine Intell.*, PAMI-11 (1989) 674–693.

26. S. Mallat, *A Wavelet Tour of Signal Processing*, Academic Press, San Diego, CA, 1998.

27. S.A. Martucci and R.M. Mersereau, The symmetric convolution approach to the nonexpansive implementations of FIR filter banks for images, *Proceedings of the IEEE International Conference on Acoustics, Speech, Signal Processing*, 1993, pp. 65–68.

28. B. Masschelein, J.G. Bormans, and G. Lafruit, Local wavelet transform: a cost-efficient custom processor for space image compression, *Applications of Digital Image Processing XXV*, A.G. Tescher, ed., Proc. SPIE, vol. 4790, 2002, pp. 334–345.

29. F. Mintzer, Filters for distortion-free two-band multirate filter banks, *IEEE Trans. Acoust., Speech, Signal Process.*, 33 (1985) 626–630.

30. G. Mongatti, L. Alparone, G. Benelli, S. Baronti, F. Lotti, and A. Casini, Progressive image transmission by content driven Laplacian pyramid encoding, *IEE Proc. I*, 139 (1992) 495–500.

31. W.B. Pennebaker and J.L. Mitchell, *JPEG: Still Image Compression Standard*, Van Nostrand Reinhold, New York, 1993.

32. K.K. Rao and J.J. Hwang, *Techniques and Standards for Image, Video, and Audio Coding*, Prentice-Hall, Englewood Cliffs, NJ, 1996.

33. R. Rinaldo and G. Calvagno, Image coding by block prediction of multiresolution subimages, *IEEE Trans. Image Process.*, 4 (1995) 909–920.

34. R.E. Roger and J.F. Arnold, Reversible image compression bounded by noise, *IEEE Trans. Geosci. Remote Sensing*, 32 (1994) 19–24.

35. A. Said and W.A. Pearlman, A new, fast and efficient image codec based on set partitioning in hierarchical trees, *IEEE Trans. Circuits Syst. Video Technol.*, 6 (1996) 243–250.

36. J.M. Shapiro, Embedded image coding using zerotrees of wavelet coefficients, *IEEE Trans. Signal Process.*, 41 (1993) 3445–3462.

37. A. Skodras, C. Christopoulos, and T. Ebrahimi, The JPEG 2000 still image compression standard, *IEEE Signal Process. Mag.*, 18 (2001) 36–58.

38. M.J.T. Smith and T. Barnwell, III, A procedure for designing exact reconstruction filter banks for tree-structured subband coders, *Proceedings of the IEEE International Conference on Acoustics, Speech, Signal Processing*, 1984, pp. 421–424.

39. M.J.T. Smith and T. Barnwell, III, Exact reconstruction techniques for tree-structured subband coders, *IEEE Trans. Acoust., Speech, Signal Process.*, 34 (1986) 434–441.

40. M.J.T. Smith and S.L. Eddins, Analysis/synthesis techniques for subband image coding, *IEEE Trans. Acoust., Speech, Signal Process.*, 38 (1990) 1446–1456.

41. W. Sweldens, The lifting scheme: a custom-design construction of biorthogonal wavelets, *Appl. Comput. Harmon. Anal.*, 3 (1996) 186–200.

42. W. Sweldens, The lifting scheme: a construction of second generation wavelets, *SIAM J. Math. Anal.*, 29 (1997) 511–546.

43. D. Taubman, High performance scalable image compression with EBCOT, *IEEE Trans. Image Process.*, 9 (2000) 1158–1170.

44. D. Taubman and M.W. Marcellin, JPEG2000: standard for interactive imaging, *Proc. IEEE*, 90 (2002) 1336–1357.

45. B.E. Usevitch, A tutorial on modern lossy wavelet image compression: foundations of JPEG 2000, *IEEE Signal Process. Mag.*, 18 (2001) 22–35.

46. P.P. Vaidyanathan, *Multirate Systems and Filter Banks*, Prentice-Hall, Englewood Cliffs, NJ, 1992.

47. M. Vetterli, Filter banks allowing perfect reconstruction, *Signal Process.*, 10 (1986) 219–244.

48. M. Vetterli and J. Kovacevic, *Wavelets and Subband Coding*, Prentice-Hall, Englewood Cliffs, NJ, 1995.

49. M.J. Weinberger, G. Seroussi, and G. Sapiro, The LOCO-I lossless image compression algorithm: principles and standardization into JPEG-LS, *IEEE Trans. Image Process.*, 9 (2000) 1309–1324.

50. P.-S. Yeh, G.A. Moury, and P. Armbruster, CCSDS data compression recommendation: development and status, in *Applications of Digital Image Processing XXV*, A.G. Tescher, ed., Proc. SPIE, vol. 4790, 2002, pp. 302–313.

2 Advanced Modeling and Coding Techniques for Image Compression

David Taubman

CONTENTS

2.1   Introduction 35
2.2   Introduction to Entropy and Coding 36
     2.2.1   Information and Entropy 36
     2.2.2   Fixed- and Variable-Length Codes 37
     2.2.3   Joint and Conditional Entropy 38
2.3   Arithmetic Coding and Context Modeling 40
     2.3.1   From Coding to Intervals on (0,1) 40
     2.3.2   Elias Coding 42
     2.3.3   Practical Arithmetic Coding 43
     2.3.4   Conditional Coding and Context Modeling 47
         2.3.4.1   Adaptive Probability Estimation 48
         2.3.4.2   Binary Arithmetic Coding Is Enough 50
     2.3.5   Arithmetic Coding Variants 50
     2.3.6   Arithmetic Coding in JBIG 51
2.4   Information Sequencing and Embedding 52
     2.4.1   Quantization 52
     2.4.2   Multiresolution Compression with Wavelets 53
     2.4.3   Embedded Quantization and Bit-Plane Coding 54
     2.4.4   Fractional Bit-Plane Coding 56
     2.4.5   Coding vs. Ordering 57
2.5   Overview of EBCOT 58
     2.5.1   Embedded Block Coding Primitives 58
         2.5.1.1   Significance Coding 60
         2.5.1.2   Sign Coding 60
         2.5.1.3   Magnitude Refinement Coding 61
     2.5.2   Fractional Bit-Plane Scan 61
     2.5.3   Optimal Truncation 62
     2.5.4   Multiple Quality Layers 63
2.6   Reflections 65
References 66

2.1 INTRODUCTION

The objective of image compression may be summarized as that of finding good approximations to the original image that can be compactly represented. The key questions are: (1) how can we choose "good" approximations? and (2) how can we represent the chosen approximations with as few bits as possible? In recent years, scalable compression has proven to be of particular interest. A scalable compressed bit-stream contains multiple embedded approximations of the same image, with the property that the bits required to represent a coarse approximation form a subset of the bits required to represent the next finer approximation. Scalable compressed bit-streams support successive refinement, progressive transmission, accurate rate control, error resilience, and many other desirable functionalities. Scalability gives rise to a third important question: (3) how can we choose a sequence of good approximations, whose representations are embedded within each other?

In this chapter, we begin by reviewing established technologies for the efficient representation of information, exploiting internal statistical redundancies in order to use as few bits as possible. Shannon referred to this as "noiseless coding" in his landmark paper, which gave birth to the field of information theory. Noiseless coding tells us only how to efficiently represent an image approximation once it has been chosen. It tells us nothing about how to choose good approximations, or how to build good embeddings. These are addressed in the second part of the chapter. Our treatment is necessarily brief, and so we alternate between a discussion of broad principles and illustration of how those principles are embodied within the specific scalable compression paradigm of the "EBCOT" algorithm found in JPEG2000.

2.2 INTRODUCTION TO ENTROPY AND CODING

2.2.1 INFORMATION AND ENTROPY

Information is a property of events whose outcomes are not certain. Information theory, therefore, is fundamentally dependent upon statistical modeling. In the simplest case, we may consider a single random variable X, having a discrete set of possible outcomes $\mathcal{A}_X$, known as the *alphabet* of X. With each $x \in \mathcal{A}_X$, we can associate a quantity $h_X(x) = \log_2 1/f_X(x)$, which can be understood as the "amount of information we receive if we learn that the outcome of X is x." Here, $f_X(x)$ is the probability mass function (PMF), representing the probability that the outcome of X is x. The average amount of information conveyed by learning the outcome of X is then the statistical expectation of the random variable $h_X(X)$, which is known as the *entropy* of X:

$$H(X) = E[h_X(X)] = E\left[\log_2 \frac{1}{f_X(X)}\right] \tag{2.1}$$

$$= \sum_{x \in \mathcal{A}_X} f_X(x) \log_2 \frac{1}{f_X(x)} \tag{2.2}$$

The fact that $H(X)$ and $h_X(x)$ are the most appropriate measures of information is not immediately obvious. This must be established by a coding theorem. It can be shown that $H(X)$ is the minimum average number of bits per sample required to code the outcomes of X. More precisely, the following can be shown: (1) among all possible schemes for representing the outcomes of a sequence of independent random variables X_n, each having the same probability distribution as X, no scheme can reliably communicate those outcomes using an average of less than $H(X)$ bits per outcome; and (2) it is possible to construct a reliable coding scheme whose bit-rate approaches the lower bound $H(X)$ arbitrarily closely, in the limit as the complexity of the scheme is allowed to grow without bound. These two statements establish $H(X)$ as the asymptotic lower bound on the average number of bits required to code outcomes of X. They were first proven by Shannon [14] in his *noiseless source coding theorem*. The proof is not complex, but will not be repeated here for brevity.

2.2.2 Fixed- and Variable-Length Codes

For a finite alphabet $\mathcal{A}_X$, with $|\mathcal{A}_X|$ elements, it is quite clear that each outcome can be represented uniquely using fixed length words with $\lceil \log_2 |\mathcal{A}_X| \rceil$ bits each.[1] This is known as *fixed-length coding*. Accordingly, it is not surprising that $H(X)$ is always upper bounded by $\log_2 |\mathcal{A}_X|$. This upper bound is achieved if and only if X is *uniformly distributed*, meaning that $f_X(x) = 1/|\mathcal{A}_X|$, $\forall x$. To be able to represent an information source more efficiently, it must produce events whose outcomes are distributed in a highly nonuniform fashion. We refer to these as *skewed sources*. Skewed sources can generally be represented more efficiently if a variable-length code is used in place of a fixed-length code. To illustrate this point, consider the following example.

EXAMPLE 2.2.1

Suppose X takes its outcomes from the set of natural numbers, i.e., $\mathcal{A}_X = \mathbb{N} = \{1, 2, 3, \ldots\}$. We can uniquely represent each outcome $x \in \mathcal{A}_X$, by sending a string of $x - 1$ "0" bits, followed by a single "1." This is known as a comma code, *since the trailing "1" of each codeword may be interpreted as a comma, separating it from the next codeword. A message consisting of the outcomes $1, 3, 2, 4$, for example, would be represented using the following string of bits:*

$$1 \mid 001 \mid 01 \mid 0001$$

The vertical bars here serve only to clarify the separation between codewords; they are not part of the encoded message.

Suppose now that X has the geometrically distributed PMF, $f_X(x) = 2^{-x}$, for which the reader can easily verify that $\sum_{x \in \mathcal{A}_X} f_X(x) = 1$. In this case, the average number of bits required to code outcomes of X is given by

$$R = \sum_{n=1}^{\infty} n \cdot f_X(n) = \sum_{n=1}^{\infty} n \cdot 2^{-n} \tag{2.3}$$

writing

$$2R = \sum_{n=1}^{\infty} n \cdot 2^{-(n-1)} = \underbrace{\sum_{n=1}^{\infty} (n-1) \cdot 2^{-(n-1)}}_{R} + \underbrace{\sum_{n=1}^{\infty} 2^{-(n-1)}}_{2} \tag{2.4}$$

we readily conclude that the code-rate is only $R = 2$ bits per sample.

The above example is illuminating from several perspectives. First, $\mathcal{A}_X$ contains an infinite number of elements, so that fixed-length coding is not even possible. Secondly, for the geometric PMF the code requires exactly $h_X(x)$ bits to represent each outcome x, so that the average number of bits required to represent the outcomes of X is exactly equal to the entropy, $H(X)$. The comma code has been widely used as a component of compression systems, including the JPEG-LS lossless image compression standard discussed in Chapter 4.

More generally, a *variable-length code* is one that assigns a unique string of bits (*codeword*) to each possible outcome, such that a message consisting of any string of such codewords can be uniquely decoded. A necessary condition for unique decodability is that the codeword lengths l_x must satisfy

$$\sum_{x \in \mathcal{A}_X} 2^{-l_x} \leq 1 \tag{2.5}$$

This is known as the *McMillan condition*. Moreover, given any set of lengths which satisfy the McMillan condition, a theorem by Kraft shows that it is always possible to construct a uniquely

---

[1] Here, $\lceil y \rceil = \min\{x \mid x \geq y\}$ denotes the value obtained by rounding y up to the nearest integer.

decodable variable length code whose codewords have these lengths. The code-rate of the code will, of course, be $R = \sum_x l_x \cdot f_X(x)$.

We are now in a position to reinforce the notion of $H(X)$ as a lower bound to the code-rate R, by observing that R and $H(X)$ would be identical if we were able to make the selection $l_x = h_X(x)$. Certainly, this choice satisfies the McMillan inequality, since

$$\sum_{x \in \mathcal{A}_X} 2^{-h_X(x)} = \sum_{x \in \mathcal{A}_X} 2^{\log_2 f_X(x)} = \sum_{x \in \mathcal{A}_X} f_X(x) = 1 \qquad (2.6)$$

Unfortunately, though, variable-length codes must have integer-length codewords. Suppose then, we choose

$$l_x = \lceil h_X(x) \rceil \qquad (2.7)$$

This choice produces integer-valued lengths in the range $h_X(x) \le l_x < h_X(x) + 1$, which satisfy the McMillan condition. Accordingly, the code-rate must lie in the range $H(X) \le R < H(X) + 1$.

EXAMPLE 2.2.2

Suppose X takes its outcomes from the alphabet $\mathcal{A}_X = \{0, 1, 2, 3\}$, with probabilities

$$f_X(0) = \frac{125}{128}, \quad f_X(1) = f_X(2) = f_X(3) = \frac{1}{128}$$

One optimal set of codewords, in the sense of minimizing the code-rate, is as follows:

$$c_0 = \text{``0,''} \quad c_1 = \text{``10,''} \quad c_2 = \text{``110,''} \quad c_3 = \text{``111''}$$

The code is uniquely decodable, since no codeword is a prefix (first part) of any other codeword, meaning that a decoder can always find the boundaries between codewords. Thus, the message

"0111010110"

is unambiguously decoded as the string of outcomes $0, 3, 0, 1, 2$. With an average bit-rate of

$$R = 1 \cdot f_X(0) + 2 \cdot f_X(1) + 3 \cdot f_X(2) + 3 \cdot f_X(3) = 1.039 \; bits \qquad (2.8)$$

the variable-length code clearly outperforms the fixed-length code-rate of 2 bits/sample. On the other hand, the source entropy of $H(X) = 0.1975$ bits suggests that we are still wasting almost 1 bit/sample.

2.2.3 JOINT AND CONDITIONAL ENTROPY

We have seen that coding can be beneficial (with respect to a naive fixed-length code) as long as the source has a skewed probability distribution. In fact, in the limit as the probability of one particular outcome approaches 1, the entropy of the source is readily shown to approach 0. Unfortunately, image samples do not typically have significantly skewed probability distributions. A typical gray-scale image, for example, may be represented by 8-bit intensity values in the range 0 (black) to 255 (white). While not all values occur with equal likelihood, there is no reason to suppose that one or even a few brightness levels will occur with much higher probability than the others. For this reason, the entropy of such an image is typically close to 8 bits.

Nevertheless, images are anything but random. To illustrate this point, suppose the image samples are encoded in raster fashion, so that at some point in the sequence the next sample to be coded is

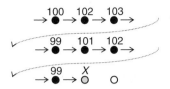

FIGURE 2.1 The advantage of context. By the time image sample "X" is to be coded, the indicated neighboring values are already known to both the encoder and the decoder, so they can be used in selecting a good code for "X."

that marked with an "X" in Figure 2.1. By the time it comes to decoding sample "X," the decoder has already observed the values of all preceding samples, some of which are also shown in the figure. Based on these previously decoded values, one (read, the decoder) would not be particularly surprised to learn that sample "X" turns out to be 100. On the other hand, one (read, the decoder) would be very surprised indeed to learn that the value of "X" is 20. Even though both values may have similar probabilities of occurring within the image as a whole, knowledge of the context in which "X" occurs can dramatically alter the amount of information conveyed by its outcome, and hence the number of bits required to encode that outcome.

The key notions here are those of *conditional probability* and *conditional entropy*. The fact that 20 seems a much less likely outcome for "X" than 100, given the surrounding values shown in Figure 2.1, arises from an underlying belief that image intensities should be somehow smooth. In fact, images are substantially smooth, having highly skewed conditional statistics.

The conditional probability distribution of a random variable X, given a context consisting of the outcomes of some other random variables $Y_1, Y_2, \ldots, Y_m$, is described by the conditional PMF

$$f_{X|Y_1,\ldots,Y_m}(x \mid y_1, \ldots, y_m) \tag{2.9}$$

or, more compactly, by $f_{X|\mathbf{Y}}(x \mid \mathbf{y})$, where $\mathbf{Y}$ is the random vector composed of the context random variables Y_1 through Y_m, and $\mathbf{y}$ is the vector of observed context outcomes. Conditional probabilities satisfy Baye's rule,

$$f_{X|\mathbf{Y}}(x \mid \mathbf{y}) = \frac{f_{X,\mathbf{Y}}(x, \mathbf{y})}{f_{\mathbf{Y}}(\mathbf{y})} = \frac{f_{X,\mathbf{Y}}(x, \mathbf{y})}{\sum_{a \in A_X} f_{X,\mathbf{Y}}(a, \mathbf{y})} \tag{2.10}$$

where $f_{X,\mathbf{Y}}(x, \mathbf{y})$ is the joint probability that the outcomes of X and $\mathbf{Y}$ are equal to x and $\mathbf{y}$, respectively.

Proceeding as before, we define

$$h_{X|\mathbf{Y}}(x \mid \mathbf{y}) = \log_2 \frac{1}{f_{X|\mathbf{Y}}(x \mid \mathbf{y})} \tag{2.11}$$

to be the amount of information communicated by the outcome $X = x$, given that we already know that the outcomes of Y_1 through Y_m are y_1 through y_m, respectively. The average amount of information communicated by learning the outcome of X, given that we already know the context $\mathbf{y}$, is the conditional entropy,

$$H(X \mid \mathbf{Y}) = E[h_{X|\mathbf{Y}}(X \mid \mathbf{Y})] = \sum_{x, y_1, \ldots, y_m} f_{X,\mathbf{Y}}(x, \mathbf{y}) \cdot \log_2 \frac{1}{f_{X|\mathbf{Y}}(x \mid \mathbf{y})} \tag{2.12}$$

Applying Baye's rule, we see that

$$H(X \mid \mathbf{Y}) = E\left[\log_2 \frac{1}{f_{X|\mathbf{Y}}(x \mid \mathbf{y})}\right] = E\left[\log_2 \frac{f_{\mathbf{Y}}(\mathbf{y})}{f_{X,\mathbf{Y}}(x, \mathbf{y})}\right] \tag{2.13}$$

which may be rewritten as

$$H(X, \mathbf{Y}) = H(X \mid \mathbf{Y}) + H(\mathbf{Y}) \tag{2.14}$$

This important result may be interpreted as follows. $H(\mathbf{Y})$ is the minimum average number of bits required to code the outcome of the random vector $\mathbf{Y}$.[2] Similarly, $H(X, \mathbf{Y})$ is the minimum average number of bits required to jointly code the outcomes of X and $\mathbf{Y}$. It follows that we can avoid the complexity and delay associated with coding the outcomes of X and $\mathbf{Y}$ jointly, while still achieving the minimum possible average number of bits, by first coding the outcome of $\mathbf{Y}$ and then coding the outcome of X, conditioned on our knowledge of the observed context $\mathbf{y}$. In fact, equation (2.14) may be recursively expanded as

$$H(X, Y_m, \ldots, Y_1) = H(X \mid Y_m, \ldots, Y_1) + H(Y_m \mid Y_1, \ldots, Y_{m-1}) + \cdots + H(Y_1) \tag{2.15}$$

This means that we can code Y_1 first, then Y_2 conditioned upon Y_1, then Y_3 conditioned upon Y_1 and Y_2, and so forth.

One particularly important observation, on which we shall later rely, is that it does not matter in what order we code the outcomes. Given a random vector $\mathbf{Y}$, consisting of m random variables Y_1 through Y_m, $H(\mathbf{Y})$ may be expanded in any of the following ways, each of which suggests a different order for coding the outcomes:

$$\begin{aligned} H(\mathbf{Y}) &= H(Y_1) + H(Y_2 \mid Y_1) + \cdots + H(Y_m \mid Y_{m-1}, \ldots, Y_2, Y_1) \tag{2.16} \\ &= H(Y_m) + H(Y_{m-1} \mid Y_m) + \cdots + H(Y_1 \mid Y_2, \ldots, Y_m) \\ &= H(Y_2) + H(Y_1 \mid Y_2) + H(Y_3 \mid Y_1, Y_2) + \cdots \\ &\quad\vdots \end{aligned}$$

To code the outcome of X with an average bit-rate approaching the conditional entropy $H(X \mid \mathbf{Y})$, it is necessary to explicitly use our knowledge of the context outcome $\mathbf{y}$. In the case of variable-length coding, we should select a different code for each possible context $\mathbf{y}$, where the codeword lengths $l_{x,\mathbf{y}}$ are as close as possible to $h_{X\mid\mathbf{Y}}(x \mid \mathbf{y}) = \log_2 1/f_{X\mid\mathbf{Y}}(x \mid \mathbf{y})$, while satisfying the McMillan condition, $\sum_x 2^{-l_{x,\mathbf{y}}} \leq 1$. If we were able to choose $l_{x,\mathbf{y}} = h_{X\mid\mathbf{Y}}(x \mid \mathbf{y})$, the average number of bits spent coding X would exactly equal $H(X \mid \mathbf{Y})$. As before, however, the need to select integer-length codewords may cost us anywhere from 0 to 1 bits per sample. Moreover, a chief difficulty with conditional variable-length coding is the need to maintain a separate set of codewords (a *codebook*) for each distinct context, $\mathbf{y}$. These difficulties are both elegantly addressed by arithmetic coding, which is the subject of the next section.

2.3 ARITHMETIC CODING AND CONTEXT MODELING

2.3.1 From Coding to Intervals on (0,1)

We have seen how variable-length codes produce a message which consists of a string of bits, whose length depends on the outcomes which were actually coded. Let $b_1, b_2, \ldots, b_{n_m}$ denote the message bits produced when coding a particular sequence of symbols (outcomes), say $x_1, x_2, \ldots, x_m$. The

---

[2] Note that a discrete random vector can always be put in 1–1 correspondence with a random variable whose alphabet contains one element for each possible outcome of the random vector. Coding a random vector then is equivalent to coding its equivalent random variable, so the interpretation of $H(X)$ as the minimum average number of bits required to code outcomes of X carries directly across to random vectors.

number of bits n_m is equal to the sum of the variable-length codes associated with each symbol, that is,

$$n_m = \sum_{k=1}^{m} l_{x_k} \tag{2.17}$$

It is instructive to think of the message bits as the fraction bits (digits following the binary point) in the binary representation of a numeric quantity β_m. Specifically, β_m has the following binary fraction representation:

$$\beta_m \equiv 0.b_1 b_2 \cdots b_{n_m} \tag{2.18}$$

with the numeric value $\beta_m = \sum_{i=1}^{n} b_i 2^{-i} \in [0, 1)$. The value of β_m uniquely identifies the m coded symbols.

Now suppose we continue coding more symbols, $x_{m+1}, x_{m+2}, \ldots$, producing more message bits and corresponding binary fractions $\beta_{m+1}, \beta_{m+2}, \ldots$. These binary fractions have the property that

$$\beta_k \in \mathcal{I}_m = [\beta_m, \beta_m + 2^{-n_m}), \quad \forall k > m \tag{2.19}$$

That is, the first m coded symbols define an interval, $\mathcal{I}_m \subset [0, 1)$, such that all messages, which commence with the same m symbols must lie within $\mathcal{I}_m$. Moreover, any message that does not commence with these same first m symbols must lie outside $\mathcal{I}_m$. It follows that

$$[0, 1) \supset \mathcal{I}_1 \supset \cdots \supset \mathcal{I}_m \supset \mathcal{I}_{m+1} \supset \cdots \tag{2.20}$$

and the coding of symbol x_{m+1} is equivalent to the process of selecting one of a number of disjoint subintervals of $\mathcal{I}_m$. These concepts are illustrated in Figure 2.2.

It is worth noting that interval length $|\mathcal{I}_m|$ is related to the number of message bits, through $n_m = \log_2 (1/|\mathcal{I}_m|)$. Equivalently, the number of bits spent for coding symbol x_m is $l_{x_m} = \log_2 (|\mathcal{I}_{m-1}|/|\mathcal{I}_m|)$. Since each distinct outcome for X_m corresponds to a disjoint subinterval of $\mathcal{I}_{m-1}$, it is clear that

$$\sum_{x \in \mathcal{A}_X} |\mathcal{I}_{m|x_m=x}| \leq |\mathcal{I}_{m-1}| \tag{2.21}$$

This reveals the origin of the McMillan condition, since

$$\sum_{x \in \mathcal{A}_X} 2^{-l_x} = \sum_{x \in \mathcal{A}_X} \frac{|\mathcal{I}_{m|x_m=x}|}{|\mathcal{I}_{m-1}|} \leq 1 \tag{2.22}$$

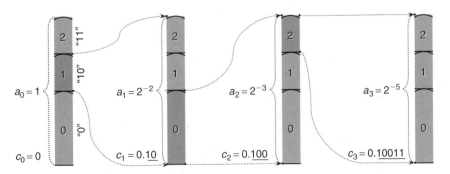

FIGURE 2.2 Variable-length encoding of the sequence $x_1 = 1$, $x_2 = 0$, $x_3 = 2$ for a ternary alphabet with codewords "0," "10," and "11," producing the encoded bit string "10011," shown as interval remapping.

The McMillan inequality is strict if and only if $\bigcup_{x \in \mathcal{A}_X} \mathcal{I}_{m|x_m=x}$ is a strict subset of $\mathcal{I}_{m-1}$; meaning that there are some strings of bits, which cannot arise from any legitimate coding of message symbols. These are the bit strings whose binary fractions β lie in the "holes" between the disjoint subintervals, $\mathcal{I}_{m|x_m=x}$. These so-called "*holes in the code-space*" represent redundancy in the coded message; they correspond to wasted bits, but many compression standards deliberately leave holes in the code-space so as to facilitate the detection of transmission errors.

2.3.2 ELIAS CODING

The connection between coding and subranging of the unit interval $[0,1)$ was recognized shortly after Shannon's original publication on information theory in 1948. The fact that this relationship can be turned around to yield an arithmetically computed coding scheme is generally attributed to Elias. The algorithm associates each m-symbol prefix of the message being encoded with a unique half-open interval $\mathcal{I}_m = [c_m, c_m + a_m) \subset [0, 1)$, such that the length of this interval satisfies

$$|\mathcal{I}_m| = a_m = \prod_{k=1}^{m} f_X(x_k) \tag{2.23}$$

This is achieved by initializing c_0 and a_0 to 0 and 1, respectively, so that $\mathcal{I}_0 = [0, 1)$, and recursively applying the following interval subranging algorithm.

Algorithm. *Elias Coding.*
For each $m = 1, 2, \ldots$

- Set $a_m = a_{m-1} \cdot f_X(x_m)$
- Set $c_m = c_{m-1} + a_{m-1} \cdot t_X(x_m)$

where $t_X(x) = \sum_{x' < x} f_X(x')$. The $t_X(x)$ are best understood as the thresholds in a partition of the unit interval, such that each element $x \in \mathcal{A}_X$ is assigned a distinct subinterval of $[0, 1)$ whose length is $f_X(x)$, as shown in Figure 2.3. The interval subranging algorithm "codes" x_m by shifting and scaling this partition to fit within $\mathcal{I}_{m-1}$ and then selects the particular shifted and scaled subinterval that corresponds to the outcome x_m.

We have seen in the case of fixed-length coding that the length of the interval $\mathcal{I}_m$ is $a_m = 2^{-n_m}$, where n_m is the length of the encoded message. In that case, the boundaries of $\mathcal{I}_m$ are aligned on exact multiples of 2^{-n_m}. In Elias coding, there is no such alignment, but it is sufficient to send to the decoder only the first

$$n_m = \left\lceil \log_2 \frac{1}{a_m} \right\rceil \tag{2.24}$$

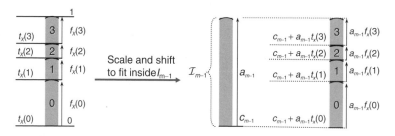

FIGURE 2.3 Subinterval thresholds and lengths for a four-symbol alphabet $\mathcal{A}_X = \{0, 1, 2, 3\}$.

bits of the binary fraction representation of c_m. To see this, observe that the decoder can always synthesize a quantity

$$\beta'_m = 0.b_1 b_2 \ldots b_{n_m} 111 \ldots \qquad (2.25)$$

by appending an infinite string of 1's to the received n_m bits. Since the first n_m fraction bits of β'_m are identical to those of c_m and the remaining, synthesized fraction bits are all 1's, it is clear that $\beta'_m \geq c_m$. It is also clear that $\beta'_m - c_m < 2^{-n_m}$, since the bit positions in which β'_m may differ from c_m correspond to multiples of $2^{-(n_m+i)}$ with $i \geq 1$. It follows that $\beta'_m \in [c_m, c_m + a_m)$, so it uniquely identifies the first m coded symbols. This means that the number of bits required to represent symbols x_1 through x_m is at most

$$\lceil \log_2 a_m \rceil = \left\lceil \sum_{k=1}^{m} \log_2 \frac{1}{f_X(x_k)} \right\rceil = \left\lceil \sum_{k=1}^{m} h_X(x_k) \right\rceil < \sum_{k=1}^{m} h_X(x_k) + 1 \qquad (2.26)$$

This, in turn, serves to emphasize the fact that $h_X(x_k)$ is the right measure for the amount of information associated with the individual outcome $X = x_k$. Evidently, the average number of bits required to encode a sequence of m symbols is within 1 bit of $m \cdot E[h_X(X)] = mH(X)$, so for long messages we essentially achieve the entropy.

Before pressing on, it is worth describing the decoding process more carefully. Suppose symbols x_1 through x_{m-1} have already been decoded. To decode the value of x_m, the decoder must read sufficient bits of the complete message β to unambiguously identify the particular subinterval of $\mathcal{I}_{m-1}$ to which it belongs. After n bits, the decoder can know that β lies within the interval $[\beta_{1:n}, \beta_{1:n} + 2^{-n})$, where

$$\beta_{1:n} = \sum_{i=1}^{n} b_i 2^{-i} \qquad (2.27)$$

It must therefore read a sufficient number of bits n, to be sure that

$$[\beta_{1:n}, \beta_{1:n} + 2^{-n}) \subseteq [c_{m-1} + a_{m-1} t_X(x), \ c_{m-1} + a_{m-1}(t_X(x) + f_X(x))] \qquad (2.28)$$

for some x, whereupon it can be certain that this $x = x_m$. Although unlikely, it is conceivable that the decoder may need to read right to the end of the encoded message before it can make such a determination. This, together with the fact that the arithmetic precision required to perform the encoding and decoding calculations grows rapidly with m, rendered Elias coding little more than an intellectual curiosity for many years. It turns out, however, that both problems can be readily solved using finite precision arithmetic, while achieving code-rates that differ negligibly from the entropy. These practical *arithmetic coding* schemes are the subject of the next subsection.

2.3.3 PRACTICAL ARITHMETIC CODING

In Elias coding, the interval $\mathcal{I}_{m-1}$, representing symbols x_1 through x_{m-1}, is updated to the interval $\mathcal{I}_m$, representing symbols x_1 through x_m, by assigning

$$a_m \leftarrow a_{m-1} f_X(x_m) \qquad (2.29)$$

$$c_m \leftarrow c_{m-1} + a_{m-1} t_X(x_m) \qquad (2.30)$$

The key observation required to bound the implementation precision is that we need not use the exact value of a_m produced by these ideal update relationships. Suppose instead that

$$0 < a_m < a_{m-1} f_X(x_m) \qquad (2.31)$$

Then the subintervals corresponding to each potential outcome of X_m remain disjoint, so that unique decoding is still guaranteed. There is, of course, some loss in coding efficiency due to the appearance of holes in the code-space; in fact, we sacrifice $\log_2 (a_{m-1}f_X(x_m))/a_m$ bits in coding the outcome $X_m = x_m$. As we shall see, however, modest arithmetic precision is sufficient to render this loss as negligible.

We are now in a position to describe a practical coding algorithm. At any point in the coding sequence, let the interval length be represented by an N-bit integer A_m, together with an exponent z_m, such that

$$z_m = \left\lfloor \log_2 \frac{1}{a_m} \right\rfloor \quad \text{and} \quad a_m = 2^{-z_m} \underbrace{(2^{-N}A_m)}_{A'_m} \tag{2.32}$$

The quantity $A'_m = 2^{-N}A_m \in [\frac{1}{2}, 1)$ is an N-bit binary fraction of the form

$$A'_m = 0.\underbrace{1aa\ldots a}_{N \text{ bits}} \tag{2.33}$$

and the quantity z_m is the number of leading 0's in the binary fraction representation of a_m, that is,

$$a_m = 0.\underbrace{00\ldots0}_{z_m \text{ zeros}}\underbrace{1aa\ldots a}_{A_m} \tag{2.34}$$

Next, we represent all probabilities approximately using P-bit integers p_x, such that

$$f_X(x) \approx p'_x = 2^{-P}p_x, \quad x \in \mathcal{A}_X \tag{2.35}$$

The interval length is then updated according to Equation (2.29) and rounded down, if necessary, to the closest representation of the form in Equation (2.34). Together, these operations are embodied by the following algorithm.

Algorithm.   *Finite Precision Interval-Length Manipulation*

- Set $T \leftarrow A_{m-1}p_{x_m}$ and $z_m \leftarrow z_{m-1}$
  Note that $T' = 2^{-(N+P)}T$ is an $(N+P)$-bit binary fraction, with $T' = A'_{m-1}p'_{x_m}$
- While $T < 2^{N+P-1}$ (i.e., while $T' < \frac{1}{2}$)
  − increment $z_m \leftarrow z_m + 1$
  − shift $T \leftarrow 2T$
- Set $A_m = \lfloor 2^{-P}T \rfloor$

Evidently, we have made two approximations that lead to slight losses in coding efficiency: (1) the probabilities are approximated with finite precision representations; and (2) the rounding operation in the last line of the algorithm generally reduces the interval length and hence increases the number of code bits by an amount strictly less than

$$\log_2 \frac{2^{N-1}+1}{2^{N-1}} \approx 2^{-(N-1)} \log_2 e \text{ bits} \tag{2.36}$$

We have now only to describe the manipulation of the interval lower bound, c_m. Since the PMF $f_X(x)$ is approximated by P-bit binary fractions p'_x, the thresholds $t_X(x)$ are also approximated by P-bit binary fractions

$$T'_X(x) = \sum_{x'<x} p'_{x'} = 2^{-P}T_X(x) \tag{2.37}$$

From the update equation (2.30) we deduce that c_m is an $(N + P + z_m)$-bit binary fraction of the form

$$c_m = 0.\underbrace{xx\ldots x}_{z_m \text{ bits}}\underbrace{cc\ldots c}_{C_m} \tag{2.38}$$

Let C_m be the integer formed from the least significant $N + P$ bits of this representation. Then the update operation consists of adding two $(N + P)$-bit integers, $A_m T_{x_m}$ and C_m, and propagating any carry bit into the initial z_m-bit prefix of c_m.

At first glance it appears that the need to resolve a carry will force us to buffer the entire z_m-bit prefix of c_m. Fortunately, however, carry propagation can affect only the least significant $r_m + 1$ bits of this prefix, where r_m is the number of consecutive least significant 1's. In fact, no future coding operations may have any effect on the more significant bits in the prefix. Consequently, the initial $z_m - r_m - 1$ bits of the codeword may be sent to the decoder immediately, so the encoder need not allocate storage for them. The binary fraction representation of the evolving codeword then consists of three key segments, as follows:

$$0.\underbrace{xxxxx\ldots x}_{z_m - r_m - 1 \text{ bits}}\underbrace{011\ldots 1}_{r_m + 1 \text{ bits}}\underbrace{cc\ldots c}_{C_m}$$

The complete encoding algorithm is shown below for the special case of a binary alphabet $\mathcal{A}_X = \{0, 1\}$. As we shall see in the next section, binary arithmetic coding is sufficient for the efficient compression of any source, regardless of its alphabet. Note that we drop the subscripts on the state variables A_m, C_m, r_m, and z_m. Also note that we need to introduce a special state, identified by $r = -1$, to deal with the possibility that a carry may occur when $r = 0$, causing the 0 bit to flip to a 1 with no subsequent 0 bits. Since future carries can never propagate this far, it is sufficient to flag the unusual condition by setting $r = -1$, which has the interpretation that the central segment in the binary fraction representation of c_m is empty and can remain empty until a zero bit is shifted out of C_m.

Algorithm. *Binary Arithmetic Encoder*

Initialize $C = 0$, $A = 2^N$, $r = -1$, $z = 0$
For each $m = 1, 2, \ldots$,

- Set $T \leftarrow A p_0$
- If $x_m = 1$,
    − set $C \leftarrow C + T$ and $T \leftarrow A p_1 = 2^P A - T$
- If $C \geq 2^{N+P}$, do "propagate carry"
  Subalgorithm. *Propagate Carry.*
  emit-bit(1)
  If $r > 0$,
      − execute $r - 1$ times, **emit-bit(0)**
      − set $r \leftarrow 0$
  else
      − set $r \leftarrow -1$
- While $T < 2^{N+P-1}$, do "renormalize once"
  Subalgorithm. *Renormalize Once.*
  Increment $z \leftarrow z + 1$
  Shift $T \leftarrow 2T$ and $C \leftarrow 2C$
  If $C \geq 2^{N+P}$ (pushing a "1" out of C)

– If $r < 0$, **emit-bit(1)**
  – else, increment $r \leftarrow r + 1$
 else
   – If $r \geq 0$, **emit-bit(0)** and execute r times, **emit-bit(1)**
   – Set $r = 0$
- Set $A \leftarrow \lfloor 2^{-P} T \rfloor$

After each iteration of the algorithm, the number of bits that have actually been output is given by $z - r - 1$, but if this quantity is of no interest the state variable z may be dropped.

We now describe a binary arithmetic decoding algorithm. The decoder maintains an N-bit state variable A, which represents the current interval width a_m, exactly as in the encoder, following identical update procedures. The decoder also maintains an $(N + P)$-bit state variable C; however, the interpretation of this quantity is somewhat different to that in the encoder.

To develop the decoding algorithm, let β denote the value represented by the entire encoded message, taken as a binary fraction. Then

$$\beta \in [c_m, c_m + a_m), \quad \forall m \tag{2.39}$$

Suppose we have correctly decoded x_1 through x_m, and the decoder has reproduced the evolution of a_m in the encoder. We could keep track of c_m in the decoder and then decode x_m, according to

$$x_m = \begin{cases} 0 & \text{if } \beta < c_m + a_{m-1}p'_0 \\ 1 & \text{if } \beta \geq c_m + a_{m-1}p'_0 \end{cases} \tag{2.40}$$

It is simpler, however, to keep track of the quantity $\beta - c_m$ and then decode x_m, according to

$$x_m = \begin{cases} 0 & \text{if } \beta - c_{m-1} < a_{m-1}p'_0 \\ 1 & \text{if } \beta - c_{m-1} \geq a_{m-1}p'_0 \end{cases} \tag{2.41}$$

To see why this is simpler note that $\beta - c_m \in [0, a_m)$, where a_m has the binary fraction representation

$$a_m = 0.\underbrace{00\ldots0}_{z_m \text{ zeros}}\underbrace{1aa\ldots a}_{A_m} \tag{2.42}$$

and $a_m p'_0$ has the binary fraction representation

$$a_m p'_0 = 0.\underbrace{00\ldots0}_{z_m \text{ zeros}}\underbrace{xx\ldots x}_{A_m p_0} \tag{2.43}$$

It follows that the z_m-bit prefix of $\beta - c_m$ is zero and the decision in Equation (2.41) may be formed using the next $N + P$ bits of $\beta - c_m$. This is the quantity managed by the decoder's state variable C_m. The binary fraction representation of $c - c_m$ has the structure

$$c - c_m = 0.\underbrace{00\ldots0}_{z_m \text{ zeros}}\underbrace{cc\ldots c}_{C_m}bbb\ldots \tag{2.44}$$

where the suffix "$bbb\ldots$," represents remaining bits in the encoded bit-stream, which have not yet been imported by the decoder. The decoding algorithm follows immediately:

Algorithm.   *Binary Arithmetic Decoder.*

Initialize $A = 2^N$, $z = 0$

Import $N + P$ bits from the bit-stream to initialize C.

For each $m = 1, 2, \ldots$,

- Set $T \leftarrow Ap_0$
- If $C < T$,
    - output $x_m = 0$
- else
    - output $x_m = 1$
    - update $C \leftarrow C - T$ and $T \leftarrow 2^P A - T$
- While $T < 2^{N+P-1}$, do "renormalize once"
  Subalgorithm.   *Renormalize Once.*
    Increment $z \leftarrow z + 1$
    Shift $T \leftarrow 2T$ and $C \leftarrow 2C$
    Load $C \leftarrow C + \textbf{load-next-bit}()$
- Set $A \leftarrow \lfloor 2^{-P} T \rfloor$

The incremental nature of the decoding procedure is particularly worth noting. Unlike the Elias algorithm, the amount by which the decoder must read ahead in the bit-stream in order to unambiguously decode the next symbol is bounded by $N + P$ bits. The decoder is also somewhat simpler than the encoder, since it need not deal with the effects of carry propagation.

2.3.4   CONDITIONAL CODING AND CONTEXT MODELING

We have seen that arithmetic coding is able to approach the entropy of a source with negligible loss, while experiencing surprisingly low complexity. As mentioned in Section 2.2.3, however, what we are really interested in is coding outcomes with the conditional entropy $H(X \mid Y)$, where $\mathbf{Y}$ represents previously encoded context information, whose outcome $\mathbf{y}$ is already known. In the context of arithmetic coding, this can be achieved with remarkable ease, by simply replacing the probabilities $f_X(x)$ with conditional probabilities $f_{X|Y}(x \mid y)$. In the particular case of the binary arithmetic encoding and decoding algorithms presented in the preceding subsection, we have only to replace the term p_0 by a position-dependent term $p_{0,m}$, which is a P-bit approximation of the conditional probability $f_{X_m|\mathbf{Y}_m}(0 \mid \mathbf{y}_m)$. Here, $\mathbf{y}_m$ denotes the observed context within which x_m is to be coded.

Since $\mathbf{y}_m$ depends upon previous outcomes, the incremental decoding process described above will have already recovered these values so that it can form exactly the same value for p_0 as that formed in the encoder. The encoder and decoder thus both form exactly the same partition of the previous interval $\mathcal{I}_{m-1}$ according to the agreed conditional probability distribution, $f_{X_m|\mathbf{Y}_m}(0 \mid \mathbf{y}_m)$. Unlike variable-length coding, where each context requires a separate codebook, arithmetic coding computes the interval subdivisions dynamically, without the need to explicitly store codebooks. This makes arithmetic coding particularly attractive for efficient, incremental conditional coding. There is, however, the question of how the encoder and decoder should decide on an agreed set of conditional probabilities, $f_{X_m|\mathbf{Y}_m}(0 \mid \mathbf{y}_m)$. Note that we are restricting our consideration here to binary alphabets, so one probability must be estimated and stored for each distinct context vector, $\mathbf{y}$.

2.3.4.1 Adaptive Probability Estimation

One of the most significant advantages of arithmetic coding is that the probabilities $p_{o,m}$ may be changed dynamically. This allows us to adaptively estimate the probability distribution associated with each context, based on information that has previously been encoded. Suppose, for example, that we maintain two counters $K_{0,\mathbf{y}}^m$ and $K_{1,\mathbf{y}}^m$, for each context $\mathbf{y}$, representing the number of times an outcome of $x_i = 0$ and $x_i = 1$ (respectively) has been seen when coding a previous sample x_i in the same context, $\mathbf{y}$. That is,

$$K_{0,\mathbf{y}}^m = \sum_{i=1}^{m-1} \delta(\mathbf{y} - \mathbf{y}_i)(1 - x_i) \quad \text{and} \quad K_{1,\mathbf{y}}^m = \sum_{i=1}^{m-1} \delta(\mathbf{y} - \mathbf{y}_i)x_i \tag{2.45}$$

Since the values of $K_{0,\mathbf{y}}^m$ and $K_{1,\mathbf{y}}^m$ depend only on previously decoded samples, the decoder can form the same counts as the encoder. Both encoder and decoder can then use the following simple estimate for $f_{X_m|\mathbf{Y}_m}(0 \mid \mathbf{y}_m)$:

$$p'_{0,m} = \frac{K_{0,\mathbf{y}}^m + \Delta}{K_{0,\mathbf{y}}^m + K_{1,\mathbf{y}}^m + 2\Delta} \tag{2.46}$$

The offset $\Delta > 0$ is included here to account for the fact that a reliable estimate of $f_{X_m|\mathbf{Y}_m}(0 \mid \mathbf{y}_m)$ cannot be formed until the counters attain values considerably greater than 0. For such *scaled-count* estimators, Δ is typically chosen in the range $\frac{1}{2}$ to 1 [20], depending on how skewed we expect the probability distribution to be.

We may think of the scaled-count estimator as a state machine, since our conditional probability model for context $\mathbf{y}$ depends on the state $(K_{0,\mathbf{y}}, K_{1,\mathbf{y}})$, of the two counters. Each time a sample is coded in context $\mathbf{y}$, the state is updated according to the rules expressed in Equation (2.45). One of the most significant developments in practical arithmetic coding has been the realization that renormalization events can be used to probabilistically gate transitions in the estimation state machine. Specifically, the state is updated immediately after any symbol encoding/decoding operation, which involves one or more calls to the "*Renormalize Once*" routine. The new state identifies the probability estimates to be used for all subsequent symbols, until a further renormalization event induces another transition in the state machine. It also turns out that the number of states associated with each probability model can be made much smaller than one might expect from our previous discussion of scaled-count estimation. This is done by collapsing states, which correspond to similar conditional probabilities.

To illustrate the above principles, Table 2.1 provides the state transition table from the MQ binary arithmetic coder, which is employed by the JPEG2000 [7] and JBIG2 [6] image compression standards. Like many other variants, the MQ coder employs a convenient trick to halve the size of the transition table. By maintaining a separate state bit $s_{\mathbf{y}}$, which provides the identity of the outcome that is currently considered most likely within context $\mathbf{y}$. We say that x_m is a most probable symbol (MPS) if it is equal to $s_{\mathbf{y}}$ when coded in context $\mathbf{y}$. Conversely, we say that x_m is a least probable symbol (LPS) if it is equal to $1 - s_{\mathbf{y}}$ when encoded in context $\mathbf{y}$. The Σ_{mps} and Σ_{lps} columns identify the new state to be used in the event that a renormalization event occurs when coding an MPS or LPS, respectively. The X_s column holds 1 if the identity of the MPS is to be switched, and the p'_{lps} column holds the estimated LPS probability associated with each state, Σ. For greater insight into the origin of this state transition table, the reader is referred to [2,18]. For our purposes here, however, the key point to observe is that the probability model associated with each context may be adaptively estimated with remarkably low complexity using a simple state machine, having

TABLE 2.1
Probability State Transition Table for the MQ Coder

Σ	Transition			Estimate	Σ	Transition			Estimate
	Σ_{mps}	Σ_{lps}	X_s	p'_{lps}		Σ_{mps}	Σ_{lps}	X_s	p'_{lps}
0	1	1	1	0.475	24	25	22	0	0.155
1	2	6	0	0.292	25	26	23	0	0.132
2	3	9	0	0.132	26	27	24	0	0.121
3	4	12	0	0.0593	27	28	25	0	0.110
4	5	29	0	0.0283	28	29	26	0	0.0993
5	38	33	0	0.0117	29	30	27	0	0.0938
6	7	6	1	0.475	30	31	28	0	0.0593
7	8	14	0	0.463	31	32	29	0	0.0499
8	9	14	0	0.397	32	33	30	0	0.0476
9	10	14	0	0.309	33	34	31	0	0.0283
10	11	17	0	0.265	34	35	32	0	0.0235
11	12	18	0	0.199	35	36	33	0	0.0145
12	13	20	0	0.155	36	37	34	0	0.0117
13	29	21	0	0.121	37	38	35	0	0.00692
14	15	14	1	0.475	38	39	36	0	0.00588
15	16	14	0	0.463	39	40	37	0	0.00287
16	17	15	0	0.447	40	41	38	0	0.00157
17	18	16	0	0.397	41	42	39	0	0.000797
18	19	17	0	0.309	42	43	40	0	0.000453
19	20	18	0	0.292	43	44	41	0	0.000194
20	21	19	0	0.265	44	45	42	0	0.000108
21	22	19	0	0.221	45	45	43	0	0.000022
22	23	20	0	0.199	46	46	46	0	0.475
23	24	21	0	0.188					

relatively few states, whose transitions occur only during renormalization events. Noting that renormalization always involves the generation of new code bits, there will be at most one transition per coded bit.

In light of the previous discussion, one might suspect that the best way to minimize the coded bit-rate of an information source is to use as many previously coded symbols as possible to form the context **y**, within which to code the next symbol. In this way, the statistical dependencies between numerous symbols can be simultaneously exploited. Of course, there will be a practical limit to the number of different contexts for which we can afford to maintain a separate probability model, but the estimator of Table 2.1 requires less than 1 byte to store the state associated with each model. Unfortunately, there is a more subtle cost to be paid by tracking too many different contexts. This cost, known as the *learning penalty*, arises from the fact that the first few symbols within any context are generally coded using inappropriate probability estimates. As more symbols are coded within that context, the probability estimates stabilize (assuming the source is approximately stationary) and coding becomes more efficient. Regardless of the specific method used to estimate context probability models, it is clear that the creation of too many contexts is undesirable. If two distinct coding contexts exhibit identical conditional PMF's, their combined learning penalty may clearly be halved by merging the contexts.

2.3.4.2 Binary Arithmetic Coding Is Enough

We have mentioned earlier that it is sufficient to consider only binary arithmetic coders. To see why this is so, suppose that $\mathcal{A}_X$ is not a binary alphabet, having 2^K entries for some integer $K > 1$.[3] Then each element of $\mathcal{A}_X$ may be represented by a K bit integer. In this way, the random variable X is equivalent to a K-dimensional random vector $\mathbf{B}$, where

$$\mathbf{B} = \begin{pmatrix} B_0 \ (\text{MSB}) \\ B_1 \\ \vdots \\ B_{K-1} \ (\text{LSB}) \end{pmatrix} \tag{2.47}$$

and the B_k are binary random variables representing the digits in the K-bit representation of X. Then

$$H(X) = H(\mathbf{B})$$
$$= H(B_0) + H(B_1 \mid B_0) + \cdots + H(B_{K-1} \mid B_0, \ldots, B_{K-2}) \tag{2.48}$$

We have thus only to encode the individual binary digits of X, being careful to condition our coding of each successive digit b_k, on the values of all previously coded digits, $b_0, \ldots, b_{k-1}$. Since a binary arithmetic coder can essentially achieve the conditional entropy, the total bit-rate associated with coding all of the digits in this fashion will essentially be equal to the entropy of the source. Of course, additional context information may be used from previously coded samples, so as to exploit the statistical dependencies between samples as well as between the digits of each sample.

2.3.5 ARITHMETIC CODING VARIANTS

In Section 2.3.3, we introduced one practical arithmetic coding algorithm. This algorithm requires one integer-valued multiplication per coded symbol, in addition to a number of addition/subtraction and comparison operations. In hardware implementations in particular, multiplication is much more expensive than addition. As a result, numerous arithmetic coding variants have been proposed, wherein the multiplication operation is approximated in one way or another. Most of these schemes were proposed in the late 1980s and early 1990s.

Multiplier-free operation and renormalization-driven probability estimation are the most distinguishing features of a broad class of arithmetic coding algorithms, which includes the Q coder [12], QM coder, [11], MQ coder and Z coder [1]. The basic idea is to replace the computation

$$T \leftarrow A_{m-1} p_{0,m} \tag{2.49}$$

found in the algorithms described hitherto, with

$$T \leftarrow \bar{p}_{\text{lps},m} = 2^N \alpha p_{\text{lps},m} \tag{2.50}$$

where $2^N \alpha \approx 2^N \frac{3}{4}$ may be interpreted as a fixed estimate of the "average" value of A_{m-1}, and the algorithm is adjusted so as to code a 1 if $x_m = s_{\mathbf{y}}$ (MPS) and a 0 if $x_m = 1 - s_{\mathbf{y}}$ (LPS). For further details regarding multiplier-free arithmetic coding and the determination of optimal values for α, the reader is referred to [8,18]. We note here, however, that the actual value of $\bar{p}_{\text{lps}}$ that is used in the MQ coder may be derived from the p'_{lps} values reported in Table 2.1 by setting $\alpha = 0.708$.

---

[3] If the size of the alphabet is not a power of 2, we can always fill it out to a whole power of 2, by adding extra symbols, each having zero probability; the extra symbols then have no impact on the entropy.

In addition to the method used to scale interval lengths (discussed above) and the method used to adaptively estimate probabilities for each context, arithmetic coder variants may also be distinguished according to the way in which they handle carry propagation effects in the encoder. There are two principal methods for handling carry propagation. The first is the method described in Section 2.3.3 above, in which a state variable r is used to keep track of the number of consecutive 1's, which are immediately more significant than the bits in the C register. This approach is taken by the Q- and QM-coder variants. A second method for handling carry propagation is to employ a *bit-stuffing* technique to deliberately open up small holes in the code-space, in such a way as to limit the number of bit positions over which a carry can propagate. This has a number of practical implementation advantages, which come at a very small cost in compression efficiency. Moreover, the holes that are introduced into the code-space represent a form of redundancy, which can readily be exploited for error detection. This approach is taken by the MQ coder, adopted by JPEG2000. In this coder, the presence of bit-stuffing means that no error-free bit-stream will ever contain two consecutive bytes whose numeric value lies in the range FF90 to FFFF. Special marker codes, with values in this range, allow a JPEG2000 decoder to resynchronize itself in the event that errors are detected. For a thorough description of the MQ coder and its bit-stuffing procedure, the reader is referred to [18].

2.3.6 ARITHMETIC CODING IN JBIG

The JBIG bi-level image compression standard [5] presents us with a relatively simple, yet effective example of the use of context-adaptive binary arithmetic coding. The Joint Bi-level Image experts Group (JBIG) committee was formed under the auspices of the International Standards Organization (ISO) to establish a standard for efficient lossless coding of bi-level, i.e., black and white images. The target source material for JBIG includes graphics and line artwork, scanned text, and halftoned images.

At the core of the JBIG algorithm is a conditional arithmetic coding scheme, based around the QM coder. The QM coder uses a related, but different probability state transition table to that shown in Table 2.1. It is a multiplier-free variant, following the principles described in the preceding subsection, with full carry propagation resolution as opposed to bit-stuffing. In JBIG, a separate context label is assigned to each of the 2^{10} possible neighborhood vectors arising from the selection of one or the other of the two neighborhood configurations shown in Figure 2.4. The first of these contexts typically gives an improvement of around 5% in bit-rate over the second; the second form is provided primarily for memory critical applications in which the need to store two preceding rows of image pixels may be prohibitive.

The JBIG standard has a number of other significant features that are of less interest to us in this present treatment, but are worth mentioning in passing. The standard provides a mechanism for modifying the context state vector from time to time by inserting certain control codes in the bit-stream. In this mode, one of the pixels used to form the context vector **y** is a *floating pixel*,

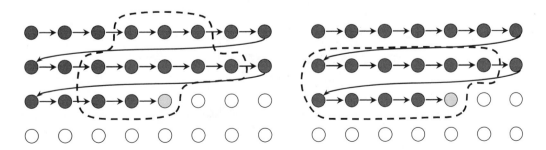

FIGURE 2.4 Neighborhoods used by the JBIG image compression standard to generate 10-bit contexts.

which has the form $x[m_1 - k_1, m_2 - k_2]$, where k_1 and k_2 are vertical and horizontal lag parameters controlling the position of this floating context pixel, relative to the location of the symbol $x[m_1, m_2]$ that is being coded. An encoder may attempt to uncover underlying periodicity in the image, such as may arise when compressing halftoned gray-level images, setting the lag terms accordingly. JBIG also contains a progressive coding mode, wherein lower resolution versions of the image may be coded first and then progressively refined. This is done by including information from the lower resolution versions within the context vector **y** used to encode higher resolution information. For a more complete summary of the technical features of the JBIG lossless binary image compression algorithm, the interested reader is referred to [4]. It is also worth noting that JBIG has been superseded by a more recent bi-level image compression standard, JBIG2 [6]. JBIG2 builds upon the approach in JBIG in many ways; interestingly, JBIG2 replaces the QM coder with the MQ variant, which is also employed by JPEG2000.

2.4 INFORMATION SEQUENCING AND EMBEDDING

In the preceding sections, we have considered only the efficient representation of information. In this section, we turn to the other questions that we posed in the abstract — namely, the selection of suitable image approximations and the creation of embedded representations.

2.4.1 QUANTIZATION

For simplicity, we restrict our attention to gray-scale images. Such images can be represented using a two-dimensional sequence of intensity values $x[n_1, n_2]$, which we will frequently write more compactly as $x[\mathbf{n}]$, where **n** is a two-dimensional sequence index. Where necessary, we suppose n_1 as the row index, working *downwards* from 0 at the top of the image, and we suppose n_2 as a column index, working *rightwards* from 0 at the left of the image. The simplest way to form an approximate representation of the image is to apply uniform scalar quantization independently to each of its samples. That is, the image is approximated by

$$y[\mathbf{n}] = \Delta q[\mathbf{n}], \quad \text{where } q[\mathbf{n}] = \left\langle \frac{x[\mathbf{n}]}{\Delta} \right\rangle \tag{2.51}$$

Here, $\langle z \rangle$ means "round z to the nearest integer," and Δ is the *quantization step size*. We refer to $q[\mathbf{n}]$ as the *quantization index* for $x[\mathbf{n}]$.

As shown in Figure 2.5, the quantization index q identifies a unique *quantization interval* (or *bin*) $\mathcal{I}_q \subseteq \mathbb{R}$. The quantizer replaces each sample value $x[\mathbf{n}]$, with the index $q[\mathbf{n}]$ of the bin to which it belongs. Associated with each bin $\mathcal{I}_q$ is a *representation level*, y_q. The dequantizer obtains the approximated image by replacing each quantization index $q[\mathbf{n}]$ with the corresponding representation level, producing $y[\mathbf{n}] = y_{q[\mathbf{n}]}$. Equation (2.51) implements these operations for the simple case of

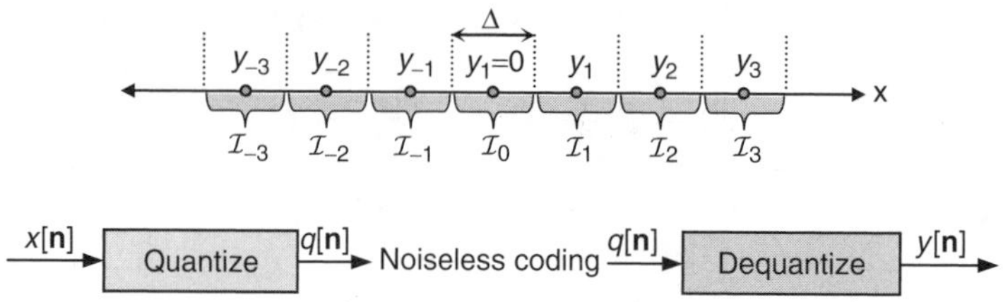

FIGURE 2.5 Quantization bins, indices, and representation levels illustrated for the simple case of uniform scalar quantization with step size Δ.

uniform quantization, wherein all bins have the same length Δ, and each representation level is at the center of its respective bin. In Section 2.4.3, however, we shall see the need to create quantization bins with different lengths.

In addition to quantization with nonuniform bins, it is possible to extend the notion of sample-by-sample quantization to that of *vector quantization*. In vector quantization, source samples are collected into groups (or vectors) and each vector is mapped to one of a number of quantization indices. Associated with each index, is a representation vector, with which the index is replaced during dequantization. Vector quantization has a number of advantages over the scalar quantization operation considered here. However, if the image is first transformed using a suitable linear transform, the advantages of vector quantization over scalar quantization become less significant. Indeed, such transforms are the subject of the next subsection. For brevity, therefore, we do not explicitly consider vector quantization in this chapter.

2.4.2 MULTIRESOLUTION COMPRESSION WITH WAVELETS

The discrete wavelet transform (DWT) provides a natural framework for scalable image compression. The DWT has already been introduced in Chapter 1, so we summarize only the most essential elements here. As shown in Figure 2.6, a first DWT stage decomposes the image into four subbands, denoted LL_1, HL_1 (horizontally high-pass), LH_1 (vertically high-pass), and HH_1. The LL_1 subband is a good low-resolution rendition of the original image, with half the width and height. The next DWT stage decomposes this LL_1 subband into four more subbands, denoted LL_2, LH_2, HL_2 and HH_2. The process continues for some number of stages D, producing a total of $3D + 1$ subbands whose samples represent the original image.

The multiresolution properties of the DWT arise from the fact that the LL_d subband is a good low-resolution image, whose dimensions are reduced by a factor of 2^d from those of the original image. The LL_d subband may be recovered from the subbands at levels $d + 1$ through D, by applying only the first $D - d$ stages of DWT synthesis. As long as the subbands are compressed independently,[4] we may convert a compressed image into a lower resolution compressed image, simply by discarding

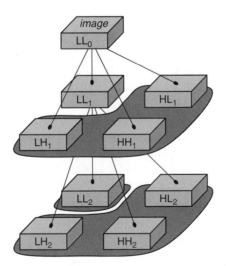

FIGURE 2.6 A $D = 2$ level DWT decomposition tree.

---

[4] Fully independent compression of the subbands is not actually required. It is sufficient to ensure that each subband from DWT stage d is compressed without reference to information from any of the subbands from DWT stages $d' < d$.

TABLE 2.2
Resolution Subsets of the "Bike" Image

Subset	% Bits	% Size	Bit-Rate (bps)
LL_0	100	100	1.00
LL_1	52.5	25.0	2.10
LL_2	20.4	6.25	3.26
LL_3	7.25	1.56	4.64

those subbands that are not required. The number of resolutions available in this way is $D + 1$. The lowest resolution is represented directly by the LL_D subband, while the highest resolution is that of the original image, which we may interpret as an LL_0 subband.

The property described above is known as *resolution scalability*. Specifically, a resolution scalable bit-stream is one from which a reduced resolution image may be obtained simply by discarding unwanted portions of the compressed data. The lower resolution representation should be identical to that which would have been obtained if the lower resolution image were compressed directly.

In addition to resolution scalability, the DWT also ensures that most of the subband samples are likely to take values which are close to zero, except in the neighborhood of image edges. This property, known as *energy compaction,* allows comparatively simple quantization and coding techniques to achieve high compression efficiency for a selected level of image quality.

Although resolution scalability (the ability to discard high-frequency subbands) provides a crude mechanism for trading image quality for compressed file size, this turns out to be much less useful than one might expect. Table 2.2 shows the reason; starting with a 2560×2048 photographic test image, "Bike," compressed to 1.0 bits/sample,[5] we might discard the higher frequency subbands to reduce the total number of compressed bits. Unfortunately, the number of compressed bits does not decrease in proportion to the number of samples. For example, discarding the two highest DWT levels divides the number of image samples by 16, but reduces the compressed file size by less than five times.

The real problem here is that the LL_2 image is available at an effective compressed bit-rate of 3.26 bits/sample, which is far more than we need to reconstruct an image at this resolution (640×480) with high visual quality. The additional bits are required only when the LL_2 image is used in the reconstruction of a higher resolution image, where small errors in the LL_2 component can be greatly magnified during the synthesis process. This example serves to highlight the fact that resolution scalability is of relatively little use unless coupled with a second type of scalability, known as *distortion scalability*. Specifically, we would like to be able to strip away some of the bits associated with the lower frequency subbands (this increases their distortion), when the image is reconstructed at a reduced resolution. Distortion scalability is the subject of the next subsection.

2.4.3 EMBEDDED QUANTIZATION AND BIT-PLANE CODING

A distortion scalable bit-stream is one in which more coarsely quantized representations are embedded within more finely quantized representations. Among other virtues, distortion scalability allows us to solve the problem presented in Table 2.2. Specifically, it allows us to reduce the accuracy (and hence compressed size) of lower resolution subbands, when the image is to be reconstructed at a reduced size, thereby yielding a reduction in file size, which is comparable to the reduction in image size.

---

[5] Details of the compression algorithm are irrelevant here, but JPEG2000 was used to obtain these results.

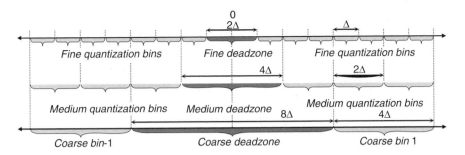

FIGURE 2.7 Three embedded deadzone quantizers.

Since prefixes of an embedded distortion scalable bit-stream must correspond to successively finer quantizations of the sample data, embedded coders are necessarily associated with a family of quantizers. In fact, these quantizers (whether scalar or vector) are inevitably embedded [19, §4B], in the sense that the quantization bins associated with a finer quantizer must be completely contained within those of each coarser quantizer. Many scalable image compressors use scalar *deadzone quantizers*, having the structure depicted in Figure 2.7. The central quantization bin, corresponding to those sample values that get quantized to 0, is known as the *deadzone*. By making the deadzone twice as large as the nonzero quantization bins, a family of embedded quantizers arises when the step size Δ is halved between successive members of the family. The enlarged deadzone, sometimes called a *fat zero*, is also helpful when coding high-frequency subband samples; these tend to be close to zero, except in the neighborhood of appropriately oriented image edges and other important features. The numerous zero-valued quantization indices, produced by samples falling within the deadzone, lead to highly skewed statistics, that are amenable to efficient coding.

The embedded quantization structure of Figure 2.7 may be conveniently associated with the bit-planes in a sign-magnitude representation of the subband samples. Let $x_b[\mathbf{n}] \equiv x_b[n_1, n_2]$ denote the two-dimensional sequence of subband samples associated with DWT subband b, and let $q_b^{(0)}[\mathbf{n}]$ denote the quantization indices (the bin labels) associated with the finest deadzone quantizer for this subband, having step size Δ_b. Then

$$q_b^{(0)}[\mathbf{n}] = \text{sign}(x_b[\mathbf{n}]) \cdot \left\lfloor \frac{|x_b[\mathbf{n}]|}{\Delta_b} \right\rfloor \qquad (2.52)$$

Letting $q_b^{(p)}[\mathbf{n}]$ denote the indices of the coarser quantizer with step size $2^p \Delta_b$, we find that

$$q_b^{(p)}[\mathbf{n}] = \text{sign}(x_b[\mathbf{n}]) \cdot \left\lfloor \frac{|x_b[\mathbf{n}]|}{2^p \Delta_b} \right\rfloor = \text{sign}(q_b^{(0)}[\mathbf{n}]) \cdot \left\lfloor \frac{\left|q_b^{(0)}[\mathbf{n}]\right|}{2^p} \right\rfloor \qquad (2.53)$$

Thus the coarser quantization indices, $q_b^{(p)}[\mathbf{n}]$, are obtained simply by discarding the least significant p bits from the binary representation of the finer quantization indices' magnitudes, $|q_b^{(0)}[\mathbf{n}]|$.

Based on the above observation, an embedded bit-stream may be formed in the manner suggested by Figure 2.8. Assuming a K-bit magnitude representation, the coarsest quantization indices $|q_b^{(K-1)}[\mathbf{n}]|$ are represented by the most significant magnitude bit of each sample, together with the signs of those samples whose magnitudes are not quantized to zero. A *bit-plane coder* walks through each of the samples, coding these bits first. If the bit-stream is truncated at this point, the decoder receives the coarsest quantization indices, $q_b^{(K-1)}[\mathbf{n}]$. The bit-plane coder then moves to the next

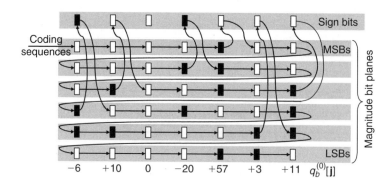

FIGURE 2.8 Bit-plane coding sequence, shown for $K = 6$ magnitude bit-planes. Black and white boxes correspond to values of "1"and "0,"respectively.

magnitude bit-plane, also coding the sign of any sample whose magnitude first becomes nonzero in this bit-plane. If the bit-stream is truncated after this point, the decoder receives the finer quantization indices $q_b^{(K-2)}[\mathbf{n}]$, and so forth.

We now remind the reader of the conclusions from Sections 2.2.3 and 2.3.4 it makes no difference in what order we code the binary digits which make up the subband samples $x_b[\mathbf{n}]$. As long as we exploit the substantial redundancy which exists between successive bit-planes (e.g., by careful use of conditional arithmetic coding), an embedded bit-stream can have the same coding efficiency as a nonembedded bit-stream in which each sample is coded completely before moving to the next. Thus, distortion scalability would appear to be achievable without any cost in coding efficiency. The reader is reminded, however, that the requirement for successive quantizers to be embedded within one another does impose a constraint on the design of the quantizer bins themselves. For many information sources, this can be shown to represent a source of suboptimality with respect to the optimal achievable nonscalable compression performance [3].

2.4.4 FRACTIONAL BIT-PLANE CODING

In the bit-plane coding procedure described above, the only natural truncation points for the embedded bit-stream are the bit-plane end points. These correspond to K different quantizations of the original subband samples, whose quantization step sizes are related by powers of 2. It is possible to achieve a more finely embedded bit-stream, with many more useful truncation points, by coding the information in each bit-plane over a series of two or more coding passes. The first coding pass for each bit-plane codes the relevant magnitude bit and any necessary sign bit, only for those samples that are likely to yield the largest reduction in distortion relative to their coding cost. Conversely, the last coding pass in each bit-plane processes those samples that are expected to be least effective in reducing distortion, relative to their cost in increased bit-rate. Together, the coding passes code exactly one new magnitude bit for every sample in the subband.

Figure 2.9 provides an illustration of these ideas, identifying the effects of different truncation lengths on the expected distortion in the reconstructed subband samples. The straight line segments between coding pass end points represent the expected mean squared error (MSE) and coded length that would result if the bit-stream were truncated part-way through a coding pass. As suggested by the figure, the operational distortion-length characteristic of a fractional bit-plane coder generally lies below (lower distortion) that of a regular bit-plane coder, except at the bit-plane end points. Importantly, this is a consequence only of information reordering. As long as the order in which samples are visited by the coding passes is dependent only on information that has already been

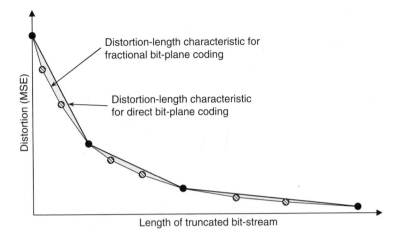

FIGURE 2.9 Effects of fractional bit-planes on the operational distortion-length characteristics of an embedded bit-stream subject to truncation. Solid dots show the bit-plane end points, while hatched dots correspond to the end points of the fractional bit-plane coding passes.

encoded, we need not send any additional bits to inform the decoder of the coding sequence. We refer to this as *context-induced* information sequencing.

The concept of fractional bit-plane coding may be found in many successful scalable image compression algorithms, albeit in different forms. The well-known SPIHT [13] algorithm builds a quad-tree structure across the DWT resolutions and uses this structure to derive the context information that determines which bits will be coded next. In Section 2.5, we describe more carefully how the fractional bit-plane procedure operates in the EBCOT [17] algorithm at the heart of JPEG2000. Indeed, fractional bit-plane coding in JPEG2000 is the culmination of several research efforts [10,15,16]. We also note that a more elaborate strategy for fractional bit-plane coding was proposed in [9], wherein information from one bit-plane might not be coded until after the coding of a less significant bit-plane has begun, depending on the expected reduction in distortion relative to coding cost.

2.4.5   CODING VS. ORDERING

The DWT is an important tool in the construction of resolution scalable bit-streams, while embedded quantization is the key to distortion scalability. A bit-stream that is both distortion and resolution scalable must contain identifiable elements, which successively reduce the distortion of the subbands in each resolution level. This is illustrated stylistically in Figure 2.10, where the relevant elements lie at the intersection of each distortion level Q_l (assume Q as standing for "quality"), with each resolution level R_r. Here, resolution level R_0 consists of the LL_D subband of a D-level DWT such as that shown in Figure 2.6, while higher resolution levels R_r each consist of the detail subbands LH_d, HL_d, and HH_d at depth $d = D + 1 - r$. The three resolution levels in Figure 2.6 are identified by shaded regions. Figure 2.10 shows two potential ways of ordering the elements in the compressed bit-stream, so as to create embedded representations that progress by resolution and quality, respectively.

It is important to observe that dependencies introduced while coding the embedded quantization indices (e.g., bit-planes) can destroy one or more degrees of scalability. For example, the tree coding structure employed by the SPIHT [13] algorithm introduces downward dependencies between resolution levels, which interfere with resolution scalability. Specifically, in SPIHT, it is not possible to discard all of the information associated with the higher resolution subbands and still decode lower resolution subbands.

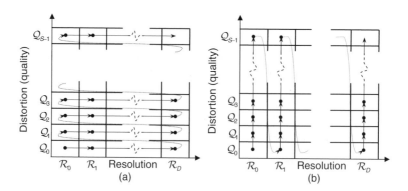

FIGURE 2.10 Elements of a distortion and resolution scalable bit-stream, with (a) quality progressive and (b) resolution progressive orderings.

We have seen that the order in which information is coded does not affect the ultimate compression efficiency we can hope to achieve. However, this is true only to the extent that we exploit the information available from previously coded symbols in the coding of later symbols. Indeed, we have seen that such context information can be important both for achieving the conditional entropy (Section 2.3.4) and for dynamically generating scanning orders with good embedding properties (Section 2.4.4). If we wish to be able to follow anything other than a predetermined one-dimensional progression through the two-dimensional scalability space depicted in Figure 2.10, we must deliberately sacrifice some of this contextual information. One particularly flexible way to do this is to partition the subband samples into small blocks and to code each block independently. Context information is exploited within a block, but not between different blocks. In the next section we shall see some of the benefits of this approach.

2.5 OVERVIEW OF EBCOT

In this section, we describe the embedded block coding with optimal truncation (EBCOT) algorithm, which lies at the heart of the JPEG2000 [7] image compression standard. The image is first subjected to a D level DWT, after which each subband is partitioned into relatively small blocks, known as *code-blocks*, having typical dimensions of 64×64 or 32×32 samples each. This is illustrated in Figure 2.11. Each code-block $\mathcal{B}_i$ is coded independently, producing its own embedded bit-stream, with many useful truncation points. In the ensuing subsections, we first describe the block coding algorithm itself; we then describe how truncation points may be selected for each individual code-block so as to minimize the overall reconstructed image distortion, for a given constraint on the bit-rate. Finally, we describe EBCOT's *abstract quality layer* paradigm.

2.5.1 EMBEDDED BLOCK CODING PRIMITIVES

In EBCOT, contexts consisting of previously coded data bits are used both for probability modeling (so that arithmetic coding can approach the relevant source conditional entropy) and for information sequencing (i.e., for defining the fractional bit-plane coding passes). Ideally, if we were able to exploit all previously coded data bits when encoding new bits, we would expect to find that the coding efficiency itself is unaffected by the coding order. Even in the practical case, where the contexts used for probability modeling must be much more heavily restricted, we find that coding efficiency is substantially independent of coding order. For this reason, we begin our description of the block coder by describing the *conditional coding primitives*; these are the same, regardless of the order in which samples are coded.

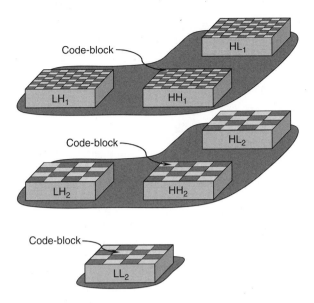

FIGURE 2.11 Division of the subbands from Figure 2.6 into code-blocks.

Following the notation developed in Section 2.4.3, let $q[\mathbf{j}] \equiv q[j_1, j_2]$ denote the sequence of quantization indices for a particular code-block, obtained after deadzone quantization with a quantization step size of Δ. Write $\chi[\mathbf{j}]$ for the sign and $v[\mathbf{j}]$ for the magnitude of $q[\mathbf{j}]$, and let $v^{(p)}[\mathbf{j}] = \lfloor 2^{-p} v[\mathbf{j}] \rfloor$ be the magnitude of the coarser quantization indices, corresponding to a step size of $\Delta 2^p$; these are obtained simply by discarding the least significant p bits from $v[\mathbf{j}]$. Finally, write $v^p[\mathbf{j}] \in \{0, 1\}$ for the least significant bit of $v^{(p)}[\mathbf{j}]$, which is also the value found in bit-plane p of $v[\mathbf{j}]$, and let K be the number of magnitude bit-planes that must be coded to capture the information in $v[\mathbf{j}]$.

Image subband samples tend to exhibit distributions that are heavily skewed toward small amplitudes. As a result, when $v^{(p+1)}[\mathbf{j}] = 0$, meaning that $x[\mathbf{j}]$ lies within the deadzone for the quantizer with step size $\Delta 2^{p+1}$, we can expect that $x[\mathbf{j}]$ is also likely to lie within the deadzone for the finer quantizer with step size $\Delta 2^p$, i.e., that $v^{(p)}[\mathbf{j}] = 0$. Equivalently, the conditional PMF, $f_{V^P | V^{(p+1)}}(v^p, 0)$, is heavily skewed toward the outcome $v^p = 0$. For this reason, an important element in the construction of efficient coding contexts is the so-called *significance* of a sample, defined by

$$\sigma^{(p)}[\mathbf{j}] = \begin{cases} 1 & \text{if } v^{(p)}[\mathbf{j}] > 0 \\ 0 & \text{if } v^{(p)}[\mathbf{j}] = 0 \end{cases} \tag{2.54}$$

To help decouple our description of the coding operations from the order in which they are applied, we introduce the notion of a binary *significance state*, $\sigma[\mathbf{j}]$. At any point in the coding process, $\sigma[\mathbf{j}]$ assumes the value of $\sigma^{(p)}[\mathbf{j}]$, where p is the most recent (least significant) bit for which information has been coded at location $\mathbf{j}$. Equivalently, we initialize the significance state of all samples in the code-block to 0 at the beginning of the coding process and then toggle the state to $\sigma[\mathbf{j}] = 1$ immediately after coding the first nonzero magnitude bit at location $\mathbf{j}$.

Given the importance of the condition $\sigma[\mathbf{j}] = 0$, we identify three different types of primitive coding operations as follows. If $\sigma[\mathbf{j}] = 0$ we refer to the task of coding $v^p[\mathbf{j}]$ as *significance coding*, since $v^p[\mathbf{j}] = 1$ if and only if the significance state transitions to $\sigma[\mathbf{j}] = 1$ in this coding step. In the event that the sample does become significant, we must invoke a *sign coding* primitive to identify $\chi[\mathbf{j}]$. For samples which are already significant, the value of $v^p[\mathbf{j}]$ serves to refine the decoder's knowledge of the nonzero sample magnitude. Accordingly, we invoke a *magnitude refinement coding* primitive.

JPEG2000 employs the binary arithmetic coder variant known as the MQ coder, with a total of 18 distinct contexts, ten of these are associated with the important significance coding primitive. We remind the reader of the learning penalty, which is incurred if too many different contexts are used. The learning penalty is particularly significant in EBCOT, since adaptive probability estimation must start from scratch within each code-block.

2.5.1.1   Significance Coding

The significance coding primitive involves a *normal mode* and a *run mode*. In the normal mode, one of nine different contexts is selected, based on the significance of the sample's eight immediate neighbors. A context index $\kappa^{\mathrm{sig}}[\mathbf{j}]$ is formed from the following three intermediate quantities:

$$\kappa^{\mathrm{h}}[\mathbf{j}] = \sigma[j_1, j_2 - 1] + \sigma[j_1, j_2 + 1] \tag{2.55}$$

$$\kappa^{\mathrm{v}}[\mathbf{j}] = \sigma[j_1 - 1, j_2] + \sigma[j_1 + 1, j_2] \tag{2.56}$$

$$\kappa^{\mathrm{d}}[\mathbf{j}] = \sum_{k_1 = \pm 1} \sum_{k_2 = \pm 1} \sigma[j_1 + k_1, j_2 + k_2] \tag{2.57}$$

Samples that lie beyond the boundaries of the relevant code-block are regarded as insignificant for the purpose of constructing these three quantities. Evidently, there are 45 possible combinations of the three quantities, $\kappa^{\mathrm{h}}[\mathbf{j}]$, $\kappa^{\mathrm{v}}[\mathbf{j}]$ and $\kappa^{\mathrm{d}}[\mathbf{j}]$. A context reduction function is used to map these 45 combinations into the nine distinct context labels, $\kappa^{\mathrm{sig}}[\mathbf{j}]$. Details of the context reduction mapping may be found in [17] or [18].

At moderate to high compression ratios, code-block samples are expected to be predominantly insignificant. For this reason, a run mode is introduced to dispatch multiple insignificant samples with a single binary symbol. The run mode serves primarily to reduce complexity, although minor improvements in compression efficiency are also typical. The run mode codes groups of four consecutive samples together, which must themselves be aligned on a four-sample boundary. As we shall see in Section 2.5.2, the scanning pattern itself works column by column on stripes of four rows at a time. This means that run-mode samples must constitute a single stripe column. The run mode is used only if all four samples have insignificant neighborhoods, meaning that no immediate neighbor of any of the four samples can be significant. In the run mode, a single binary digit is coded within its own context κ^{run}, to identify whether or not any of the samples in the run become significant in the current bit-plane. If so, the index of the first such sample is signalled using what essentially amounts to a 2-bit fixed length code.

2.5.1.2   Sign Coding

The sign coding primitive is invoked at most once in any given location $\mathbf{j}$, immediately after the significance coding operation in which that location first becomes significant. Most algorithms proposed for coding subband sample values, whether embedded or otherwise, treat the sign as an independent, uniformly distributed random variable, devoting 1 bit to coding its outcome. It turns out, however, that the signs of neighboring sample values exhibit significant statistical redundancy. Some arguments to suggest that this should be the case are presented in [17].

The JPEG2000 sign coding primitive employs five contexts. Context design is based upon the relevant sample's immediate four neighbors, each of which may be in one of three states: significant and positive, significant and negative, or insignificant. There are thus 81 unique neighborhood configurations. For details of the symmetry conditions and approximations used to map these 81 configurations to one of five context labels, $\kappa^{\mathrm{sign}}[\mathbf{j}]$, the reader is referred to [17].

2.5.1.3 Magnitude Refinement Coding

The magnitude refinement primitive is used to code the next magnitude bit $v^p[\mathbf{j}]$ of a sample, which was previously found to be significant. As already noted, subband samples tend to exhibit probability distributions $f_X(x)$, which are heavily skewed toward $x = 0$. In fact, the conditional PMF $f_{V^p|Q^{(p+1)}}(v^p \mid q^{(p+1)})$ typically exhibits the following characteristics: (1) it is independent of the sign of $q^{(p+1)}$; (2) $f_{V^p|Q^{(p+1)}}(0 \mid q^{(p+1)}) > \frac{1}{2}$ for all $q^{(p+1)}$; and (3) $f_{V^p|Q^{(p+1)}}(0 \mid q^{(p+1)}) \approx \frac{1}{2}$ for large $|q^{(p+1)}|$.

As a result, it is beneficial to condition the coding of $v^p[\mathbf{j}]$ upon the value of $v^{(p+1)}[\mathbf{j}]$ only when $v^{(p+1)}[\mathbf{j}]$ is small. We also find that it can be useful to exploit redundancy between adjacent sample magnitudes when $v^{(p+1)}[\mathbf{j}]$ is small. These observations serve to justify the assignment of one of three coding contexts κ^{mag}, as follows:

$$\kappa^{\mathrm{mag}}[\mathbf{j}] = \begin{cases} 0 & \text{if} \quad v^{(p+1)}[\mathbf{j}] = 1 \quad \text{and} \quad \kappa^{\mathrm{h}}[\mathbf{j}] + \kappa^{\mathrm{v}}[\mathbf{j}] + \kappa^{\mathrm{d}}[\mathbf{j}] = 0 \\ 1 & \text{if} \quad v^{(p+1)}[\mathbf{j}] = 1 \quad \text{and} \quad \kappa^{\mathrm{h}}[\mathbf{j}] + \kappa^{\mathrm{v}}[\mathbf{j}] + \kappa^{\mathrm{d}}[\mathbf{j}] > 0 \\ 2 & \text{if} \quad v^{(p+1)}[\mathbf{j}] > 1 \end{cases} \qquad (2.58)$$

2.5.2 FRACTIONAL BIT-PLANE SCAN

In JPEG2000, three fractional bit-plane coding passes are employed. Each coding pass traverses the sample locations within its code-block, following the stripe-oriented scan depicted in Figure 2.12. The first coding pass in each bit-plane, $\mathcal{P}_1^p$, includes any sample location $\mathbf{j}$, which is itself insignificant but has a significant neighborhood, that is, at least one of its eight neighbors is significant. Membership in $\mathcal{P}_1^p$ may be expressed by the conditions $\sigma[\mathbf{j}] = 0$ and $\kappa^{\mathrm{h}}[\mathbf{j}] + \kappa^{\mathrm{v}}[\mathbf{j}] + \kappa^{\mathrm{d}}[\mathbf{j}] > 0$. These conditions are designed to include those samples that are most likely to become significant in bit-plane p. Moreover, for a broad class of probability models, the samples in this coding pass are likely to yield the largest decrease in distortion relative to the increase in code length [10]. Once a sample becomes significant, the four neighbors that have not yet been visited in the scan also have significant neighborhoods, and will be included in $\mathcal{P}_1^p$ unless they were already significant. This pass is called the *significance propagation pass*, due to this tendency to dynamically propagate its membership.

Figure 2.12 provides an example of the significant propagation pass for one stripe of the code-block. In the figure, empty circles identify samples that are insignificant in bit-plane p; shaded circles identify samples that become significant during the pass, $\mathcal{P}_1^p$; and solid dots indicate samples that were already significant in bit-plane $p - 1$. Crosses are used to mark those samples that do not belong to the significance propagation pass, either because they were already significant, or because their neighborhood is entirely insignificant at the point in the scan when they are visited.

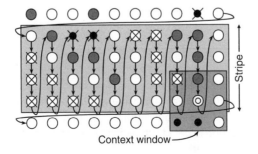

FIGURE 2.12 Stripe-oriented scan of the JPEG2000 block coder, illustrating the 3×3 context window used both for probability modeling and coding pass membership selection.

In the second pass of each bit-plane, $\mathcal{P}_2^p$, the magnitude refinement primitive is used to code magnitude bit $v^p[\mathbf{j}]$ of any sample that was already significant in the previous bit-plane. Equivalently, $\mathcal{P}_2^p$ includes any sample whose significance state is $\sigma[\mathbf{j}] = 1$, which was not already included in $\mathcal{P}_1^p$. The final coding pass, $\mathcal{P}_3^p$, includes all samples for which information has not already been coded in bit-plane p. From the definitions of $\mathcal{P}_1^p$ and $\mathcal{P}_2^p$, we see that samples coded in this pass must be insignificant. We also note that the conditions for the run mode may occur only in this coding pass.

2.5.3   OPTIMAL TRUNCATION

In the previous subsections, we have described the mechanisms whereby each code-block $\mathcal{B}_i$ is converted into a single embedded bit-stream. The natural truncation points for this embedded bit-stream are the coding pass end points, of which there are three per bit-plane. For each such truncation point z, we can identify the length L_i^z, of the smallest prefix of the embedded bit-stream from which all of the arithmetically encoded symbols in the first z coding passes can be unambiguously decoded. We assume that the overall reconstructed image distortion can be represented (or approximated) as a sum of distortion contributions from each of the code-blocks and let D_i^z denote the distortion contributed by block $\mathcal{B}_i$, if only the first z coding passes are decoded. The calculation or estimation of D_i^z depends upon the subband to which block $\mathcal{B}_i$ belongs, as well as the particular model of distortion that is deemed relevant. For our present purposes, it is sufficient to note that both MSE and the so-called *visually weighted MSE* measures satisfy our additivity assumption, provided the DWT basis functions are approximately orthogonal and the quantization errors can be considered uncorrelated. Visually weighted MSE refers to the weighting of distortion contributions from different subbands in accordance with "average" visual detection thresholds for the corresponding spatial frequencies.

Since the code-blocks are compressed independently, we are free to use any desired policy for truncating their embedded bit-streams. If the overall length of the final compressed bit-stream is constrained by $L_{\max}$, we are free to select any set of truncation points $\{z_i\}$, such that

$$L = \sum_i L_i^{z_i} \leq L_{\max} \tag{2.59}$$

Of course, the most attractive choice is that which minimizes the overall distortion,

$$D = \sum_i D_i^{z_i} \tag{2.60}$$

The selection of truncation points may be deferred until after all of the code-blocks have been compressed, at which point the available truncation lengths L_i^z, and the associated distortions D_i^z, can all be known. For this reason, we refer to the optimal truncation strategy as one of *postcompression rate-distortion optimization* (PCRD-opt).

Let $\{z_{i,\lambda}\}$ be any set of truncation points that minimizes

$$D(\lambda) + \lambda L(\lambda) = \sum_i (D_i^{z_{i,\lambda}} + \lambda L_i^{z_{i,\lambda}}) \tag{2.61}$$

for some $\lambda > 0$. It is easy to see that these truncation points are optimal in the sense that the distortion D cannot be further reduced without also increasing the length L, or vice versa. Thus, if we can find a value of λ such that the $\{z_{i,\lambda}\}$, which minimize Equation (2.61) yield $L(\lambda) = L_{\max}$, this set of truncation points must be a solution to our optimization problem. Since the set of available truncation points is discrete, we shall not generally be able to find such a λ. Nevertheless, since the code-blocks are relatively small and there are typically many truncation points, it is sufficient in practice to find the smallest value of λ such that $L(\lambda) \leq L_{\max}$.

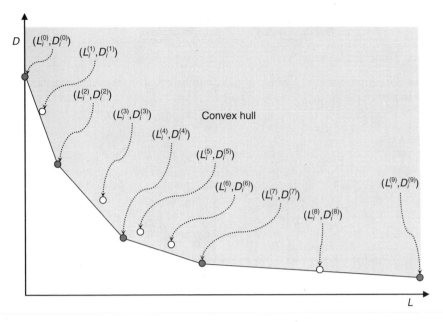

FIGURE 2.13 Upper convex hull produced by the distortion-length coordinates of a set of possible truncation points. Solid dots show those truncation points that can arise as solutions to Equation (2.62).

For any given λ, minimization of Equation (2.61) reduces to an independent minimization task for each code-block, i.e.,

$$z_{i,\lambda} = \text{argmin}_z \, (D_i^z - \lambda L_i^z) \tag{2.62}$$

Interestingly, some truncation points z can never arise as solutions to this simple optimization problem, regardless of the selected value for λ. These "useless" truncation points are those whose (D_i^z, L_i^z) coordinates fall strictly inside the *upper convex hull* of the set $\{(D_i^z, L_i^z)\}_z$, as illustrated in Figure 2.13. A good embedded block coder should have the property that most of its coding passes produce truncation points that lie on the boundary of this convex hull. Studies of the EBCOT coding passes, such as one found in [18], suggest that all of its coding passes produce truncation points that have a high likelihood of contributing to the convex hull boundary.

2.5.4 MULTIPLE QUALITY LAYERS

Since each code-block has its own embedded bit-stream, it is convenient to use a separate term to refer to the overall compressed image bit-stream. We shall call this a *pack-stream*, because it is inevitably constructed by packing contributions from the various code-block bit-streams together in some fashion. The simplest pack-stream organization consistent with the EBCOT paradigm is illustrated in Figure 2.14. In this case, the optimally truncated block bit-streams are simply concatenated, with length tags inserted to identify the contribution from each code-block.

This simple pack-stream is resolution-scalable, since each resolution level consists of a well-defined collection of code-blocks, explicitly identified by the length tags. The pack-stream also possesses a degree of spatial scalability. As long as the subband synthesis filters have finite support, each code-block influences only a finite region in the reconstructed image. Thus, given a spatial region of interest, the relevant code-blocks may be identified and extracted from the pack-stream.

Interestingly, the simple pack-stream of Figure 2.14 is not distortion-scalable, even though its individual code-blocks have embedded representations. The problem is that the pack-stream offers

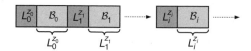

FIGURE 2.14 Simple pack-stream, created by concatenating optimally truncated code-block bit-streams together with length tags.

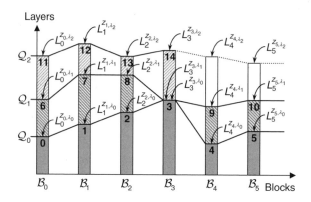

FIGURE 2.15 $S = 2$ EBCOT quality layers, indicating the order in which code-block contributions should appear for a quality-progressive pack-stream.

no information to assist in the construction of a smaller pack-stream whose code-block contributions are optimized in any way.

Figure 2.15 illustrates the quality layer abstraction introduced by the EBCOT algorithm [17] to resolve the above difficulty. Only six code-blocks are shown for illustration. There are a total of S quality layers, labeled Q_0 through Q_{S-1}. The first layer Q_0 contains optimized code-block contributions, having lengths $L_i^{z_i,\lambda_0}$, which minimize the distortion $D^0 = \sum_i D_i^{z_i,\lambda_0}$, subject to a length constraint $\sum_i L_i^{z_i,\lambda_0} \leq L_{\max}^0$. Subsequent layers Q_l contain additional contributions from each code-block, having lengths $L_i^{z_i,\lambda_l} - L_i^{z_i,\lambda_{l-1}}$, which minimize the distortion $D^l = \sum_i D_i^{z_i,\lambda_l}$, subject to a length constraint $\sum_i L_i^{z_i,\lambda_l} \leq L_{\max}^l$.

Although each quality layer notionally contains a contribution from every code-block, some or even all of these contributions may be empty. In Figure 2.15, for example, code-block $\mathcal{B}_3$ makes no contribution to layer Q_1. A distortion-scalable pack-stream may be constructed by including sufficient information to identify the contribution made by each code-block to each quality layer. Moreover, quality progressive organizations are clearly supported by ordering the information in the manner suggested by the numbering in Figure 2.15.

If a quality progressive pack-stream is truncated at an arbitrary point then the decoder can expect to receive some number of complete quality layers, followed by some fraction of the blocks from the next layer. In the example shown in Figure 2.15, the third quality layer Q_2 is truncated before code-block $\mathcal{B}_4$. In this case, the received prefix will not be strictly optimal in the PCRD-opt sense. However, this suboptimality may be rendered negligible by employing a large number of layers. On the other hand, more layers imply a larger overhead to identify the contributions made by each block to each layer. The EBCOT algorithm provides a *second tier* coding strategy for this type of information. For further details of the second tier coding strategy, the reader is referred to [17,18].

2.6 REFLECTIONS

We have seen that the information in any given approximation of an image can be quantified, at least in theory, using the notion of joint entropy. Moreover, we have seen that the joint entropy of an information source can be expanded in terms of conditional entropy, according to Equation (2.16). This expansion suggests that the average overall coded bit-rate can attain its minimum possible value, as long as we are able to code each outcome with a bit-rate equal to its conditional entropy, conditioned on the values of all previously coded outcomes. Importantly, this is true regardless of the order in which the information is coded. We have also seen that highly practical arithmetic encoding techniques exist, which are essentially able to achieve the relevant conditional entropies, subject to the provision of suitable probability models. Together, these observations reduce the problem of efficient image coding to that of conditional probability modeling.

We have briefly investigated the problem of image approximation through quantization. More specifically, we have noted that the DWT provides an excellent framework, in which to quantize an image's information. The DWT also provides a multiresolution representation, such that the omission of higher frequency DWT subbands does not interfere with the reconstruction of lower resolution versions of the image. At the same time, embedded quantization and coding techniques allow each subband to be coded progressively, in coarse to fine fashion. Together, these properties introduce the possibility of scaling both the resolution and the quality (e.g., signal-to-noise ratio) of the reconstructed image, by selectively discarding elements from the compressed bit-stream. Indeed, in Section 2.4.2, we have shown that the multiresolution properties of the DWT cannot be fully exploited unless it is combined with embedded quantization and coding. This is because high-resolution imagery requires the lower resolution subbands to be quantized with much higher fidelity than would otherwise be warranted when reconstructing the image at a lower resolution.

Fortunately, since the order in which information is coded need not affect the efficiency with which it is encoded, it is possible to construct embedded coders that yield high coding efficiency at all of their natural truncation points. In fact, the number of natural truncation points may be increased by coding information in an order that is dynamically generated on the basis of previously coded outcomes, so as to first encode those bits that are likely to provide larger decreases in distortion relative to their coding cost. We demonstrated these various principles through the fractional bit-plane embedded coding algorithm, which lies at the heart of the JPEG2000 image compression standard. This algorithm yields state-of-the-art compression performance, together with a highly scalable compressed representation.

It is interesting to note that in JPEG2000, the DWT is essentially playing three different roles. In the first instance, the DWT provides the multiresolution representation required for resolution scalability. In the second instance, the DWT produces subbands that are largely uncorrelated; this simplifies the probability modeling problem, allowing us to construct good conditional coding contexts that consist only of a sample's immediate neighbors. Finally, the DWT provides an excellent environment for designing dynamic scanning algorithms with good embedding properties. That is, the subband statistics are such that it is possible to predict which data bits are likely to produce the largest decrease in distortion relative to their coding cost, based only on those data bits that have already been encoded.

In this chapter, we have also investigated some aspects of the interaction between coding and ordering. Contexts formed from previously encoded data bits are useful both for conditional probability modeling (i.e., for coding efficiency) and for information ordering (i.e., for defining the membership of fractional bit-plane coding passes). On the other hand, the use of contextual information imposes constraints on our ability to discard elements of the compressed bit-stream. The EBCOT algorithm in JPEG2000 provides an excellent compromise between the objectives of an efficient finely embedded bit-stream (i.e., good context modeling) and a highly scalable and accessible bit-stream (i.e., one with few constraints). By producing an independent finely embedded bit-stream for each block of subband samples, EBCOT allows for both resolution scalability and spatial region-of-interest accessibility. Moreover, EBCOT's abstract quality layer paradigm provides a coarse mechanism for applications to explicitly control the order in which information is embedded. This is quite different to the ordering of information within each code-block's embedded bit-stream, which is an intrinsic part of the block coding algorithm. By explicitly adjusting the truncation points associated with each code-block, within each of a number of quality layers, JPEG2000 streams can be optimized to exploit the actual measured distortion properties of the image coder within each code-block. When done using the PCRD-opt approach, this often results in coding gains that outweigh the small losses due to limiting the coding contexts to the scope of each code-block. Moreover, quality layers can be constructed so as to optimize application-specific or sophisticated visual measures of distortion, which could not otherwise have been envisaged within the design of the low-level block coding algorithm. In this way, EBCOT provides a combination of intrinsic and extrinsic methods for ordering information, so as to provide as much scalability as possible while maintaining high compression efficiency.

REFERENCES

1. L. Bottou, P. Howard, and Y. Bengio, The z-coder adaptive binary coder, *Proceedings of the IEEE Data Compression Conference*, Snowbird, 1998, pp. 13–22.
2. D. Duttweiler and C. Chamzas, Probability estimation in arithmetic and adaptive-huffman entropy coders, *IEEE Trans. Image Proc.*, 4 (1995), 237–246.
3. H. Equitz and T. Cover, Successive refinement of information, *IEEE Trans. Inform. Theory*, 37 (1991), 269–275.
4. H. Hampel, R. Arps, D.D., Chamzas, D. Dellert, D.L. Duttweiler, T. Endoh, W. Equitz, F. Ono, R. Pasco, I. Sebestyen, C.J. Starkey, S.J. Ulban, Y. Yamazaki, and T. Yoshida, Technical features of the JBIG standard for progressive bi-level image compression, vol. 4, April 1992.
5. ISO/IEC 11544, JBIG bi-level image compression standard, 1993.
6. ISO/IEC 14492, JBIG2 bi-level image compression standard, 2000.
7. ISO/IEC 15444-1, Information technology – JPEG 2000 image coding system – Part 1: Core coding system, 2000.
8. G. Langdon, Probabilistic and q-coder algorithms for binary source adaptation, *Proceedings of the IEEE Data Compression Conference*, Snowbird, 1991, pp. 13–22.
9. J. Li and S. Lei, Rate-distortion optimized embedding, *Proceedings of the Picture Coding Symposium*, Berlin, 1997, pp. 201–206.
10. E. Ordentlich, M. Weinberger, and G. Seroussi, A low-complexity modeling approach for embedded coding of wavelet coefficients, *Proceedings of the IEEE Data Compression Conference*, Snowbird, 1998, pp. 408–417.
11. W. Pennebaker and J. Mitchell, *JPEG: Still Image Data Compression Standard*, Van Nostrand Reinhold, New York, 1992.
12. W. Pennebaker, J. Mitchell, G. Langdon, and R. Arps, An overview of the basic principles of the q-coder adaptive binary arithmetic coder, *IBM J. Res. Develop.*, 32 (1988), 717–726.
13. A. Said and W. Pearlman, A new, fast and efficient image codec based on set partitioning in hierarchical trees, *IEEE Trans. Circuit Syst. Video Tech.*, 6 (1996), 243–250.
14. C. Shannon, A mathematical theory of communication, *Bell Syst. Tech. J.*, 27 (1948), 379–423 (Part I); 623–656 (Part II). Reprinted in book form with postscript by W. Weaver, University of Illinois Press, Urbana, 1949.
15. F. Sheng, A. Bilgin, P. Sementilli, and M. Marcellin, Lossy and lossless image compression using reversible integer wavelet transforms, *Proceeding of the IEEE International Conference on Image Processing*, 1998, pp. 876–880.
16. D. Taubman, Embedded, Independent Block-Based Coding of Subband Data, Technical Report N871R, ISO/IEC JTC1/SC29/WG1, July 1998.
17. D. Taubman, High performance scalable image compression with EBCOT, *IEEE Trans. Image Proc.*, 9 (2000), 1158–1170.

18. D. Taubman and M. Marcellin, *JPEG2000: Image Compression Fundamentals, Standards and Practice*, Kluwer, Boston, 2002.
19. D. Taubman and A. Zakhor, A common framework for rate and distortion based scaling of highly scalable compressed video, *IEEE Trans. Circuit Syst. Video Tech.*, 6 (1996), 329–354.
20. A. Zandi and G. Langdon, Bayesian approach to a family of fast attack priors for binary adaptive coding, *Proceedings of the IEEE Data Compression Conference*, Snowbird, 1992.

3 Perceptual Aspects of Image Coding

Alessandro Neri, Marco Carli, and Sanjit K. Mitra

CONTENTS

3.1 Introduction 69
3.2 The Human Vision System 70
3.3 Physiological Models 72
3.4 Perceptual Distortion Metrics 76
3.5 Evaluation of the JND Threshold 78
3.6 Effects of Perception in DCT Domain 79
3.7 Perception Metrics in the Wavelet Domain 81
3.8 Conclusions 84
References 84

3.1 INTRODUCTION

Modeling-perceived distortion is a central aspect of modern lossy image and video coding systems. In fact, the limitations of the *Human Visual System* (HVS) in distinguishing two signals with similar spectral, temporal, or spatial characteristics can be exploited to reduce the amount of resources necessary to store or to transmit a multimedia content [21].

The two basic mechanisms characterizing the perception of still images that have attracted the attention of many researchers over the last few decades are the *frequency sensitivity* and the *local contrast sensitivity*. Frequency sensitivity is usually described by the *modulation transfer function* (MTF), characterizing the response of the human eye to sine-wave gratings at various frequencies. Contrast sensitivity models the local inhibition of the perception of weak features induced by the presence of strong patterns like edges and lines. More recently, mathematical models accounting for the masking effects introduced by complex textures have also been proposed [24].

Based on these researches, the models for the analytical assessment of the *just noticeable distortion* (JND) [37,38], that defines the error profile whose associated distortion is perceived just by a small fraction of the potential users, have been refined. The JND is in fact the cornerstone of any perceptual distortion metric.

In this chapter, after a short overview of the basic characteristics of the HVS, we discuss several neurophysiological models that enlighten the basic biological mechanisms of perception. We then discuss the description of the macroscopic aspects of the behavior of the HVS, which have been investigated through psychophysical subjective experiments, to quantify a distortion metric based on the HVS sensitivity and masking properties. Techniques for direct evaluation of the perceptual distortion metric in the discrete cosine transform (DCT) and wavelet transform domains are finally presented in the last section. These techniques constitute the basis for coding algorithms that minimize the perceptual entropy.

3.2 THE HUMAN VISION SYSTEM

Even now, it is still not exactly known how the HVS works. Several studies have been carried out pointing out its main features and the differences existing among individuals.

The eye is the interface between the input signal (the visual information) and the processing unit (the brain). It can be compared to a self-focusing camera: it automatically adjusts the amount of light intensity entering into the eye and it delivers the information carried by the light intensity to the optic nerve. Each eye bulb is anchored in its position by six extraocular muscles whose function is to keep it aligned and synchronized to the system. The light, going through the *cornea* (the transparent front part of the eye), *aqueous-humor*, and *lens*, is projected through the *vitreous-humor* onto the *retina*. Figure 3.1 shows a section of the eye.

The largest change (two third) in bend occurring to light, which is necessary for focusing, takes place at the air–cornea interface. The successive lens of the eye provides the remaining third of the focusing power; its main purpose is to adjust the focusing on objects at various distances. The eye changes the focus not varying the distance between lens and retina but by changing the shape of the lens by using the *ciliary muscles*. The adjustment capability varies from person to person and with the age of the person.

The amount of the light entering into the eye is determined by the size of the *pupil*. The *iris* controls the size of the pupil and lies between the aqueous-humor and the lens. The pupil diameter varies between 2 and 8 mm, changes according to the amount of light falling into the eye.

The retina converts the light into neurosignals, allowing the vision under different conditions ranging for very low-intensity light (night or scotopic vision) to sunlight (photopic vision).

Investigation is being carried out continuously to understand the operation of the retina and its building blocks. Thanks to these efforts our knowledge about the treatment of light signal in the eye and in the visual cortex has increased dramatically. The retina, has the shape of a plate 1/4-mm thick. It consists of three layers of cells bodies separated by two layers containing synapses made by the axons and dendrites of these cells. These cells (horizontal cells, bipolar cells, amacrine cells, etc.) gather and process the optical signal.

The most important part of the retina is constituted by the photosensitive (*photoreceptors*) cells at the lower part of it. In the human eye, there are four different types of receptors. These cells are blocks in which the light signal is converted into neurosignal. In short, the receptors respond to light through the *bleaching* process. A photon of light is absorbed by the pigment and is then chemically transformed.

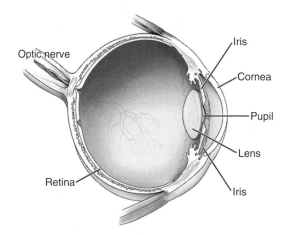

FIGURE 3.1 Diagram of the eye showing the macula and fovea (black and white). Courtesy of U.S.A. National Eye Institute, National Institutes of Health.

The central area of the retina, called *fovea*, can be seen as a packed array of photoreceptors and provides the highest spatial resolution. There are two kinds of photoreceptors: *rods* and *cones*. These cells have different distribution in the retina, present different shape and size, and are differently interconnected.

Rods are responsible for our vision under very low-light condition. They are long and slender and contain the same pigment, the photochemical rhodopsin. There are about 100 million rods in an eye, uniformly distributed in the fovea, which has a diameter of about 0.5 mm. The light sensitivity is achieved by combining the output of several rods into one ganglion cell neuron, penalizing the spatial resolution. The adaptation of our vision system to the dark is obtained by increasing slowly the amount of rhodopsins. Ganglion cells are of two types: on-center cells and off-center cells. An on-center cell responds to a simple stimulus consisting of a bright spot on a dark background, while an off-center cell responds to a stimulus consisting of a dark spot on a bright background. The size of the receptive field of ganglion cells varies in a systematic way: the size is smaller in the fovea, becoming progressively larger toward the retinal periphery. In addition, the response of a ganglion cell is independent of the stimulus orientation.

Cones do not respond under low-light conditions and are responsible for fine details and color vision. Cones, shorter than rods, are present throughout the retina but are most densely packed in the fovea. The rods, there are three different types of cones, each one characterized by a different photochemical pigment. These three pigments are sensitive to different light wavelengths. Cones are classified according to the wavelength of their sensitivity peak as follows: short-wavelength (S) (*blue* cones), medium-wavelength (M) (*green* cones) and long-wavelength (L) (*red* cones). The number of cones in the retina is much smaller than the number of rods and is estimated to be around 5 millions and most concentrated in the fovea; there are more red and green cones than blue cones and this fact can explain the lower sensitivity of HVS to the blue component. Consequently, the maximum spatial resolution that can be perceived is higher for red and green components than for the blue one. In the fovea, red and green cones are separated by approximately 2.5 µm, resulting in a maximum resolution of 60 cycles/deg. The blue cones are spaced at intervals of about 50 µm, resulting in a maximum resolution of 3 cycles/deg. Our perception of colors depends on the responses of these three set of cones.

By analyzing how the signals are treated by our vision system, we get an idea on what is really relevant and what can instead be considered useless in image and video coding. For example, two signals with spectral power density distribution producing the same response in all three kind of cones are perceived as the same color. The trichromatic nature of the visual system is exploited by modern imaging and display systems. Almost all the display systems use three color phosphors, the emission of each kind of phosphor matched to the response of a particular cone type.

The color opponent theory [4] states that not all the colors can co-exist (e.g., greenish, blue, and greenish red). It means that the output of the cones are encoded in three channels. One channel correspond to the black and white (luminance) perception, while the other two channels encode chrominance signals around red–green and blue–yellow axes.

Color video coding standards exploit this fact to encode the tristimulus color information resulting in a smaller amount of data to be stored or transmitted. In addition, the spatial resolution of color channels is lower than the resolution in the black and white, or luminance, channel. As a consequence, in image and video coding, the chromatic bands are usually downsampled by a factor 2, to increase the compression rate.

Behind the photoreceptors are cells containing mainly a black pigment called *melanin*. A possible function of this dark layer is to absorb the light that has passed through the retina, avoiding in this way its reflection and scattering inside the eye. It corresponds to the black wall inside a photo camera.

The optic nerve, carrying the retina's entire output, is a bundle of axons of the ganglion cells. It passes through the optic chiasm. From there the fibers of the optic nerve continue to several areas of the brain, mostly terminating to the two lateral geniculate bodies. Each lateral geniculate body consists essentially of one synaptic stage whose elements have on-center and off-center receptive fields and respond to colored light in a way similar to the response of the retinal ganglion cells. The lateral geniculate bodies send their output to the primary visual, or striate cortex, a complex structure with a greater variety of cell types, each tuned to an elaborate stimulus, organized into three or four stages. Cells of the visual cortex are classified as simple or complex based on the complexity of their behavior.

Each simple cell has a small, clearly delineated spatial receptive field so that it generates either on or off responses when a specific pattern falls inside. While the retinal and the geniculate cells exhibit an isotropic behavior, with circularly symmetric excitatory and inhibitory domains, the domains of the cortical simple cells are always separated by a straight line or by two parallel lines. The most common receptive map consists of a long, narrow excitatory region flanked, on both sides, by larger inhibitory regions. Both simple and complex cells present an anisotropic response, reacting to specifically oriented patterns over a limited region of the visual field. However, the behavior of complex cells cannot be explained by a neat subdivision of the receptive field into excitatory and inhibitory regions.

3.3 PHYSIOLOGICAL MODELS

Although many aspects of the behavior of simple cells are consistent with linear models, there are relevant exceptions, like output saturation and masking effects that cannot be explained in the linear context. Recently, several nonlinear computational models of the cortical visual processing have been proposed. These models attempt to describe in a compact and abstract mathematical form the transformations performed on the image signals by the HVS. They do not necessarily copy, in detail, the biological mechanism of the HVS [16]. Even though a complete and systematic mathematical description of the whole set of simple and complex cells of the visual cortex is far away from being known, the study of the partial models, identified up to now, allows an understanding of the basic mechanisms of perception, thus offering a consistent background for the design of subjective psychophysical experiments. With this aim, in this section, we review two multistage nonlinear models that illustrate the frequency sensitivity and the local contrast sensitivity behavior from a physiological point of view.

Let us first consider the two-stage neuron model for V1 simple cells presented in [11]. As illustrated in Figure 3.2, the first stage consists of a linear combination of the neuron inputs. In the second stage, the signal is normalized with respect to the pooled activities of a group of neurons, presumed to be near neighbors in the visual cortex, and then rectified. The normalization signal pooled over a large set of neurons with a wide variety of tuning properties produces a variety of phenomena, collectively described as *nonspecific suppression*. An example of this kind of phenomena is the response of an orientation-selective cell that presents a strong response in the presence of an isolated grating stimulus at the preferred orientation, and a null response when the orientation of the stimulus is perpendicular to the preferred direction. In fact, the response to the superimposed pair of stimuli, preferred plus perpendicular, is typically about half the response to the rightward stimulus alone, as observed in experiments. In addition, normalization captures the amplitude saturation at high contrast.

This physiological model has then inspired the psychophysical *divisive normalization model*, which will be discussed later in the section devoted to JND evaluation.

A rather similar approach is followed in [32] to model simple cells. In this case, the first stage consist of a linear, orientation-selective filter performing a space–frequency analysis whose point spread function (PSF) is a Gabor function cascaded with a memoryless nonlinearity acting as a half-wave rectifier.

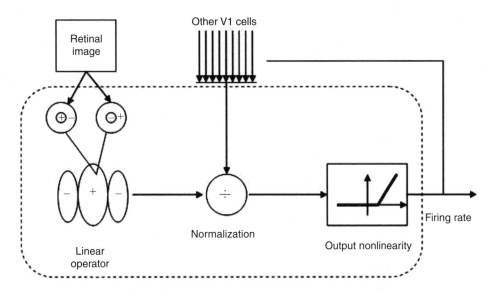

FIGURE 3.2 V1 model.

Let us denote by $\mathbf{x} = (x_1, x_2)$ the points on the image plane. Let $h^{\mathcal{G}}_{f_0, \Theta, \phi}(\mathbf{x})$ be the 2-D Gabor function with spatial frequency f_0 and orientation Θ:

$$h^{\mathcal{G}}_{f_0, \Theta, \phi}(\mathbf{x}) = \exp\left\{-\frac{1}{2\sigma^2}\mathbf{x}^T R_\Theta^T \Gamma R_\Theta \mathbf{x}\right\} \cos(2\pi f_0 \mathbf{u}_1^T R_\Theta \mathbf{x} + \phi) \tag{3.1}$$

where R_Θ is the rotation operator

$$R_\Theta = \begin{bmatrix} \cos\Theta & -\sin\Theta \\ \sin\Theta & \cos\Theta \end{bmatrix} \tag{3.2}$$

$\mathbf{u}_1$ is the column vector

$$\mathbf{u}_1 = \begin{bmatrix} 1 \\ 0 \end{bmatrix} \tag{3.3}$$

and Γ is the diagonal eccentricity matrix with spatial aspect ratio ρ,

$$\Gamma = \begin{bmatrix} 1 & 0 \\ 0 & \rho^2 \end{bmatrix} \tag{3.4}$$

The spatial bandwidth of $h^{\mathcal{G}}_{f_0, \Theta, \phi}(\mathbf{x})$ is controlled by the product σf_0. Specifically, half-response spatial frequency bandwidth of one octave approximately corresponds to $\sigma f_0 = 0.56$. It has been observed that the spatial aspect ratio usually varies in the range $0.23 < \rho < 0.92$. Finally, the parameter ϕ determines the symmetry of the PSF. In particular, $\phi = 0$ for center-on, symmetric receptive fields; $\phi = \pi$ for center-off, symmetric receptive fields; and $\phi = \pi/2$ and $\phi = -\pi/2$ for antisymmetric receptive fields with opposite polarities. Thus, for a given luminance distribution $g(\mathbf{x})$ the output of the first stage of a grating cell oriented along a direction Θ is given by

$$r_{f, \Theta, \phi}(\mathbf{x}) = u_{-1}[g(\mathbf{x}) * h^{\mathcal{G}}_{f, \Theta}(\mathbf{x})] \tag{3.5}$$

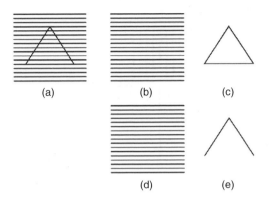

FIGURE 3.3 Triangle over a bar grating: (a) complex pattern; (b, c) original elements composing the image; and (d, e) decomposition more likely perceived.

where u_{-1} is the Heaviside step function and $*$ the convolution operator. The output of the linear system is then normalized with respect to the average gray level within the receptive field $m_\Theta(\mathbf{x})$, given by

$$m_\Theta(\mathbf{x}) = g(\mathbf{x}) * \exp\left\{-\frac{1}{2\sigma^2}\mathbf{x}^\mathsf{T} R_\Theta^\mathsf{T} \Gamma R_\Theta \mathbf{x}\right\} \tag{3.6}$$

Finally, the output of the simple cell is set equal to the hyperbolic ratio given by

$$s_{f,\Theta,\phi}(\mathbf{x}) = \begin{cases} 0, & m_\Theta(\mathbf{x}) = 0 \\ \chi\left(\dfrac{\frac{r_{f,\Theta,\phi}(\mathbf{x})}{m_\Theta(\mathbf{x})}R}{\frac{r_{f,\Theta,\phi}(\mathbf{x})}{m_\Theta(\mathbf{x})} + C}\right), & \text{otherwise} \end{cases} \tag{3.7}$$

where R and C are the maximum response level and the semisaturation constant, respectively.

Until now we have considered the perception produced by several simple stimuli over a uniform background. More complex masking mechanisms affect the perception of edges and object contours in the presence of structured textures. As an example, let us consider the image of Figure 3.3, which consists of a triangle superimposed to a bar grating. Usually, only two sides of the triangle are clearly distinguished from the background, while the third side is perceived as a bar of the grating.

A similar situation arises with the superposition of the rectangle illustrated in Figure 3.4. Even if in this case the fourth side should be easily perceived due to partial overlapping.

Physiological experiments evidenced that while complex cells are able to detect textures, edges, and contours, discrimination between textures and form information is carried out by the grating cells and the bar cells of the visual cortex. Grating cells react to periodic-oriented patterns, while bar cells have a complementary behavior, with a peak in the response in the presence of isolated bars.

The computational models for grating and bar cells have been investigated [24]. In essence, the grating cell model consists of two stages. The first stage is composed of grating subunits whose responses are computed using as input the responses to center-on and center-off simple cells with symmetrical receptive fields. The model of a grating subunit is conceived in such a way that the unit is activated by a set of three bars with appropriate periodicity, orientation, and position. In the second stage, the responses of grating subunits with preferred orientation Θ and frequency f are added together to compute the response of a grating cell. The activity of a grating subunit can be computed from the activities of simple cells with symmetric receptive fields along a line segment of length 3λ, where $\lambda = 1/f$ is the spatial wavelength, with orientation Θ passing through the point

FIGURE 3.4 Rectangle over a bar grating.

$\mathbf{x} = (x_1, x_2)$. Let us denote by $M_{f,\Theta,n}(\mathbf{x})$ the maximum output of a simple cell, alternatively center-on or center-off, over the nth subinterval $E_{\Theta,n}(\mathbf{x})$ of length $\lambda/2$ of the said interval:

$$M_{f,\Theta,n}(\mathbf{x}) = \underset{\mathbf{y} \in E_{\Theta,n}(\mathbf{x})}{\text{Max}} \left\{ s_{f,\Theta,(n+1)\pi}(\mathbf{y}) \right\} \tag{3.8}$$

where $n \in \{-3, -2, -1, 0, 1, 2\}$, and

$$E_{\Theta,n}(\mathbf{x}) = \left\{ \mathbf{y} \mid n\tfrac{\lambda}{2} \cos \Theta \leq y_1 - x_1 \leq (n+1)\tfrac{\lambda}{2} \cos \Theta \right.$$
$$\left. n\tfrac{\lambda}{2} \sin \Theta \leq y_2 - x_2 \leq (n+1)\tfrac{\lambda}{2} \sin \Theta \right\} \tag{3.9}$$

Let us denote by $\mathcal{M}_{f,\Theta}(\mathbf{x})$ the maximum value of $M_{f,\Theta,n}(\mathbf{x})$ over the whole line segment of length 3λ, i.e.,

$$\mathcal{M}_{f,\Theta}(\mathbf{x}) = \underset{n \in \{-3,-2,-1,0,1,2\}}{\text{Max}} \{M_{f,\Theta,n}(\mathbf{x})\} \tag{3.10}$$

Then, each grating subunit is considered activated if $M_{f,\Theta,n}(\mathbf{x})$, normalized with respect to its maximum value $\mathcal{M}_{f,\Theta}(\mathbf{x})$, exceeds an activation threshold $\rho < 1$, or, equivalently, if $M_{f,\Theta,n}(\mathbf{x})$ is greater than ρ times $\mathcal{M}_{f,\Theta}(\mathbf{x})$. The grating cell is activated if center-on and center-off cells, with the same preferred orientation Θ and spatial frequency f, are alternatively activated in intervals of length $\lambda/2$, along a segment of length 3λ centered on $\mathbf{x}$. Thus, denoting by $q_{f,\Theta}(\mathbf{x})$ the activity of a grating cell whose receptive field is centered on $\mathbf{x}$, we have

$$q_{f,\Theta}(\mathbf{x}) = \begin{cases} 1, & \text{if } \forall n, \ M_{f,\Theta,n}(\mathbf{x}) \geq \rho \mathcal{M}_{f,\Theta}(\mathbf{x}) \\ 0 & \text{otherwise} \end{cases} \tag{3.11}$$

Finally, in the second stage of the model, the response $w_{f,\Theta}(\mathbf{x})$ is computed as the weighted sum of the contribution of the grating subunits with orientations Θ and $\Theta + \pi$:

$$w_{f,\Theta}(\mathbf{x}) = \iint \exp\left(-\frac{||\mathbf{x} - \mathbf{y}||^2}{2\beta^2\sigma^2}\right)[q_{f,\Theta}(\mathbf{y}) + q_{f,\Theta+\pi}(\mathbf{y})] \, dy_1 \, dy_2 \tag{3.12}$$

The β parameter determines the effective area of the weighted summation. A value of $\beta = 5$ results in good approximation with observed values.

The computational model of a bar cell is based on the thresholding of the difference between the activity of a simple or a complex cell and the activity of a grating cell with the same preferred orientation and spatial frequency.

3.4 PERCEPTUAL DISTORTION METRICS

Although the mathematical model of the HVS at microscopic level is still incomplete, the previous sections emphasized the fact that the HVS is a complex system characterized by a collection of individual mechanisms, each selective in terms of frequency and orientation. The visual cortex decomposes the retinal image into frequency- and orientation-selective band-limited components, providing some kind of multichannel representation of the original image [12–15,22]. Several basis sets, such as Gabor [29,45], Cauchy [10], and Gaussian differences [23], have been proposed in the recent literature to fit the visual cortex decomposition [17–20]. However, while these representations are useful to devise specific algorithms for image analysis and understanding, the HVS properties, whose exploitation has a higher impact on perceptual redundancy reduction, are better investigated by means of complementary psychophysical studies aimed at the identification of the global characteristics of the HVS from a macroscopic point of view.

With the intent of illustrating how perceptual aspects can impact on the design of image coders, let us first formalize a general framework for lossy entropy coding.

Given a complete basis $\{\varphi_k(x_1, x_2), k \in K, \mathbf{x} \in T \subseteq R^2\}$, where K is the index set, an image $g(\mathbf{x})$ can be represented as a linear combination of the basis elements:

$$g(\mathbf{x}) = \sum_{k \in K} c_k \varphi_k(\mathbf{x}) \tag{3.13}$$

Usually, maximization of coding efficiency with reasonable computational complexity asks for a basis leading to a sparse representation with statistically independent coefficients. In this case, a lossy entropy coding is obtained by applying thresholding and scalar quantization to the representation coefficients $\{c_k\}$. The quantized coefficients $\{\widehat{c}_k\}$ are then employed to reconstruct the decoded image $\{\widehat{g}(\mathbf{x})\}$.

The loss associated to the encoding/decoding process can be described by the difference $e(\mathbf{x})$ between the original and the decoded image:

$$e(\mathbf{x}) = \sum_{k \in K} [c_k - \widehat{c}_k] \varphi_k(\mathbf{x}) \tag{3.14}$$

However, the distortion $\mathcal{D}$ perceived by a subject is directly related to the amount of artifacts contained in $e(\mathbf{x})$ as well as to their nature. Moreover, psychophysical tests have verified that their perception is affected by the image content itself. Specifically, the presence of strong patterns such as edges, lines, and textured regions, can inhibit the perception of weaker features. As a consequence, when the differences are small, the HVS is unable to distinguish the original and the encoded image. A similar situation applies when an additional signal, like a watermark, is hidden in the original image.

As argued in the original works of Watson and others (see, for instance, [31,39,46–48], three major phenomena affect the visibility of artifacts in achromatic images: contrast sensitivity, luminance masking, and contrast masking. The contrast sensitivity (CS) describes the ability of the HVS to perceive an isolated stimulus presented over a flat, neutral gray background. The luminance masking, also known as light adaptation [46], describes the variation of the contrast sensitivity vs. the background brightness. Finally, the contrast masking accounts for the capability of the HVS in perceiving a target pattern in the presence of a textured background.

Factors like contrast sensitivity and light adaptation capability differ from user to user, and may change during time, based on the current psychological and emotional status.

To build an analytical model of the masking effect accounting for the inter-subject variability, the concept of JND has been introduced in the literature. In essence, the JND defines the characteristics of the artifacts produced by the encoding process whose distortion is not perceived by a fraction P of the test population. Values of $P = 0.50$, 0.75, and 0.95 are typically employed, depending on how conservative the assessment has to be.

Since the HVS performs a frequency- and orientation-selective processing, the assessment of the perceived distortion is usually built upon the experiments on the effects produced by simple stimuli constituted by localized grating patterns characterized by a spatial frequency f and an orientation Θ.

Following [33,44], the probability of perceiving a distortion produced by such a pattern located in $\mathbf{x}$, with amplitude $q_{f,\Theta}(\mathbf{x})$, presented over a background with average intensity equal to $m(\mathbf{x})$, can be modeled as follows:

$$P_{f,\Theta}^{D}(\mathbf{x}) = 1 - \exp\left\{ -\left| \frac{q_{f,\Theta}(\mathbf{x})}{\gamma t^{\mathrm{JND}}[f, \Theta, m(\mathbf{x})]} \right|^{\beta_{f,\Theta}} \right\} \tag{3.15}$$

where $t^{\mathrm{JND}}[f, \Theta, m(\mathbf{x})]$ denotes the JND threshold at frequency f and orientation Θ.

As detailed in the following, the JND threshold depends on the energy of the neighbors in space, orientation, and scale. This formulation is consistent with the physiological models previously discussed, where a space, scale, and orientation linear decomposition are cascaded with a nonlinear gain control that normalizes the output of the decomposition with respect to the linear combination of the energy of neighbors.

Since statistical independence between perceived distortions caused by different patterns can be reasonably assumed, the overall probability of perceiving at least one distortion produced by a set of grating patterns with amplitude $q_{f,\Theta}(\mathbf{x})$ and characterized by a spatial frequency f and an orientation Θ, in a given region of interest (ROI) centered in $\mathbf{x}$, is

$$P_{f,\Theta}^{\mathrm{ROI}}(\mathbf{x}) = 1 - \exp\left\{ -\sum_{\mathbf{y} \in \mathrm{ROI}(\mathbf{x})} \left| \frac{q_{f,\Theta}(\mathbf{y})}{\gamma t^{\mathrm{JND}}[f, \Theta, m(\mathbf{y})]} \right|^{\beta_{f,\Theta}} \right\} \tag{3.16}$$

Note that the above equation can also be written as

$$P_{f,\Theta}^{\mathrm{ROI}}(\mathbf{x}) = 1 - \exp\left\{ -|\mathfrak{D}(\mathbf{x}, f, \Theta)|^{\beta_{f,\Theta}} \right\} \tag{3.17}$$

where $\mathfrak{D}(\mathbf{x}, f, \Theta)$ is the Minkowski metric with exponent $\beta_{f,\Theta}$ of the normalized error, i.e.,

$$\mathfrak{D}(\mathbf{x}, f, \Theta) = \left[\sum_{\mathbf{y} \in \mathrm{ROI}(\mathbf{x})} \left| \frac{q_{f,\Theta}(\mathbf{y})}{\gamma t^{\mathrm{JND}}[f, \Theta, m(\mathbf{y})]} \right|^{\beta_{f,\Theta}} \right]^{1/\beta_{f,\Theta}} \tag{3.18}$$

We can finally combine the effects generated by pooling the errors produced by the grating patterns with different frequencies and orientations. As reported in [46], the overall distortion $\mathfrak{D}$ can be approximated by the Minkowski metric with exponent $\widetilde{\beta}$ associated to $\mathfrak{D}(\mathbf{x}, f, \Theta)$,

$$\mathfrak{D}(\mathbf{x}) = \left[\sum_{f} \sum_{\Theta} |\mathfrak{D}(\mathbf{x}, f, \Theta)|^{\beta} \right]^{1/\beta} \tag{3.19}$$

In practice, $\widetilde{\beta} = \infty$ is adopted. Consequently, the probability of perceiving a distortion is given by the highest value $P_{f,\Theta}^{\text{ROI}}$.

3.5 EVALUATION OF THE JND THRESHOLD

Analytical models for the JND threshold $t^{\text{JND}}[f, \Theta, m(\mathbf{x})]$, appearing in Equation (3.15), have been investigated by means of subjective experiments, each focused on a specific aspect of the HVS. Based on their results, the JND threshold is usually factored as follows:

$$t^{\text{JND}}(f, \Theta, m) = t^D(f, \Theta, m)a_C(f, \Theta, m) = t^{CS}(f, \Theta)a_D(m)a_C(f, \Theta, m) \qquad (3.20)$$

where

- $t^D(f, \Theta, m) = t^{CS}(f, \Theta)a_D(m)$ is the contrast sensitivity threshold
- $t^{CS}(f, \Theta)$ is the inverse of the smallest contrast producing a noticeable difference when an isolated sinusoidal stimulus at a frequency f with orientation Θ is presented over a uniform gray background corresponding to a luminance of about 5.28 cd/m^2
- $a_D(m)$ is the luminance masking factor modeling the variation of the contrast sensitivity vs. the background brightness
- $a_C(f, \Theta, m)$ is the contrast-masking factor accounting for the capability of the HVS of perceiving a target pattern in presence of a masking pattern with identical frequency and orientation, as a function of the masker contrast [3,5,7–9,25]

Equation (3.20) allows us to compute the JND threshold, which represents the cornerstone in the optimization of lossy coders driven by perceptual aspects, from the knowledge of the partial factors. Before discussing the relationships between the contrast sensitivity threshold, the luminance and the contrast-masking functions, and the quantization steps in the DCT and in the wavelet domain, it should be noted that Equation (3.20) is a special case of a more general divisive normalization scheme suggested by the recent nonlinear physiological models of neuron in visual cortex, [28,50]. The divisive normalization is constituted by a linear transform $h_{f,\Theta}(\mathbf{x})$ performing a space–scale–orientation decomposition, as the Gabor decomposition, cascaded with a nonlinear stage normalizing the stimulus with respect to the energy of the neighbors in space, scale, and orientation.

Then, the output $q_{f,\Theta}(\mathbf{x})$ of the first stage of the divisive normalization scheme at frequency f, orientation Θ, and location $\mathbf{x}$ is given by the spatial convolution between the input image $g(\mathbf{x})$ and the linear kernel $h_{f,\Theta}(\mathbf{x})$:

$$q_{f,\Theta}(\mathbf{x}) = g(\mathbf{x}) * h_{f,\Theta}(\mathbf{x}) \qquad (3.21)$$

Let $E_s(\mathbf{x})$, $E_r(\Theta)$, and $E_f(f)$, respectively, be the neighborhoods of $\mathbf{x}$, Θ, and f. The output of the gain control stage $r_{f,\Theta}(\mathbf{x})$ is then computed as follows:

$$r_{f,\Theta}(\mathbf{x}) = \frac{\text{sign}[q_{f,\Theta}(\mathbf{x})]|q_{f,\Theta}(\mathbf{x})|^{\delta_e}}{b_{f,\Theta} + \displaystyle\sum_{\mathbf{y} \in E_s(\mathbf{x})} \sum_{\alpha \in E_r(\Theta)} \sum_{v \in E_f(f)} w_{f,v,\Theta,\alpha}|q_{v,\alpha}(\mathbf{y})|^{\delta_i}} \qquad (3.22)$$

where $w_{f,v,\Theta,\alpha}$ are the weights specifying the entity of the interactions between the grating stimulus centered in $\mathbf{x}$, with spatial frequency f, and orientation Θ, and the grating stimulus centered in $\mathbf{y}$, with spatial frequency v and orientation α. Finally, δ_e and δ_i are the excitatory and the inhibitory exponents, respectively, and $b_{f,\Theta}$ is the saturation coefficient. Thus, denoting by $\widehat{q}_{f,\Theta}(\mathbf{x})$ and $\widehat{r}_{f,\Theta}(\mathbf{x})$ the output of the first and the second stage, respectively, corresponding to the encoded image, and with $e_{f,\Theta}(\mathbf{x}) = q_{f,\Theta}(\mathbf{x}) - \widehat{q}_{f,\Theta}(\mathbf{x})$ the error at the output of the first stage, the distortion component $\mathfrak{D}(\mathbf{x}, f, \Theta)$ can be computed as follows [50]:

$$\mathfrak{D}(\mathbf{x}, f, \Theta) = \left[\sum_{\mathbf{y} \in \text{ROI}(\mathbf{x})} \left| r_{f,\Theta}(\mathbf{y}) - \widehat{r}_{f,\Theta}(\mathbf{y}) \right|^{\beta_{f,\Theta}} \right]^{1/\beta_{f,\Theta}} \tag{3.23}$$

Since the JND is usually small, the above equation can be approximated as follows:

$$\mathfrak{D}(\mathbf{x}, f, \Theta) \cong \left[\sum_{\mathbf{y} \in \text{ROI}(\mathbf{x})} \left| \frac{\left| e_{f,\Theta}(\mathbf{x}) \right|^{\delta_e}}{b_{f,\Theta} + \sum_{\mathbf{y} \in E_s(\mathbf{y})} \sum_{\alpha \in E_r(\Theta)} \sum_{\nu \in E_f(f)} w_{f,\nu,\Theta,\alpha} |q_{\nu,\alpha}(\mathbf{y})|^{\delta_i}} \right|^{\beta_{f,\Theta}} \right]^{1/\beta_{f,\Theta}} \tag{3.24}$$

As reported in [29], the parameters of the divisive normalization scheme can be identified by means of psychophysical experiments.

3.6 EFFECTS OF PERCEPTION IN DCT DOMAIN

Direct use of the divisive normalization scheme for distortion assessment requires the computation of the image representation that accurately describes the psychophysical behavior of the HVS. Nevertheless, fast techniques for coder optimization require exploitation of the effects produced by quantization in the encoder transform domain. Therefore, in the following, we analyze how the JND is related to quantization in the DCT and in the wavelet domains.

In DCT-based still image coding, the original image is first partitioned into square blocks of $N_{\text{DCT}} \times N_{\text{DCT}}$ pixels. Let us denote by D the one-dimensional DCT operator, and by $\mathbf{Z}(n_1, n_2)$ the 2-D array corresponding to the (n_1, n_2) block, and by $\mathbf{C}(n_1, n_2) = D\mathbf{Z}(n_1, n_2)D^{\mathsf{T}}$ its DCT. To determine the magnitude $t_{i,j}^{\text{JND}}(n_1, n_2)$ of the quantization error of the DCT coefficient $c_{i,j}(n_1, n_2)$ of block (n_1, n_2), corresponding to the index pair (i, j), producing a JND, on the basis of Equation (3.20), we first observe that the spatial bandwidth $f_{i,j}$ and the orientation Θ of the DCT basis function $\phi_{i,j}$ can be computed as

$$f_{i,j} = \frac{1}{2N_{\text{DCT}}} \sqrt{\frac{i^2}{w_x^2} + \frac{j^2}{w_y^2}} \tag{3.25}$$

and

$$\Theta_{i,j} = \arcsin \frac{2f_{i,0} f_{0,j}}{f_{i,j}^2} \tag{3.26}$$

where w_x and w_y denote the width and the height of a pixel, respectively, expressed in degree of visual angle.

Let us denote by $L_{\min}$ and $L_{\max}$ the minimum and the maximum display luminance, and by M the number of gray levels. Then, following Ahumada and Peterson [1], the contrast sensitivity threshold $t_{i,j}^D(n_1, n_2)$ can be computed as

$$t_{i,j}^D(n_1, n_2) = \frac{MT_{i,j}(n_1, n_2)}{2\alpha_i \alpha_j (L_{\max} - L_{\min})} \tag{3.27}$$

where $T_{i,j}(n_1, n_2)$ is the background luminance-adjusted DCT coefficient contrast sensitivity, and α_i and α_j are the DCT normalizing factors, i.e.,

$$\alpha_h = \begin{cases} \dfrac{1}{\sqrt{N_{\text{DCT}}}}, & h = 0 \\[3mm] \sqrt{\dfrac{2}{N_{\text{DCT}}}}, & h \neq 0 \end{cases} \tag{3.28}$$

As detailed in [1], the background luminance-adjusted DCT coefficient contrast sensitivity $T_{i,j}(n_1, n_2)$ can be approximated by a parabola in log spatial frequency, i.e.,

$$T_{i,j}(n_1, n_2) = 10^{g_{i,j}(n_1, n_2)} \tag{3.29}$$

with

$$g_{i,j}(n_1, n_2) = \log_{10} \frac{T_{\min}(n_1, n_2)}{r + (1-r)\cos^2 \Theta_{i,j}} + K(n_1, n_2)(\log_{10} f_{i,j} - \log_{10} f_{\min}(n_1, n_2))^2 \tag{3.30}$$

where

$$f_{\min}(n_1, n_2) = \begin{cases} \left[\dfrac{L(n_1, n_2)}{L_f} \right]^{\alpha_T} f_0, & L(n_1, n_2) \leq L_f, \\[3mm] f_0, & L(n_1, n_2) > L_f, \end{cases} \tag{3.31}$$

$$K(n_1, n_2) = \begin{cases} \left[\dfrac{L(n_1, n_2)}{K_0} \right]^{\alpha_T} f_0, & L(n_1, n_2) \leq L_K, \\[3mm] K_0, & L(n_1, n_2) > L_K, \end{cases} \tag{3.32}$$

$$T_{\min}(n_1, n_2) = \begin{cases} \left[\dfrac{L(n_1, n_2)}{L_T} \right]^{\alpha_T} \dfrac{L_T}{S_0}, & L(n_1, n_2) \leq L_T, \\[3mm] \dfrac{L(n_1, n_2)}{S_0}, & L(n_1, n_2) > L_T. \end{cases} \tag{3.33}$$

In the above equations, $L(n_1, n_2)$ is the local background luminance given by

$$L(n_1, n_2) = L_{\min} + \frac{L_{\max} - L_{\min}}{M} \left[\frac{\sum_{(0, m_1, m_2) \in F_{0, n_1, n_2}} c_{0, m_1, m_2}}{N_{\text{DCT}} \mathcal{N}(F_{0, n_1, n_2})} + m_g \right] \tag{3.34}$$

where $\mathcal{N}(F_{0, n_1, n_2})$ is the number of subband coefficients contained in the foveal region corresponding to block (n_1, n_2) in subband zero and m_g is the global mean of the image.

We note that $\mathcal{N}(F_{0, n_1, n_2})$ can be computed, based on the viewing distance D, the display resolution R, and the visual angle $\Delta\theta$ sustained by a foveal region ($\Delta\theta \cong 2°$) as follows:

$$\mathcal{N}(F_{0, n_1, n_2}) = \left\lfloor \frac{2DR \tan(\Delta\theta/2)}{N_{\text{DCT}}} \right\rfloor^2 \tag{3.35}$$

where $\lfloor \; \rfloor$ means rounding to the nearest smallest integer. The remaining model parameters, whose typical values are listed in Table 3.1, have been identified by means of psychophysical experiments [15].

Let us now examine the behavior of the contrast-masking factor that models the reduction in the visibility of one image component induced by the presence of other nonuniform elements. The contrast-masking adjustment in DCT domain is of the form [6]:

$$a_{CM}(b, n_1, n_2) = \begin{cases} \max\left\{ 1, \left| \dfrac{c_{F_{(b, n_1, n_2)}}}{t_{\text{DCT}}(b, n_1, n_2)} \right|^{0.6} \right\}, & b \neq 0 \\[3mm] 1 & b = 0 \end{cases} \tag{3.36}$$

where $c_{F_{(b, n_1, n_2)}}$ is the average number of DCT coefficients in the foveal region.

TABLE 3.1
Parameters for DCT Contrast Sensitivity Threshold Model

Parameter	Value
r	0.7
L_T	13.45 cd/m^2
S_0	94.7
α_T	0.649
f_0	6.78 cycles/deg
α_f	0.182
L_f	300 cd/m^2
K_0	3.125
α_k	0.0706
L_k	300 cd/m^2

Finally, direct extension of Equation (3.15) yields the following expression for probability of perceiving a distortion in the block (n_1, n_2) due to the quantization error $q_{i,j}(n_1, n_2)$ of $c_{i,j}(n_1, n_2)$:

$$P_{i,j}^D(n_1, n_2) = 1 - \exp\left\{ -\left| \frac{q_{i,j}(n_1, n_2)}{t_{i,j}^{\text{JND}}(n_1, n_2)} \right|^{\beta_{i,j}} \right\} \qquad (3.37)$$

where $t_{i,j}^{\text{JND}}(n_1, n_2)$ can be computed as product of the contrast sensitivity threshold $t_{i,j}^D(n_1, n_2)$ given by Equation (3.27) and the contrast-masking adjustment $a_{CM}(b, n_1, n_2)$ given by Equation (3.36). Incidentally, we observe that Equation (3.37) implies that $t_{i,j}^{\text{JND}}(n_1, n_2)$ is the magnitude of the quantization error of the (i, j) DCT coefficient that produces a perception of distortion in one block in about 63 users out of 100.

Under the assumption of statistical independence among perceived distortions caused by quantization of different coefficients, the overall probability of perceiving a distortion in any block of a given ROI is

$$P_{i,j}^{\text{ROI}}(n_1, n_2) = 1 - \exp\left\{ -\sum_{(n_1, n_2) \in \text{ROI}} \left| \frac{q_{i,j}(n_1, n_2)}{t_{i,j}^{\text{JND}}(n_1, n_2)} \right|^{\beta_{i,j}} \right\} \qquad (3.38)$$

The above equations that directly relate perceived distortion to quantization factor in the DCT domain are on the basis of several algorithms for the enhancement of the JPEG standard performance. Among them, we cite the DCTune technique, [34,35,46–48] and the adaptive perceptual distortion control proposed by Hoentsch and Karam [14,43], which overcome the overhead of storing and transferring the quantization thresholds by re-estimating the amount of masking at the decoder site.

3.7 PERCEPTION METRICS IN THE WAVELET DOMAIN

The multiresolution analysis decomposes a signal at different scales or resolutions by using a basis whose elements are localized in both time and frequency, so that the representation of short-duration, nonstationary signals focuses on few components that immediately enlighten predominant frequencies and the location of abrupt changes.

In principle, the *continuous wavelet transform* (CWT) can be computed by correlating the image $g(\mathbf{x})$ with a scaled and translated versions of the *mother wavelet* $\psi(\mathbf{x})$, usually localized in both the time and the frequency domains. More useful in image coding are the orthogonal wavelet series expansions, named *discrete wavelet transforms* (DWT), derived by the CWT when scale and translation factors are constrained to discrete values.

For clarity, we first summarize the basic properties of one-dimensional DWT. Let ϕ be a (smooth) basic scaling function such that the discrete set $\{2^{-i/2}\phi(2^{-i}x - k)\}$ forms an orthonormal basis for a subspace $\mathbf{V}_i \in \mathbf{L}^2(\mathbf{R})$. Let ψ be an admissible wavelet such that (1) the discrete set $\{2^{-i/2}\psi(2^{-i}x-k)\}$ forms an orthonormal basis for a subspace $\mathbf{W}_i \in \mathbf{L}^2(\mathbf{R})$; (2) the two subspaces $\mathbf{V}_i$ and $\mathbf{W}_i$ are mutually orthogonal, i.e., $\mathbf{W}_i \perp \mathbf{V}_i$; and (3) the subspace $\mathbf{V}_{i-1}$ can be expressed as direct sum of $\mathbf{V}_i$ and $\mathbf{W}_i$, namely,

$$\mathbf{V}_{i-1} = \mathbf{V}_i \oplus \mathbf{W}_i \tag{3.39}$$

Then, a signal $s(x) \in \mathbf{V}_0$ is represented from a smoothed approximation at resolution $i = M$ obtained by combining translated versions of the basic scaling function $\phi(x)$, scaled by a factor 2^M, and M details at the dyadic scales $a = 2^\ell$ obtained by combining shifted and dilated versions of the mother wavelet $\psi(x)$:

$$s(x) = \sum_k 2^{-M/2}c_M[k]\phi(2^{-M/2}x - k) + \sum_{\ell=1}^{M}\sum_k 2^{-\ell/2}d_\ell[k]\psi(2^{-\ell/2}x - k) \tag{3.40}$$

The wavelet and the basic scaling function satisfy the *dilation equations*

$$\phi(x) = \sqrt{2}\sum_k h[k]\phi(2x - k), \quad \psi(x) = \sqrt{2}\sum_k g[k]\phi(2x - k) \tag{3.41}$$

where $g[k]$ and $h[k]$ are the coefficients of two *quadrature mirror filters* (QMFs)

$$H(\omega) = \frac{1}{\sqrt{2}}\sum_k h[k]e^{-j\omega k}, \quad G(\omega) = \frac{1}{\sqrt{2}}\sum_k g[k]e^{-j\omega k} \tag{3.42}$$

The DWT coefficients can be iteratively computed by means of the Mallat's pyramidal algorithm [26].

Specifically, the scaling coefficients $c_\ell[k]$ and the detail coefficients $d_\ell[k]$, at resolution 2^ℓ, can be computed by filtering the scaling coefficients $c_{\ell-1}[k]$ with the low-pass $h[n]$ and the high-pass $g[n]$ filters, respectively, followed by a factor-of-2 downsampling

$$c_\ell[n] = \sum_k c_{\ell-1}[k]\,h[2n - k], \quad d_\ell[n] = \sum_k c_{\ell-1}[k]\,g[2n - k] \tag{3.43}$$

The DWT recursively filters only the scaling coefficients $c_\ell[n]$, whereas the detail coefficients $d_\ell[n]$ are never reanalyzed and constitute the DWT outputs. In the *wavelet packet decomposition*, instead, the detail coefficients $d_\ell[n]$ are also recursively decomposed, following the same filtering and subsampling scheme: this approach is a generalization of the DWT and offers a richer signal analysis [2].

In image coding, a separable 2D DWT obtained by applying first a 1D DWT with respect to one coordinate (e.g., horizontal) cascaded with a 1D DWT with respect to the other coordinate (e.g., vertical) is employed. Since each resolution level of the 2D DWT decomposition is obtained by the four possible combinations of low- and high-pass filters along the horizontal and vertical directions, in the following, the orientation Θ of the elements of the wavelet basis will be indexed by $\Theta_{DWT} \in \{LL, HL, LH, HH\}$. In addition, for a display resolution of r pixels/deg, the frequency bandwidth f associated with the Lth level of the decomposition is

$$f = r2^{-L} \tag{3.44}$$

As verified by empirical evidence, although the wavelet transform approximates an ideal "decorrelator" quite well, a statistical dependency between wavelet transform coefficients still remains. This dependency possesses the two properties of clustering within the same scale and persisting across different scales. In other words, in the WT domain, the wavelet coefficients across scales tend to coagulate in correspondence of visually significant spatial events (edges, corners, lines, etc.) so that nonzero higher-resolution coefficients are usually attached to nonzero lower-resolution values [27]. This observation has inspired quite efficient image coding techniques. Several algorithms have been proposed during the last decade to encode the DWT coefficients [30]. We cite here Shapiro's embedded zero-tree wavelet compression (EZW), where wavelet coefficients are encoded orderly following the three ramifications starting from the lowest resolution plane of the wavelet coefficients [36], Said and Pearlman's spatial partitioning of images into hierarchical trees (SPHIT), Taubman and Zakhor's layered zero coding (LZC), and Taubman's embedded block coding with optimized truncation (EBCOT). Among them, EBCOT, which as been incorporated in the JPEG2000 standard, offers quality and resolution scalability, and random access in a single bit-stream [43].

The effects of the uniform quantization of a single band of coefficients as a function of resolution level, orientation, and display visual resolution have been measured by Watson et al. [51] (Table 3.2).

The experimental results evidenced many similarities with the behavior of the JND threshold in the DCT domain. In fact, the relationship between $\log t^{CS}(f, \Theta)$ and $\log f$ is still parabolic:

$$\log t^{CS}(f, \Theta) = \log \alpha + k(\log f - \log g_{\Theta_{DWT}} f_0)^2 \tag{3.45}$$

The parabola has the minimum at $g_{\Theta_{DWT}} f_0$, which is shifted by an amount $g_{\Theta_{DWT}}$, varying with the orientation.

Amplitude thresholds increase rapidly with spatial frequency and with orientation from low-pass to horizontal/vertical to diagonal. Thresholds also increase by a factor 2 from Y to C_r and by a factor 4 from Y to C_b.

The uniform quantization factor $Q(\ell, \Theta_{WDT})$ of the subband Θ_{WDT} of the ℓth resolution level can be computed from the JND threshold in the spatial domain. Let $A(\ell, \Theta_{WDT})$ be the amplitude of the basis function that results from a unit WDT coefficient $d_\ell[n_1, n_2] = 1$ of subband Θ_{WDT} of the ℓth resolution level. Since for a uniform quantization factor $Q(\ell, \Theta_{WDT})$, the largest possible coefficient error is $Q(\ell, \Theta_{WDT})/2$, we can set

$$Q(\ell, \Theta_{WDT}) = 2 \frac{t^{JND}(r^{-2\ell}, \Theta_{WDT})}{A(\ell, \Theta_{WDT})} \tag{3.46}$$

TABLE 3.2
Parameters for DWT YC_bC_r Threshold Model for Observer sfl

Color	α	k	f_0	g_{LL}	g_{HL}	g_{LH}	g_{HH}
Y	0.495	0.466	0.401	1.501	1	1	0.534
C_r	0.944	0.521	0.404	1.868	1	1	0.516
C_b	1.633	0.353	0.209	1.520	1	1	0.502

Thus, substitution of Equation (3.45) into Equation (3.46) yields

$$Q(\ell, \Theta_{WDT}) = \frac{2}{A(\ell, \Theta_{WDT})} \alpha 10^{k \log \left(\frac{2^\ell f_0 g \Theta_{WDT}}{r} \right)^2} a_D(m) a_C(r^{-\ell, \Theta_{WDT}, m}) \qquad (3.47)$$

As suggested in [51], the effects of light adaptation and texture masking in the wavelet domain could be deducted from those produced by quantization in the DCT domain. In practice, the lack of direct experimental observations suggests the application of more conservative coding schemes obtained from Equation (3.47) by setting $a_D(m) = 1$ and $a_C(r^{-\ell, \Theta_{WDT}, m}) = 1$ (see [37,40–42]).

3.8 CONCLUSIONS

In this chapter, we have illustrated those aspects of the visual perception that can be fruitfully exploited in order to reduce the spatial redundancy presented by still images without impairing the perceived quality.

The physiological models at microscopic level have demonstrated that the visual system is characterized by a rich collection of individual mechanisms, each selective in terms of frequency and orientation. The local inhibition of the perception of weak features induced by the presence of strong patterns like edges and lines, associated with those mechanisms, explains the local contrast sensitivity characterizing the human perception of still images. Although a systematic investigation of the physiological mechanisms is still incomplete, the experience and the knowledge acquired till now are on the basis of the macroscopic psychophysical models of the global characteristics of the HVS. Since trade-off between coding efficiency and perceived quality requires the knowledge of the impact produced by quantization on the perceived distortion, the second part of this chapter has been devoted to the illustration of the relationships between just noticeable distortion and quantizers operating in the DCT and in the wavelet domains derived from the analysis of the results of the psychophysical experiments. As a matter of fact, due to the wide diffusion of the JPEG standard, past research has been mainly focused on the visibility of artifacts produced by coders based on the DCT operator. Only a few results reported in the current literature address the problem of perceived distortion, when the coder operates on a wavelet representation of the image.

Future work could address the systematic analysis of the visibility of impairments produced by more general coding schemes in the presence of complex stimuli as well as the impact on perception of the terminal capabilities (e.g., size, spatial resolution, and viewing distance).

REFERENCES

1. A.J. Ahumada and H.A. Peterson, Luminance-model-based DCT quantization for color image compression, *Proceedings of Human Vision, Visual Processing, Digital Display III*, 1992, pp. 365–374.
2. J.J. Benedetto and M.W. Frazier, *Wavelets: Mathematics and Applications*, CRC Press, Boca Raton, FL, 1994.
3. J.M. Foley, Binocular distance perception: egocentric distance tasks, *J. Exp. Psychol.: Human Perception Performance*, 11, 133–148, 1985.
4. E. Hering, Zue Lehre vom Lichtsinn, Vienna, 1878.
5. J.M. Foley, Human luminance pattern vision mechanisms: masking experiments require a new model, *J. Opt. Soc. Am. A*, 11, 1710–1719, 1994.
6. J.M. Foley and G.M. Boynton, A new model of human luminance pattern vision mechanisms: analysis of the effects of pattern orientation, spatial phase and temporal frequency, *Proc. SPIE*, 2054, 32–42, 1994.

7. J.M. Foley and C.C. Chen, Pattern detection in the presence of maskers that differ in spatial phase and temporal offset: threshold measurements and a model, *Vision Res.*, 39, 3855–3872, 1999.

8. J.M. Foley and G.E. Legge, Contrast detection and near-threshold discrimination in human vision, *Vision Res.*, 21, 1041–1053, 1981.

9. J.M. Foley and W. Schwarz, Spatial attention: the effect of position uncertainty and number of distractor patterns on the threshold versus contrast function for contrast discrimination, *J. Opt. Soc. Am. A*, 15, 1036–1047, 1998.

10. N. Graham, Neurophysiology and psychophysics, in *Visual Pattern Analyzers*. Academic, New York, 1989, chap. 1, p. 334.

11. D.J. Heeger, E.P. Simoncelli, and J.A. Movshon, Computational models of cortical visual processing, *Proc. Natl. Acad. Sci. USA*, 93, 623–627, 1996. Colloquium paper.

12. I. Hoentsch and L.J. Karam, Locally-adaptive perceptual quantization without side information for DCT coefficients, *Conference Record of the 31st Asilomar Conference on Signals, Systems and Computers*, Vol. 2, 2–5 Nov. 1997, pp. 995–999.

13. I. Hoentsch and L.J. Karam, Locally-adaptive image coding based on a perceptual target distortion, *Proc. IEEE Int. Conf. On Acoustics, Speech, and Signal Processing, 1998, ICASSP '98*, Vol. 5, 12–15, May 1998, pp. 2569–2572.

14. I. Hoentsch and L.J. Karam, Locally adaptive perceptual image coding, *IEEE Trans. Image Process.*, 9, 1472–1483, 2000.

15. I. Hoentsch and L.J. Karam, Adaptive image coding with perceptual distortion control, *IEEE Trans. Image Process.*, 11, 213–222, 2002.

16. D.N. Hubel, *Eye, Brain, and Vision*, Vol. 22 of Scientific American Library, W. H. Freeman and Company, New York, 1988.

17. D.N. Hubel and T.N. Wiesel, Receptive fields, binocular interaction and functional architecture in the cats visual cortex, *J. Physiol.*, 160, 106–154, 1962.

18. D.N. Hubel and T.N. Wiesel, Receptive fields and functional architecture of money striate cortex, *J. Physiol.*, 198, 215–243, 1968.

19. D.N. Hubel and T.N. Wiesel, Sequence regularity and orientation columns in the monkey striate cortex, *J. Compar. Neurol.*, 158, 295–306, 1974.

20. D.N. Hubel and T.N. Wiesel, Functional architecture of macaque visual cortex, *Proc. R. Soc. Lond. B*, 198, 159, 1977.

21. N. Jayant, J. Johnston, and R. Safranek, Signal compression based on models of human perception, *Proc. IEEE*, 81, 1385–1422, 1993.

22. L.J. Karam, An analysis/synthesis model for the human visual based on subspace decomposition and multirate filter bank theory, *Proceedings of IEEE International Symposium on-Time-Frequency Time-Scale Analysis*, Oct. 1992, pp. 559–562.

23. S.A. Klein and D.N. Levi, Hyperacuity thresholds of 1 sec: theoretical predictions and empirical validation, *J. Opt. Soc. Am.*, 2, 1170–1190, 1985.

24. P. Kruizinga and N. Petkov, Nonlinear operator for oriented texture, *IEEE Trans. Image Process.*, 8, 1395–1407, 1999.

25. G.E. Legge and J.M. Foley, Contrast masking in human vision, *J. Opt. Soc. Am.*, 70, 1458–1471, 1980.

26. S. Mallat, A theory for multiresolution signal decomposition: The wavelet representation, *IEEE Trans. Pattern Mach. Intelligence*, 11, 674–693, 1989.

27. S. Mallat and S. Zhong, Characterization of signal from multiscale edges, *IEEE Trans. Pattern Anal. Mach. Intelligence*, 14, 710–732, 1992.

28. J. Malo, I. Epifanio, R. Navarro, and E.P. Simoncelli, Non-linear image representation for efficient perceptual coding, *IEEE Trans. On Image Process.*, 15, 68–80, 2006.

29. S. Marcelja, Mathematical description of the responses of simple cortical cells, *J. Opt. Soc. Am.*, 70, 1297–1300, 1980.

30. F. Marino, T. Acharya, and L.J. Karam, A DWT-based perceptually lossless color image compression architecture, *Conference Record of the Thirty-Second Asilomar Conference on Signals, Systems and Computers, 1998*, Vol. 1, Nov. 1998, pp. 149–153.

31. H.A. Peterson, A.J. Ahumada, and A.B. Watson, An improved detection model for DCT coefficient quantization, *Proceedings of Human Vision, Visual Processing, Display VI: SPIE*, 1993, pp. 191–201.

32. N. Petkov and P. Kruizinga, Computational models of visual neurons specialised in the detection of periodic and aperiodic oriented visual stimuli: bar and grating cells, *Biol. Cybern.*, 76, 83–96, 1997. Computing Science, University of Groningen.

33. J.G. Robson and N. Graham, Probability summation and regional variation in contrast sensitivity across the visual field, *Vision Res.*, 21, 409–418, 1981.

34. R. Rosenholtz and A.B. Watson, Perceptual adaptive JPEG coding, *Proceedings of IEEE International Conference on Image Processing*, 1996, pp. 901–904.

35. R.J. Safranek and J.D. Johnston, A perceptually tuned sub-band image coder with image dependent quantization and post-quantization data compression, *International Conference on Acoustics, Speech, and Signal Processing 1989, ICASSP-89*, Vol. 3, May 1989, pp. 1945–1948.

36. J. Shapiro, Embedded image coding using zerotrees of wavelet coefficients, *IEEE Trans. Signal Proc.*, 41, 3445–3462, 1993.

37. D.-F. Shen and J.-H. Sung, Application of JND visual model to SPHIT image coding and performance evaluation, *Proceedings of 2002 International Conference on Image Processing*, Vol. 3, June 2002, pp. III-249–III-252.

38. D.-F. Shen and L.-S. Yan, JND measurements and wavelet-based image coding, *Input/Output and Imaging Technologies, SPIE Proceedings*, Vol. 3422, 1998, pp. 146–156.

39. J.A. Solomon, A.J. Ahumada, and A.B. Watson, Visibility of DCT basis functions: effects of contrast masking, *Proceedings of Data Compression Conference*, 1994, pp. 361–370.

40. D.M. Tan, H.R. Wu, and Z. Yu, Vision model based perceptual coding of digital images, *Proceedings of 2001 International Symposium on Intelligent Multimedia, Video and Speech Processing*, May 2001, pp. 87–91.

41. D.M. Tan, H.R. Wu, and Z.H. Yu, Perceptual coding of digital monochrome images, *IEEE Signal Process. Lett.*, 11, 239–242, 2004.

42. D. Taubman, High performance scalable image compression with EBCOT, *IEEE Trans. Image Process.*, 9, 1158–1170, 2000.

43. L. Tuyet-Trang, L.J. Karam, and G.P. Abousleman, Robust image coding using perceptually-tuned channel-optimized trellis-coded quantization, *Proceedings of 42nd Midwest Symposium on Circuits and Systems*, Vol. 2, 8–11 Aug. 1999, pp. 1131–1134.

44. A.B. Watson, Probability summation over time, *Vision Res.*, 19, 515–522, 1979.

45. A.B. Watson, The cortex transform: rapid computation of simulated neural images, *Comput. Vis., Graph., Image Process.*, 39, 311–327, 1987.

46. A.B. Watson, DCTune: a technique for visual optimization of DCT quantization matrices for individual images, *Society of Information Display Digital Technical Papers XXIV*, 1993, pp. 946–949.

47. A.B. Watson, DCT quantization matrices visually optimized for individual images, *Proceedings of Human Vision, Visual Processing, Digital Display IV*, B.E. Rogowitz, Ed., 1993, pp. 202–216.

48. A.B. Watson, Visually optimal DCT quantization matrices for individual images, *Proceedings of Data Compression Conference* 1993, pp. 178–187.

49. A.B. Watson, *Image data compression having minimum perceptual error*, U.S. Patent 5-426–512, 1995.

50. A.B. Watson and J.A. Solomon, A model of visual contrast gain control and pattern masking, *J. Opt. Soc. A*, 14, 2379–2391, 1997.

51. A.B. Watson, G.Y. Yang, J.A. Solomon, and J. Villasenor, Visibility of wavelet quantization noise, *IEEE Trans. Image Process.*, 6, 1164–1175, 1997.

4 The JPEG Family of Coding Standards

Enrico Magli

CONTENTS

4.1 Introduction 88
4.2 A Brief History of the JPEG Family of Standards 88
4.3 The JPEG Standard 89
    4.3.1 Transform 89
    4.3.2 Quantization 90
    4.3.3 Entropy Coding 91
        4.3.3.1 Huffman Coding 91
        4.3.3.2 Arithmetic Coding 93
    4.3.4 Lossless Mode 94
    4.3.5 Progressive and Hierarchical Encoding 95
    4.3.6 Codestream Syntax 95
4.4 The JPEG-LS Standard 96
    4.4.1 Context-Based Prediction 97
    4.4.2 Entropy Coding 98
        4.4.2.1 Golomb coding 98
        4.4.2.2 Run Mode 99
    4.4.3 Near-Lossless Mode 99
    4.4.4 JPEG-LS Part 2 99
    4.4.5 Codestream Syntax 100
4.5 The JPEG 2000 Standard 101
    4.5.1 Transform and Quantization 101
        4.5.1.1 DC Level Shifting 101
        4.5.1.2 Multicomponent Transformation 102
        4.5.1.3 Wavelet Transformation 102
        4.5.1.4 Quantization 103
    4.5.2 Data Organization 103
    4.5.3 Entropy Coding 104
    4.5.4 Codestream Syntax, Progression Orders, and Codestream Generation 105
    4.5.5 Advanced Features 106
        4.5.5.1 Region of Interest Coding 106
        4.5.5.2 Error Resilience 107
    4.5.6 Other Parts of the Standard 108
        4.5.6.1 Part 2 — Extensions 108
        4.5.6.2 Part 3 — Motion JPEG 2000 108
        4.5.6.3 Other Parts 108
4.6 Advanced Research Related to Image-Coding Standards 109
    4.6.1 DCT-Based Coding 109
    4.6.2 Wavelet-Based Coding and Beyond 109

4.7   Available Software 110
       4.7.1   JPEG 110
       4.7.2   JPEG-LS 110
       4.7.3   JPEG 2000 110
References 110

4.1   INTRODUCTION

The term "JPEG," which stands for joint photographic experts group, has become very popular since Internet pages have started to contain "JPEG" images, i.e., images that can be downloaded in a shorter time than uncompressed or losslessly compressed images. Over the years, the committee that standardized the JPEG image compression algorithm has developed and issued other standards, such as JPEG-LS, JBIG, JBIG2, and JPEG 2000. This chapter addresses the standards related to compression of continuous-tone images; it does not cover compression of bilevel images. In particular, a research-oriented overview of the JPEG, JPEG-LS, and JPEG 2000 standards is provided. As for JPEG, the baseline compression process is described in detail, including the lossless mode, and the progressive and hierarchical modes are surveyed. Then, the recent JPEG-LS standard is reviewed, focusing on lossless compression and then describing the near-lossless mode. Finally, the new JPEG 2000 lossy and lossless compression standard is described. Pointers to reference software are given, and a discussion of recent advances and new research directions that are related to the technology contained in the standards is provided in the final section.

4.2   A BRIEF HISTORY OF THE JPEG FAMILY OF STANDARDS

As early as 1982, a working group was formed by the International Standardization Organization (ISO) to design a color image compression standard [22]; the initial target application was progressive image transmission at ISDN rates (64 kbit/sec). The JPEG group was established in 1986 as a collaborative effort from ISO and CCITT. In February 1989, the committee selected a set of system configurations based on the "adaptive DCT" (discrete cosine transform). A baseline system was defined, which achieved lossy compression by means of sequential DCT and Huffman coding; an extended DCT-based system was also defined, including features such as extended Huffman coding, progressive coding, and hierarchical coding. Further refinements were made, leading to the Committee Draft that was issued in April 1990; after incorporating comments, the draft international standard was balloted and approved in July 1992 [10].

After the success of the JPEG standard, it was decided to revisit the lossless coding mode in JPEG. A new standardization activity was started, the main target being effective lossless and near-lossless compression of continuous-tone, grey scale, and color still images. A number of proposals were submitted, and the LOCO algorithm [37] was selected as basis for further development. The JPEG-LS standard was eventually issued in 1998 [12].

Later on, driven by recent research results in the field of image coding, a call for technical contributions was issued in 1997 as a preliminary step to a novel compression standard, called JPEG 2000, and featuring improved low bit-rate performance with respect to JPEG, as well as embedded lossy and lossless compression, random codestream access, and robustness to bit errors. Based on the first core experiments, a wavelet-based technique followed by arithmetic coding was selected in 1998. Further refinements led to the final version of the algorithm, which was accepted as international standard in December 2000 [14].

4.3 THE JPEG STANDARD

JPEG is said to be "colorblind," in that each color component of the input image is treated independently; no information about the colorspace is written in the basic compressed file, although file formats like JFIF may contain colorspace details. JPEG can handle images of size up to $2^{16} \times 2^{16}$ with up to 255 components, each sample being represented with N bits/sample. N can be 8 or 12 in the lossy modes, and from 2 to 16 in the lossless modes.

JPEG supports four modes of operation: sequential, progressive, hierarchical, and lossless. In sequential coding, each image component is encoded in raster order. In progressive coding, the image is encoded in multiple passes, and each additional pass provides refinement data in order to achieve coarse-to-fine image buildup. In hierarchical encoding, the image is encoded at multiple resolutions; this allows to access a low-resolution version of the image without having to decode the whole codestream. In lossless mode, the decoded image is identical to the original one. The terms "baseline" and "extended" are also often used with regard to JPEG encoding; however, they actually refer to the required capabilities that a decoder must have in terms of the above modes of operation (and related parameters), and not specifically to a coding mode. A baseline process shall support sequential coding for $N = 8$ with Huffman coding. An extended process shall support both sequential and progressive encoding, for $N = 8$ and 12, using Huffman and arithmetic coding. A lossless process shall support the lossless sequential mode, with $2 \leq N \leq 16$, using both Huffman and arithmetic coding. A hierarchical process shall support the extended and lossless processes, along with the hierarchical mode.

In the following we describe the processing steps applied to each component in the sequential mode (see [35] for an excellent survey); the decoder performs exactly the same operations in reverse order. The lossless mode is described in Section 4.3.4, while the progressive and hierarchical modes are described in Section 4.3.5.

Note that the JPEG standard defines both the decoding process and the compressed file format, but not the encoding process, which must only fulfill the compliance tests in [11].

4.3.1 TRANSFORM

A block diagram of JPEG sequential coding is shown in Figure 4.1. JPEG follows the common transform coding paradigm, according to which the image undergoes a linear invertible transform, and then the transform coefficients are quantized and entropy-coded. The first operation performed on each component is level shifting, i.e., the unsigned integer sample values are subtracted 2^{N-1}, where N is the number of bits/sample on which the samples are represented; then, they are input to the DCT.

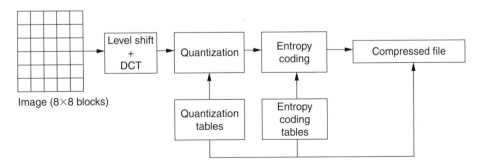

FIGURE 4.1 Block diagram of JPEG sequential coding. The image is divided in 8×8 blocks; for each block, the DCT is computed, its coefficients are quantized, and then entropy-coded. The quantization and entropy coding tables are written in the compressed file along with the entropy coded data.

In JPEG, each block of 8×8 samples is independently transformed using the two-dimensional DCT. This is a compromise between contrasting requirements; larger blocks would provide higher coding efficiency, whereas smaller blocks limit complexity.

The definition of the forward and inverse DCT used in JPEG can be found in [22]. It is worth noticing that the standard does not specify the DCT implementation to be used. Furthermore, since there is no point in requiring that the DCT be computed with greater accuracy than necessary for the subsequent quantization, the compliance testing defined in Part 2 of the standard [11] refers to the combined DCT and quantization, and not to the DCT alone. This leaves room for the implementation of faster (though possibly less accurate) DCTs.

4.3.2 QUANTIZATION

The 8×8 DCT coefficients of each block have to be quantized before entropy coding. Quantization is a crucial step, in that it allows to reduce the accuracy with which the DCT coefficients are represented, opening the door to different rate/quality trade-offs. The choice of the quantization step size for each DCT coefficient is a bit allocation problem. In principle, the step sizes could be computed and optimized for each 8×8 block so as to achieve minimum distortion for a given target bit-rate; however, the selected step sizes for each block would also have to be signaled to the decoder in addition to the entropy coded data. JPEG adopts a simpler approach, which consists in performing uniform scalar quantization with different step size for each DCT coefficient; this accounts for the fact that few low-frequency coefficients usually contain most of the signal energy, and have to be represented with higher accuracy than the high-frequency (AC) coefficients.

Letting $s_{i,j}$ $(i,j = 0, \ldots, 7)$ be the 64 DCT coefficients of each image block, the quantized coefficients are obtained by

$$s_{i,j}^* = \text{round}\left(\frac{s_{i,j}}{q_{i,j}}\right) \qquad (4.1)$$

where $q_{i,j}$ is the quantization step size for coefficient $s_{i,j}$, and round ( ) denotes rounding to the nearest integer. The collection of quantization step sizes is organized in an 8×8 quantization table. At the decoder, the dequantizer simply computes the reconstructed coefficient as $s_{i,j}^r = s_{i,j}^* \cdot q_{i,j}$. Note that this quantizer does not have a finite number of levels, as no clipping is performed to keep the quantized DCT coefficients to a maximum number of bits. In fact, this resembles an entropy-constrained quantizer [15], in that the quantizer overload error is avoided by not limiting the maximum number of bits; the impact on the bit-rate is minimum, since the entropy coder can easily accommodate coefficients that need many bits to be represented, but occur with very low probability.

After quantization, the $s_{0,0}$ coefficient (also called the DC coefficient, as its value is proportional to the average value of the image samples in the 8×8 block) is treated differently than the other AC coefficients. Specifically, the DC coefficient is differentially encoded by subtracting a predicted value from it. This predicted value is the quantized DC coefficient of the preceding block of the same component. This leads to a more compact representation of the DC coefficient, which often carries most of the energy in the transform domain.

JPEG defines an example quantization table (see [22]), which was selected experimentally so as to account for the visibility of errors in addition to minimizing distortion. However, for certain types of images the example table may not be optimal; hence, it is possible to use other custom quantization tables, which can be signaled in the compressed file.

When using JPEG, it is desirable to be able to select not one, but several different rate/quality trade-offs. If one employs the example quantization table on a given image, this will result in a certain bit-rate and a certain peak signal-to-noise ratio (PSNR) of the decoded image with respect to the original one. So how does one achieve other trade-offs? This is where the "quality factor" comes into play. Setting a given quality factor amounts to employing a scalar coefficient α that scales the

0	1	5	6	14	15	27	28
2	4	7	13	16	26	29	42
3	8	12	17	25	30	41	43
9	11	18	24	31	40	44	53
10	19	23	32	39	45	52	54
20	22	33	38	46	51	55	60
21	34	37	47	50	56	59	61
35	36	48	49	57	58	62	63

FIGURE 4.2 Zig-zag scanning of DCT coefficients; the scanning order follows increasing numbering in the figure.

quantization matrix to obtain a modified set $\alpha \cdot q_{i,j}$ of quantization step sizes. If this coefficient is greater than 1, the new step sizes will be larger than the original ones; the larger they are, the smaller the compressed file size, and the lower the PSNR, and vice versa. There is however a caveat for this procedure, as scaling all step sizes by the *same* coefficient is in general suboptimal from the rate-distortion standpoint, and better results may be obtained by *ad hoc* modifications of the quantization table.

4.3.3   ENTROPY CODING

The 8×8 matrix of transform coefficients for each block, including the predicted DC coefficient, is sent as input to the entropy-coding stage. Before entropy coding, the quantized samples are rearranged into a one-dimensional vector. The data are read in a zig-zag order, in such a way that frequency components[1] that are "close" in the two-dimensional DCT matrix are also close in the one-dimensional vector. The zig-zag scan also has the nice feature that the samples are ordered by decreasing probability of being 0 [22]. The zig-zag order is depicted in Figure 4.2, where coefficients are scanned from number 0 to 63.

The one-dimensional vector obtained after zig-zag scanning is then fed to the entropy coder for further compression. In fact, since many high-frequency components happen to be quantized to zero, especially at low bit-rates, long runs of zeros can occur in this vector. To take advantage of this, the entropy coder incorporates a run-length mode, in which zero runs are encoded by specifying the number of consecutive zeros in the vector. JPEG defines two entropy-coding modes, namely Huffman and arithmetic coding. The Huffman-coding mode must be implemented in all JPEG processes, whereas the arithmetic mode is not required in a baseline process.

4.3.3.1   Huffman Coding

Huffman coding [4] is a technique that explicitly maps input symbols to codewords in such a way that the most probable symbols are described with very short codewords, and the least probable ones with longer codewords; on an average, this coding procedure provides a more compact description of the source.

Huffman coding in JPEG is done in slightly different ways for the DC and the AC coefficients. The (predicted) DC coefficient, denoted by D, is always encoded first; for data with up to 12 bits/sample, the encoding is done by subdividing the input alphabet $[-32767, 32768]$ into

---

[1] This is a somewhat incorrect wording, in that the DCT does not perform a harmonic decomposition like the Fourier transform.

TABLE 4.1
Huffman Coding of the DC Coefficient

S	DC Coefficient
0	0
1	$-1, 1$
2	$-3, -2, 2, 3$
3	$-7, \ldots, -4, 4, \ldots, 7$
4	$-15, \ldots, -8, 8, \ldots, 15$
...	...
15	$-32767, \ldots, -16384, 16384, \ldots, 32767$
16	32768

Note: The right column contains logarithmically increasing intervals, whereas the left column contains the number of additional bits required to specify the quantized coefficient. Each row in the table is a symbol and needs to be assigned a Huffman code.

intervals with logarithmically increasing size, as shown in Table 4.1. Note that, since Huffman coding is also used in lossless mode, the full 16-bit precision range is considered as input alphabet; this also accommodates the increased DCT dynamic range for lossy coding. Each of the intervals in the left column of Table 4.1 is a symbol that must be assigned a Huffman code. As can be seen, this Huffman coder has a very small number of symbols, which makes it somewhat suboptimal in terms of entropy; on the other hand, this results in a very compact Huffman code specification. Each element of the input array is specified by the Huffman code corresponding to the interval it belongs to, plus S additional bits specifying the sign and magnitude of the coefficient D within the interval (except if the coefficient is 0 or 32,768, for which additional bits are not necessary).

The Huffman coding procedure is slightly more complicated for the AC coefficients, since it is necessary to compactly encode runs of zeros. To this end, the basic symbol is defined to be a pair (R, S), where R is the length of a run of zeros interrupted by a nonzero coefficient A, which shall fall in one of the intervals specified in Table 4.1; R can be no larger than 15. Encoding is done by outputting the Huffman code for the symbol (R, S), plus S additional bits that completely specify the coefficient A. There are two special cases in this procedure. The first one is the end-of-block condition, which signals that all the remaining AC coefficients are 0; this is very useful, since blocks are often ended by long sequences of 0s. The second one is the zero run length, which serves to encode a run of 16 zeros in the rare case that the run is larger than 15. Special symbols are defined for either case.

So far, nothing has been said as to how the Huffman tables are generated. The standard [10] provides example Huffman tables that can be used to encode an image. However, one may be willing to generate their own optimized tables; the motivation is that, on a specific set of images (e.g., medical images) an optimized table may perform better than the example one. Even better performance can be obtained if the Huffman tables are optimized for the specific image to be coded; however, this requires that the whole image be pre-encoded, so that symbol occurrence frequencies can be counted to compute the optimized Huffman tree. The syntax allows one to specify the Huffman table that has been used (DHT marker segment; see Section 4.3.6). The Huffman code is described by a sequence of fields containing (1) the number of codes of each code length (from 1 to 16 bits), and (2) the list of values for each code, in addition to the specification of whether the table is to be used for DC or

AC coefficients. The standard provides procedures for unambiguously constructing Huffman tables from the content of the DHT marker segment.

4.3.3.2 Arithmetic Coding

In addition to Huffman coding, JPEG also defines an arithmetic-coding mode to obtain improved coding efficiency; in fact, it has been found [22] that the arithmetic coder provides about 13% bit-rate saving with respect to Huffman coding with fixed tables, and about 8% saving with respect to Huffman coding with tables optimized for each image. Arithmetic coding [4] is based on recursive subdivision of the unit interval proportionally to the cumulative distribution function of the input symbols. In a binary arithmetic coder, as is used in JPEG, each new input bit generates a subdivision of the current interval into two subintervals whose size is proportional to the probability that the input bit is 0 or 1, and a selection of the subinterval corresponding to the actual value of the input bit. A major difference in the JPEG arithmetic coding mode with respect to Huffman coding is that the symbol probabilities are *conditional* probabilities depending on past values of the to-be-coded data (more on conditional coding in Section 4.4); apart from conditioning, the arithmetic coder uses the same coding model as the Huffman coder, which is based on the interval subdivisions in Table 4.1. JPEG does not use a "true" arithmetic coder, but rather a multiplierless approximation called QM-coder [22]; a description of the QM-coder is omitted here, since the QM-coder is identical to the MQ-coder employed in JPEG 2000 (except for the state transition tables and the carry propagation handling), which is described in another chapter of this book.

On the other hand, in the following we describe the statistical model used for arithmetic coding of DC and AC coefficients. The model contains two parts; in the first part, the input symbols (up to 16 bit/sample) are binarized into decision bits, while in the second part conditional probabilities are computed for each decision bits. The decision bits and conditional probabilities are used as input to the QM-coder.

The DC coefficient D is binarized as a sequence of decision outcomes, each of which is accompanied with a "context," i.e., a similarity class of past outcomes for which conditional probabilities can be defined. The first decision is $D = 0$, with context S0. If $D \neq 0$, the sign of D is coded with context SS. The next decision is $|D| - 1 > 0$ with context SN or SP depending on the sign of D. If $|D| - 1 \neq 0$, a sequence of binary decisions follows, to identify the logarithm of the magnitude of D; this is akin to Huffman coding of the symbol S using logarithmically spaced intervals (see Table 4.1). In particular, the sequence of decisions X_n are made, consisting of the comparisons $|D| > 2^n$; the sequence is stopped upon the first negative outcome X_L of the comparison. The arithmetic coding mode in JPEG codes the same values as the Huffman mode, including some extra bits required to identify $|D|$ within the selected interval, but using different contexts.

The contexts S0, SS, SP, and SN are conditioned on the value of the previous predicted DC coefficient D_p in the same component. D_p is classified into five categories, namely zero, small positive, small negative, large positive, and large negative; the boundaries to select between a small positive/negative and a large positive/negative can be determined by the user, and in general reflect the fact that the predicted DC coefficients should be tightly distributed around 0. For each of the five categories defined for D_p, a state-transition machinery is defined to update the conditional probabilities for each context S0, SS, SP, and SN. Each of the decision bits $X_1, \ldots, X_{15}$ and $M_2, \ldots, M_{15}$ also have their own contexts.

Coding of the AC coefficients follows a similar scheme. Since the end-of-block condition is very important to generate a compact codestream, this condition is retained as separate part of the statistical model, and is checked separately before encoding each AC coefficient A; if it holds, i.e., A is zero and there is no other later coefficient different from zero in the array, a 1-decision is coded with context SE, and this completes coding of the 8×8 block. Otherwise, a decision tree equal to that described for the DC coefficient D is entered.

In the AC case, three conditional probabilities are defined for SE, S0, and the union of SP/SN/X_1. However, unlike coding of the DC coefficient, conditioning is not based on past values, but only on the index K of the coefficient A in the zig-zag sequence; this implies that 63 counters are updated for each of the three conditional probabilities. Notice that conditioning does not depend on the values of neighboring AC coefficients, since the DCT yields transform samples that are almost completely uncorrelated. The sign bits are encoded with a fixed probability estimate of 0.5; this is easily justified by intuition, as well as the fact that SP and SN are in the same "bin" for conditioning, for there should be no statistical difference between positive and negative number in the AC coefficients of the DCT. Merging X_1 with SP and SN stems from empirical considerations. The decisions $X_2, \ldots, X_{15}$ and $M_2, \ldots, M_{15}$ have their own contexts; different contexts are defined for low- and high-frequency DCT coefficients.

Note that, unlike Huffman coding, when arithmetic coding is used there is no run mode.

4.3.4 LOSSLESS MODE

JPEG also foresees a lossless coding mode, which is not based on the DCT, but rather on linear prediction. The standard defines a set of seven predictors, plus a no-prediction mode reserved for the differential coding in hierarchical progressive mode; the predictor is fixed within a given data portion, but can be changed between different portions, as the predictor identifier is written in the scan header (see Section 4.3.6). Each pixel is predicted from its causal neighborhood of order 1. Let a, b, and c be respectively the left, top, and top-left neighbors of the current pixel x, as in Figure 4.3; the predicted value for the pixel can be found, for each predictor, using the formulas in Table 4.2. The prediction error is calculated modulo 65536.

Boundary pixels are treated as follows. Predictor 1 is used for the first raster line at the start of a new data portion and at the beginning of each restart interval, whereas the elected predictor is used for all subsequent lines. The first sample of each of these lines uses predictor 2.

If Huffman coding is used, the entropy coding mode employed for the DC coefficient in the DCT domain is used without modification to code the prediction error samples. If arithmetic coding is used, the DC mode is also used, with a few modifications to extend the contexts to a 2D form.

c	b
a	x

FIGURE 4.3 Neighborhood for prediction in JPEG lossless mode.

TABLE 4.2
Definition of Predictors for Lossless Coding

ID	Prediction
0	No prediction
1	a
2	b
3	c
4	$a + b - c$
5	$a + (b - c)/2$
6	$b + (a - c)/2$
7	$(a + b)/2$

4.3.5 PROGRESSIVE AND HIERARCHICAL ENCODING

In many applications, it is desirable that the receiver does not have to wait to download the whole compressed file to display the image; for example, a coarse version of the image may be sufficient to decide ether to download the complete image, or to switch to a new image. To enable this functionality, JPEG provides the progressive and hierarchical modes, in which coarse image information is stored at the beginning of the compressed file, while refinement information is available later on. The ability of a compression technique to provide images at different quality or resolution levels is also known as scalability. As will be seen in Section 4.5.4, this concept is developed in a very sophisticated way in the JPEG 2000 standard; however, several building blocks are already available in JPEG.

The *progressive* mode applies to lossy DCT compression, and is based on the use of multiple encoding passes (also called scans) to convey successive refinements of the 8×8 blocks of DCT coefficients. The DCT decomposes each 8×8 block into a sequence of coefficients that roughly correspond to low to high frequencies of each 8×8 image block. As a consequence, it is straightforward to use this spectral decomposition to send first the low-frequency components and then the high-frequency ones, obtaining the progressive feature. JPEG allows to do this in two ways, namely spectral selection and successive approximation.

In spectral selection, contiguous DCT coefficients in the zig-zag sequence are grouped into subbands of "neighboring" frequencies. Rather than containing all DCT coefficients for one block, each encoding pass contains only the coefficients of a given set of spatial frequencies for *all* blocks; multiple scans are used to encode all subbands. By doing so, the receiver can start decoding and displaying the coarse image obtained from the low-frequency subbands, and then refine it when high-frequency subbands are received; this is somewhat equivalent to the resolution-progressive mode in JPEG 2000. In successive approximation, the DCT coefficients are not divided into sets of spatial frequencies; instead, all coefficients are encoded at progressively higher accuracy by transmitting the most significant bits first. Multiple scans are used to send refinement information for all coefficients; this is equivalent to the quality-progressive mode in JPEG 2000, and typically provides better quality than baseline transmission and spectral selection at low bit-rates. In both modes, the first scan must contain a band with all (and only) the predicted DC coefficient of each block. Note that JPEG also allows to use flexible combinations of spectral selection and successive approximation.

The hierarchical mode employs progressive coding at different stages, in which *spatial* resolution is increased at each stage; each version of the image is encoded in a different frame. In the first stage, a possibly lower resolution version of the image is encoded using JPEG in sequential or progressive mode. The output image is then upsampled (if necessary) and used as prediction for the next stage, which shall encode the difference between the image and the reference frame. For these differential frames, since prediction is already performed from a reference image, level-shifting and differential coding of the DC coefficient are not carried out.

4.3.6 CODESTREAM SYNTAX

Besides the encoding and decoding processes, JPEG also defines an interchange file format, as well as abbreviated file formats. A basic description of the interchange file format is provided hereafter.

The compressed file is a sequence of marker segments, which contain general information about the image (e.g., size, number of components, downsampling factors for each component, and so on) as well as coding parameters and entropy coded segments; each marker segment begins with a unique 2-byte code identifying its function. The compressed data are placed in one or more "frames" (more than one frame is allowed only for hierarchical processes). Each frame consists of one or more "scans" through the data, and each scan can contain data from a single component or interleaved data from different components. Frames and scans are preceded by their own headers that contain parameters needed for decoding; they can also be preceded by marker segments that specify or modify other parameters, e.g., quantization tables (DQT marker segments) and entropy-coding tables (DHT marker segments). Scan headers are always followed by entropy coded segments. To achieve error resilience, the scan data may be divided into restart intervals; the data in each restart interval are stuffed into one entropy coded segment, and the entropy coded segments (if there is more than one in the scan) are separated by a 2-byte periodic code called a restart marker. Note that, to achieve error resilience, the data in each entropy coded segment must be decodable without knowledge of other entropy coded segments. This implies that the entropy coder statistics must be reset at the beginning of each entropy coded segment.

Coding parameters can be varied through frames and scans. The frame header contains the quantization table selector for the current frame. The scan header contains the quantization table selector and the entropy coding table selectors for the current scan, as well as the parameters that specify the progressive and hierarchical modes.

4.4 THE JPEG-LS STANDARD

The lossless mode in JPEG is based on the use of a fixed predictor for each scan, followed by Huffman or arithmetic coding. However, while a fixed predictor provides a very simple compression strategy, it is often suboptimal as to coding efficiency.

A lossless compression scheme can attain an average codeword length as low as the entropy rate of the sequence of input samples $\{X_n\}$, which can be written [4] as $\mathcal{H}(X) = \lim_{n \to \infty} H(X_n \mid X_{n-1}, X_{n-2}, \ldots, X_1)$. Achieving the entropy rate implies that coding of a generic image sample X_n must conditionally depend on all the other samples in the sequence. Therefore, adaptive entropy coding should be performed so as to take into account not only the symbol occurrence probabilities $P(X_n)$, but rather the conditional probabilities $P(X_n \mid X_{n-1}, X_{n-2}, \ldots, X_1)$. Turning from theory to practice, one has to define:

- a reasonably small neighborhood of the current sample, which is used to condition the encoding of that sample to the values assumed by its neighbors;
- a finite number of "contexts", i.e., statistically homogeneous classes on which description of conditional dependencies can be based.

The statistical model is described in terms of a total number of parameters that depends on the number of contexts and the number of selectable coding parameters for each context. Even though in causal encoding the encoder may not need to send these parameters along with the data, there is a cost incurred by choosing too many contexts or too many parameters. In fact, the statistics computed on-the-fly by the encoder and the decoder may be inaccurate if the sample set for each parameter is not large enough; this effect is known as "context dilution."

A simplified block diagram of the JPEG-LS coding process is shown in Figure 4.4. The basic element of a encoding pass (also called a scan, similar to JPEG) is a line of image samples.

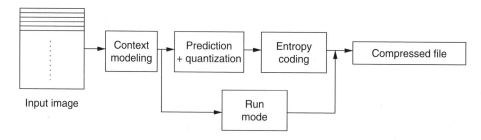

FIGURE 4.4 Basic block diagram of JPEG-LS.

At each sample, a context is determined, and used to condition prediction and entropy coding. Then the current sample is predicted from a causal neighborhood, and a prediction error is computed; if near-lossless compression is used, the prediction error is also quantized. The contexts are then used to switch between two different coding modes according to the local image characteristics. A regular mode is used, unless it is found that all image samples in the current neighborhood have the same value; this triggers a run-length coding mode.

4.4.1 CONTEXT-BASED PREDICTION

The basic set of image samples used to condition the prediction contains four neighbors of the current sample x; we will denote these samples as a, b, c, and d, respectively, the left, top, top-left, and top-right neighbors of x, as in Figure 4.5.

The predictor consists of a fixed and an adaptive part. The fixed part has the objective of detecting edges, while the adaptive part adds a context-dependent integer term. Specifically, the fixed part provides a prediction of x as follows:

$$\hat{x} = \begin{cases} \min(a,b) & \text{if } c \geq \max(a,b) \\ \max(a,b) & \text{if } c \leq \min(a,b) \\ a+b-c & \text{otherwise} \end{cases} \tag{4.2}$$

The motivation behind this predictor is to avoid picking as prediction for x a pixel lying on a edge close to x, but to which x does not belong, as this would lead to a large prediction error. Instead, the predictor performs a sort of rough-edge detection. If $c \geq \max(a,b)$, then c and b may be lying on a positive edge; the predictor then selects as estimate the pixel *not* lying on the edge, i.e., $\min(a,b)$. The same reasoning applies to negative edges. If edges are not likely, then the prediction is a linear combination of the neighbors; note that this is equal to predictor 4 in the lossless JPEG mode (see Table 4.2). The prediction error e has a larger range of admissible values than the original signal. To reduce this range, the prediction error is computed modulo 2^N; this operation does not significantly affect the statistical distribution of prediction error samples.

Prediction errors generated by fixed predictors typically follow a two-sided geometric distribution with zero mean value. However, context conditioning often results in biased prediction errors, where the bias μ is in general a noninteger value; it is useful to separately consider the integer and fractional parts of μ, which can be rewritten as $\mu = R - s$, with R as an integer number and $0 \leq s < 1$. The purpose of the adaptive part of the predictor in JPEG-LS is to cancel R. To do so, one can count the number N_C of occurrences of each context, along with the sum of the prediction errors contributed by the fixed predictor for each context, and estimate R for each context as the mean value of the prediction error; JPEG-LS actually implements a simplified version of this technique that avoids divisions [37]. In the following we will denote by e the prediction error samples after bias correction. Note that bias correction can also be viewed as a part of the entropy-coding process instead of the decorrelation.

Context modeling is based on gradient information in the neighborhood of x. In particular, three local gradients are estimated as $g_1 = d - b$, $g_2 = b - c$, and $g_3 = c - a$. Since all possible combinations of values would lead to too high a number of contexts, gradient values are quantized into a fixed number of symmetric and roughly equiprobable regions, indexed as $-T, \ldots, 0, \ldots, T$, leading to

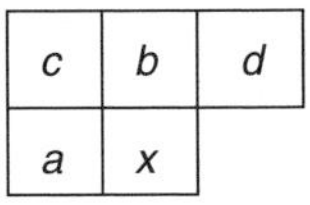

c	b	d
a	x	

FIGURE 4.5 Neighborhood for prediction in JPEG-LS.

$(2T + 1)^3$ different contexts $\mathcal{C} = [g_1 \cdot g_2 \ g_3]$. Since for symmetry the probabilities $P(\mathcal{C})$ and $P(-\mathcal{C})$ are equal, contexts $\mathcal{C}$ and $-\mathcal{C}$ are merged. Upon finding the first negative value in $\mathcal{C}$, the encoder uses context $-\mathcal{C}$ and flips the sign of the prediction error; this is undone by the decoder as it computes the context for the current sample. The total number of contexts thus becomes $((2T + 1)^3 + 1)/2$; since $T = 4$ is used in JPEG-LS, 365 contexts are employed. Default quantization thresholds are defined for 8-bit samples (see [12]), which can be changed by writing the desired thresholds in the codestream.

4.4.2 ENTROPY CODING

Entropy coding in JPEG-LS can be done in two different ways, according to whether context information indicates that the neighborhood of the current pixel contains details, or is very smooth. In the former case, entropy coding is carried out employing Golomb codes, whereas in the latter case run-length coding is performed so as to take advantage of long sequences of image samples with the same value in smooth image regions. The decision is made by checking the current context for $g_1 = g_2 = g_3 = 0$; this condition triggers the run mode for the sample value $x = x^*$, while Golomb coding is used otherwise.

4.4.2.1 Golomb Coding

JPEG-LS uses a class of codes, namely Golomb codes, which are optimal for an input alphabet following a one-sided geometric distribution. A Golomb code G_m of order m represents a nonnegative integer $x \geq 0$ as the concatenation of (i) the unary representation of $\lfloor x/m \rfloor$, i.e., as many 0s as $\lfloor x/m \rfloor$, followed by a 1, and (ii) a binary representation of $x \bmod m$. Since the input samples x follow a stationary distribution with probability density function proportional to $(1 - \lambda)\lambda^x$ (with $0 < \lambda < 1$), for each λ there is an optimal m that provides optimal compression.

The special case of m being a power of 2 leads to a very simple coding procedure; these G_{2^k} codes are also known as Golomb–Rice codes, and have been used in the CCSDS lossless data compression standard [3]. In this case, coding x with G_{2^k} is simply done by appending the k least significant bits of x to the unary representation of the number obtained considering all other most significant bits of x (from bit $k + 1$ up to the most significant bit).

Golomb codes must be applied to samples with one-sided distribution. However, prediction error samples have a two-sided distribution; therefore, before encoding, these samples have to be mapped onto nonnegative integers. In particular, instead of the prediction error e, a mapped error $e' = M'(e) = 2|e| - u(e)$ is coded, where $u(e) = 1$ if $e < 0$, and 0 otherwise. Note that this mapping merely outputs the index of e in the sequence $0, -1, +1, -2, +2, \ldots$, and has the property that, if $s \leq 1/2$ (where s is the fractional part of the bias, see Section 4.4.1), then prediction error samples are sorted according to nonincreasing probability. If $s > 1/2$, the mapping $e'' = M''(e) = 2|-e-1| - u(-e-1)$ is used.

For selecting the optimal parameter k of the Golomb–Rice code, JPEG-LS uses an approximated version of the maximum likelihood procedure. In particular, *for each context* an additional counter is updated, which stores the accumulated magnitude A of the prediction error samples. Letting N_C be the number of occurrences of the context for the current sample, the optimal k is yielded by $k = \min\{k' \mid 2^{k'} N_C \geq A\}$. (If $k = 0$ it is still necessary to select between the two mappings M' and M''.)

This general encoding procedure has the drawback that it can occasionally expand the signal representation instead of compressing it. To avoid this, a maximum codeword length $L_{\max}$ is fixed (e.g., $L_{\max} = 32$); the coding procedure for encoding x, which can take on values in $[0, \alpha]$ (e.g., $\alpha = 255$), follows the steps described above if $\lfloor 2^{-k} x \rfloor < L_{\max} - \lceil \log \alpha \rceil - 1$, as this condition guarantees that the total codeword length does not exceed $L_{\max}$ bits. Otherwise, the number $L_{\max} - \lceil \log \alpha \rceil - 1$ is coded in unary to act as "escape" code, and is followed by the binary representation of $x - 1$ using $\lceil \log \alpha \rceil$ bits; this requires a total of L_{max} bits.

4.4.2.2   Run Mode

The run mode is started upon detection of a zero context with image sample value x^*. For the next samples, the encoder only looks for image samples that have value x^*, regardless of whether the gradients are still zero and encodes the run length for x^*. As soon as the encoder detects a sample whose value is different from x^*, it encodes the difference $x - b$ and exits the run mode. The run mode is also interrupted upon detection of ends of lines; note that the termination conditions are based on past samples, and can be exactly reproduced at the decoder.

Coding of run lengths is also based on Golomb–Rice codes; however, the code adaptation strategy differs from that for encoding of prediction error samples (see [36] for details).

4.4.3   Near-Lossless Mode

Besides lossless compression, JPEG-LS also offers a particular lossy compression mode, named "near-lossless," in which it is guaranteed that the maximum absolute difference between each pixel of the decoded and the original image does not exceed a user-defined value δ. Notice that near-lossless compression is very difficult to achieve by means of transform coding (e.g., the lossy version of JPEG), since the inverse transform propagates the quantization error of a transformed coefficient across a number of image samples; this makes it difficult to adjust the quantization so as to guarantee a maximum error on a per-pixel basis. Conversely, predictive schemes are suitable for near-lossless compression, since quantization in the DPCM feedback loop [15] produces the same maximum absolute error on each pixel, which is equal to half of the quantizer step size. Near-lossless compression can hence be achieved by means of a uniform scalar quantizer with step size $2\delta + 1$ and midpoint reconstruction [38], such that, letting e be the prediction error, its reconstructed value $\hat{e}$ is given by

$$\hat{e} = \text{sign}(e)(2\delta + 1) \left\lfloor \frac{|e| + \delta}{2\delta + 1} \right\rfloor \tag{4.3}$$

Since the reconstructed and not the original samples are now available at the decoder, some modifications have to be made to the encoder. In particular, context modeling and prediction must be based on the reconstructed values. Moreover, the condition for entering the run mode is relaxed to $|g_i| < \delta$, as values lower than δ can result from quantization errors; once in run mode, the termination conditions now depend on $|a - b| \leq \delta$. Other modifications are listed in [37]. Note that, while we have described the near-lossless case as an add-on to the lossless one, the standard [12] defines the lossless case as a special case of near-lossless compression with $\delta = 0$.

4.4.4   JPEG-LS Part 2

Part 2 of JPEG-LS [13] provides a number of extensions to the baseline compression algorithm, and namely:

- arithmetic coding (in addition to Golomb coding);
- variable near-lossless coding;
- modified prediction for sources with sparse distributions;
- modified Golomb coding;
- fixed-length coding;
- modified sample transformation process.

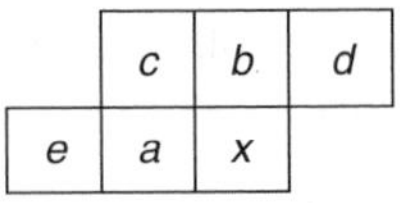

FIGURE 4.6 Neighborhood for context modeling in arithmetic coding mode for Part 2 of JPEG-LS.

While Part 1 of JPEG-LS uses Golomb coding, for certain types of images (e.g., compound documents) this may lead to suboptimal performance. Therefore, an arithmetic coding mode is foreseen to overcome this problem. In particular, in this mode no difference is made between the regular and the run mode; the prediction error samples are binarized and sent to the binary arithmetic coder. Unlike the baseline algorithm, the context for arithmetic coding employs an additional pixel, as shown in Figure 4.6. Four local gradients are computed as $g_1 = d - b$, $g_2 = b - c$, $g_3 = c - a$, and $g_4 = a - e$, and used to form contexts; however, the predictor for the current sample is the same as in the baseline algorithm.

The variable near-lossless coding extension allows to vary the δ parameter during the encoding process. In the "visual quantization" mode, δ can change according to the context of the current pixel. In particular, if $|g_1| + |g_2| + |g_3| > T_v$, $\delta + 1$ is used for maximum error, otherwise δ is used; the motivation is that a higher error can be tolerated in very high-activity regions, where the error is less visible, whereas the default error is used in normal regions. The threshold T_v can be selected by the user.

Moreover, the δ parameter can be re-specified along the vertical direction; this allows to control the output bit-rate, in that if the compressed-data buffer is getting full, the encoder can switch to a larger δ, which will provide a higher degree of compression.

Sources with sparse distribution are known to present difficulties for predictive compression schemes [23]. Such sources can be encountered in several applications; an example is given by multispectral images represented with say 12 bit/sample, whose samples undergo an offset/gain radiometric correction and are eventually represented with 16 bits/sample. If the offset and gain are the same for each sample, only 2^{12} of the total 2^{16} levels are used, leading to a sparse histogram. It has been shown [6] that histogram packing leads to images with lower total variation, which can be losslessly compressed at lower bit-rates. JPEG-LS specifies a procedure for histogram packing.

Fixed-length coding may be used to prevent the compressed data from having larger size than the original data. This is achieved by simply writing the binary representation of the input sample x (or of the quantizer index in the near-lossless compression case) using $\lceil \log_2 ([2^N + \delta/2\delta + 1]) \rceil$ bits/sample.

4.4.5 CODESTREAM SYNTAX

Similar to JPEG, the JPEG-LS standard defines both the encoding and decoding processes, along with the compressed file format; there is no requirement that an encoder shall be able to operate for all ranges of parameters allowed for each encoding process, provided that it meets the compliance requirements in [12].

JPEG-LS supports images with up to 255 components, of size up to $2^{64} \times 2^{64}$, with sample precision of 2 to 16 bits as in the lossless JPEG mode.

JPEG-LS uses a file format that is similar to the JPEG interchange format, as it consists of frames, scans, and restart intervals. In fact, JPEG-LS uses the same markers as JPEG (except for a few that do not apply). Moreover, it adds new marker segments containing JPEG-LS specific information, namely specific start-of-frame and start-of-scan marker segments, and an LSE marker segment for preset parameters; in fact, unlike JPEG, parameters have default values that can be overridden by inserting other marker segments. JPEG-LS supports single- and multicomponent scans; in this latter case, a *single* set of context counters is used throughout all components, whereas prediction and context determination are done independently on each component. The data in a component scan can be interleaved either by lines or by samples. Unlike JPEG, JPEG-LS also supports compression of palettized images.

4.5 THE JPEG 2000 STANDARD

More than 10 years after the standardization of JPEG, it seems that its success is still growing, with new applications making use of JPEG lossy-compressed images. So why is there the need of a new compression standard?

As a matter of fact, JPEG has a number of limitations that were hardly relevant at the time of its definition, but are becoming increasingly important for new applications. There are plenty of novel aspects to JPEG 2000; perhaps the single most innovative paradigm is "compress once: decompress many ways" [31]. This implies that, once a codestream has been properly generated, selecting a maximum quality or rate (possibly including lossless compression) and resolution, a user can:

- decompress the image at the maximum quality and resolution;
- decompress at a lower rate with optimal rate-distortion performance;
- decompress at reduced resolution with optimal performance;
- decompress only selected spatial regions of the image; if region-of-interest (ROI) coding has been performed at the encoder side, improved quality can be selected for certain spatial regions;
- decompress a number of selected components;
- extract information from a codestream to create a new codestream with different quality/resolution *without* the need of decompressing the original codestream; other transformations, like cropping and geometric manipulations, require minor processing and do not need full decompression.

On top of that, JPEG 2000 provides several additional features such as improved error resilience, visual weighting, image tiling, support for higher bit depths, and number of components, just to mention a few. A thorough discussion of the JPEG 2000 functionalities is provided in [27,30].

In the following we describe Part 1 of JPEG 2000 ("core coding system," [14]); a brief summary of the other parts of the standard is given in Section 4.5.6. JPEG 2000 can handle images with up to 16,384 components; each component can be divided into up to 65,535 tiles whose maximum size is $2^{32} \times 2^{32}$ samples; each sample can be represented by up to 38 bits/sample. It is not required that different components have the same bit depth. As can be seen, this overcomes the 8 to 12 bits/sample limitation of JPEG, and extends the maximum image size and bit depth well beyond previous standards, allowing application of JPEG 2000 to images with high sample accuracy and very large size (e.g., scientific images or digital cinema video sequences).

JPEG 2000 can be operated in *irreversible* and *reversible* mode. The irreversible mode corresponds to lossy compression, and allows to generate an embedded codestream that can be decoded up to the maximum available rate; however, owing to the finite precision computer arithmetic, it does not support perfectly lossless compression. The reversible mode provides an embedded codestream that can be decoded at any lossy rate up to lossless compression. It is worth noticing that, apart from minor differences in the transform, the two modes are exactly alike; this is a major advance with respect to JPEG, in which the lossy and lossless modes are utterly different compression techniques.

4.5.1 TRANSFORM AND QUANTIZATION

4.5.1.1 DC Level Shifting

Figure 4.7 shows the sequence of operations that generate the compressed image. Each image component is considered separately, and is first divided into rectangular tiles. If the samples are unsigned

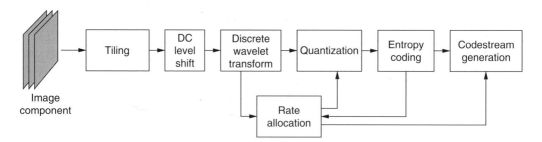

FIGURE 4.7 Basic block diagram of JPEG 2000. After the optional multicomponent transformation, each component is tiled, wavelet-transformed, quantized, and entropy-coded; the rate allocator decides which compressed data portions will be inserted in the codestream.

integers, level shifting is performed as in JPEG, by subtracting 2^{N-1} from each image sample, where N is the bit depth.

4.5.1.2 Multicomponent Transformation

Unlike JPEG, the JPEG 2000 standard foresees the optional use of a color transformation, which applies to the first three components of the image.[2] In fact, two multicomponent transformations are defined, according to whether the reversible or irreversible modes are employed.

The reversible component transformation can only be used in reversible mode. For illustration purposes, let R, G, and B be samples in the same spatial position of the first, second, and third component (although they need not be red, green, and blue), and I_0, I_1, and I_2 the corresponding samples after color transformation. The reversible component transformation is defined as

$$
\begin{aligned}
I_0 &= \left\lfloor \frac{R+2G+B}{4} \right\rfloor \\
I_1 &= R - G \\
I_2 &= B - G
\end{aligned}
\tag{4.4}
$$

Its corresponding inverse transform is

$$
\begin{aligned}
G &= I_0 - \left\lfloor \frac{I_1+I_2}{4} \right\rfloor \\
R &= I_1 + G \\
B &= I_2 + G
\end{aligned}
\tag{4.5}
$$

and ensures exact reconstruction of the original components. The irreversible component transformation can only be used in irreversible mode, and is defined as

$$
\begin{aligned}
I_0 &= 0.299R + 0.587G + 0.114B \\
I_1 &= -0.16875R - 0.33126G + 0.5B \\
I_2 &= 0.5R - 0.41869G - 0.08131B
\end{aligned}
\tag{4.6}
$$

4.5.1.3 Wavelet Transformation

A L-level wavelet transform [24,33] is applied separately to each component of each tile, generating a set of subbands of wavelet coefficients. JPEG 2000 employs one out of two wavelet filters, namely the (9,7) and (5,3) biorthogonal filters. The (9,7) filter was first proposed in [1], and has been found

---

[2] In case the multicomponent transformation is used, the first three components *must* have the same bit depth.

to be the overall best wavelet filter for image coding [34]; on the other hand, the (5,3) filter [16] has rational coefficients, which make it more convenient for low-precision implementations and for reversible transforms.

Two different wavelet transformations are defined in JPEG 2000, corresponding to the irreversible and reversible modes. The irreversible transformation is a classical wavelet transform employing the (9,7) filter, with symmetric extension at image boundaries. The reversible transformation is based on filter decomposition into lifting steps [8,29]; by rounding the intermediate results after each lifting step, this procedure allows to obtain a slightly nonlinear wavelet transform that maps integers to integers [2]. Since the wavelet coefficients are represented as integer values, as opposed to the irreversible transformation that yields real values, lossless compression is possible by avoiding quantization of the transform coefficients, and simply performing entropy coding of the transform coefficients. The reversible mode employs this integer-to-integer transform along with the (5,3) filter.

The wavelet transform can be implemented by buffering the whole image to allow for the horizontal and vertical convolutions to take place. However, in some applications it is desirable to avoid excessive buffering. To this end, it is possible to compute a line-based transform. This transform only requires to buffer a few lines of image samples; horizontal convolutions are performed in the usual way, while vertical convolution samples output one sample at a time, every time a new image line is acquired. This transform is exactly equivalent to the case of full image buffering. However, the rate allocation process works differently due to the memory constraints in the line-based case (note, however, that the rate-distortion optimization is not part of the standard). In the full-buffering case, it is possible to encode the whole image at a high rate and, after the encoding, to select, for a given target rate, the data portions that contribute most to image quality, leading to globally optimal rate-distortion performance. In the line-based case, rate allocation must be done on-the-fly, picking portions only of those code blocks available in the buffer. This may lead to locally optimal performance, in that a later data portion that yields a high-quality contribution may be discarded in favor of a less important earlier data portion.

4.5.1.4 Quantization

After the wavelet transformation, the transform coefficients need to be quantized to an integer representation. JPEG 2000 uses a uniform scalar quantizer with deadzone, which is known to be very efficient especially at low bit rates [15,17]. A different quantization step size is used for the coefficients of each subband; the deadzone is twice the step size. Quantization of each coefficient $x_b(i,j)$ of subband b is done with step size Δ_b according to the following formula:

$$x_b^*(i,j) = \text{sign}(x_b(i,j))\text{round}\left(\frac{|x_b(i,j)|}{\Delta_b}\right) \qquad (4.7)$$

In reversible mode, Δ_b must be equal to 1.

In implicit mode, Δ_b is taken proportionally to the nominal dynamic range of each subband, starting from a user-specified global initial step size. In explicit mode, the user can specify all step sizes Δ_b, which are written in the codestream.

4.5.2 DATA ORGANIZATION

The quantized wavelet coefficients of each tile are conceptually structured in a way that allows flexible generation of the codestream from the compressed data; this organization is sketched in Figure 4.8. The wavelet coefficients of each subband are partitioned into rectangular nonoverlapping sets called *code-blocks*. Each code-block has a limited scope of influence in the decoded image, due to the finite length of the wavelet synthesis filters. As a consequence, if one wants to decode a portion of

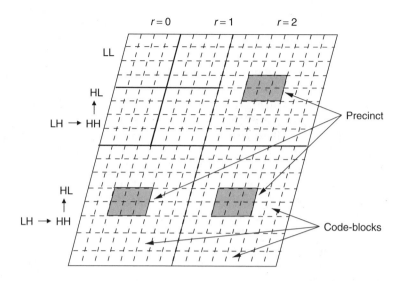

FIGURE 4.8 Data organization in the wavelet domain: subbands, precincts, and code-blocks.

a tile instead of the complete tile, it is possible to identify and decode only those code-blocks that contribute to that portion; thus, code-blocks also provide a sort of random access to the image.

Code-blocks can be grouped according to location and resolution into the so-called *precincts*. Precincts are defined for each resolution level r, with $r = 0, \ldots, L + 1$, and $\tau = 0$ indicating the LL subband after L transform levels; in particular, a precinct is a rectangular collection of whole code-blocks, replicated in each of the three high-frequency subbands of resolution level τ. The usefulness of precincts is that it can be decided to perform entirely independent encoding of each precinct, in such a way that all code-blocks in a precinct can be recovered without reference to any other precinct.

4.5.3 ENTROPY CODING

JPEG 2000 employs bit-plane-based binary arithmetic coding of the quantized wavelet coefficients of each code-block. Within each subband, code-blocks are scanned in raster order, and the coefficients of the code-block are encoded by bit planes from the most to the least significant bit. Within each code-block, bit-planes are not formed by scanning all coefficients in raster order, but rather considering the concatenation of groups of four columns, each of which is raster-scanned. Each bit of the bit-plane is coded in only one out of three *coding passes*, called significance propagation, magnitude refinement, and clean-up; coding passes are the "smallest" data units in JPEG 2000, and are the bricks for building up the codestream. Each coding pass is encoded by means of an arithmetic coder, which exploits contextual information to estimate conditional symbol probabilities. Arithmetic coding is performed by the MQ coder; since this coder is extensively treated in another chapter of this book, along with context modeling, its description is omitted here.

During the encoding process, it is possible to compute the distortion reduction obtained by decoding each coding pass; moreover, after arithmetic coding the rate necessary to encode the coding pass is also known. This pair of distortion and rate values can be used during the codestream formation stage in order to obtain the best quality for a given rate and progression order. It is worth noticing that, since JPEG 2000 standardizes the syntax and the decoder, the rate allocation process is not a mandatory part of the standard, but different trade-offs can be achieved among quality, delay, and complexity. In fact, optimal rate allocation requires to encode the whole image, generating and coding even those coding passes that will not be inserted in the codestream; suboptimal strategies can be devised to limit memory usage or complexity.

FIGURE 4.9 Structure of JPEG 2000 codestream.

4.5.4 CODESTREAM SYNTAX, PROGRESSION ORDERS, AND CODESTREAM GENERATION

The JPEG 2000 codestream syntax differs from JPEG and JPEG-LS in many aspects; the concepts of frames and scans are substituted by more general and flexible data structures. JPEG 2000 also defines an optional JP2 file format, which can be used to provide additional information about the image, such as metadata, colorspaces, information about intellectual property rights, and so on; in the following we will exclusively refer to the codestream syntax.

The basic structure of a JPEG 2000 codestream is depicted in Figure 4.9. Conceptually, in JPEG 2000 an image component can be divided into smaller, independently coded subimages known as "tiles"; the *compressed* data related to each tile can be further divided into tile-parts. The codestream contains a main header, followed by tile-part headers and tile-part data, and is terminated by an end-of-codestream marker. The main header contains ancillary information that applies to the whole codestream, whereas the tile-part headers contain information specific to each tile-part; this mechanism allows to vary coding parameters between different tile-parts. Tile-part compressed data are organized as a sequence of "packets," which can contain very flexible combinations of data; each packet is preceded by a packet header specifying which information is contained in the packet.

The organization of the compressed data as a sequence of packets is functional to the desired results in terms of rate, quality, and scalability. In JPEG 2000, this is achieved by employing the concept of *layers*; a layer is simply an "increment" in image quality. For instance, one can encode an image at 1 bpp using three layers at 0.25, 0.5, and 1 bpp; the rate allocator can perform the encoding in such a way that decoding at those rates provides optimal rate-distortion performance, whereas in general decoding at other rates is possible but may incur a slight performance decrease. An increment with respect to the current quality can be achieved by decoding "more data," and these data can be selected very flexibly among all available coding passes of the data in all subbands. The way coding passes are picked up and put into layers determines the final results in terms of rate, quality, and scalability.

All compressed data representing a *specific* tile, layer, component, resolution level, and precinct appears in one *packet*; this does not mean that a complete code-block is coded into one single packet, since the code-block compressed data can be fragmented into different layers. In a packet, contributions from the LL, HL, LH, and HH subbands appear in that order (resolution level $r = 0$ only contains the LL subband, whereas all other resolution levels only contain the HL, LH, and HH subbands); within a subband, code-block contributions appear in raster order within the boundaries of a precinct.

A layer is nothing but a collection of packets, one from each precinct of each resolution level. Therefore, it is easy to see that a packet can be roughly interpreted as a quality increment for a specific resolution level and "spatial location," and a layer as a quality increment for the entire image. In the codestream, data from different components are interleaved layer-wise. The final codestream consists of the headers, followed by a succession of layers terminated by the end-of-codestream marker.

The rate allocator decides which coding passes of which code-block go in every layer (or are not included in the codestream in case of lossy compression). This is done so as to obtain a specific user-specified data progression order. JPEG 2000 defines five different progression orders that interleave layers, resolution levels, components, and positions (i.e., precincts) in various ways:

- layer–resolution level–component–position (also known as quality-progressive);
- resolution level–layer–component–position;

- resolution level–position–component–layer;
- position–component–resolution level–layer (also known as spatially progressive);
- component–position–resolution level–layer (also known as component-progressive);

As an example, quality-progressive encoding amounts to interleaving the packets in the following order:

for each layer $y = 0, \ldots, Y - 1$,
   for each resolution level $r = 0, \ldots, L + 1$,
      for each component $c = 0, \ldots, C_{max} - 1$,
         for each precinct $p = 0, \ldots, P_{max} - 1$,
            packet for component c, resolution level r, layer y, and
               precinct p.

Note that the progression order may be changed within the codestream.

The flexible data organization described above has one drawback, i.e., each packet must be preceded by a header that specifies what information is contained in the packet body. In particular, the following information is contained in each packet header:

(1) If the current packet is an empty packet; if this is the case, this is the only information contained in the packet header.
(2) Which code-blocks are included in the packet.
(3) For each code-block included for the first time, the actual number of bit-planes used to represent the coefficients in the code-block.
(4) The number of coding passes included in this packet from each code-block.
(5) The number of bytes contributed from each included code-block.

Signaling of empty/nonempty packets obviously requires one bit. For indicating the number of coding passes, a simple variable-length code is used. The number of bytes for each code-block are represented in binary format. Code-block inclusion in the packet and the number of bit-planes for each code-block are signaled by means of tag tree data structures; tag trees allow to represent two-dimensional arrays of nonnegative integers in a hierarchical way, so that only the tag tree information necessary to decode the code-blocks in the current packet is included in the packet header.

As can be seen, packet headers require some overhead information to specify what is included in each packet. Note that to generate an extremely finely scalable codestream, one would like to form a very high number of quality layers. However, all other things being equal, in quality-progressive mode the total number of packets equals the number of packets in case of no layering (i.e., only one layer) times the number of layers. Since more layers require more packets, extremely fine scalability is achieved at the expense of a slight performance decrease owing to the packet header overhead. However, reasonably high numbers of layers can be generated with only marginal performance decrease, so that in practice this is not an issue.

4.5.5 ADVANCED FEATURES

4.5.5.1 Region of Interest Coding

In many applications, different portions of an image may have unequal importance from a perceptual or objective point of view; as a consequence, it can be interesting to code one or more ROIs with higher quality with respect to the background.

In JPEG 2000, the MAXSHIFT method is employed to encode ROIs with higher quality than the background, and to transmit them first. The basic idea of this method is to identify those wavelet coefficients that would contribute to the ROI reconstruction,[3] and to upscale those coefficients by a user-defined scaling value S. By doing so, the most significant bit-planes will mostly contain wavelet coefficients pertaining to the ROI, which will be encoded prior to the background ones.

Note that if one chooses S in such a way that the minimum nonzero wavelet coefficient magnitude belonging to the ROI is larger than the maximum coefficient in the background, there is no need to explicitly signal which coefficients belong to the ROI, as the decoder can identify them based on the quantization indices; this allows implicit usage of arbitrarily shaped ROIs. Since bit-planes containing information related to the ROI and the background are separated, the rate allocator can independently select coding rates for the two regions.

4.5.5.2   Error Resilience

Besides storage and TCP/IP lossless image transmission, JPEG 2000 also targets image transmission applications over channels or networks potentially prone to errors. However, as is well known, entropy-coded data are very sensitive to errors, and the effect of an erroneous codeword can propagate throughout the codestream, with dramatic effects on image quality. As a consequence, error resilience tools are necessary to combat the effect of such errors. Recognizing this issue, in JPEG, restart markers were defined as a way to restart the decoder after a synchronization loss due to a codestream error.

JPEG 2000 takes a much more sophisticated approach to the problem of transmission errors, and defines several error resilience tools at the syntax level, the wavelet transform level, and the entropy coding level. These tools are the following:

- code-blocks and precincts;
- arithmetic coder predictable termination and segmentation symbol;
- resynchronization markers;
- packed packet headers.

In general, the aim of error resilience tools is to limit the scope of transmission errors by allowing the decoder to (1) detect errors, and (2) restart decoding at the next error-free boundary. Code-block size in the wavelet domain is a user-selectable parameter that can reduce error propagation. Normally, code-block size is chosen to be not too small, so as to allow better coding efficiency by providing the arithmetic coder with larger input data blocks; however, since code-blocks are independently entropy coded, smaller code-blocks provide limited error propagation. To further strengthen the decoder's resynchronization ability, precincts can be employed; in fact, tag trees in the packet headers are applied independently within each precinct, facilitating decoder resynchronization.

The most powerful error resilience tool in JPEG 2000 is the arithmetic coder predictable termination. This procedure can be used to bring the arithmetic coder into a known state at the end of each coding pass. Upon occurrence of a bit error in a given coding pass, the decoder will likely find itself in the wrong state; this causes declaration of an error. A "segmentation symbol" can also be used for error detection. It amounts to encoding a fictitious input symbol at the end of each bit-plane; erroneous decoding of this symbol causes declaration of an error.

Errors detected by means of arithmetic coder termination and/or segmentation symbol can be concealed by simply skipping the decoding of the erroneous coding passes (and possibly a few previous ones to account for error detection delays). This procedure avoids the decoding of erroneous data, and provides greatly improved image quality in error-prone environments.

In addition to the error resilience tools described above, JPEG 2000 defines syntactical elements that can be used to help decoder resynchronization. In fact, decoding restart on a new code-block is only possible if the header information is intact. In order to make it easier to protect packet

---

[3] This can be done by looking at the wavelet cones of influence defined by the synthesis filters.

header information, a particular mode is foreseen, which avoids the fragmentation of packet headers throughout the codestream. Namely, all packet headers can be extracted from the codestream and grouped in specific marker segments (PPM/PPT) in the main or tile-part headers. Obviously, if *all* header data are contiguous, they can be easily assigned a higher level of error protection, e.g., by using a stronger channel code for the rest of the codestream, or by assigning a higher number of retransmission attempts, and so forth. Moreover, the start-of-packet (SOP) and end-of-packet-header (EPH), two-byte markers can be used to delimit the packet headers; this is especially useful if the packet headers are fragmented in the codestream, as it facilitates the decoder in locating the next packet in case of errors.

4.5.6 OTHER PARTS OF THE STANDARD

Part 1 of JPEG 2000 defines the wavelet-based core coding system; however, there is much more to this standard than Part 1. At the time of this writing, the JPEG committee has defined 14 parts of the standard, most of which are still ongoing; a brief summary is given hereafter.

4.5.6.1 Part 2 — Extensions

Some of the advanced features that were studied during the development of the standard could not find place in the core coding system, but have been integrated in Part 2. These include (not exclusively) the following:

- Arbitrary DC offset values can be used for any component.
- The quantizer deadzone can be set to an arbitrary value.
- A trellis-coded quantizer [26] can be used instead of the uniform scalar quantizer to achieve improved coding efficiency.
- The wavelet decomposition is no longer bounded to be dyadic, but arbitrary horizontal/ vertical subsampling and low/highpass iteration schemes can be used.
- Arbitrary wavelet filters can be used in addition to the (9,7) and (5,3) filters.
- For multicomponent images, besides "spatial" per-component decorrelation the standard also supports transformations across components; an inverse component transformation can be defined in a specific marker segment.
- ROI coding also supports the scaling method in addition to the MAXSHIFT method.
- A flexible extended file format (JPX) is defined to provide advanced color space, composition, and animation facilities.

4.5.6.2 Part 3 — Motion JPEG 2000

After the standardization of JPEG, it was found that many applications, especially in the field of videosurveillance and low-cost consumer electronics, can effectively employ JPEG-based intraframe video compression (Motion JPEG). However, although most existing implementations of Motion JPEG are based on the same technology, they may differ in the file format, as Motion JPEG is not an international standard. To avoid the same problem, Part 3 of JPEG 2000 standardized Motion JPEG 2000, which is a file format allowing to embed a sequence of intraframe JPEG 2000 coded pictures along with metadata.

4.5.6.3 Other Parts

In addition, several new parts of JPEG 2000 have been or are being developed and standardized.

Part 4 of JPEG 2000 defines conformance testing procedures to validate implementations of the standard. Part 5 deals with reference software (more on this in Section 4.7). Part 6 defines a file format for compound images. Part 8 (JPSEC) addresses security aspects for JPEG 2000 imagery. Part 9 (JPIP) defines an interactive protocol whereby clients and servers can exchange images in a very flexible way. Part 10 (JP3D) defines improved coding modes for 3D and volumetric data. Part 11 (JPWL) defines improved error protection tools for JPEG 2000 image and video transmission over wireless and error-prone environments. Part 12 defines an ISO base media file format, which is common with MPEG-4.

4.6 ADVANCED RESEARCH RELATED TO IMAGE-CODING STANDARDS

International standards for image compression freeze the best available technology at a given time; however, research progresses after the definition of the standard, as new results are made available by the scientific community. The purpose of this section is to list a number of research topics and results that have appeared after the standardization of the techniques described above. In-depth description of many of these topics is provided in other chapters of this book.

4.6.1 DCT-BASED CODING

After publication of the JPEG standard, users started to notice that, at very low bit-rates, blockiness appears in the decoded images due to the 8×8 tiling for the DCT. This is also a common problem to all DCT-based video coders such as MPEG-2, MPEG-4, H.261, and H.263. Therefore, a large body of research has been devoted to the design of filters that aim at reducing blocking artifacts (see, e.g., [40]). In hybrid video coding, a common solution is to insert the deblocking filter directly in the DPCM feedback loop at the encoder. In JPEG, deblocking is usually a post-processing step on the decoded image, even though it can be embodied in the decoder by performing the filtering operation directly in the DCT domain.

In parallel, lapped orthogonal transforms were developed for the same purpose (see, e.g., [19]). These transforms are based on the idea of computing the DCT (or other local trigonometric transforms) on *overlapping* data blocks, which can help avoid blocking artifacts. However, the advent of wavelet-based coders, which exhibit higher coding efficiency than DCT-based coders and do not suffer from blocking artifacts, has somewhat diverted attention from lapped transforms in the field of image processing.

To come up with fast and efficient JPEG implementations, there has been a lot of work on the design of fast algorithms for DCT computation. Interestingly, it has been shown that the lifting scheme can be used to obtain integer-to-integer versions of the DCT, as well as other transforms [9,32]. An integer DCT is also used in the novel H.264/AVC video coding standard; this transform operates on 4×4 blocks instead of the typical 8×8 size [20].

4.6.2 WAVELET-BASED CODING AND BEYOND

Rate allocation for JPEG 2000 is not defined in the standard. Most existing software implementations support the so-called postcompression rate-distortion optimization; this technique provides optimal performance, but requires that the whole image be coded to a rate significantly higher than the final rate, in order to pick the most significant coding passes to be embedded in the codestream. Research is on-going on the development of rate allocation techniques that require less computations and less memory. An example is given in [39].

After the huge success of wavelets as building blocks for image compression, researchers have investigated new transforms in an attempt to achieve improved energy compaction. In particular, transform that are able to effectively describe both contours and texture would be a perfect fit for image compression. Several such transforms have been proposed, including ridgelets, wedgelets, curvelets, contourlets, and bandelets (see, e.g., [5,28]). The method of matching pursuits [18] has also been studied in the field of image and video coding [7,21]. However, no technique has been shown to significantly and consistently outperform state-of-the-art wavelet-based image coders such as JPEG 2000 and SPIHT [25].

4.7   AVAILABLE SOFTWARE

The reader interested in learning more and practicing with JPEG, JPEG-LS, and JPEG 2000 is encouraged to download and use publicly available software implementing these standards. At the time of this writing, the links provided in this section are working; for obvious reasons, the author cannot guarantee that they will continue to work indefinitely.

4.7.1   JPEG

The most popular JPEG software is that developed by the Independent JPEG Group, which can be downloaded from www.ijg.org. Because of patent issues, this software does *not* implement the arithmetic coding options of JPEG.

4.7.2   JPEG-LS

JPEG-LS software can be downloaded from the Hewlett-Packard web site at www.hpl.hp.com/loco/.

4.7.3   JPEG 2000

JPEG 2000 has two official reference softwares, namely Jasper and JJ2000. Jasper is written in C language, and can be obtained at http://www.ece.uvic.ca/~mdadams/jasper/. JJ2000 is written in Java language, and can be obtained at jj2000.epfl.ch. David Taubman's Kakadu software is written in C++; it is not freely available, but a demo can be obtained at www.kakadusoftware.com OpenJPEG is a new implementation of JPEG 2000 in C language. It can be obtained at www.openjpeg.org

REFERENCES

1. M. Antonini, M. Barlaud, P. Mathieu, and I. Daubechies, Image coding using wavelet transform, *IEEE Trans. Image Process.*, 1, 205–220, 1992.
2. R.C. Calderbank, I. Daubechies, W. Sweldens, and B. Yeo, Wavelet transforms that map integers to integers, *Appl. Comput. Harmonic Anal.*, 5, 332–369, 1998.
3. CCSDS 121.0-B-1 Blue Book, *Lossless Data Compression*, May 1997.
4. T.M. Cover and J.A. Thomas, *Elements of Information Theory*, Wiley, New York, 1991.
5. M.N. Do and M. Vetterli, The finite ridgelet transform for image representation, *IEEE Trans. Image Process.*, 12, 16–28, 2003.
6. J.S.G. Ferreira and A.J. Pinho, Why does histogram packing improve lossless compression rates?, *IEEE Signal Process. Lett.*, 9, 259–261, 2002.
7. P. Frossard, P. Vandergheynst, R.M. Ventura, and M. Kunt, A posteriori quantization of progressive matching pursuit streams, *IEEE Trans. Signal Process.*, 52, 525–535, 2004.
8. M. Grangetto, E. Magli, M. Martina, and G. Olmo, Optimization and implementation of the integer wavelet transform for image coding, *IEEE Trans. Image Process.*, 11, 596–604, 2002.
9. P. Hao and Q. Shi, Matrix factorizations for reversible integer mapping, *IEEE Trans. Signal Process.*, 49, 2314–2324, Oct. 2001.

10. ISO/IEC 10918-1 and ITU-T T.81, *Information Technology — Digital Compression and Coding of Continuous-tone Still Images: Requirements and Guidelines*, 1992.

11. ISO/IEC 10918-2 and ITU-T T.83, *Information technology — Digital Compression and Coding of Continuous-tone Still Images: Compliance Testing*, 1994.

12. ISO/IEC 14495-1 and ITU-T T.87, *Information Technology — Lossless and Near-lossless Compression of Continuous-tone Still Images — Baseline*, 1998.

13. ISO/IEC 14495-2 and ITU-T T.870, *Information Technology — Lossless and Near-lossless Compression of Continuous-tone Still Images: Extensions*, 2003.

14. ISO/IEC 15444-1 and ITU-T T.800, *Information Technology — JPEG 2000 Image Coding System — Part 1: Core Coding System*, 2000.

15. N.S. Jayant and P. Noll, *Digital Coding of Waveforms*, Prentice-Hall, Englewood Cliffs, NJ, 1984.

16. D. Le Gall and A. Tabatabai, Sub-band coding of digital images using symmetric short kernel filters and arithmetic coding techniques, *Proceedings of IEEE International Conference on Acoustics, Speech and Signal Processing (ICASSP)*, New York, 1988.

17. S. Mallat and F. Falzon, Analysis of low bit rate image transform coding, *IEEE Trans. Signal Process.*, 46, 1027–1042, 1998.

18. S.G. Mallat and Z. Zhang, Matching pursuits with time-frequency dictionaries, *IEEE Trans. Signal Process.*, 41, 3397–3415, 1993.

19. H.S. Malvar, Biorthogonal and nonuniform lapped transforms for transform coding with reduced blocking and ringing artifacts, *IEEE Trans. Signal Process.*, 46, 1043–1053, 1998.

20. H.S. Malvar, A. Hallapuro, M. Karczewicz, and L. Kerofsky, Low-complexity transform and quantization in H.264/AVC, *IEEE Trans. Circuits Syst. Video Technol.*, 13, 598–603, 2003.

21. R. Neff and A. Zakhor, Matching pursuit video coding. I. Dictionary approximation, *IEEE Trans. Circuits Syst. Video Technol.*, 12, 13–26, 2002.

22. W.B. Pennebaker and J.L. Mitchell, *JPEG Still Image Data Compression Standard*, Van Nostrand Reinhold, New York, 1993.

23. A.J. Pinho, An online preprocessing technique for improving the lossless compression of images with sparse histograms, *IEEE Signal Process. Lett.*, 9, 5–7, 2002.

24. O. Rioul and M. Vetterli, Wavelets and signal processing, *IEEE Signal Process. Mag.*, 8, 14–38, 1991.

25. A. Said and W.A. Pearlman, A new, fast, and efficient image codec based on set partitioning in hierarchical trees, *IEEE Trans. Circuits Syst. Video Technol.*, 6, 243–250, 1996.

26. P.J. Sementilli, A. Bilgin, J.H. Kasner, and M.W. Marcellin, Wavelet TCQ: submission to JPEG 2000, *Proceedings of SPIE*, San Diego, 1998.

27. A. Skodras, C. Christopoulos, and T. Ebrahimi, The JPEG 2000 still image compression standard, *IEEE Signal Process. Mag.*, 18, 36–58, 2001.

28. J.-L. Starck, E.J. Candes, and D.L. Donoho, The curvelet transform for image denoising, *IEEE Trans. Image Process.*, 11, 670–684, 2002.

29. W. Sweldens, The lifting scheme: a construction of second generation wavelets, *SIAM J. Math. Anal.*, 29, 511–546, 1997.

30. D.S. Taubman and M.W. Marcellin, JPEG 2000: standard for interactive imaging, *Proc. IEEE*, 90, 1336–1357, 2002.

31. D.S. Taubman, and M.W. Marcellin, *JPEG 2000 — Image Compression Fundamentals, Standards and Practice*, Kluwer, Dordrecht, 2002.

32. T.D. Tran, The binDCT: fast multiplierless approximation of the DCT, *IEEE Signal Process. Lett.*, 7, 141–144, 2000.

33. M. Vetterli and J. Kovačević, *Wavelets and Subband Coding*, Prentice-Hall PTR, Englewood Cliffs, NJ, 1995.

34. J.D. Villasenor, B. Belzer, and J. Liao, Wavelet filter evaluation for image compression, *IEEE Trans. Image Process.*, 4, 1053–1060, 1995.

35. G.K. Wallace, The JPEG still picture compression standard, *IEEE Trans. Consumer Electron.*, 38, xviii–xxxiv, 1992.

36. M.J. Weinberger and G. Seroussi, Sequential prediction and ranking in universal context modeling and data compression, *IEEE Trans. Inf. Theory*, 43, 1697–1706, 1997.

37. M.J. Weinberger, G. Seroussi, and G. Sapiro, The LOCO-I lossless image compression algorithm: principles and standardization into JPEG-LS, *IEEE Trans. Image Process.*, 9, 1309–1324, 2000.
38. X. Wu and P. Bao, L_∞ constrained high-fidelity image compression via adaptive context modeling, *IEEE Trans. Image Process.*, 9, 536–542, 2000.
39. Y.M. Yeung, O.C. Au, and A. Chang, An efficient optimal rate control scheme for JPEG 2000 image coding, *Proceeding of IEEE International Conference on Image Processing (ICIP)*, Barcelona, Spain, 2003.
40. L. Ying and R.K. Ward, Removing the blocking artifacts of block-based DCT compressed images, *IEEE Trans. Image Process.*, 12, 838–842, 2003.

5 Lossless Image Coding

Søren Forchhammer and Nasir Memon

CONTENTS

5.1 Introduction 113
5.2 General Principles 114
    5.2.1 Prediction 115
    5.2.2 Context Modeling 116
    5.2.3 Entropy Coding 117
        5.2.3.1 Huffman Coding 117
        5.2.3.2 Arithmetic Coding 119
5.3 Lossless Image Coding Methods 121
    5.3.1 JPEG Lossless 121
        5.3.1.1 Huffman Coding Procedures 121
        5.3.1.2 Arithmetic Coding Procedures 122
    5.3.2 Context-Based Adaptive Lossless Image Coding 123
        5.3.2.1 Gradient-Adjusted Predictor 124
        5.3.2.2 Coding Context Selection and Quantization 125
        5.3.2.3 Context Modeling of Prediction Errors and Error Feedback 126
        5.3.2.4 Entropy Coding of Prediction Errors 126
    5.3.3 JPEG-LS 127
    5.3.4 Reversible Wavelets — JPEG2000 128
    5.3.5 Experimental Results 129
5.4 Optimizations of Lossless Image Coding 129
    5.4.1 Multiple Prediction 130
    5.4.2 Optimal Context Quantization 131
5.5 Application Domains 134
    5.5.1 Color and Multiband 134
        5.5.1.1 Predictive Techniques 134
        5.5.1.2 Band Ordering 135
        5.5.1.3 Interband Prediction 135
        5.5.1.4 Error Modeling and Coding 136
        5.5.1.5 Hyperspectral Images 136
        5.5.1.6 Reversible Transform-Based Techniques 137
    5.5.2 Video Sequences 137
    5.5.3 Color-Indexed Images and Graphics 139
References 140

5.1 INTRODUCTION

Image compression serves the purpose of reducing bandwidth or storage requirement in image applications. Lossy image compression is widely deployed, e.g., using the classic JPEG standard [21]. This standard also has a less-known lossless version. Lossless compression has the advantage of avoiding the issue whether the coding quality is sufficient. In critical applications, lossless coding may be mandatory. This includes applications where further processing is applied to the images.

Examples are medical imaging, remote sensing, and space applications, where scientific fidelity is of paramount importance. In other areas such as prepress and film production, it is the visual fidelity after further processing which is of concern.

In this chapter, we present an overview of techniques for lossless compression of images. The basis is techniques for coding gray-scale images. These techniques may be extended or modified in order to increase performance on color and multiband images, as well as image sequences.

Lossless coding is performed in a modeling and a coding step. The focus of this text is on the paradigm of modeling by prediction followed by entropy coding. In JPEG, a simple linear prediction filter is applied. In recent efficient schemes, nonlinear prediction is applied based on choosing among a set of linear predictors. Both the predictors and the selection is based on a local neighborhood. The prediction residuals are coded using context-based entropy coding. Arithmetic coding provides the best performance. Variable-length codes related to Huffman coding allow faster implementations.

An interesting alternative to predictive coding is established by the use of reversible wavelets. This is the basis of the lossless coding in JPEG2000 [50] providing progression to lossless. For color-mapped images having a limited number of colors per pixel, coding directly in the pixel domain may be an efficient alternative.

This chapter is organized as follows. First, the general principles are introduced. In Section 5.3, the lossless version of the classic JPEG exemplifies the basic techniques. More recent techniques are represented by the lossless compression of the JPEG-LS [22] and JPEG2000 standards. Context-based adaptive lossless image coding (CALIC) [61] is presented as state of the art at a reasonable complexity. The focus is on coding gray-scale natural images, but the techniques may also be applied to color images coding the components independently. Section 5.4 presents recent techniques for optimizing the compression using multiple prediction and optimal context quantization. These techniques are complex but they may lead to new efficient coding schemes. In Section 5.5, important application domains are treated. Single image coding is extended to encompass color and multiband images as well as coding image sequences including video sequences. An important issue is how to capture the correlations of the extra dimensions and what the benefit of the increased complexity is. Finally, color-indexed images and pixel-based graphics are considered. This is a class of images which does not comply well with models aimed at locally smooth images. Therefore, the traditional image coding techniques do not perform optimally on these. Newer schemes often apply a special mode to improve performance, but it is demonstrated that there are images within the class for which significantly better compression may be achieved by pixel-domain-oriented techniques.

5.2   GENERAL PRINCIPLES

The pixels of an image, or the symbols resulting from the processing of an image, are scanned in a sequential order prior to coding. We shall now consider the sequence of pixels obtained by a raster scan of an image as a sequence of random variables.

Let $x_1, \ldots, x_T$ denote a finite sequence of random variables $X_1, \ldots, X_T$ over a discrete alphabet $\mathcal{A}$. In the statistical approach to lossless compression, a conditional probability mass function (pmf) $p_t(\cdot)$ is assigned to each variable X_t, prior to the encoding, given the outcomes of the past variables, $X_1, \ldots, X_{t-1}$.

A lower bound of the average code length is given by the entropy. If we just consider one variable X over the discrete alphabet $\mathcal{A}$, the entropy of the source S is defined by [10]

$$H(S) = \sum_{i=1}^{|\mathcal{A}|} P_i \log_2 \frac{1}{P_i} \tag{5.1}$$

where $|\mathcal{A}|$ is the size of the alphabet.

Considering the whole string of variables X^T as one variable, the probability $P(X^T)$ may be decomposed by

$$p_t(x_t) = \Pr(X_t = x_t | X_1 = x_1, X_2 = x_2, \ldots, X_{t-1} = x_{t-1}), \quad \text{for all } x \in \mathcal{A} \quad (5.2)$$

in which case the entropy may be written as

$$\frac{1}{T} H(X^T) = \frac{1}{T} \sum_{t=1}^{T} H(X_t | X_1, \ldots, X_{t-1}) \text{ bits/symbol} \quad (5.3)$$

which is the Shannon entropy rate of the sequence and $H(\cdot|\cdot)$ is the conditional entropy function. If the variables are independent and identically distributed (iid), the entropy per symbol of the sequence is expressed by the entropy of one element, $H(X)$.

Given the sequence of pmfs $p_t(\cdot)$, there are coding techniques, e.g. arithmetic coding [43], that will encode $X_t = x_t$ using approximately $-\log_2 p_t(x_t)$ bits. On the average, the sequence $X_1, \ldots, X_T$ will then be encoded using approximately

$$E\left[\frac{1}{T} \sum_{t=1}^{T} -\log_2 p_t(X_t)\right] \text{ bits/sample} \quad (5.4)$$

Accordingly, to encode X_t, we should ideally produce the conditional pmf in (5.2) or an approximation or estimate thereof. The encoding of a pixel depends on the values of previously scanned pixels, where the ordering is given by the scan order.

Clearly, with the above (optimal) approach, there could be a different pmf for every value of t and every sequence of values of $X_1, \ldots, X_{t-1}$ leading to an intractable estimation problem. Universal source coders, e.g. [44], provide solutions that are asymptotically optimal, but using domain knowledge less complex efficient coders may be specified. Hence, practical lossless compression techniques use a model to capture the inherent structure in the source. This model can be used in a number of ways. One approach is to use the model to generate a residual sequence, which is the difference between the actual source output and the model predictions. If the model accurately reflects the structure in the source output the residual sequence can be considered to be iid. Often, a second stage model is used to further extract any structure that may remain in the residual sequence. The second stage modeling is often referred to as *error modeling*. Once we get (or assume that we have) an iid sequence, we can use entropy coding to obtain a coding rate close to the entropy as defined by (5.4). Another approach is to use the model to provide a context for the encoding of the source output, and encode sequences by using the statistics provided by the model. The sequence is encoded symbol by symbol. At each step, the model provides a probability distribution for the next symbol to the encoder, based on which the encoding of the next symbol is performed. These approaches separate the task of lossless compression into a *modeling* task and a *coding* task. As we shall see in the next subsections, encoding schemes for iid sequences are known which perform optimally and, hence, the critical task in lossless compression is that of modeling. The model imposed on the source determines the rate at which we would be able to encode a sequence emitted by the source. Naturally, the model is highly dependent on the type of data being compressed. Later in this chapter we describe some popular modeling schemes for image data.

5.2.1 PREDICTION

Prediction essentially attempts to capture the intuitive notion that the intensity function of typical images is usually quite "smooth" in a given local region and hence the value at any given pixel is quite similar to its neighbors. In any case, if the prediction made is reasonably accurate, then the prediction error has significantly lower magnitude and variance when compared with the original signal and it can be encoded efficiently with a suitable variable-length coding technique.

Linear predictive techniques (also known as lossless DPCM) usually scan the image in raster order, predicting each pixel value by taking a linear combination of pixel values in a casual neighborhood. For example, we could use $(I[i-1,j]+I[i,j-1])/2$ as the prediction value for $I[i,j]$, where $I[i,j]$ is the pixel value in row i, column j. Despite their apparent simplicity, linear predictive techniques are quite effective and give performance surprisingly close to more state-of-the-art techniques for natural images. If we assume that the image is being generated by an autoregressive (AR) model, coefficients that best fit given data in the sense of the L2 norm can be computed. One problem with such a technique is that the implicit assumption of the data being generated by a stationary source is, in practice, seldom true for images. Hence, such schemes yield very little improvement over the simpler linear predictive techniques such as the example in the previous paragraph. Significant improvements can be obtained for some images by adaptive schemes that compute optimal coefficients on a block-by-block basis or by adapting coefficients to local changes in image statistics. Improvements in performance can also be obtained by adaptively selecting from a set of predictors. An example scheme that adapts in presence of local edges is given by the median edge detection (MED). MED detects horizontal or vertical edges by examining the north, west, and northwest neighbors of the current pixel. The north (west) pixel is used as a prediction in the case of a vertical (horizontal) edge. In case of neither, planar interpolation is used to compute the prediction value. The MED predictor has also been called the median adaptive predictor (MAP) and was first proposed by Martucci [30] as a nonlinear adaptive predictor that selects the median of a set of three predictions in order to predict the current pixel. One way of interpreting such a predictor is that it always chooses either the best or the second-best predictor among the three candidate predictors. Martucci reported the best results with the following three predictors: (1) $I[i,j-1]$, (2) $I[i-1,j]$, and (3) $I[i,j-1]+I[i-1,j]-I[i-1,j-1]$. In this case, it is easy to see that MAP turns out to be the MED predictor. The MED predictor gives superior or almost as good a performance as many standard prediction techniques, many of which being significantly more complex.

5.2.2 CONTEXT MODELING

For the optimal approach (5.2) to lossless coding, there could be a different pmf, $p_t(\cdot)$, for every value of t and every sequence of values of $X_1, \ldots, X_{t-1}$. This is clearly not tractable. A widely used approach to probability estimation for practical lossless image compression is to reduce the dependency to some limited measurement of the past, called a *context*. This is called context modeling [44]. Prediction may also be seen as a means to reduce the number of pmfs, which need to be specified.

Associated with a given context model is a finite set of contexts $\mathcal{C}$, given by functions or mappings that assigns a context $C \in \mathcal{C}$ to each element in the data sequence $x_1, \ldots, x_t, 0 < t < \infty$. For example, the contexts might consist of $\{X_{i-2} = a, X_{i-1} = b\}$ for all $a, b \in \mathcal{A}$. This is an example of a template-based context: The term context in relation to a symbol x_t is used to refer to the context assigned by a mapping of $x_1, \ldots, x_{t-1}$.

Viewed as a model, the context model assumes that the distribution of the current symbol depends only on some limited context. That is, given its context, the current symbol is conditionally independent of past data symbols.

Associated with each context C is the conditional pmf $p(\cdot|C)$, and the conditional entropy is given by

$$H(X|C) \equiv -\sum_x p(x|C) \log_2 p(x|C) \text{ bits/symbol} \tag{5.5}$$

when X appears in context C, and the overall conditional entropy is

$$H(X|\mathcal{C}) \equiv \sum_{C \in \mathcal{C}} p(C)H(X|C) \text{ bits/symbol} \tag{5.6}$$

where $p(C)$ is the probability of context C occurring.

An encoder based on $p(\cdot|C)$ will (approximately) achieve the rate $H(X|\mathcal{C})$ even if the current symbol is not conditionally independent of the past, given the contexts.

If there is conditional independence, $H(X|\mathcal{C})$ equals the entropy of the source (5.3), and thus the least possible rate. In practical lossless image compression, the pmf $p(\cdot|C)$ is estimated either beforehand or adaptively and the question of conditional independence is not addressed.

Usually, the mapping defining the number of contexts, is chosen such that this number is much less than the length of the image data sequence. Contemporary lossless image coder takes an approach where the pmfs, $p(\cdot|C)$, are unknown *a priori*, but estimated adaptively. They can be estimated by maintaining counts of symbol occurrences within each context (as in CALIC [61]) or by estimating the parameters of an assumed pmf (as in LOCO-I [54] and JPEG-LS [22]). The coding efficiency relies on obtaining good estimates for the pmf within each context.

5.2.3 ENTROPY CODING

If the probability of the symbols being encoded is not uniform then it is clearly advantageous to assign shorter codewords to the more frequently occurring symbols. Suppose a codeword of length L_i is assigned to the symbol a_i (that occurs with probability P_i). The average (expected) length of the codeword is

$$L_{av} = \sum_{i=1}^{|\mathcal{A}|} P_i L_i$$

From a result in information theory [10], the average length L_{av} is bounded from below by the entropy (5.1) of the source S.

Clearly, the goal of the symbol-coding unit is to achieve an average codeword length as close to the entropy as possible. There are systematic procedures of coding that perform very close to the entropy bound. These coding procedures include Huffman coding and arithmetic coding. We will elaborate on Huffman coding and arithmetic coding, which are commonly used in image compression.

5.2.3.1 Huffman Coding

Huffman coding is based on the knowledge of the probabilities of occurrence of symbols. It leads to minimum average codeword length under the condition that no codeword is a prefix of another. Huffman coding is optimal (achieves entropy bound) in the case where all symbol probabilities are integral powers of 1/2. Given the probabilities of occurrence of symbols, the following procedure can be used to construct a Huffman code:

1. Arrange the symbols in a rank-ordered list according to the probability of occurrence.
2. Perform the following iterations to reduce the size of the list by creating composite symbols until a reduced list with only two composite symbols is reached.
   (a) Combine the two symbols with the lowest probability to form a composite symbol, and add the probabilities.
   (b) Create a new list from the old by deleting the two symbols that were combined, and adding the composite symbol in a rank-ordered manner.

TABLE 5.1
Huffman Coding

Stage 1 (5 Symbols)			Stage 2 (4 Symbols)			Stage 3 (3 Symbols)			Stage 4 (2 Symbols)		
Prob	S	C	Prob	S	C	Prob	S	C	Prob	S	C
0.500	a_1	0	0.500	a_1	0	0.500	a_1	0	0.500	a_1	0
0.125	a_2	110	0.250	a_{45}	10	0.250	a_{45}	10	0.500	a_{2-5}	1
0.125	a_3	111	0.125	a_2	110	0.250	a_{23}	11			
0.125	a_4	100	0.125	a_3	111						
0.125	a_5	101									

Note: S denotes symbols and C denotes codewords.

(c) Construct the binary tree in which each node represents the probability of all nodes beneath it.
(d) Following a convention of assigning a "0" or "1" to the direction of movement in the tree, assign a code by traversing a path to each leaf.

To illustrate the procedure, consider the case of coding the output of a source with $|\mathcal{A}| = 5$ possible symbols $\{a_i, i = 1, \ldots, 5\}$, with probabilities $P_1 = \frac{1}{2}$, $P_2 = P_3 = P_4 = P_5 = \frac{1}{8}$. The entropy in this case is given by

$$H(S) = \sum_{i=1}^{|\mathcal{A}|} P_i \log_2 \frac{1}{P_i} = \frac{1}{2} \cdot 1 + 4 \cdot \frac{1}{8} \cdot 3 = 2 \text{ bits/symbol}$$

Huffman coding is performed as shown in Table 5.1. In stage 1, the symbol probabilities are arranged in decreasing order. The lowest two probabilities corresponding to symbols a_4 and a_5 are added and assigned to a new symbol a_{45}, which then appears in the shorter list of stage 2. Note that the symbols a_4 and a_5 are deleted, while a_{45} is inserted in the correct rank-ordered position. The procedure is repeated until there are only two symbols. These two symbols are assigned bits 0 and 1 to represent them. Next, the component symbols a_{45} and a_{23} of the last new symbol a_{2-5} are assigned an additional bit 0 and 1, respectively. The procedure is repeated to traverse back to the first stage.

In the above case, it can be verified that the average length of the codeword is $L_{av} = 2$ bits. A rate of 2 bits/symbol is equal to the entropy, and this is achieved because the symbol probabilities are integral powers of 1/2.

Now, consider a case where the condition that the probabilities are integral powers of 1/2 does not hold. Assume that a source generates $|\mathcal{A}| = 3$ symbols $\{a_i, i = 1, \ldots, 3\}$, with probabilities $P_1 = P_2 = P_3 = 1/3$. In this case, the source entropy is 1.585 bits/symbol. With Huffman coding, it can be verified that a code $\{0, 10, 11\}$ is generated to represent the symbols with an average length

$$L_{av} = \tfrac{1}{3}(1 + 2 + 2) = 1.67 \text{ bits/symbol}$$

A question that can be asked is whether one can improve over this performance of 1.67 bits/symbol. This improvement can be achieved by jointly coding a block of symbols. Toward this end, let us define a new symbol as a block of N occurrences of original symbols. Let us denote these $|\mathcal{A}|^N$ possible new symbols by $\{b_i, i = 1, \ldots, |\mathcal{A}|^N\}$. By applying Huffman coding to the new symbols, it is possible to reduce the average number of bits per original symbol. For example, in the above case of $|\mathcal{A}| = 3$, consider a block of $N = 5$ occurrences of original symbols $\{a_i, i = 1, \ldots, 3\}$, generating new symbols $\{b_i, i = 1, \ldots, 243\}$. Even without formally applying the Huffman coding procedure, we note a fixed codeword length of $L = 8$ bits can be used to uniquely represent the 243 possible blocks of $N = 5$ original symbols. This gives us a length of $8/5 = 1.6$ bits/original symbol, significantly closer to the bound of 1.585 bits/symbol.

In this simple case it was possible to use a suitable block size to get close to the bound. With a large set of symbols and widely varying probabilities, an alternative method called arithmetic coding, which we describe below proves to be more effective in approaching the bound in a systematic way.

5.2.3.2   Arithmetic Coding

Huffman coding, introduced in the previous section, was the entropy coder of choice for many years. Huffman coding in the basic form has some deficiencies, namely that it does not meet the entropy unless all probabilities are integral powers of 1/2, it does not code conditional probabilities, and the probabilities are fixed. All these deficiencies may be addressed by alphabet extension, having multiple Huffman codes (one for each set of probabilities) and using adaptive Huffman coding evaluating and possibly redesigning the Huffman tree when the (estimated) probabilities change. All approaches are used in practice, but especially if all are to be used at once the complexity becomes prohibitive. An alternative is given by arithmetic coding [43] which elegantly deals with all these issues.

Arithmetic coding is often referred to as coding by interval subdivision. The coding specifies a real number (pointing) within a subinterval of the unit interval [0;1[. The codeword is a binary representation of this number. Let $x^n = (x_1, x_2, \ldots, x_n)$ denote a sequence of symbols which is to be coded; x^n has length n and runs from 1 to n. Let x_i be the symbol at i. Let $P(\cdot)$ denote a probability measure on all strings of length n and $P(x^n)$ denote the probability of the string x^n. First, we disregard the restriction of finite precision. We shall associate the string of symbols x^n with an interval $I(x^n)$ of width equal to the probability of the string $P(x^n)$, $I(x^n) = [F(x^n); F(x^n) + P(x^n)[$. Later we shall use (a binary representation of) $F(x^n)$ as the codeword. Ordering all the possible sequences of length n alphabetically and locating the intervals $I(x^n)$ one after the other, we see that we have partitioned the unit interval [0;1[into intervals of lengths equal to the probabilities of the strings. (Summing the probabilities of all possible sequences adds to unity as we have presumed a probability measure.) This means $F(x^n)$ is the cumulative probability of all the strings prior in ordering to x^n. This interval subdivision may instead be done sequentially in such a way that we only need to calculate the subinterval associated with the sequence we are coding.

The codeword subinterval (of x^{n+1}) is calculated by the following recursion:

$$F(x^n 0) = F(x^n) \tag{5.7}$$

$$F(x^n i) = F(x^n) + \sum_{j < i} P(x^n j), \quad i = 0, \ldots, |\mathcal{A}| - 1 \tag{5.8}$$

$$P(x^n i) = P(x^n) P(i|x^n), \quad i = 0, \ldots, |\mathcal{A}| - 1 \tag{5.9}$$

where $|\mathcal{A}|$ is the size of the symbol alphabet.

The width of the interval associated with x^n in (5.9) equals its probability $P(x^n)$. Using pointers with a spacing of powers of 1/2, at least one pointer of length $\lceil - \log_2 (P(x^n)) \rceil$ will point to the interval. ($\lceil y \rceil$ denotes the ceiling or round-up value of y.) This suggests we may come arbitrarily close to the per symbol entropy.

The major drawback of the recursions is that even with finite precision conditional probabilities $P(i|j)$ in (5.9), the left-hand side will quickly require a precision beyond any register width we choose.

(We note that it is mandatory that the encoder and decoder can perform exactly the same calculations.) There are several remedies to this. The most straightforward is to approximate the multiplication of (5.9). For simplicity, an approximative solution is given for the binary case

$$\bar{P}(x^n i) = \lfloor \bar{P}(x^n)\bar{P}(i|x^n) \rfloor_q, \quad i = 0, 1 \tag{5.10}$$

where $\lfloor z \rfloor_q$ denotes the truncation of a fractional binary number z to q digits precision.

The decoding is performed reversing the process. Consider the binary case having decoded the data string up to $t-1$ as $x^{t-1} = u$; now the question is whether the next symbol x_t is 0 or 1. This is resolved by observing that

$$F(x^n) - F(u) \geq \bar{P}(u0) \quad \Leftrightarrow \quad x_t = 1 \tag{5.11}$$

which, in turn, can be decided looking at the q next bits of the code string and carrying out the subtraction which is the inverse of (5.8). (Again, fixed register implementations can perform this task.)

A first step to solving the practical problems in the binary case is given by (5.10). The finite precision problem is similar but more complicated in the m-ary alphabet case. Furthermore, in both cases there are issues of carry overflow and decodability including stopping. A popular algorithm for an m-ary alphabet is given in [55].

Binary arithmetic coding is relatively simpler than m-ary arithmetic coding, so even in the nonbinary case, binary arithmetic coding may be applied by decomposing the values of the alphabet in a binary tree [14].

The arithmetic coding in current lossless image coding standards are all binary with roots in the Q coder [40], which in turn is an extension of the skew coder [25]. The Q coder avoids multiplication by approximation (5.9). A modified version, the QM coder, is used in JPEG with the arithmetic coding option and in the JBIG standard [39]. Instead of dealing directly with the 0s and 1s put out by the source, the QM coder maps them into a more probable symbol (MPS) and less probable symbol (LPS). If 1 represents black pixels, and 0 represents white pixels, then in a mostly black image, 1 will be the MPS, while in an image with mostly white regions 0 will be the MPS. In order to make the implementation simple, several deviations from the standard arithmetic coding algorithm has been adopted. The update equations in arithmetic coding that keep track of the subinterval to be used for representing the current string of symbols involve multiplications which are expensive in both hardware and software. In the QM coder expensive multiplications are avoided. Let A_n refer to the approximation of $I(x_n)$ with endpoints l_n and u_n. Let q refer to the LPS probability. The recursions are given by

$$l_n = \begin{cases} l_{n-1} & \text{for } x_n \text{ } MPS \\ l_{n-1} + A_{n-1} - q & \text{for } x_n \text{ } LPS \end{cases} \tag{5.12}$$

$$A_n = \begin{cases} A_{n-1} - q & \text{for } x_n \text{ } MPS \\ q & \text{for } x_n \text{ } LPS \end{cases} \tag{5.13}$$

The approximation applied to avoid multiplications is that $qA_{n-1} \approx q$. The QM coder scales A_n to lie in the interval of 0.75 to 1.5. The rescalings of the interval take the form of repeated doubling, which corresponds to a left shift in the binary representation.

The QM coder also achieves speedup by applying an approximation in the adaptive probability estimate. The basis of the estimate is the following estimator, which may be interpreted as a Bayesian estimator. Consider k observations of a sequence of independent binary variables taken from the same distribution, i.e., a Bernoulli process. Let $n_0(k)$ denote the number of 0s and $n_1(k)$ the number of 1s out of the k symbols. If the probability p_1 itself is considered a stochastic variable with Beta distribution ($p_1 \in B(\delta, \delta)$ [1]) specified by the parameter δ, the Bayes estimate of p_1 given k observations is

$$\hat{p}_1 = (n_1(k) + \delta)/(n_1(k) + n_0(k) + 2\delta). \tag{5.14}$$

$\delta = 1$ is optimal if the prior distribution of p_1 is uniform. Choosing $\delta = 0.5$ is better for a more skewed symmetric distribution. $\delta = 0.45$ is used in the estimate the QM coder is based on. The estimate based on $n_0(k)$ and $n_1(k)$ is used when the context appears the next time, i.e., when coding instance number $k+1$ occurs in that context.

The QM coder [39] implements an approximation of the estimate using a finite-state machine (FSM). Each state of the FSM is assigned a probability estimate as well as the new state transitions. The probability q of the LPS for context C is updated each time the context C is active and a rescaling takes place, i.e., a bit is output. Rescalings occur for all LPS instances, whereas for MPS instances the rescaling will occur with a certain probability, which is incorporated in the update of the estimate. In a nonstationary situation, it may happen that the symbol assigned to LPS actually occurs more often than the symbol assigned to MPS. In this situation, the assignments are reversed, i.e., the symbol assigned the LPS label is assigned the MPS label and vice versa. The test is conducted every time a rescaling takes place. The decoder for the QM coder operates in much the same way as the encoder — by mimicking the encoder operation.

The use of conditional probabilities increases the modeling power. The conditional probabilities are fed to the arithmetic coder performing the actual entropy coding. For binary images (or bit-planes of an image) the context given by a template may directly define the coding model. In the classic bi-level image coding standard, JBIG [19], a 10-pixel template is used to define the conditioning states. For each of these conditioning states, the state of the FSM is stored holding the current probability estimate. The application of context-based arithmetic coding to transformed pixel values in a lossless image coder reflects that the first step of prediction (or transformation) does not provide perfect decorrelation.

As we shall see, context-based arithmetic coding is a very powerful tool coding a one-dimensional sequence of conditional probabilities, but defining the conditioning context in 2D or generally in the dimensionality the data set may represent.

5.3   LOSSLESS IMAGE CODING METHODS

This section presents how the principles are applied in a suite of well-known lossless image coders. JPEG (lossless) combines simple linear prediction with entropy coding. CALIC and JPEG-LS apply more advanced prediction and effectively capture the redundancy due to the fact that the prediction does not provide perfect decorrelation. This redundancy is addressed using context-based entropy coding. Neighboring pixels within a template of two-dimensional causal pixels are used for prediction. The notation of the pixels is shown in Figure 5.1.

Lossless coding based on reversible wavelets is briefly mentioned with JPEG2000 as the primary example. Finally, results are presented for a set of gray scale and a set of color images. The latter includes other images than just natural images. The overview [9] presents more coders in the evolution of lossless image coding.

5.3.1   JPEG LOSSLESS

The well-known JPEG standard [21] has a lesser known lossless version which employs a predictive approach. The prediction is followed by coding of the prediction errors using Huffman coding or as an option arithmetic coding. The standard allows the user to choose between eight simple predictors (Table 5.2). The coding of the prediction errors $e = I - \hat{I}$, according to the standard, is described in the rest of this subsection.

5.3.1.1   Huffman Coding Procedures

In the Huffman coding version, the prediction errors are coded using the Huffman table provided in the bit stream using the specified syntax. The errors are coded independently using the table given. Thus the prediction errors are assumed to be iid.

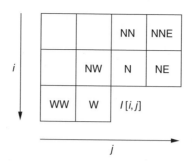

FIGURE 5.1 Notation used for specifying neighborhood pixels of current pixel $I[i, j]$.

TABLE 5.2
JPEG Predictors for Lossless Coding

Mode	Prediction for $I[i, j]$
0	0 (no prediction)
1	N
2	W
3	NW
4	N + W − NW
5	W + (N − NW)/2
6	N + (W − NW)/2
7	(N + W)/2

The alphabet size for the prediction errors is twice the original alphabet size. In JPEG, a special length-limited Huffman code is used. Using large Huffman tables leads to increased complexity in hardware implementation. To reduce the complexity, the coding is split into two, the first part represented by a Huffman table, followed by uncoded bits in the second part. In the first part, each prediction error is classified into a "magnitude category," k, which is coded using a Huffman table. Table 5.3 shows the 17 different categories that are defined. Each category k, except for the last, contains 2^k members $\{\pm 2^{k-1}, \ldots, \pm 2^k - 1\}$. The element within the category is coded directly using k bits in the second part. The bits coding the specific prediction error e within category k are given by a k-bit number

$$n = \begin{cases} e & \text{if } e \geq 0 \\ 2^k - 1 + e & \text{if } e < 0 \end{cases}$$

5.3.1.2   Arithmetic Coding Procedures

Arithmetic coding may be used as an option. The arithmetic coding version uses quantized prediction errors at neighboring pixels as contexts for conditional coding of the prediction error. This is a simple form of error modeling. Thus the arithmetic coding version attempts to capitalize on the remaining structure in the prediction residual, as opposed to the Huffman version which assumes the prediction error probability distribution to be iid. Encoding within each context is carried out by a binary arithmetic coder by decomposing the prediction error into a sequence of binary decisions. The first binary decision codes whether the prediction error is zero. If the error is nonzero, the sign of the error is determined in the second step. The subsequent steps classify the magnitude of the prediction error

TABLE 5.3
Mapping of Prediction Errors to Magnitude Category and Extra Bits

Category	Symbols	Extra Bits
0	0	—
1	$-1, 1$	0, 1
2	$-3, -2, 2, 3$	00,01,10,11
3	$-7, \ldots, -4, 4, \ldots, 7$	$000, \ldots, 011, 100, \ldots, 111$
4	$-15, \ldots, -8, 8, \ldots, 15$	$0000, \ldots, 0111, 1000, \ldots, 1111$
$\vdots$	$\vdots$	$\vdots$
16	32768	

into one of a set of ranges and the final uncoded bits determine the exact prediction error magnitude within the range. The binary decomposition allows using a fast binary arithmetic coder, which in JPEG is the QM coder (see Section 5.2.3). For a detailed description of this coder and the standard refer to [39].

5.3.2 CONTEXT-BASED ADAPTIVE LOSSLESS IMAGE CODING

The simple modeling in JPEG may be further developed for better performance. The universal context modeling (UCM) scheme proposed by Weinberger et al. [53] introduced advanced modeling for adaptive prediction and context formation. This is a principled but highly complex context-based image coding technique. A solution of lower complexity was presented by CALIC [61]. The complexity was later further reduced in JPEG-LS [22]. The modeling is composed of prediction followed by a context-based model of the prediction error.

In this subsection, we give a brief description of CALIC and its major components (Figure 5.2) as an example of a highly efficient context-based image coder. For a more detailed description, the reader is referred to [61]. In the prediction step, CALIC employs a simple gradient-based nonlinear prediction scheme called gradient-adjusted predictor (GAP), which adjusts prediction coefficients based on estimates of local gradients. Prediction is then made context-sensitive and adaptive by modeling of prediction errors and feedback of the expected error conditioned on properly chosen modeling contexts. The modeling context is a combination of quantized local gradient and texture pattern, two features that are indicative of the error behavior. The net effect is a nonlinear, context-based, adaptive prediction scheme that can correct itself by learning from its own past mistakes under different contexts. The context-based error modeling is done at a low model cost. By estimating expected prediction errors rather than error probabilities in different modeling contexts, CALIC can afford a large number of modeling contexts without suffering from context dilution problems or from excessive memory use. This is a key feature of CALIC.

Another innovation of CALIC was the introduction of two modes, distinguishing between binary and continuous-tone types of images on a local, rather than a global, basis. This distinction is important because the compression methodologies are very different in the two modes. The former codes the pixel values directly, whereas the latter uses predictive coding. CALIC selects one of the two modes based on a local causal template without using any side information. The two-mode design contributes to the robustness of CALIC over a wide range of images, including the so-called multimedia images that mix text, graphics, line art, and photograph.

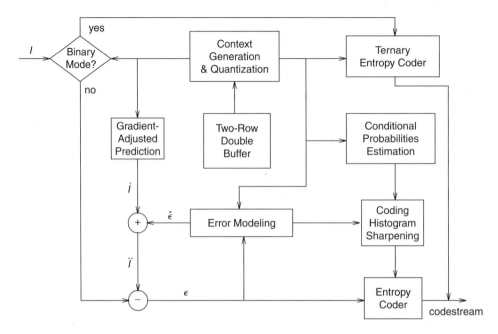

FIGURE 5.2 Block diagram of CALIC.

The CALIC codec was proposed as a candidate algorithm for the international standardization effort on lossless compression of continuous-tone images leading to the JPEG-LS standard. CALIC achieved an average lossless bit rate of 2.99 bits per pixel on the 18 8-bit test images selected by ISO for proposal, thus providing a significant gain compared to an average bit rate of 3.98 bits per pixel for lossless JPEG (Huffman coding) on the same test set. Section 5.3.5 presents bit rates that CALIC obtained on ISO test images and makes comparisons with some popular lossless image coders (Tables 5.4 and 5.5).

CALIC introduced new efficient algorithmic techniques for context formation, quantization, and modeling. Although conceptually elaborate, CALIC is algorithmically quite simple, involving mostly integer arithmetic and simple logic.

CALIC encodes and decodes images in raster scan order with a single pass through the image. The coding process uses prediction templates that involve only the previous two scan lines of coded pixels. CALIC adapts between the two modes of coding: binary and continuous-tone modes (Figure 5.2). The system automatically selects one of the two modes during the coding process, depending on the context of the current pixel. Binary mode is applied when no more two intensity values are present locally. For each pixel a ternary decision is coded including an escape symbol besides the two local symbols. Values at the six nearest-neighboring pixels are used as the context (Figure 5.1).

In the continuous-tone mode, the system has four components (Figure 5.2): prediction, context selection and quantization, context modeling of prediction errors, and entropy coding of prediction errors. We briefly describe each below.

5.3.2.1 Gradient-Adjusted Predictor

GAP is a simple, adaptive, nonlinear predictor. Robust prediction in areas with strong edges is obtained by local adaption to the intensity gradients. Out of a set of predefined linear predictors, for each pixel, GAP makes a choice of which to use based on estimated gradients of the image.

In GAP, the gradient of the intensity function at the current pixel I is estimated by computing the following quantities:

$$d_h = |I[i-1,j] - I[i-2,j]| + |I[i,j-1] - I[i-1,j-1]|$$
$$+ |I[i+1,j-1] - I[i,j-1]|$$
$$d_v = |I[i-1,j] - I[i-1,j-1]| + |I[i,j-1] - I[i,j-2]|$$
$$+ |I[i+1,j-1] - I[i+1,j-2]|. \tag{5.15}$$

A prediction is then made by the following procedure [61]:

IF $(d_v - d_h > 80)$ {sharp horizontal edge}
   $\dot{I}[i,j] = w$
ELSE IF $(d_v - d_h < -80)$ {sharp vertical edge}
   $\dot{I}[i,j] = n$
ELSE {
   $\dot{I}[i,j] = (w+n)/2 + (ne - nw)/4;$
   IF $(d_v - d_h > 32)$ {horizontal edge}
      $\dot{I}[i,j] = (\dot{I}[i,j] + w)/2$
   ELSE IF $(d_v - d_h > 8)$ {weak horizontal edge}
      $\dot{I}[i,j] = (3\dot{I}[i,j] + w)/4$
   ELSE IF $(d_v - d_h < -32)$ {vertical edge}
      $\dot{I}[i,j] = (\dot{I}[i,j] + n)/2$
   ELSE IF $(d_v - d_h < -8)$ {weak vertical edge}
      $\dot{I}[i,j] = (3\dot{I}[i,j] + n)/4$
}

where

$$n = I[i,j-1], w = I[i-1,j], ne = I[i+1,j-1],$$
$$nw = I[i-1,j-1], nn = I[i,j-2], ww = I[i-2,j]. \tag{5.16}$$

The thresholds given in the above procedure are for eight bit data and are adapted on the fly for higher-resolution images. They can also be specified by the user if offline optimization is possible.

5.3.2.2 Coding Context Selection and Quantization

The predicted value $\dot{I}$ is further adjusted via an error feedback loop of one-step delay. This results in an adaptive, context-based, nonlinear prediction $\ddot{I}$ as explained shortly. The residue obtained from the final prediction $\ddot{I}$ is entropy-coded based on eight estimated conditional probabilities in eight different contexts. These eight contexts, called error energy contexts, are formed by quantizing an error energy estimator Δ, a random variable, into eight bins. The quantizer bins partition prediction error terms into eight classes by the expected error magnitude.

The error energy estimator is computed as follows:

$$\Delta = ad_h + bd_v + c|e_w| \tag{5.17}$$

where d_h and d_v are defined in (5.15) and $e_w = I[i-1,j] - \dot{I}[i-1,j]$. The coefficients a, b, and c can be optimized based on training data. In [61], $a = b = 1$ and $c = 2$ is recommended for the coefficients in (5.17).

To improve coding efficiency, prediction errors are mapped into classes of different variances by conditioning the error distribution on Δ. Context quantization is applied to Δ in the estimation conditional probabilities $p(e|\Delta)$ for time and space efficiency. The error energy estimator Δ is quantized to $M = 8$ levels, as larger values of M will only improve coding efficiency marginally. Let $Q(\Delta) \in \{0, 1, \ldots, 7\}$ denote the Δ quantizer. The context quantizer, $Q(\Delta)$, is a scalar quantizer, which may be determined by minimizing the total entropy of the errors based on $p(e|Q(\Delta))$. Based on a training set of (e, Δ) pairs from test images, standard dynamic programming technique was used to choose $0 = q_0 < q_1 < \cdots < q_{M-1} < q_M = \infty$ to partition Δ into M intervals such that

$$-\sum_{d=0}^{M-1} \sum_{q_d \leq \Delta < q_{d+1}} p(e) \log p(e) \tag{5.18}$$

is minimized. An image-independent Δ quantizer given by the thresholds

$$q_1 = 5, \quad q_2 = 15, \quad q_3 = 25, \quad q_4 = 42, \quad q_5 = 60, \quad q_6 = 85, \quad q_7 = 140$$

was found to work almost as well as the optimal image-dependent Δ quantizer [61].

5.3.2.3 Context Modeling of Prediction Errors and Error Feedback

Performance of the GAP predictor can be significantly improved as the predictor does not provide perfect decorrelation of the data. Context modeling of the prediction error $e = I - \dot{I}$ using a combination of texture and energy information is used to exploit higher-order structures such as texture patterns. A total of 576 compound contexts are defined by 144 texture contexts and 4 error energy contexts ($\Delta/2$). The texture context information is obtained by a quantization of local values as follows:

$$C = \{x_0, \ldots, x_6, x_7\} = \{n, w, nw, ne, nn, ww, 2n - nn, 2w - ww\} \tag{5.19}$$

C is then quantized to an 8-bit binary number $B = b_7 b_6 \cdots b_0$ using the prediction value $\dot{I}$ as the threshold,

$$b_k = \begin{cases} 0 & \text{if } x_k \geq \dot{I}[i, j] \\ 1 & \text{if } x_k < \dot{I}[i, j] \end{cases}, \quad 0 \leq k < K = 8 \tag{5.20}$$

Thus a texture pattern relative to the prediction value $\dot{I}$ is obtained. The compound contexts are denoted $C([Q(\Delta)/2], B)$.

Bias cancellation is introduced in the prediction for each context, $C(\delta, \beta)$, by feeding back the conditional mean $\bar{e}(\delta, \beta)$ and adjusting the prediction $\dot{I}$ to $\ddot{I} = \dot{I} + \bar{e}(\delta, \beta)$. The new prediction error $\epsilon = I - \ddot{I}$ is used in context-based error modeling. This results in an improved predictor for I: $\ddot{I} = \dot{I} + \bar{\epsilon}(\delta, \beta)$, where $\bar{\epsilon}(\delta, \beta)$ is the sample mean of ϵ conditioned on compound context $C(\delta, \beta)$. As a last modeling step, sign flipping is introduced. Prior to encoding, the encoder checks whether $\bar{\epsilon}(\delta, \beta) < 0$. If so, $-\epsilon$, is coded instead of ϵ. The decoder has the information to reverse the sign, if necessary.

5.3.2.4 Entropy Coding of Prediction Errors

The sign-predicted prediction error is encoded using the eight primary contexts Δ described above. CALIC allows separation of the modeling and entropy coding stages; thus any entropy coder can be interfaced with the CALIC modeling. A simple ternary adaptive arithmetic coder is used in binary mode to encode one of the two values or an escape symbol. Thirty-two different contexts are used each updating occurrence counts. In the continuous-tone mode, an adaptive m-ary arithmetic coder is used based on [18].

TABLE 5.4
Bit Rates (bpp) for a Test Set of 8-Bit Natural Gray-Scale Images

Image	JPEG-LS	CALIC	Wavelet (2+2,2)	JPEG2000
Balloon	2.90	2.78	2.78	3.07
Barb 1	4.53	4.33	4.34	4.65
Zelda	3.84	3.72	3.70	3.92
Hotel	4.53	4.22	4.30	4.63
Barb 2	4.71	4.49	4.55	4.84
Board	3.82	3.50	3.53	3.81
Girl	3.96	3.71	3.74	4.10
Gold	4.56	4.38	4.37	4.65
Boats	4.03	3.77	3.79	4.11
Average	4.06	3.88	3.90	4.20

TABLE 5.5
Lossless Bit Rates (bpp) for a Test Set of Color Images

	LJPG	LJPG-A	JPEG-LS	PNG	CALIC	LOCO-A
Natural	4.60	4.11	3.87	4.17	3.73	3.78
Graphics	3.13	2.09	1.95	2.14	1.83	1.75
Average	4.08	3.40	3.19	3.46	3.06	3.06

5.3.3   JPEG-LS

JPEG-LS [22] is the best-performing standard for lossless image compression at present (Tables 5.4 and 5.5). UCM [53] introduced efficient modeling at high complexity. JPEG-LS was later developed based on a framework similar to that of CALIC, but with even more emphasis on low complexity. The baseline version is based on LOCO-I and the extension in part two on LOCO-A [54]. An important difference is that the extension may apply arithmetic coding whereas the baseline does not. The baseline version achieves high compression performance without the use of arithmetic coding at significantly less complexity than CALIC. LOCO-A almost achieves the performance of CALIC on natural images (Table 5.5).

LOCO-I is outlined briefly. A more detailed description including theoretical background is given in [54]. The prediction is a MED predictor (see Section 5.2.1) using the N, W, and NW pixels (Figure 5.1). The prediction is modified by a context-dependent integer bias-cancellation step, which also involves the NE pixel. The coding contexts are determined by quantizing each of the three differences between neighboring context pixels. The vector of the three quantized differences is further reduced by a sign-flipping step. The total number of contexts is 365. A Golomb-type entropy coder [54] is used based on a two-sided geometric distribution (TSGD) modeling, which has two free parameters. The mean value is used for the bias cancellation. A simple technique for estimating the parameters is applied. To improve compression on highly compressible images including graphics, a run mode is introduced. For lossless coding, the mode is entered when the four coding context pixels have the same value. The run of pixels having the value of the W pixel is coded using another adaptive Golomb-type entropy coder [54].

LOCO-A adds arithmetic coding to LOCO-I. The TSGD assumption is relaxed by only using the parameter information to quantize contexts into conditioning states. Binary arithmetic coding is applied to a binary tree structure describing the prediction error values.

JPEG-LS also provides an option for near-lossless coding. A parameter k specifies that each pixel in the reconstructed image is within $\pm k$ of the original [54]. This is readily achieved by applying quantization of the prediction residuals. This approach was also introduced in CALIC.

5.3.4 Reversible Wavelets — JPEG2000

While spatial domain techniques using prediction have dominated lossless image coding, frequency domain techniques such as discrete cosine transform (DCT) and wavelets dominate lossy image coding. An inherent problem of frequency domain techniques w.r.t. lossless coding is how to control the pixel errors efficiently. The decorrelating transforms used for lossy compression usually cannot be used for lossless compression as they often require floating point computations, which result in loss of data when implemented with finite precision arithmetic. This is especially true for "optimal" transforms like the KL (Karhunen–Loeve) transform and the DCT. Also, techniques purely based on vector quantization are of little utility for lossless compression.

Integer-to-integer wavelets [8], naturally described using the lifting technique, provides an efficient frequency domain solution to lossless image coding. Combining the reversible wavelet transform with advanced context modeling and coding of transform coefficients provides results almost matching those of highly efficient spatial domain compression algorithms for natural images. There are several advantages offered by a wavelet (or subband) approach for lossless image compression. The most important of which is perhaps the natural integration of lossy and lossless compression that becomes possible. By transmitting entropy-coded wavelet coefficients in an appropriate manner, one can produce an embedded bit stream that permits the decoder to extract a lossy reconstruction at a desired bit rate. This enables progressive decoding of the image that can ultimately lead to lossless reconstruction. The image can also be recovered at different spatial resolutions. These features are of great value for specific applications in, e.g., remote sensing and "network-centric" computing in general.

The JPEG2000 [50] lossless coding option is briefly described below. It is based on reversible wavelets for lossless coding. The wavelet coefficients are coded in bit-planes within the subbands for efficient as well as progressive (or embedded) coding. The context modeling plays an important role.

The default JPEG2000 setting for lossless compression is based on a dyadic decomposition using a 5/3[1] bi-orthogonal wavelet decomposition [50]. The entropy coding uses the correlation of the magnitude of neighboring wavelet coefficients, e.g., the clustering of high-magnitude coefficients in the high-frequency bands. The coding is performed in multiple passes. In each pass, the bit-planes of the subbands of wavelet coefficients are coded. The bit-planes of a subband are coded from most significant bit (msb) to least significant bit (lsb) being dependent only on higher bit-planes of the same subband. In each pass, the encoder uses three coding primitives: the significant bit coding, the sign bit coding, and the refinement coding. For coefficients which are insignificant so far, i.e., have 0s in higher bit-planes, it is coded whether they become significant, i.e., have a 1 in the current bit-plane. After a coefficient becomes significant, the sign bit is coded. Refinement coding is applied to the (less significant) bits of the coefficients already significant. JPEG2000 applies two (heuristic) models, using context quantization, adopted from EBCOT [49] for coding the significance information (plus a run-mode context). These models use the current significance information of the eight neighboring coefficients which is mapped into nine conditioning states. In all, only 18 conditioning states are used in the entropy coding. Using few states enables quick learning of parameters so that the independent conditional probabilities may efficiently be learned within even small blocks. As shown in Section 5.4.2, the mapping is well chosen and little is to be

---

[1] The two filters have five and three taps, respectively.

gained by mapping to more states for natural images. The modeling in the entropy coding is crucial for obtaining good results. The most is to be gained in the significance pass as the entropy is fairly high in the other two passes.

In the embedded wavelet entropy coders, EZW [47] and SPIHT [46], predating JPEG2000, the correlation of coefficients, at the same spatial position of subbands with the same orientation, is captured in a tree structure. Improving upon SPIHT, a reversible wavelet coder was presented [35], which almost meets the performance of CALIC (within 1% on the test set of natural images; Table 5.4). In this coder, the (2+2,2) lifting wavelet filters [8] were used, followed by bit-plane coding of the wavelet coefficients forming conditioning states similar to CALIC. The energy level was estimated by a linear combination of neighboring coefficients as well as parent and sister band coefficients. Bias cancellation was also used. The LL band was coded using CALIC. The correlation between father and sons is not as important as the wavelet decomposition suggests though once efficient modeling is performed within the band. In a fast version [59] of ECECOW [58], a linear combination of near-by significance values is applied to a larger neighborhood of values within the subband. For speed, this linear combination is calculated using convolution. High performance (within 2% of CALIC on the test set of natural images; see Table 5.4) is achieved without the use of parent subband information. Efficient scalable wavelet-based coding allowing both lossy and lossless decoding using a single bit stream is presented in [5].

5.3.5 EXPERIMENTAL RESULTS

Results for the methods presented in this section are listed in Tables 5.4 and 5.5. Table 5.4 presents the results for JPEG-LS, CALIC, JPEG2000, and the more complex lossless wavelet-based codec [35]. (The JPEG2000 results are obtained using InfranView, the others are from [35].) The JPEG-LS results given here are based on adaptive Golomb coding. The test set is the luminance component, i.e., 8-bit gray-scale of classic test images. Table 5.5 presents the results for lossless JPEG, both Huffman-based (LJPG) and with arithmetic coding (LJPG-A). The table also presents the results for JPEG-LS (Huffman), PNG, CALIC, and LOCO-A, which is equivalent to JPEG-LS with arithmetic coding. The test set is a subset of the newer JPEG test images used for JPEG-LS evaluations. The results [54] are accumulated and reported in the two subsets of 11 natural images and 6 computer-generated images.[2] The natural images are mostly color images in different color spaces (YUV, RGB, and CMYK). The computer-generated images are represented in the color spaces RGB or CIElab. The average values given are averages over the bit rates (in bits per pixel [bpp]) of the individual images. It may be noted that LOCO-A matches the performance of CALIC overall. This is due to better performance on the computer graphic images while CALIC performs about 1% better on the natural images, the class of images for which it was primarily optimized.

5.4 OPTIMIZATIONS OF LOSSLESS IMAGE CODING

CALIC and JPEG-LS/LOCO provide very efficient coding at a reasonable complexity. A natural question is how much improvement is possible if more complex methods are used, including optimization for the individual images. As there is no method for measuring the entropy of a natural image, the approach is to construct better and better coding schemes.

Once arithmetic coding is chosen for the entropy coding, it is the modeling step which becomes the focal point of the optimization. This section describes methods to optimize the prediction and context modeling steps, which constitutes the dominating paradigm. Dynamic programming is one optimization technique which may be applied with the code length as the cost function. A number

---

[2] One of these, areial2, is actually an image with a sparse histogram.

of schemes have been introduced using multiple prediction. There are well-known techniques, e.g., linear regression for optimizing linear predictors in a two-pass scheme. The following step of context modeling has been dominated by heuristic approaches. In this section, a general design of optimal context quantization is introduced and applied to the context formation for reversible wavelets. The focus is still on natural gray-scale images. Improvements related to utilizing the correlations of multiple components or in image sequences as well as improvements for graphic images are treated in Section 5.5.

5.4.1 MULTIPLE PREDICTION

Choosing a predictor which achieves small prediction errors is a good start, but the predictor should be optimized with respect to the following context modeling. Especially for adaptive prediction the issue is subtle. While there is a vast literature on adaptive prediction in general, JPEG-LS and CALIC apply simple prediction of filter values that do not adapt in time but are given by the pixels of the prediction template.

Recently, schemes such as [3,31,36] based on adaptive or classification-based use of multiple predictors have been able to improve upon the results of CALIC. These schemes apply a two-pass or rather two-stage approach. The individual image is analyzed first optimizing the coding. The parameters are coded in the header of the file and in a final pass the image is coded.

In TMW [36], multiple linear predictors are applied. The (estimated) error probability distributions for the predictors are blended. Besides the pixel predictor, a sigma predictor and a blending predictor are adaptively used for coding each pixel. The sigma predictor predicts the magnitude of the error of each pixel predictor, based on the errors of pixels within a causal template. This in turn determines the parameter for modeling the predictors' error distribution, for which a variation of the t-distribution is used. The weights for blending the error distributions are determined based on the code length required by the predictor within a template window. Thus, the blending is the new element of the modeling in TMW. Large templates are used. Twelve pixels are found to be efficient for the pixel predictor and 30 pixels are used for the sigma predictor. Arithmetic coding is used for the actual coding. In [37], the TMW coder is optimized. For the test set of Table 5.4, an average code length of 3.63 bpp was reported. TMW demonstrates that improvement using large neighborhoods and multiple predictors is possible. The price is a dramatic increase in complexity at both encoder and decoder due to the blending and adaption. An improvement of only 7% over CALIC does suggest that a level of diminishing returns has been reached for natural images.

Multiple predictors are also considered in [31], but chosen on a block basis leading to faster decoding while the encoding is still slow. Linear predictors are used. For each (8×8) block a predictor is chosen. The coding context information is based on error information for pixels within a local neighborhood. The error is mapped to an error index. The error indices are summed and then quantized using nonlinear quantization. The quantization thresholds are determined for each class using dynamic programming. A generalized Gaussian function is used to model the error distribution. This includes the Gaussian and Laplacian distributions. The distribution is determined by two parameters. The standard deviation for each context is fixed for all images. The so-called shape parameter of the distribution is determined for each context for the individual images. The parameters are iteratively optimized for the individual images.

The resulting side information, which is stored in the header part of the file contains information about

- Class label for each block specifying the predictor
- Prediction coefficients for each class
- Thresholds for context quantization for each class
- Probability distribution parameter for each context.

Using arithmetic coding, results slightly better than TMW and comparable to TMW^{Lego} were reported [17,31]. A similar less complex scheme achieved results 3% better than CALIC and within 1% of TMW [3] on a set of 24 gray-scale images. Good results are also reported in [16], although little information is given about this technique.

5.4.2 OPTIMAL CONTEXT QUANTIZATION

The best-performing lossless image compression algorithms introduced in Section 5.3.2, CALIC [61] and JPEG-LS [22], as well as the JPEG2000 entropy-coding algorithm EBCOT [49] all quantize the contexts C_t into a relatively small number of conditioning states. These context quantizers are heuristic but they have produced some of the best-performing image compression algorithms, by developing the quantizers using training data of the type of material they are targeted for, e.g. images [58]. As discussed in Section 5.2.2 and exemplified by the coders mentioned above, the design of the context quantizer Q, as part of the context mapping, is of crucial importance. An approach to optimizing context quantization is introduced below. It may be used both to design and analyze context quantizers.

The above-mentioned coders have all recognized that estimating the conditional probabilities $P(X_t|C_t)$ directly using count statistics defining the context C_t by the combination of values of a set of past samples can lead to the so-called context dilution problem, especially if the number of symbols in the context is large, or if the symbol alphabet is large. Instead $P(X_t|Q(C_t))$ is estimated after context quantization, thus reducing the number of parameters to estimate (or learn). Reducing the number of conditioning states may also be motivated by the advantage that less memory is required to store the estimates.

Algorithm context [44] is a universal coding scheme using a tree-structured context quantizer. UCM [53] combined prediction and a tree structure for image coding. Inspired by the high performance of more heuristic quantizers, formal optimization and training of general context quantizers become interesting when it is possible to integrate domain knowledge in the coder, e.g., based on a training set.

Two recent concepts of optimal context quantization are introduced below. They are based on minimizing the conditional entropy [60] and minimizing the adaptive code length [12], respectively.

The minimum conditional entropy context quantization (MCECQ), [60] is based on minimizing the conditional entropy, $H(X|Q(\mathcal{C}))$, given the probability density functions, $P(X|\mathcal{C})$, for a specific number of quantized contexts, M. This approach addresses the problem of minimizing the increase in conditional entropy for a given limited number of context classes, M. That the conditional entropy will not decrease by decreasing M is given by the convexity of the entropy function, H. The resulting quantization regions, A_m, defined on C may be quite complex. However, their associated sets, B_m, defined on the probability density functions, $P_{X|C}$, are simple convex sets [60]. In [13], the minimum conditional entropy criterion given above was applied to compute optimal context quantizers for the application of bi-level image compression.

The solution is especially simple when estimating the probability of a binary random variable, Y, as $P(Y|C)$ is expressed by one parameter and therefore the quantization regions B_m, defined on the pdf, are simple intervals. The binary variables could be binary decisions decomposing a variable used in an image compression scheme. For the binary variable, Y, we describe the quantization interval, B_m, by the random variable $P(Y=1|C)$ (the posterior probability that $Y=1$ as a function of C). The conditional entropy $H(Y|Q(\mathcal{C}))$ of the optimal context quantizer can be expressed by

$$H(Y|Q(\mathcal{C})) = \sum_{m=1}^{M} Pr\{P(Y=1|C) \in [q_{m-1},q_m)\}H(Y|P(Y=1|C) \in [q_{m-1},q_m)) \quad (5.21)$$

for some set of thresholds $\{q_m\}$ dividing the unit probability interval into M contiguous quantization intervals B_m. Posing the optimal MCECQ this way enables the solution to be found by searching over the set of interval end points, $\{q_m\}$, which is a scalar quantization problem. The optimal solution can be obtained using dynamic programming [56]. Once the end points defining the probability intervals, B_m, are found, the optimal MCECQ regions A_m are given by

$$A_m = \{\mathbf{c} : P(Y = 1|\mathbf{c}) \in [q_{m-1}, q_m)\} \qquad (5.22)$$

The MCECQ approach provides a theoretical solution, but if the probability density functions $P(X|C)$ were known, context dilution would not be a problem and it would be more efficient (compression-wise) to use $P(X|C)$ directly when coding rather than applying context quantization. This leads to modifying the view on the context quantizer design. In practice, a training set or the data sequence to be coded is used to estimate $P(X|C)$, i.e., the design is based on a given data set rather than a probability density function. As described, the best lossless image coding techniques apply adaptive coding. Thus the natural question is how to design the context quantizer to minimize the actual adaptive code length of the given training set, x^T [12]. This approach is formalized below.

The presentation is restricted to the case where the variable to be coded x comes from a binary alphabet, $\mathcal{A} = \{0, 1\}$. The adaptive code length is based on entropy coding of the sequence of adaptive probability estimates based on occurrence counts given by (14). Let n_i denote the count of the symbol i in context $\mathbf{c}$.

The ideal adaptive code length for the data set $x^T = x_1, \ldots, x_T$ is given by

$$L(x^T|\mathcal{C}) = \sum_{t=1}^{T} -\log \hat{p}_t(x_t|\mathbf{c}_t) \qquad (5.23)$$

where $\mathbf{c}_t$ is the context x_t appears in and $\hat{p}_t(x_t|\mathbf{c}_t)$ is given by (5.14) based on the counts of the causal part, x^{t-1}, of the data. This is easily implemented by sequentially updating the counters.

Owing to properties of the adaptive code length, a fast solution may again be obtained in the binary case using dynamic programming. We consider the data set x^T. As the order of the symbol appearance within a context class does not influence the ideal adaptive code length (5.23), this code length may be computed based on the set of counts over the contexts $\mathbf{c}$. Likewise, the ideal code length after context quantization may be computed just based on adding the counts of contexts which are quantized to the same CQ cell.

Let $N_i(\mathbf{c})$ be the occurrence count for symbol i in context $\mathbf{c}$ in x^T. For a given context quantizer, $Q(\mathbf{c})$, defined by the quantization regions $A_m = \{\mathbf{c} : Q(\mathbf{c}) = m\}$, $m = 1, \ldots, M$, the counts after quantization are simply given by adding the counts of contexts mapped to the same CQ cell, i.e.,

$$N_i(m) = \sum_{\mathbf{c} \in A_m} N_i(\mathbf{c}) \qquad (5.24)$$

Let L_m denote the corresponding ideal adaptive code length based on the counts, $N_i(m)$, accumulated in A_m. Let T_m be the total number of occurrences in A_m. Reorganizing the terms of (5.14) used in (5.23) gives, in the binary case

$$L_m = \sum_{k=0}^{T_m-1} \log(k + 2\delta) - \sum_{i=0}^{1} \sum_{j=0}^{N_i(m)-1} \log(j + \delta) \qquad (5.25)$$

where the last term only sums over values of i for which $N_i(m) > 0$. Now the context quantizer $Q(\mathbf{c})$ minimizing the total code length may be expressed based on computations of the counts in the original contexts, $\mathbf{c}$,

$$\min_{M,Q(\mathbf{c})} \sum_{m=1}^{M} L_m = \min_{M,Q(\mathbf{c})} L(x^T|Q(\mathbf{c})) \qquad (5.26)$$

Dynamic programming may be applied to the adaptive code length criteria assuming that the cells are contiguous intervals of the probability estimate simplex. Fast computation of the count-based adaptive code lengths, L_m (5.25) is possible using look-up tables for the terms in the sum (5.25), which may be computed once and for all for any given δ. The size of these look-up tables may be bounded using Stirlings approximation for large counts [27]. Both techniques for fast computation of L_m based on counts have constant complexity. Thus, the suggested use of dynamic programming may also be efficiently applied to the adaptive code length criteria.

If a training set is used to estimate the probabilities used in the CQ design it is natural also to use this information in the actual coding. One way is to initialize the adaptive probability estimates, e.g., for each i, by replacing δ with $\delta_i = \delta + N_i^{init}$ in (5.14), where N_i^{init} is the initialization of the count for symbol i. The initialization may be included in the dynamic coding loop by searching for the optimal values of N_i^{init} along the ratio given by the counts [24].

The optimal context quantization based on adaptive code lengths was applied to the entropy coding model of the significance information of wavelet coefficients as in JPEG2000 (Section 5.3.4). The JPEG2000 EBCOT models for the significance information and the optimized models were used in the same settings: the same dyadic reversible wavelet transform using the 5/3 filter [50], the same arithmetic coder from [55], and the same (raster) scanning order of the wavelet transform subbands. For simplicity, the coding passes were separated by the subbands.

In the experiment, 60 natural (8-bit) images were used for training the context quantizer, which was applied to the raw contexts of the EBCOT model. The adaptive code length criteria using initialization was used. The result of applying the optimized models to the nine test images of Table 5.4 is given in Table 5.6. Quantizing to nine conditioning states, the improvement is only 0.21% and using 64 states (or the optimal number) the improvement is 0.35%. Even increasing the raw context size by including more information about the energy level of the significant coefficients before quantization only gave a 0.43% improvement (Table 5.7). These results can be interpreted as a justification of the heuristically designed JPEG2000 context quantization for coding the significance information in the entropy coding of natural images. The best-performing lossless wavelet coders mentioned in Section 5.3.4 do however demonstrate that better performance is possible. Optimal quantization was also applied to the entropy coding in lossless compression of image sequences [12]. Applied to image data sequences as alpha plane sequences (defining video

TABLE 5.6
Average Bit Rates (in Bits per Pixel) for Coding the Significance Information Using the JPEG2000 Models and the Models of Different Sizes Obtained by Optimizing the JPEG2000 Raw Model (256 Raw Contexts) for the Set of Test Images and the Training Set

	Models		9 States,	9 States	64 States
	No init.		No init.		
Average bit rates	2.1302	2.1290	2.1257	2.1243	2.1227
Average bit rates, training set	2.1104	2.1089	2.1049	2.1032	2.1012

TABLE 5.7
Average Bit Rates (in Bits per Pixel) for Coding the Significance Information Using the Initialized JPEG2000 Models and the High-Order Optimized Models of Different Size for the Set of Test Images and the Training Set

Models	9 States	32 States	64 States	Optimal
Average bit rates	2.1290	2.1223	2.1211	2.1210	2.1214
Average bit rates, training set	2.1089	2.0990	2.0968	2.0965	2.0958

objects) and sequences of street maps, significant improvement was achieved using optimal context quantization.

5.5 APPLICATION DOMAINS

Important domains include medical applications, space application including satellite images, high-quality film and TV production as well as computer graphics. The coding schemes presented in Section 5.3 present efficient solutions for two-dimensional data. In many of the applications the data sets have a third (and possibly a fourth) dimension. In these cases, the higher-dimensional correlations may be utilized to improve compression. This section investigates the potentials for color and multiband images as well as for video sequences. Full color images are multiband images, but color information may also be represented by a color-mapped index. In this case, the characteristics deviate from those of natural images which may be characterized as being locally "smooth." Application of other coding techniques are investigated for color-mapped images and computer graphic images.

5.5.1 Color and Multiband

The best way of exploiting spatial and spectral redundancies for lossy and lossless compression is usually quite different. In this section, we primarily focus on how prediction-based techniques can be extended to provide lossless compression of multispectral data. The wavelet approach is also briefly commented. Irrespective of the transform used, there is a significant amount of redundancy that remains in the data after decorrelation. For high performance, the modeling and capturing of the residual redundancy constitutes a crucial step in lossless compression.

5.5.1.1 Predictive Techniques

Predictive techniques may naturally be extended to exploit interband correlations, but new issues such as the following arise:

- *Band ordering.* Using interband correlation, raises the questions of which bands are best as reference band(s) for predicting and modeling intensity values in a given band. This in turn raises the question of the best ordering of the bands.
- *Interband prediction.* What is the best way to incorporate pixels located in previously encoded spectral bands to improve prediction?
- *Interband error modeling.* What is the best way to utilize the residual redundancy after prediction across bands?

The issues of band ordering is specific to multiband coding, whereas the two other issues relate to the basic prediction and error modeling in lossless image coding. Typical approaches that have been taken to address these questions are presented below.

5.5.1.2 Band Ordering

Wang et al. [52] analyzed correlations between the seven bands of LAND-SAT TM images, and proposed an order and a mix of intra- and interband prediction, based on heuristics. Different orders of the bands have been devised, e.g., in [2].

The optimal ordering may be solved using graph-theoretic techniques [48]. When only one reference band is used for each band, an $O(N^2)$ solution exists for an N-band image, when no restrictions are imposed as to how many bands need to be decoded prior to decoding any given band. A single reference band was found to be sufficient [48]. A drawback of the technique is that it is two-pass. Thus the main use is by offline processing determining an optimal ordering for different types of images. A limitation compression-wise is that reordering entire bands disregards that spatial variation may lead to different spectral relationships.

In the remainder of this subsection, we assume that the band ordering has been decided. The band ordering is one aspect of the scan ordering of the 3D data set composed by multiband images. Generally, the pixels may be pixel interleaved, line interleaved, or one band at a time. The original ordering may be altered using buffering of data. Obviously, the pixels of the 3D data set used for prediction must be causal relative to the scan order used in the encoding.

5.5.1.3 Interband Prediction

In order to exploit interband correlations, it is natural to generalize (linear) predictors from two- to three-dimensional prediction. Let Y denote the current band and X the reference band. A 3D prediction of the current pixel $Y[i, j]$ is given by

$$\hat{Y}[i,j] = \sum_{a,b\in N_1} \theta_{a,b} Y[i-a, j-b] + \sum_{a',b'\in N_2} \theta'_{a',b'} X[i-a', j-b'] \tag{5.27}$$

where N_1 and N_2 are appropriately chosen causal neighborhoods with respect to the scan. The coefficients $\theta_{a,b}$ and $\theta'_{a',b'}$ can be optimized by minimizing $||Y-\hat{Y}||$, given multispectral image data.

Roger and Cavenor [45] performed a detailed study on prediction of AVIRIS[3] images. Their conclusion was that a third-order spatial-spectral predictor based on the immediate two neighbors $Y[i, j-1]$, $Y[i-1, j]$ and the corresponding pixel $X[i, j]$ in the reference band is sufficient. They further suggested the use of forward prediction, transmitting optimal coefficients for each row. The reasons were twofold. It provided a locally adaptive predictor and the AVIRIS image data is acquired in a line interleaved fashion. Compressionwise, a block-based adaption of prediction coefficients would be better.

To avoid the overhead cost of coding frequently changing prediction coefficients, Wu and Memon [62] proposed a (backward) adaptive interband predictor. Extending ideas in CALIC, relationships between local gradients of adjacent spectral bands are utilized. The following *difference-based interband interpolation*, was proposed as means of interpolating the current pixel based on local gradients in the reference band. The estimated value $\hat{Y}[i, j]$ is given by

$$\frac{Y[i-1,j] + (X[i,j] - X[i-1,j]) + Y[i,j-1] + (X[i,j] - X[i,j-1])}{2} \tag{5.28}$$

---

[3] Airborne visible infrared imaging spectrometer. It delivers calibrated images in 224 contiguous spectral bands with wavelengths from 400 to 2500 nm.

Owing to the difficulties in finding one global interband predictor, Wu and Memon proposed a predictor switching between inter- and intraband prediction. Interband prediction is only used if the interband correlation in a local window is strong enough; otherwise intraband prediction is used. Simple heuristics to approximate this correlation were used to reduce complexity. They reported that switched inter-/intraband prediction gives significant improvement over optimal predictors using inter- or intraband prediction alone.

5.5.1.4   Error Modeling and Coding

As in the lossless coding of single images, additional coding gain may be obtained by using an error model that captures the structure that remains after prediction. In Section 5.3.2, details of error modeling techniques were given. Here, examples of error modeling used for compression of multispectral images is given.

Roger and Cavenor [45] investigated two different variations. One technique was to determine the optimal Rice–Golomb code in a row by an exhaustive search over the parameter set, i.e., modelwise the assumption is that the prediction errors follow a single geometric pmf. In the second technique, the variance of prediction errors for each row was used to choose one of eight predesigned Huffman codes. Tate [48] used a context based on quantizing the prediction error in the corresponding location in the reference band and used this as a conditioning state for arithmetic coding. Since this involves estimating the pmf in each conditioning state, only a small number of states (4 to 8) are used. An example of a hybrid approach is given by Wu and Memon [62] (3D-CALIC) who propose an elaborate context formation scheme that includes gradients, prediction errors, and quantized pixel intensities from the current and reference band. They select between one of eight different conditioning states for arithmetic coding based on a prediction error variance estimate.

A simple solution of subtracting one band from another was considered in [9]. For R,G,B (red, green, and blue) images, the green component was coded first and then subtracted (modulo 256) from the red and blue component. Applying JPEG-LS to the difference bands gave results close to those of interband CALIC. It was noted though that in case the cross-band correlation was low, taking the differences would make the compression worse.

Just taking a simple difference between the prediction error in the current and reference band provides another simple technique for exploiting relationships between prediction errors in adjacent bands. The approach can be further improved by conditioning the differencing operation. The prediction errors would still not provide perfect correlation, but contain enough structure to benefit from the use of error modeling.

5.5.1.5   Hyperspectral Images

Having up to more than 2000 closely spaced bands in hyperspectral images makes the use of high-order interband prediction profitable. Aiazzi et al. [2] considered up to three causal bands in the prediction step. Weighted linear predictors were used. The linear predictors were optimized within classes determined by fuzzy logic. Using some of these ideas, a global approach to prediction was taken in [38] for hyperspectral images. A high-order linear predictor was applied at each spatial position over the spectral bands. The pixels are clustered using vector quantization and an optimized predictor is calculated for each cluster. The prediction errors are coded with an adaptive entropy coder for each cluster with the band as additional context information. The coder was applied to AVIRIS images in 224 bands, reporting 69% higher compression ratios than applying JPEG-LS to the individual images. This advantage was reduced to 19% by applying simple differencing between bands prior to JPEG-LS. In a recent study [51], a fast correlation-based band ordering heuristic combined with dynamic programming is introduced. Applied to hyperspectral sounder data, a 5% compression gain is achieved compared with natural ordering. This result is within a few percent of what is achieved by the optimal ordering. CALIC in the 3D version was modified for hyperspectral images [26]. The modifications include always using the interband predictor due to the high correlation, combining two bands in the interband prediction and a different optimization of parameters. On data compressed in band-interleaved-by-line format, it outperforms state-of-the-art algorithms including 3D-CALIC. In band-sequential format it was outperformed by, e.g., the 3D fuzzy prediction technique [2].

5.5.1.6 Reversible Transform-Based Techniques

Transform-based lossless image compression may be extended to multispectral images. Bilgin et al. [6] extend the well-known zero tree algorithm for compression of multispectral data. They perform a 3D dyadic subband decomposition of the image and encode transform coefficients by using a zero tree structure extended to three dimensions. They report an improvement of 15 to 20% over the best 2D lossless image compression technique. Lossless compression based on 3D wavelets was also presented in [63]. The scheme provides efficient lossy-to-lossless compression. Applied to 3D medical images, an average gain of 35% over the best 2D lossless image coders was reported.

5.5.2 VIDEO SEQUENCES

Lossless coding of video is of interest in the line of production and contribution of video for film and broadcast applications. Digital cinema is one important example. For distribution the video is mostly lossy coded, notably using one of the MPEG standards which apply hybrid coding, combining motion-compensation and 2D-DCT-based coding. For lossy coding, the use of block-based motion compensation is quite simple and effective.

For lossless coding, techniques building on the techniques for lossless image coding are the most effective. Extending these to efficiently utilize the temporal correlations is more difficult though. Video adds the temporal dimension to the spatial and spectral dimensions of interband coding. In broad terms, the motion is related to image objects (projected onto the image plane), which in turn leads to a spatial and temporal change of the temporal correlation. Thus adaptive mechanisms such as, e.g., switching between prediction methods become important for effective coding.

Video is as color images often captured in red, green, and blue (R,G,B) components. An early work on lossless video coding considered video in an (R,G,B) representation [32]. Owing to the high correlation between the color bands it was concluded that using interband information in the prediction scheme was of prime importance. Using spectral correlation exceeded the effectiveness of solely using the temporal correlation. Block-based techniques for capturing the temporal correlation was explored, block-based differencing being the simplest case. This may be extended by estimating and coding motion vectors for each block as in MPEG coding. An alternative approach is to design 3D predictors. A 3D predictor may be designed based on a 2D predictor by averaging the results of using the 2D predictor in the three coordinate planes passing through the pixel to be coded. Using the JPEG predictor 4 (Table 5.2) yields the 3D predictor

$$\frac{2(I_k[i,j-1] + I_k[i-1,j] + I_{k-1}[i,j])}{3} - \frac{I_{k-1}[i-1,j] + I_{k-1}[i,j-1] + I_k[i-1,j-1]}{3} \quad (5.29)$$

where I_k and I_{k-1} refer to pixels in the current and the previous frame, respectively.

For the (R,G,B) material, the best solution was found to be a block-based switching between simple differencing for temporal prediction and spectral prediction [32]. The spectral correlation was exploited by using the best predictor for the red band on the green band and the best predictor for the green band on the blue band. The predictors considered were spatial predictors.

Utilizing spectral and temporal information in the context definition for coding the residual was also considered. Again the spectral approach was found to give the best result based on the magnitude of prediction residuals in the previous spectral band within a block around the current pixel. On the whole, the bit rate was reduced by 2.5 bits per (R,G,B) pixel (measured by the residual entropy) compared with lossless JPEG [32].

Subsequently, a study concluded that for luminance–chrominance representations, the interband correlation provides minor improvements to lossless coding [34]. In JPEG2000, a reversible color transform is introduced to convert (R,G,B) images to a luminance–chrominance representation such that the lossless coding could be performed in the latter, but still provide a perfect reconstruction of the (R,G,B) images. Such a transform may also be used to decorrelate an (R,G,B) video sequence.

In [7,28,29], lossless coding of video in the standard CCIR luminance–chrominance format for TV was considered. Motivated by the findings in [34], the components were proposed to be coded independently and only the luminance was considered for the experiments. Thus the focus was on utilizing the temporal redundancies in the coding process. In [28,29], a lossless video coder was designed based on extending JPEG-LS. (This was later referred to as LOCO 3D [7].) The basic elements of JPEG-LS, including context-based prediction, bias cancellation, and Golomb coding, were combined with multiple predictors and coding of side information for the prediction. The extensions were high-resolution motion fields, 3D predictors, prediction using one or multiple previous images, predictor dependent modeling, and motion field selection by code length. The framework was based on combining motion estimation with multiple predictors, coding the motion vector and the predictor chosen as side information on a block basis. The predictors chosen were [29] as follows:

- *Predictor 0.* Intra prediction using the JPEG-LS predictor.
- *Predictor 1.* The motion-compensated version of the 3D predictor in (29), i.e., the pixels from the other band is replaced by the motion compensated pixels of the reference frame.
- *Predictor 2.* A DC shift predictor in 3D, which calculates the difference of the JPEG mode 4 predictor in the (motion-compensated) reference image and the current image. The motion-compensated pixel at the current position is modified by this difference.
- *Predictors 3 and up.* Motion compensated prediction using bi-linear interpolation for subpixel accuracy.

Predictor 0 is an intra-predictor. Predictors 1 and 2 are motion-compensated 3D filters where only integer pixel motion vectors are considered. Each subpixel position is represented by a predictor with an index of 3 or higher. For all predictors bias correction is applied as in JPEG-LS.

The decision of which predictor to use is determined on a block basis by deriving the incremental code length based on the Golomb coding parameters. As side information, the integer part of the motion vectors is coded as an image for each coordinate. The choice of predictor is also coded as an image and this is also the case for the reference image identity if more than one reference image is considered. These three- or four-side information images have a lower resolution, namely the image resolution divided by the block size. The context for coding is determined by the differences of the eight causal neighbors as in JPEG-LS combined with the motion compensation difference. These differences are quantized as in JPEG-LS.

The motion-compensated coder described above was applied to interlace and progressive material [28,29]. For the interlace material, each frame is composed of two image fields of 720×288 pixels. For the progressive material each image is 360×288 pixels. The results were compared with the results of applying JPEG-LS to each of the images in the sequence. Using one reference frame, the average saving was 1.1 bpp corresponding to 20%. For large numbers of reference images (10 frame images or 12 fields), additional savings of 10 to 20% were achieved. There were significant differences from (sub)sequence to (sub)sequence. Example given, in parts where there was no global motion and only smaller moving objects, savings up to 40% improvement were achieved even with just one reference frame. In this case the improvement of using multiple reference frames was insignificant. For slow pan and zoom, the overall gain was smaller but significant improvement was achieved increasing the number of reference frames.

The same scheme was used in [7] and extended with adaptive filtering. The conclusions were that the gain obtained by the adaptive filtering was very small. Images of 720×576 were obtained by interleaving the two fields of each frame of the interlace material. Savings of about 0.8 bpp compared to using JPEG-LS for each image were reported for this test material.

5.5.3 COLOR-INDEXED IMAGES AND GRAPHICS

The application of (linear) prediction or frequency transformation in image coding is based on an assumption of smoothness of the image function. For color-indexed images and images having limited bits per pixel this assumption is questionable. Examples are typically seen on web pages depicting icons, logos, maps, graphics in general, and palletized images. Mobile phone is another important application area for such images. These images are often coded using GIF or PNG, which are based on the one- dimensional Lempel–Ziv coding leading to fast but nonoptimal coding for images. We shall consider the class of color-indexed images where the pixel values are defined by an index combined with a color table that holds the color value of each index. Further we restrict the range of index values to lie within 0 to 255, i.e., up to 8 bits per pixel.

There is no standard really efficiently targeting these color-indexed images. Existing compression schemes may be optimized for these images though. Given a lossless image coder, say JPEG-LS, the indices of the color map of a palletized image or computer graphics may be reordered to minimize the code length. Determining the optimal reordering is intractable (NP-complete), but heuristic approaches have been devised and combined with lossless JPEG [33] as well as JPEG-LS and JPEG2000 [65]. Significant improvements compared with GIF were obtained. For color-mapped images with 256 colors, JPEG2000 gave the best results. For fewer colors, either JPEG-LS or JPEG2000 provided the best results. A survey of reordering methods is given in [41]. The reordering methods were combined with JPEG-LS and JPEG2000. In these tests, JPEG-LS consistently provided the best results, both for 256 and fewer colors. (These test images with 256 colors were smaller than those in [65].) The conclusion was that the method of Memon and Venkateswaran [33] is the most effective. The second most effective method was a modified version of Zeng's reordering [41,65], which was 3 to 5% worse, but much faster.

For images with limited bits per pixel, efficient coding may also be obtained by coding each bit-plane using a bi-level image coder. JBIG [19] recommends using Gray coding of the pixel values in order to minimize the number of bits in the binary representation which change their value when two neighboring values differ by one. More generally the binary representation of the pixel values may be optimized to reduce the code length of bit-plane coding.

A recent efficient coding scheme for color-indexed images is provided by the piecewise-constant (PWC) codes [4]. PWC decomposes the coding into binary decisions and chooses a different context for each of the decisions. For each pixel, PWC applies arithmetic coding to the binary decisions until the value is determined. First it is coded whether the current pixel has the same value as one of the four neighbors in four steps. An edge map provides the context information for coding whether the current pixel value is equal to the west neighbor. If not the north neighbor is considered. Thereafter the north-west and north-east neighbors are considered if necessary. If the pixel value is different from the four causal neighbors, PWC resorts to a list of good guesses and finally if necessary the actual value is coded possibly using the predictor of JPEG-LS [22]. A special skip-innovation technique is also applied to code runs of pixels in uniform areas. A related scheme is runs of adaptive pixel patterns (RAPP) [42], which also codes whether the current pixel has the same value as one of the four causal neighbors and if not it codes an escape. The context for coding this decision is quantized by labeling each pixel according to whether it is a new color or identical to one of the others within the template. The resulting pattern is used as context. This may readily be generalized for larger templates, e.g., in a scheme called template relative pixel patterns (TRPP) [11]. Even templates up to a size of 10 become manageable with respect to memory regardless of the alphabet size. An efficient indexing based on enumerative coding techniques may be used for TRPP contexts.

TABLE 5.8
Lossless Code Lengths (Bytes) for a (Composite) Street Map

GIF	PNG	JPEG-LS	JPEG-LS Reorder	JBIG Gray	RAPP ($M = 5$)	PWC (Flip, Collapse)	Layer [11]
49248	43075	66778	46742	30426	26472	20600	14770

A variation of bit-plane coding is to code binary layers corresponding to specific index values. This is efficient for coding maps which are composed of different layers. Coding bit-planes or binary layers, the basic bi-level coding can be extended by using cross-layer context information. In JBIG-2 [15,20], lossy bi-level images may be refined to lossless images using template pixels in both the lossy (reference) image and the final lossless image. This was also applied to coding layers of street maps [11]. In EIDAC [64], bit-plane coding was applied placing template pixels at the current position in all the higher bit-planes. From a graphics image as a map, layers may be defined by the colors. In this case, a so-called skip coding may be applied [11], where in a given layer the coding of pixels already colored in a previous layer is skipped. These skipped values may be mapped to the background value or an extra skip value as part of a context quantization. Once interlayer dependency is introduced in the coding of a layered image the coding order of the layers may influence the coding efficiency. This is amenable to optimization and especially for single reference layer coding an efficient dynamic programming technique may be applied [23], similar to the problem of interband ordering [48] (see Section 5.5.1). In cross-layer coding, various techniques of context optimization may be applied, e.g. searching for good context pixels and performing context quantization [23].

Table 5.8 [11] gives the results of coding a street map using existing coding methods. The map has 723×546 pixels and 12 different layers or colors. It is seen that direct use of JPEG-LS gives the highest bit rate even being outperformed by GIF and PNG. Reordering the indices [32] prior to JPEG-LS coding improves the performance to the level of GIF and PNG. JBIG coding of bit-planes (after Gray coding) achieved the best results among the standard methods. The gain using Gray coding prior to coding was almost 10%. Even with reordering, it is significantly outperformed by PWC. PWC is almost 3.2 times more efficient than JPEG-LS (without index reordering). Using layered coding even better results may be obtained and at the same time provide progression [11]. Given the full layers using template coding and with reference coding for some layers a bit rate of 16.797 bytes was obtained. Replacing the coding of some of the layers by free tree coding [27] gave 14.770 bytes as the overall best result. (Both of these results are for coding all but the last layer enabling reconstruction of the composite image.) Given only the composite image and extracting the layers, a code length of 20.700 bytes was obtained using skip coding of the binary layers. A combination of two binary layers and coding the residual (using TRPP) yielded 19.109 bytes.

As more information actually is coded when coding the full layers, it is seen that there is definitely room for improving the lossless coding of color-indexed images as the (composite) map image.

REFERENCES

1. A. Agresti, *Categorical Data Analysis*, Wiley, New York, 1990.
2. B. Aiazzi, P. Alba, L. Alparone, and S. Baronti, Lossless compression of multi/hyper-spectral imagery based on a 3-D fuzzy prediction, *IEEE Trans. Geosci. Remote Sensing*, 37 (1999) 2287–2294.
3. B. Aiazzi, L. Alparone, and S. Baronti, Near-lossless image compression by relaxation-labelled prediction, *Signal Process.*, 82 (2002) 1619–1631.
4. P.J. Ausbeck, Jr., The piecewise-constant image model, *Proc. IEEE*, 88 (2000) 1779–1789.

5. A. Bilgin, P.J. Sementilli, F. Sheng, and M.W. Marcellin, Scalable image coding using reversible integer wavelet transforms, *IEEE Trans. Image Process.*, 9 (2000) 1972–1975.

6. A. Bilgin, G. Zweig, and M.W. Marcellin, Three-dimensional image compression using integer wavelet transforms, *Appl. Opt.*, 39 (2000) 1799–1814.

7. D. Brunello, G. Calvagno, G.A. Mian, and R. Rinaldo, Lossless compression of video using temporal information, *IEEE Trans. Image Process.*, 12 (2003) 132–139.

8. R. Calderbank, I. Daubechies, W. Sweldens, and B.-L. Yeo, Wavelet transforms that map integers to integers, *Appl. Comput. Harmon. Anal.*, 5 (1998) 332–369.

9. B. Carpentieri, M.J. Weinberger, and G. Seroussi, Lossless compression of continuous-tone images, *Proc. IEEE*, 88 (2000) 1797–1809.

10. T.M. Cover and J.A. Thomas, *Elements of Information Theory*, Wiley, New York, 1991.

11. S. Forchhammer and O.R. Jensen, Content layer progressive coding of digital maps, *IEEE Trans. Image Process.*, 11 (2002) 1349–1356.

12. S. Forchhammer, X. Wu, and J.D. Andersen, Optimal context quantization in lossless compression of image data sequences, *IEEE Trans. Image Process.*, 13 (2004) 509–517.

13. D. Greene, F. Yao, and T. Zhang, A linear algorithm for optimal context clustering with application to bi-level image coding, *Proceedings of the International Conference on Image Processing*, 1998, pp. 508–511.

14. P.G. Howard, Lossless and lossy compression of text images by soft pattern matching, *Proceedings of the IEEE Data Compression Conference, DCC '96,* Salt Lake City, UT, March 1996, pp. 210–219.

15. P.G. Howard, F. Kossentini, B. Martins, S. Forchhammer, and W.J. Rucklidge, The emerging JBIG2 standard, *IEEE Trans. Circuits Syst. Video Tech.*, 8 (1998) 838–848.

16. http://compression.ru/ds

17. http://itohws03.ee.noda.sut.ac.jp/~matsuda/mrp/

18. A. Moffat, R. Neal, and I. Witten, Arithmetic coding revisited, *Proceedings of the IEEE Data Compression Conference, DCC '98,* Salt Lake City, UT, March 1995, pp. 202–211.

19. JBIG, Progressive bi-level image compression, *ISO/IEC Int. Standard 11544*, 1993.

20. JBIG2, Lossy/Lossless coding of bi-level images (JBIG2), *ISO/IEC Int. Standard 14492*, 2000.

21. JPEG, JPEG technical specification, *ISO/IEC Int. Standard 10918-1*, 1994.

22. JPEG-LS, Lossless and near-lossless compression of continuous-tone still images, *ISO/IEC Int. Standard 14495*, 1999.

23. P. Kopylov and P. Fränti, Compression of map images by multi-layer context tree modeling, *IEEE Trans. Image Process.*, 14 (2005) 1–11.

24. A. Krivoulets, X. Wu, and S. Forchhammer, On optimality of context modeling for bit-plane entropy coding in the JPEG2000 standard, *Visual Content Processing and Representation, Proceedings of the 8th VLBV*, N. Garcia, J.M. Martnez, and L. Salgado, Eds., Springer, Heidelberg, September 2003.

25. G.G. Langdon, Jr. and J.J. Rissanen, A simple general binary source code, *IEEE Trans. Info. Theory*, 28 (1982) 800–803.

26. E. Magli, G. Olmo, and E. Quacchio, Optimized onboard lossless and near-lossless compression of hyperspectral data using CALIC, *IEEE Geosci. Remote Sensing Lett.*, 1 (2004) 21–25.

27. B. Martins and S. Forchhammer, Tree coding of bilevel images, *IEEE Trans. Image Process.*, 7 (1998) 517–528.

28. B. Martins and S. Forchhammer, Lossless compression of video using motion compensation, Proceedings of the *IEEE Data Compression Conference, DCC '98*, Salt Lake City, UT, March 1998, p. 560.

29. B. Martins and S. Forchhammer, Lossless compression of video using motion compensation, *Proceedings of the 7th Danish Conference Pattern Rec. Image Analysis*, Copenhagen, Denmark, July 1998, pp. 59–67.

30. S.A. Martucci, Reversible compression of HDTV images using median adaptive prediction and arithmetic coding, *Proc. IEEE Int'l. Symp. Circuits and Systems*, IEEE Press, Piscataway, NJ, 1990, pp. 1310–1313.

31. I. Matsuda, N. Shirai, and S. Itoh, Lossless coding using predictors and arithmetic code optimized for each image, *Visual Content Processing and Representation*, *Proceedings of the 8th VLBV*, N. Garcia, J.M. Martnez, and L. Salgado, Eds., Springer, Heidelberg, Sept. 2003.

32. N. Memon and K. Sayood, Lossless compression of video sequences, *IEEE Trans. Commun.*, 44 (1996) 1340–1345.

33. N. Memon and A. Venkateswaran, On ordering color maps for lossless predictive coding, *IEEE Trans. Image Process.*, 5 (1996) 1522–1527.

34. N. Memon, X. Wu, V. Sippy, and G. Miller, An interband coding extension of the new lossless JPEG standard, *Proceedings of IS&TSPIE Symposium on Elec. Im.*, 3024, 1997, pp. 47–58.

35. N. Memon, X. Wu, and B.-L. Yeo, Improved techniques for lossless image compression with reversible integer wavelet transforms, *Proceedings of the IEEE International Conference on Image Processing, ICIP'98*, 1998, pp. 891–895.

36. B. Meyer and P. Tischer, TMW — a new method for lossless image compression, *Proceedings of the Picture Coding Symposium, PCS'97*, 1997, pp. 533–538.

37. B. Meyer and P. Tischer, TMW^{Lego}- an object oriented image modelling framework, *Proceedings of the IEEE Data Compression Conference, DCC'01*, Salt Lake City, UT, 2001, p. 504.

38. J. Mielikainen and P. Toivanen, Clustered DPCM for the lossless compression of hyperspectral images, *IEEE Trans. Geosci. Remote Sensing*, 41 (2003) 2943–2946.

39. W.B. Pennebaker and J.L. Mitchell, *JPEG: Still Image compression Standard*, Van Nostrand Reinhold, New York, 1992.

40. W.B. Pennebaker, J.L. Mitchell, G.G. Langdon, Jr., and R.B. Arps, An overview of the basic principles of the Q-coder adaptive binary arithmetic coder, *IBM J. Res. Dev.*, 32 (1988) 717–752.

41. A.J. Pinho and A.J.R. Neves, A survey on palette reordering methods for improving the compression of color-indexed images, *IEEE Trans. Image Process.*, 13 (2004) 1411–1418.

42. V. Ratnaker, RAPP: Lossless image compression with runs of adaptive pixel patterns, *Proceedings of the 32nd Asilomar Conference on Signals, Systems and Computers*, Nov. 1998.

43. J. Rissanen, Generalized Kraft inequality and arithmetic coding, *IBM J. Res. Dev.*, 20 (1976) 198–203.

44. J. Rissanen, A universal data compression system, *IEEE Trans. Info. Theory*, 29 (1983) 656–664.

45. R.E. Roger, and M.C. Cavenor, Lossless compression of AVIRIS images, *IEEE Trans. Image Process.*, 5 (1996) 713–719.

46. A. Said, and W.A. Pearlman, An image multiresolution representation for lossless and lossy compression, *IEEE Trans. Image Process.*, 5 (1996) 1303–1310.

47. J.M. Shapiro, Embedded image coding using zerotrees of wavelet coefficients, *IEEE Trans. Signal Process.*, 41 (1993) 3445–3462.

48. S.R. Tate, Band ordering in lossless compression of multispectral images, *IEEE Trans. Comput.*, 46 (1997) 477–483.

49. D. Taubman, High performance scalable image compression with EBCOT, *IEEE Trans. Image Process.*, 9 (2000) 1158–1170.

50. D.S. Taubman and M.W. Marcellin, *JPEG2000*, Kluwer, Boston, 2002.

51. P. Toivanen, O. Kubasova, and J. Mielikainen, Correlation-based band-ordering heuristic for lossless compression of hyperspectral sounder data, *IEEE Geosci. Remote Sensing Lett.*, 2 (2005) 50–54.

52. J. Wang, K. Zhang, and S. Tang, Spectral and spatial decorrelation of Landsat-TM data for lossless compression, *IEEE Trans. Geosci. Remote Sensing*, 303 (1995) 1227–1285.

53. M.J. Weinberger, J. Rissanen, and R.B. Arps, Applications of universal context modeling to lossless compression of gray-scale images, *IEEE Trans. Image Process.*, 5 (1996) 575–586.

54. M.J. Weinberger, G. Seroussi, and G. Sapiro, The LOCO-I lossless image compression algorithm: Principles and standardization into JPEG-LS, *IEEE Trans. Image Process.*, 9 (2000) 1309–1324.

55. I.H. Witten, R.M. Neal, and J.G. Cleary, Arithmetic coding for data compression, *Commun. ACM*, 30 (1987) 520–540.

56. X. Wu, Optimal quantization by matrix-searching, *J. Algorithms*, 12 (1991) 663–673.

57. X. Wu, Lossless compression of continuous-tone images via context selection and quantization, *IEEE Trans. Image Process.*, 6 (1997) 656–664.

58. X. Wu, High-order context modeling and embedded conditional entropy coding of wavelet coefficients for image compression, *Proceedings of the 31st Asilomar Conference on Signals, Systems, and Computers*, November 1997, pp. 1378–1382.

59. X. Wu, Compression of wavelet transform coefficients, in *The Transform and Data Compression Handbook*, K.R. Rao and P.C. Yip, Eds., CRC Press, Boca Raton, FL, 2001, Chap. 8, pp. 347–378.

60. X. Wu, P.A. Chou, and X. Xue, Minimum conditional entropy context quantization, *Proceedings of the International Symposium on Information Theory*, 2000, pp. 43.

61. X. Wu and N. Memon, Context-based, adaptive lossless image coding, *IEEE Trans. Commun.*, 45 (1997) 437–444.

62. X. Wu and N. Memon, Context-based lossless interband compression extending CALIC, *IEEE Trans. Image Process.*, 9 (2000) 994–1001.

63. Z. Xiong, X. Wu, S. Cheng, and J. Hua, Lossy-to-lossless compression of medical volumetric data using three-dimensional integer wavelet transforms, *IEEE Trans. Med. Imaging*, 22 (2003) 459–470.

64. Y. Yoo, Y. Kwon, and A. Ortega, Embedded image-domain adaptive compression of simple images, *Proceedings of the 32nd Asilomar Conference on Signals, Systems and Computers*, November 1998, pp. 1256–1260.

65. W. Zeng, J. Li, and S. Lei, An efficient color re-indexing scheme for palette-based compression, *Proceedings of the IEEE International Conference on Image Processing, ICIP'00*, 2000, pp. 476–479.

6 Fractal Image Compression

Raouf Hamzaoui and Dietmar Saupe

CONTENTS

6.1   Introduction 145
6.2   The Fractal Image Model 148
    6.2.1   Variants 153
6.3   Image Partitions 154
    6.3.1   Quadtrees 155
    6.3.2   Other Hierarchical Partitions 155
    6.3.3   Split–Merge Partitions 156
6.4   Encoder Complexity Reduction 157
    6.4.1   Feature Vectors 157
    6.4.2   Classification Schemes 159
        6.4.2.1   Jacquin's Approach 159
        6.4.2.2   Classification by Intensity and Variance 159
        6.4.2.3   Clustering Methods 160
    6.4.3   Tree-Structured Methods 160
    6.4.4   Multiresolution Approaches 160
    6.4.5   Fast Search via Fast Convolution 161
    6.4.6   Fractal Image Compression without Searching 162
6.5   Decoder Complexity Reduction 162
    6.5.1   Fast Decoding with Orthogonalization 162
    6.5.2   Hierarchical Decoding 162
    6.5.3   Codebook Update 163
    6.5.4   Other Methods 164
6.6   Attractor Coding 164
6.7   Rate-Distortion Coding 166
6.8   Extensions 167
    6.8.1   Hybrid Methods 167
    6.8.2   Channel Coding 167
    6.8.3   Progressive Coding 167
    6.8.4   Postprocessing 167
    6.8.5   Compression of Color Images 168
    6.8.6   Video Coding 168
    6.8.7   Applications 168
6.9   State of the Art 168
6.10 Conclusion 170
Acknowledgments 171
References 172

6.1   INTRODUCTION

In the mid-1980s fractal techniques were introduced in computer graphics for modeling natural phenomena [75]. One of the new ideas came from a mathematical theory called iterated function systems (IFS) [3]. This theory had previously been developed in 1981 by John Hutchinson, however, without any practical applications in mind [49]. It was Michael Barnsley and his research group from the Georgia Institute of Technology who first realized the potential of iterated function systems for modeling, e.g., clouds, trees, and leaves. While other modeling techniques in computer graphics such as procedural modeling and L systems were dominating the IFS approach, one of the visions of Barnsley — namely that of encoding real-world images using IFS — turned into an innovative technique in the image compression field.

The innovation that distinguishes fractal image coding is that an image is not *explicitly* coded, e.g., as a list of quantized coefficients for a suitable set of image basis vectors in transform coding. Instead, the fractal image code consists of a description of a contractive image operator, and the decoder recovers the fixed point of this operator as an approximation of the original image to be encoded. Therefore, fractal image compression defines an image reconstruction *implicitly* as a fixed point of an image operator, i.e., as the solution of a system of equations.

This design principle has far-reaching consequences. On the encoder side the image operator should be constructed choosing the best one (or perhaps a suboptimal one) in a suitable class of image operators. Since one can hardly expect the encoder to come up with the image operator for the whole image in one batch, construction of the operator must proceed piece by piece. But then, during the encoding, one has only a partial definition of an image operator, and it is not sensible to assume a "current" fixed point as an intermediate image approximation. Thus, it is intrinsically difficult to define the pieces of the operator such that at the end its fixed point is as close to the given image as possible.

Also the decoder is fundamentally different from traditional image compression decoders. Its job is to solve a system of equations in a finite-dimensional vector space of gray-scale (or color) images. Moreover, the dimension of the system is huge, namely equal to the number of pixels in the original image. Therefore, the solution method typically is iterative and the number of iterations may depend on the many choices made in the construction of the particular image operator as well as on the original image. In contrast, decoding in traditional compression systems is not iterative, and the number of operations (or an upper bound) can easily be calculated.

These new challenges made the approach a very attractive research topic. The problem at the core is that of finding a suitable fractal image model, in other words, a suitable class of image operators. The class should have several desirable properties. To name the most important ones, the operators should exhibit sufficient structure allowing their piecewise construction as well as fast iterative fixed-point computation. Moreover, it must be possible to efficiently encode the operators in the form of a bit stream. Last, it is desirable to be able to interpret the image operator at arbitrary image resolution, thereby yielding a resolution-independent image coding method. It is probably not surprising to learn that the class of image operators that was identified as useful for fractal compression does not consist of highly nonlinear operators, but rather of a set of "least nonlinear" operators, namely a particular subset of affine mappings. Moreover, the requirements of a simple operator structure enforces that the corresponding matrices are sparse.

After solving this fundamental problem, two further very important issues needed to be addressed. The first one is of an algorithmic or engineering type. For the basic class of fractal image operators, how does one best design the algorithms for the encoder and decoder. In particular, how can one best trade off the three basic commodities that must be considered in any compression–decompression system: (1) bit rate (or, equivalently, compression ratio), (2) approximation quality usually measured in peak-signal-to-noise ratio (PSNR), and (3) time and space complexity of both the encoder and the decoder as well as algorithmic complexity. The second issue is an evaluation of the approach in relation to other competing image compression methods.

Before giving an overview of this paper we present a short synopsis of the history of fractal coding. Back in 1987, Michael Barnsley and Alan Sloan [5] speculated about very high fractal compression ratios and announced that it was possible to transmit such compressed image files at video rates over regular telephone lines. However, even though a few encoded images were published (without the originals), at that time nobody seemed to know exactly how to faithfully reproduce images at reasonable compression ratios with IFSs. What was the problem? The class of image operators defined by IFS is simply not appropriate. The fractals that one can easily generate with an IFS are all of a particular type. They are images that can be seen as collages of deformed and intensity transformed copies of themselves [3]. Thus, e.g., in an IFS encoding of a picture of an object one should see miniature distorted copies of the same object everywhere. This seemed not only unreasonable but also technically infeasible. Human interaction was required to first segment an image into a few more or less homogeneous regions, each of which was then covered by a collection of rescaled copies of itself, where scaling factors, translation, and other mapping parameters were far from obvious. The bulk of the work had to be done manually, and others irreverently called the method the "graduate student algorithm."

However, in 1989, Arnaud Jacquin, one of the graduate students of Barnsley, was the first to report an automatic fractal encoding system in his Ph.D. thesis, in a couple of technical reports, and shortly after in a conference paper [52]. In his work, he proposed a nonlinear IFS image operator that works locally by combining transformed pieces of the image. By leaving behind the rigid thinking in terms of global IFS mappings, the approach broke the ice for a novel direction of research in image coding.

The basic new idea was very simple. An image should not be thought of as a collage of copies of the entire image, but as a collage of copies of smaller parts of it. For example, a part of a cloud does not look like an entire landscape with clouds, but it does not seem so unlikely to find another section of some cloud or some other structure in the image that looks like the given cloud section. Thus, the general approach is to first subdivide the image into a partition — fixed size square blocks in the simplest case — and then to find a matching image portion for each part. Jacquin provided a workable image compression implementation. However, there were many open questions: for example, how should the image be partitioned? where should one search for matching image portions? how should the intensity transformation be designed? Moreover, the algorithm as proposed was slow. Thus, methods for acceleration were urgently needed.

One of the good things of standards is that people can build further research and applications on them, thereby accelerating scientific progress. This is what happened after Yuval Fisher made his well-written C code for an adaptive quadtree based fractal encoder available on the world wide web along with a thorough theoretical and practical documentation [30].

Fractal image compression quickly ripened to the state where it could easily outperform the JPEG standard. However, over the years, new image compression techniques — predominantly those based on subband and wavelet transform coding — were pushing the compression performance far ahead. This competition and the first promising results of fractal compression provided a lasting motivation for research in fractal methods. Our survey describes the main results of these efforts. We present the theory specifically for coding digital gray-scale images. Moreover, we essentially discuss fractal image coding in the "pure" spatial domain.

In Section 6.2, we lay out the basic fractal image model, some of its variants and extensions. In Section 6.3, methods for partitioning the image are discussed. Section 6.4 presents an overview of the many complexity reduction methods for fractal image encoding, whereas Section 6.5 discusses the corresponding complexity reduction in the fractal decoder. Section 6.6 presents coding techniques that aim at overcoming the limitations of the suboptimal algorithm used in practice (collage coding). In Section 6.7, we treat fractal rate-distortion coding whose aim is to optimally trade off compression vs. image reconstruction fidelity. In Section 6.8, we discuss hybrid methods using fractal transforms to improve other encoding methods and vice versa, error protection for the transmission of fractal codes over noisy channels, progressive fractal codes in which the order of transmitted parameters needs to be organized to maximize reconstruction quality at intermediate points of the bit stream, postprocessing to reduce disturbing blocking artefacts, fractal video coding, and briefly mention various applications. Section 6.9 compares the rate-distortion performance of a state-of-the-art fractal coder in the spatial domain to one of the best wavelet coders [79].

6.2 THE FRACTAL IMAGE MODEL

In this section, we introduce our terminology and explain the principles of fractal image compression. We try to provide a consistent and rigorous description. Sometimes this was not possible without formalism.

Let $\mathcal{F}_d$ be the set of monochrome digital images of size $M \times N, M, N \geq 1$, which we see as functions

$$f : \mathcal{X} = \{0, \ldots, M-1\} \times \{0, \ldots, N-1\} \to \{0, \ldots, N_q - 1\} \subset \mathbb{R} \qquad (6.1)$$

Thus, the gray value, or intensity, of pixel $(i, j) \in \mathcal{X}$ is $f(i, j)$, and f can be stored by using $MN\lceil \log_2 N_q \rceil$ bits. We also introduce the space $\mathcal{F}$ of functions $f : \mathcal{X} \to \mathbb{R}$, which is a Banach space over $\mathbb{R}$ for any vector norm. We will often consider $f|B$, the restriction of f to a nonempty subset B of $\mathcal{X}$. In this context, $\mathbf{x}_{f|B}$ will denote the column vector formed by stacking the pixel intensities of B, row by row, left to right, and top to bottom. For example, if $B \subset \mathcal{X}$ is an $n \times n$ square block, that is, $B = \{i, i+1, \ldots, i+n-1\} \times \{j, j+1, \ldots, j+n-1\}$, then

$$\begin{aligned}
\mathbf{x}_{f|B} = {}& (f(i,j), f(i, j+1), \ldots, f(i, j+n-1), \\
& f(i+1, j), f(i+1, j+1), \ldots, f(i+1, j+n-1), \ldots, \\
& f(i+n-1, j), f(i+n-1, j+1), \ldots, f(i+n-1, j+n-1))^{\mathrm{T}}
\end{aligned}$$

Definition 6.2.1 (Contraction).

An image operator $T : \mathcal{F} \to \mathcal{F}$ is said to be contractive (or a contraction) for the vector norm $\| \cdot \|$ if there exists $s \in (0, 1)$ such that for all $f, g \in \mathcal{F}$,

$$\|T(f) - T(g)\| \leq s\|f - g\| \qquad (6.2)$$

Definition 6.2.2 (Fractal transform).

Let $T : \mathcal{F} \to \mathcal{F}$ be an image operator. If there exists a vector norm for which T is contractive, then T is called a fractal transform.

Definition 6.2.3 (Fractal encoder).

Let $\mathcal{T}$ be a set of fractal transforms. A fractal encoder is a function from $\mathcal{F}_d$ to $\mathcal{T}$.

Since $\mathcal{F}$ is a Banach space, the contraction mapping principle states that a fractal transform T has a unique fixed point $f_T \in \mathcal{F}$, which can be obtained as the limit of the sequence $\{f_k\}_{k \geq 0}$, where $f_{k+1} = T(f_k)$, and $f_0 \in \mathcal{F}_d$ is an arbitrary starting image.

Definition 6.2.4 (Fractal decoder).

Let $\mathcal{T}$ be a set of fractal transforms. A fractal decoder is a function from $\mathcal{T}$ to $\mathcal{F}$, which maps $T \in \mathcal{T}$ to its fixed point f_T.

We ask now how to construct a set $\mathcal{T}$ of fractal transforms, a fractal encoder $\mathcal{F}_d \to \mathcal{T}$, and a mapping $\omega : \mathcal{T} \to \{0, 1\}^*$, which associates to a fractal transform T a binary codeword $\omega(T)$ (*fractal code*), such that for any image $f^* \in \mathcal{F}_d$, the fractal encoder can quickly find a fractal transform $T \in \mathcal{T}$ satisfying the following two conditions:

1. the fixed point of T is close to f^*.
2. $|\omega(T)|$, the length of $\omega(T)$, is smaller than $MN\lceil \log_2 N_q \rceil$.

Jacquin [53] and Barnsley [4] gave an answer to this problem by showing experimentally that a special class of fractal transforms can meet the above requirements for natural images. In the following, we describe the Jacquin–Barnsley approach. For simplicity, we assume that $M = N = 2^k$, $k \geq 1$.

- Let $\mathcal{R} = \{R_1, \ldots, R_{n_R}\}$ be a partition of $\mathcal{X}$ into $2^n \times 2^n$ square blocks called *ranges*. Thus, $\mathcal{X} = \bigcup_{i=1}^{n_R} R_i$ and $R_i \cap R_j = \emptyset$, $i \neq j$.
- Let $\mathcal{D} = \{D_1, \ldots, D_{n_D}\}$ be a set of $2^{n+1} \times 2^{n+1}$ square blocks $D_i \subset \mathcal{X}$ called *domains*.
- Let $\mathcal{S} = \{s_1, \ldots, s_{n_s}\}$ be a set of real numbers called *scaling factors*.
- Let $\mathcal{O} = \{o_1, \ldots, o_{n_o}\}$ be a set of real numbers called *offsets*.
- Let $\mathcal{P} = \{\mathbf{P}_1, \ldots, \mathbf{P}_{n_p}\}$ be a set of permutation matrices of order $2^{2(n+1)}$.

Then the partition $\mathcal{R}$ together with any n_R-tuple

$$((D_T(1), s_T(1), o_T(1), \mathbf{P}_T(1)), \ldots, (D_T(n_R), s_T(n_R), o_T(n_R), \mathbf{P}_T(n_R))) \tag{6.3}$$

where $(D_T(i), s_T(i), o_T(i), \mathbf{P}_T(i)) \in \Pi = \mathcal{D} \times \mathcal{S} \times \mathcal{O} \times \mathcal{P}$ yields an image operator T called *Jacquin–Barnsley operator*, defined piecewise by

$$\mathbf{x}_{T(f)|R_i} = s_T(i)\hat{\mathbf{D}}\mathbf{P}_T(i)\mathbf{x}_{f|D_T(i)} + o_T(i)\mathbf{1}, \quad i = 1, \ldots, n_R \tag{6.4}$$

Here $\mathbf{1} = (1, \ldots, 1)^T \in \mathbb{R}^{2^{2n}}$, and $\hat{\mathbf{D}}$ is the $2^{2n} \times 2^{2(n+1)}$ *downsampling matrix*

$$\hat{\mathbf{D}} = \frac{1}{4} \begin{pmatrix} \mathbf{Q} & \mathbf{Q} & 0 & 0 & \ldots & 0 & 0 \\ 0 & 0 & \mathbf{Q} & \mathbf{Q} & 0 & \ldots & 0 \\ \vdots & & & & & & \vdots \\ 0 & \ldots & 0 & \ldots & 0 & \mathbf{Q} & \mathbf{Q} \end{pmatrix} \tag{6.5}$$

where $\mathbf{Q}$ is the $2^n \times 2^{n+1}$ submatrix

$$\mathbf{Q} = \begin{pmatrix} 1 & 1 & 0 & 0 & \ldots & 0 & 0 & 0 & 0 \\ 0 & 0 & 1 & 1 & 0 & \ldots & 0 & 0 & 0 \\ \vdots & & & & & & & \vdots \\ 0 & \ldots & 0 & 0 & 0 & \ldots & 0 & 1 & 1 \end{pmatrix} \tag{6.6}$$

Equality (6.4) has a simple interpretation. The permutation matrix $\mathbf{P}_T(i)$ shuffles the intensities of the pixels in the domain $D_T(i)$, the matrix $\hat{\mathbf{D}}$ maps the vector $\mathbf{P}_T(i)\mathbf{x}_{f|D_T(i)} \in \mathbb{R}^{2^{2(n+1)}}$ to a vector of dimension 2^{2n} by averaging the new intensities of pairwise disjoint groups of four neighboring pixels of $D_T(i)$. Finally, the scaling factor $s_T(i)$ and the offset $o_T(i)$ change the contrast and the brightness, respectively. Note that usually only eight permutation matrices corresponding to the eight isometries of the square are used [30].

The following proposition gives a sufficient condition for a Jacquin–Barnsley operator to be a contraction.

Proposition 6.2.5.

Let T be a Jacquin–Barnsley operator. If $\max_{1 \leq i \leq n_R} |s_T(i)| < 1$, then T is a fractal transform.

Proof

Let $f, g \in \mathcal{F}$. Then

$$
\begin{aligned}
\|T(f) - T(g)\|_\infty &= \max_{1 \leq i \leq n_R} \|\mathbf{x}_{T(f)|R_i} - \mathbf{x}_{T(g)|R_i}\|_\infty \\
&= \max_{1 \leq i \leq n_R} \|s_T(i)\hat{\mathbf{D}}\mathbf{P}_T(i)(\mathbf{x}_{f|D_T(i)} - \mathbf{x}_{g|D_T(i)})\|_\infty \\
&\leq \left(\max_{1 \leq i \leq n_R} |s_T(i)| \right) \\
&\qquad \max_{1 \leq i \leq n_R} \|\hat{\mathbf{D}}\|_\infty \|\mathbf{P}_T(i)\|_\infty \|\mathbf{x}_{f|D_T(i)} - \mathbf{x}_{g|D_T(i)}\|_\infty \\
&= \left(\max_{1 \leq i \leq n_R} |s_T(i)| \right) \max_{1 \leq i \leq n_R} \|\mathbf{x}_{f|D_T(i)} - \mathbf{x}_{g|D_T(i)}\|_\infty \\
&\leq \left(\max_{1 \leq i \leq n_R} |s_T(i)| \right) \|f - g\|_\infty
\end{aligned}
$$

which shows that T is a contraction for the l_∞ norm.

Let T be a Jacquin–Barnsley operator given by a partition $\mathcal{R}$ and the *code parameters* $D_T(i), s_T(i), o_T(i), \mathbf{P}_T(i)$, $i = 1, \ldots, n_R$. Then we say that R_i is *encoded* by $(D_T(i), s_T(i), o_T(i), \mathbf{P}_T(i))$.

Let $\mathcal{T} = \mathcal{T}(\mathcal{R}, \mathcal{D}, \mathcal{S}, \mathcal{O}, \mathcal{P})$ be a set of contractive Jacquin–Barnsley operators and suppose that we assign to each $T \in \mathcal{T}$ a codeword $\omega(T)$ with a fixed length. Given an original image $f^* \in \mathcal{F}_d$, a fractal encoder will be optimal if it finds a fractal transform $T_{\text{opt}} \in \mathcal{T}$ that solves the minimization problem

$$
\min_{T \in \mathcal{T}} \Delta(f^*, f_T) \tag{6.7}
$$

where $\Delta(f^*, f_T) \geq 0$ is the reconstruction error, typically $\Delta(f^*, f_T) = \|f^* - f_T\|_2^2$. Since there are $(n_D n_s n_o n_P)^{n_R}$ solutions, finding T_{opt} by enumeration is impractical for large n_R. Usually, a suboptimal solution is found by a greedy algorithm known as *collage coding* [53]. The idea consists of minimizing the *collage error* $\|f^* - T(f^*)\|_2^2$ instead of the reconstruction error $\|f^* - f_T\|_2^2$. The motivation for collage coding is the following theorem.

Theorem 6.2.6 (Collage Theorem).

Suppose that the fractal transform T is a contraction for the Euclidean norm with contraction factor $c(T) \in (0, 1)$. Then

$$
\|f^* - f_T\|_2 \leq \frac{1}{1 - c(T)} \|f^* - T(f^*)\|_2 \tag{6.8}
$$

Proof

$$
\begin{aligned}
\|f^* - f_T\|_2 &\leq \|f^* - T(f^*)\|_2 + \|T(f^*) - f_T\|_2 \\
&\leq \|f^* - T(f^*)\|_2 + c(T)\|f^* - f_T\|_2
\end{aligned}
$$

which gives the desired result.

The theorem says that if $c(T)$ is not too close to 1, then a small collage error yields a small reconstruction error. It is easier to minimize the collage error because

$$\|f^* - T(f^*)\|_2^2 = \sum_{i=1}^{n_R} \|\mathbf{x}_{f^*|R_i} - \mathbf{x}_{T(f^*)|R_i}\|_2^2$$

$$= \sum_{i=1}^{n_R} \|\mathbf{x}_{f^*|R_i} - (s_T(i)\hat{\mathbf{D}}\mathbf{P}_T(i)\mathbf{x}_{f^*|D_T(i)} + o_T(i)\mathbf{1})\|_2^2$$

Hence collage coding reduces the minimization of the collage error to solving the n_R independent minimization problems

$$\min_{(D,s,o,\mathbf{P})\in\Pi} \|\mathbf{x}_{f^*|R_i} - (s\hat{\mathbf{D}}\mathbf{P}\mathbf{x}_{f^*|D} + o\mathbf{1})\|_2^2, \quad i = 1, \ldots, n_R \tag{6.9}$$

Each of these minimization problems can be efficiently solved as follows. For a given $(D, \mathbf{P}) \in \mathcal{D} \times \mathcal{P}$, let s and o denote the solutions of the least squares problem

$$\min_{s,o\in\mathbb{R}} \|\mathbf{x}_{f^*|R_i} - (s\hat{\mathbf{D}}\mathbf{P}\mathbf{x}_{f^*|D} + o\mathbf{1})\|_2^2 \tag{6.10}$$

If we denote the vector $\mathbf{x}_{f^*|R_i}$ by $\mathbf{r}_i$ and the vector $\hat{\mathbf{D}}\mathbf{P}\mathbf{x}_{f^*|D}$ by $\mathbf{c}$, then the least-squares solution is given by

$$s = \frac{2^{2n}\langle\mathbf{c}, \mathbf{r}_i\rangle - \langle\mathbf{c}, \mathbf{1}\rangle\langle\mathbf{r}_i, \mathbf{1}\rangle}{2^{2n}\langle\mathbf{c}, \mathbf{c}\rangle - \langle\mathbf{c}, \mathbf{1}\rangle^2} \tag{6.11}$$

and

$$o = \frac{1}{2^{2n}}[\langle\mathbf{r}_i, \mathbf{1}\rangle - s\langle\mathbf{c}, \mathbf{1}\rangle] \tag{6.12}$$

if $\mathbf{x}_{f^*|D}$ is not in the linear span of $\mathbf{1}$. Otherwise, $s = 0$ and $o = (1/2^{2n})\langle\mathbf{r}_i, \mathbf{1}\rangle$. Here $\langle,\rangle$ denotes the standard scalar product in $\mathbb{R}^{2^{2n}}$. Note that $(1/2^{2n})\langle\mathbf{r}_i, \mathbf{1}\rangle$ is the average of the pixel intensities in the set R_i and will be denoted by $\mu(R_i)$.

Next, we quantize s and o to their nearest neighbors in $\mathcal{S}$ and $\mathcal{O}$, respectively, obtaining a scaling factor s^* and an offset o^*. Finally, we select a pair $(D, \mathbf{P})$ in $\mathcal{D} \times \mathcal{P}$ that minimizes the *local collage error* $\|\mathbf{x}_{f^*|R_i} - (s^*\hat{\mathbf{D}}\mathbf{P}\mathbf{x}_{f^*|D} + o^*\mathbf{1})\|_2^2$, which is equal to

$$\langle\mathbf{c}, \mathbf{c}\rangle(s^*)^2 + 2\langle\mathbf{c}, \mathbf{1}\rangle s^*o^* + 2^{2n}(o^*)^2 - 2\langle\mathbf{r}_i, \mathbf{c}\rangle s^* - 2\langle\mathbf{r}_i, \mathbf{1}\rangle + \langle\mathbf{r}_i, \mathbf{r}_i\rangle \tag{6.13}$$

Whereas the Collage theorem gives an upper bound on the reconstruction error, the following result gives a lower bound on the same error.

Proposition 6.2.7 (Anti-Collage Theorem) [92].

Suppose that the fractal transform T is a contraction for the Euclidean norm with contractivity factor $c(T) \in [0, 1)$. Then

$$\|f^* - f_T\|_2 \geq \frac{1}{1 + c(T)}\|f^* - T(f^*)\|_2 \tag{6.14}$$

Proof [92]

$$\|f^* - T(f^*)\|_2 \leq \|f^* - f_T\|_2 + \|T(f^*) - f_T\|_2$$
$$\leq \|f^* - f_T\|_2 + c(T)\|f^* - f_T\|_2$$

which gives the desired result.

The above model can be extended by allowing ranges and domains to have arbitrary size and shape. For example, the ranges may be squares of various sizes, given by a quadtree partition. They may also be rectangles, triangles, or unions of edge-connected square blocks (see Section 6.3). Also, the ratio of the domain size to the range size may be other than 4.

A generalized Jacquin–Barnsley operator $T : \mathcal{F} \rightarrow \mathcal{F}$ is characterized by $\mathcal{R} = \{R_1, \ldots, R_{n_R}\}$, a partition of $\mathcal{X}$ into sets (ranges) of size $|R_i| = n_i$, and for each $i = 1, \ldots, n_R$:

- a domain $D_T(i) \in \mathcal{D} = \{D_1, \ldots, D_{n_D}\} \subset 2^{\mathcal{X}}$, with $|D_T(i)| = m_i n_i$, $m_i \geq 1$,
- a scaling factor $s_T(i) \in \mathcal{S} = \{s_1, \ldots, s_{n_s}\} \subset \mathbb{R}$,
- an offset $o_T(i) \in \mathcal{O} = \{o_1, \ldots, o_{n_o}\} \subset \mathbb{R}$,
- and a permutation matrix $\mathbf{P}_T(i)$ of order $m_i n_i$,

which define T through

$$\mathbf{x}_{T(f)|R_i} = s_T(i)\hat{\mathbf{E}}(i)\mathbf{P}_T(i)\mathbf{x}_{f|D_T(i)} + o_T(i)\mathbf{1} \tag{6.15}$$

where $\hat{\mathbf{E}}(i)$ is an $n_i \times m_i n_i$ matrix. The most common choice for $\hat{\mathbf{E}}(i)$ is the averaging matrix

$$\frac{1}{m_i} \begin{pmatrix} \mathbf{1} & \mathbf{0} & \ldots & \mathbf{0} \\ \mathbf{0} & \mathbf{1} & \ldots & \mathbf{0} \\ \vdots & \vdots & \ddots & \vdots \\ \mathbf{0} & \mathbf{0} & \ldots & \mathbf{1} \end{pmatrix} \tag{6.16}$$

where $\mathbf{1} = (1 \ldots 1)$ and $\mathbf{0} = (0 \ldots 0)$ are $1 \times m_i$ submatrices. We now give a sufficient condition for a generalized Jacquin–Barnsley operator to be a fractal transform.

Proposition 6.2.8.

Let T be a generalized Jacquin–Barnsley operator. Suppose that $|s_T(i)|\|\hat{\mathbf{E}}(i)\|_\infty < 1$ for all $i = 1, \ldots, n_R$. Then T is a fractal transform.

Proof

The proof is similar to that of Proposition 6.2.

If $\hat{\mathbf{E}}(i)$ is the averaging matrix, then $\|\hat{\mathbf{E}}(i)\|_\infty = 1$ and the generalized Jacquin–Barnsley operator is a fractal transform when $|s_T(i)| < 1$ for all $i = 1, \ldots, n_R$.

We conclude this section by defining the *dependence graph* of a Jacquin–Barnsley operator [27], which will be needed later.

Definition 6.2.9 (Dependence graph).

Let $T \in \mathcal{T}$ be a Jacquin–Barnsley operator with set of ranges $\mathcal{R}$. A range $R_j \in \mathcal{R}$ is called a child of a range $R_i \in \mathcal{R}$ if the domain that encodes R_j overlaps R_i. The dependence graph of T is a directed graph $\mathcal{G}(T) = (V, E)$, where $V = \mathcal{R}$ and $(R_i, R_j) \in E$ if R_j is a child of R_i.

6.2.1 VARIANTS

For simplicity, the description in this section is restricted to the Jacquin–Barnsley operators given by (6.4).

Øien [72] proposed to modify the Jacquin–Barnsley operator such that

$$\mathbf{x}_{T(f)|R_i} = s_T(i)\hat{\mathbf{D}}\mathbf{P}_T(i)(\mathbf{x}_{f|D_T(i)} - \mu(D_T(i))\mathbf{1}) + o_T(i)\mathbf{1} \tag{6.17}$$

With this setting, the least-squares optimal scaling factor is unchanged, while the optimal offset is the mean of the range block. Moreover, one can show that under certain conditions on the range partition and the domains, fast decoding can be achieved (see Section 6.5).

An extension of the Jacquin–Barnsley operator consists of including more fixed vectors [67,72]. This corresponds to

$$\mathbf{x}_{T(f)|R_i} = s_T(i)\hat{\mathbf{D}}\mathbf{P}_T(i)\mathbf{x}_{f|D_T(i)} + \sum_{k=1}^{p} o_{T,k}(i)\mathbf{u}_k \tag{6.18}$$

where $\mathbf{u}_1, \ldots, \mathbf{u}_p$ ($p < 2^{2n}$) is an orthonormal basis of a p-dimensional subspace of $\mathbb{R}^{2^{2n}}$. For a fixed domain D and permutation matrix $\mathbf{P}$, the encoding problem can then be stated as the least-squares problem

$$E(\mathbf{c}, \mathbf{r}_i) = \min_{s, o_1, \ldots, o_p \in \mathbb{R}} \|\mathbf{r}_i - (s\mathbf{c} + \sum_{k=1}^{p} o_k \mathbf{u}_k)\|_2 = \min_{\mathbf{x} \in \mathbb{R}^{p+1}} \|\mathbf{r}_i - \mathbf{A}\mathbf{x}\|_2 \tag{6.19}$$

where the codebook vector $\mathbf{c}$ is equal to $\hat{\mathbf{D}}\mathbf{P}\mathbf{x}_{f^*|D}$, $\mathbf{x}_{f^*|R_i}$ is denoted by $\mathbf{r}_i$, $\mathbf{A}$ is a $2^{2n} \times (p+1)$ matrix whose columns are $\mathbf{c}, \mathbf{u}_1, \ldots, \mathbf{u}_p$, and $\mathbf{x} = (s, o_1, \ldots, o_p) \in \mathbb{R}^{p+1}$ is a vector of coefficients. If $\mathbf{c}$ is not in the linear span of $\mathbf{u}_1, \ldots, \mathbf{u}_p$, then the minimization problem (6.19) has the unique solution

$$\bar{\mathbf{x}} = (\mathbf{A}^T \mathbf{A})^{-1} \mathbf{A}^T \mathbf{r}_i \tag{6.20}$$

where the matrix $\mathbf{A}^+ = (\mathbf{A}^T \mathbf{A})^{-1} \mathbf{A}^T$ is also known as the pseudo-inverse of $\mathbf{A}$. Thus, $\mathbf{r}_i$ is approximated by the vector $\mathbf{A}\mathbf{A}^+ \mathbf{r}_i$, where $\mathbf{A}\mathbf{A}^+$ is the orthogonal projection matrix onto the column space of $\mathbf{A}$. Now let P be the orthogonal projection operator which projects $\mathbb{R}^{2^{2n}}$ onto the subspace $\mathcal{B}$ spanned by $\mathbf{u}_1, \ldots, \mathbf{u}_p$. By orthogonality, we have for $\mathbf{r}_i \in \mathbb{R}^{2^{2n}}$

$$P\mathbf{r}_i = \sum_{k=1}^{p} \langle \mathbf{r}_i, \mathbf{u}_k \rangle \mathbf{u}_k \tag{6.21}$$

Then $\mathbf{r}_i$ has a unique orthogonal decomposition $\mathbf{r}_i = O\mathbf{r}_i + P\mathbf{r}_i$ where the operator $O = I - P$ projects onto the orthogonal complement $\mathcal{B}^\perp$. For $\mathbf{z} \in \mathbb{R}^{2^{2n}} - \mathcal{B}$, we define the operator

$$\phi(\mathbf{z}) = \frac{O\mathbf{z}}{\|O\mathbf{z}\|_2} \tag{6.22}$$

Now for $\mathbf{c} \notin \mathcal{B}$, we have

$$\mathbf{A}\mathbf{A}^+ \mathbf{r}_i = \langle \mathbf{r}_i, \phi(\mathbf{c}) \rangle \phi(\mathbf{c}) + \sum_{k=1}^{p} \langle \mathbf{r}_i, \mathbf{u}_k \rangle \mathbf{u}_k \tag{6.23}$$

To get the least-squares error, we use the orthogonality of $\phi(\mathbf{r}_i), \mathbf{u}_1, \ldots, \mathbf{u}_p$ to express $\mathbf{r}_i$ as

$$\mathbf{r}_i = \langle \mathbf{r}_i, \phi(\mathbf{r}_i) \rangle \, \phi(\mathbf{r}_i) + \sum_{k=1}^{p} \langle \mathbf{r}_i, \mathbf{u}_k \rangle \mathbf{u}_k \tag{6.24}$$

We insert the result for $\mathbf{r}_i$ in the first part of $\mathbf{AA}^+\mathbf{r}_i$ in (6.23) and find that

$$\langle \mathbf{r}_i, \phi(\mathbf{c}) \rangle \phi(\mathbf{c}) = \langle \mathbf{r}_i, \phi(\mathbf{r}_i) \rangle \, \langle \phi(\mathbf{c}), \phi(\mathbf{r}_i) \rangle \, \phi(\mathbf{c}) \tag{6.25}$$

Thus

$$\mathbf{AA}^+\mathbf{r}_i = \langle \mathbf{r}_i, \phi(\mathbf{r}_i) \rangle \, \langle \phi(\mathbf{c}), \phi(\mathbf{r}_i) \rangle \, \phi(\mathbf{c}) + \sum_{k=1}^{p} \langle \mathbf{r}_i, \mathbf{u}_k \rangle \mathbf{u}_k \tag{6.26}$$

Using (6.24) and (6.26) we get

$$E(\mathbf{c}, \mathbf{r}_i) = \langle \mathbf{c}, \phi(\mathbf{r}_i) \rangle \sqrt{1 - \langle \phi(\mathbf{c}), \phi(\mathbf{r}_i) \rangle^2} \tag{6.27}$$

Other researchers use in addition to the fixed vectors several domains $D_{T,1}(i), \ldots, D_{T,m}(i)$ [35,91]. Thus, we have

$$\mathbf{x}_{T(f)|R_i} = \sum_{k=1}^{m} s_{T,k}(i)\hat{\mathbf{D}}\mathbf{P}_{T,k}(i)\mathbf{x}_{f|D_{T,k}(i)} + \sum_{k=1}^{p} o_{T,k}(i)\mathbf{u}_k \tag{6.28}$$

However, there is here one major complication because it is not clear how to guarantee the contractivity of T [35,91].

6.3   IMAGE PARTITIONS

There are two problems when the Jacquin–Barnsley operator is based on uniform image partitions. The first one is encoding complexity. For good image reconstruction quality generally small image range block sizes must be chosen. This leads to a large number of range blocks, requiring an equal number of time-expensive searches in the domain pool. The other problem concerns rate-distortion performance. An adaptive partitioning of an image may hold strong advantages over encoding range blocks of fixed size. There may be homogeneous image regions in which a good collage can be attained using large blocks, while in high contrast regions smaller block sizes may be required to arrive at the desired quality. For typical fractal codecs the bulk of the rate of the code of the fractal operator is given by the specification of domain blocks, scaling, and offset factors, thus, is roughly proportional to the number of range blocks. Therefore, adaptivity of the image partition allows to reduce the rate without loss in image reconstruction fidelity.

However, while adaptive partitions allow to reduce the number of range blocks they also require additional code for the specification of the range blocks, i.e., for the partition, which is an integral part of the fractal code. Thus, methods for efficient image partition encoding must be considered. Moreover, the tradeoff between the overhead in rate induced by having to encode adaptive partitions and the achievable gain in rate by reducing the number of range blocks must be considered. These aspects set up the stage for a variety of approaches ranging from simple quadtree to irregular image partitions. There is no underlying theory that would be able to guide the research for the best methods. Therefore, the different image partition methods can only be compared empirically using, e.g., rate-distortion curves.

The use of adaptive partitions makes it easy to design a variable rate fractal encoder. The user may specify goals for either the image quality or the compression ratio. The encoder can recursively break up the image into suitable portions until either criterion is reached.

6.3.1 QUADTREES

The first approach for adaptive partitions (already taken by Jacquin) was to consider square blocks of varying sizes, e.g., being 4, 8, and 16 pixels wide. This idea leads to the general concept of using a quadtree partition, first explored in the context of fractal coding in [10,51].

A quadtree fractal encoding algorithm targeting fidelity proceeds as follows:

1. Define a minimum range block size.
2. Define a tolerance for the root-mean-square error (the collage error divided by the square root of the number of pixels in the range).
3. Initialize a stack of ranges by pushing a collection of large range blocks of size $2^n \times 2^n$ partitioning the entire image onto it.
4. While the stack is not empty carry out the following steps:
    4.1. Pop a range block R from the stack and search the code parameters $(D, s, o, \mathbf{P}) \in \Pi$ yielding the smallest local collage error.
    4.2. If the corresponding root-mean-square error is less than the tolerance or if the range size is less than or equal to the minimum range size, then save the code for the range, i.e., the address of D and the index of s, o, and $\mathbf{P}$.
    4.3. Otherwise partition R into four quadrants and push them onto the stack.

By using different fidelity tolerances for the collage error one obtains a series of encodings of varying compression ratios and fidelities [30]. The quadtree algorithm can be improved in several ways. For example, in the split-decision criterion to break up a range block into four quadrants if the root-mean-square error for the range block is larger than the threshold, one may make the threshold adapt to the range block size, improving the rate-distortion performance [25]. Alternatively, one may save encoding time by not using time-expensive domain pool searches. Then one cannot base the split-decision function on collage error. Instead, variance [84] or entropy [25] may be used with good results.

6.3.2 OTHER HIERARCHICAL PARTITIONS

For quadtree partitions the size of the square blocks is image-adaptive but the block shapes are not. A first step toward adaptive block shapes is given by hierarchical rectangular partitions in which the aspect ratio of the rectangles can be adapted to image content. In *horizontal–vertical (HV) partitioning* [31] the image is segmented into such rectangles. If for a given rectangular range block no acceptable domain match is found, the block is split into two rectangles either by a horizontal or a vertical cut. The choice of the split direction and location is based on block uniformity and incorporates a rectangle degeneration prevention mechanism as follows.

For the range block $R = \{0, \ldots, n-1\} \times \{0, \ldots, m-1\}$, the biased differences of vertical and horizontal pixel sums, respectively, are computed:

$$h_j = \frac{\min(j, m-j-1)}{m-1} \left(\sum_i f(i,j) - \sum_i f(i,j+1) \right) \tag{6.29}$$

$$v_i = \frac{\min(i, n-i-1)}{n-1} \left(\sum_j f(i,j) - \sum_j f(i+1,j) \right) \tag{6.30}$$

For other range blocks corresponding differences are defined similarly. The maximal value of these differences determines splitting direction and position. A decision tree containing this information

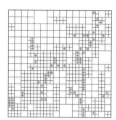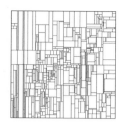

FIGURE 6.1 Hierarchical partitions for fractal image coding: a quadtree, a horizontal–vertical, and a polygonal partition for the *Lenna* image.

has to be stored. In spite of the higher cost for storing the partition information, the simulation results show a considerable rate-distortion improvement over the quadtree scheme.

A further step in adaptivity is *polygonal partitioning* [77] in which additionally cutting directions at angles of 45° and 135° are allowed.

Figure 6.1 shows some examples of hierarchical partitions for fractal image compression.

6.3.3 SPLIT–MERGE PARTITIONS

The flexibility of hierarchical partitions can be increased by allowing neighboring regions to merge. Davoine et al. [19,20] advocate the use of Delaunay triangulations for split–merge partitioning. The advantage of triangulations is the unconstrained orientation of edges. Delaunay triangulations maintain a certain regularity avoiding long and skinny triangles. Starting with a Delaunay triangulation of a set of regularly distributed grid points, the partition is refined by splitting nonuniform triangles (as measured by standard deviation of pixel image intensities). This splitting step is done by adding an additional vertex at the center of gravity of the triangle and by recomputing the Delaunay triangulation for this enlarged set of vertices. The splitting is stopped when the image intensity is sufficiently uniform in each triangle. In the merging pass, vertices p are iteratively removed if all triangles sharing a vertex p have approximately the same mean value. After each removal action the Delaunay triangulation of the reduced set of vertices is computed. Davoine et al. [21] also allow the merging of two triangles when the resulting quadrilateral is convex and both triangles have more or less the same gray value distribution.

Using Delauney triangulations instead of simple rectangles causes a number of new technical issues. The triangles need to be scan-converted, and an appropriate pool of corresponding domain blocks needs to be defined. When mapping triangles to triangles, the corresponding downsampling matrices $\hat{\mathbf{D}}$ must be generated, which depend on the shapes of the triangles.

In region-based fractal coding, the shapes of range blocks can be defined arbitrarily allowing for maximal adaptivity of the image partition. The regions (ranges) are usually defined with a split–merge approach, again. First, the image is split in atomic square blocks, e.g., of size 4×4 or just individual pixels in the extreme case. Then neighboring blocks are merged successively to build larger ranges of irregular shapes. Since one ends up with only a few large ranges, there are only a few transformations to store. The irregular shapes of the ranges imply that for each range a different pool of domain blocks (with the same shape) must be defined, which requires a more complex fractal encoder. Different merging strategies may be used.

Thomas and Deravi [89] introduced region-based fractal coding and used heuristic strategies in which a range block is merged iteratively with a neighboring block without changing the corresponding domain block as long as a distortion criterion is satisfied. Then the procedure is repeated for other atomic blocks.

Hartenstein et al. [44] improved the strategy by merging pairs of neighboring range blocks that lead to the least increase in collage error (Figure 6.2). This approach requires maintaining fractal

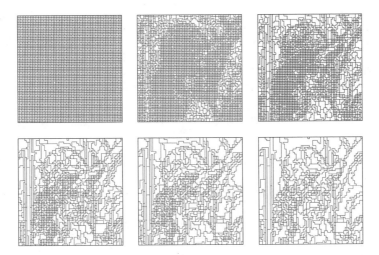

FIGURE 6.2 Split–merge partitions for a region-based fractal coder. Starting from a uniform partition, blocks are iteratively merged based on the least increase of collage error.

codes for all possible merged range pairs. The full search for matching domain blocks for all range pairs is prohibitive in time complexity and, thus, a reduced search is done using only appropriately extended domain blocks that are known for the two constituting range blocks.

In the above two approaches, the atomic blocks are taken as same size square blocks, which are composed in the merging phase to form irregular image regions. The partition needs to be encoded and sent to the decoder as side information. For the encoding of such partitions one may adapt algorithms based on chain coding, which encode a trace of the boundaries of the regions, or one can use region edge maps that encode for each atomic block whether it has a boundary edge along its northern and its western border (see [85] for a comparison). Additional gains can be achieved by using a simple image adaptive partition such as a quadtree partition for the split phase and then merging the blocks as before. The side information for the image partition then consists of two parts, the initial (quadtree) partition and the merging information. A coder based on this approach was proposed in [14] and improved in [71] using context modeling for region edge maps.

6.4 ENCODER COMPLEXITY REDUCTION

Fractal image compression suffers from long encoding times if implemented without acceleration methods. The time-consuming part of the encoding is the search for an appropriate domain for each range. The number of possible domains that theoretically may serve as candidates is prohibitively large. Thus, one must impose certain restrictions in the specification of the allowable domains. In a simple implementation, one might consider as domain pool, e.g., only subsquares of a limited number of sizes and positions. Now for each range in the partition of the original image all elements of the domain pool are inspected. If the number of domains in the pool is n_D, then the time spent for each search is linear in n_D, $O(n_D)$. Several methods have been devised to reduce the time complexity of the encoding. As in Section 6.2.1, we call the domain vectors after pixel shuffling and downsampling codebook vectors $\mathbf{c} = \hat{\mathbf{D}}\mathbf{P}\mathbf{x}_{f^*|D}$ and range blocks are given by vectors $\mathbf{r}_i = \mathbf{x}_{f^*|R_i}$.

6.4.1 FEATURE VECTORS

In the feature vector approach [81], a small set of d real-valued keys is devised for each domain, which make up a d-dimensional feature vector. These keys are constructed such that searching in the domain pool can be restricted to the *nearest neighbors* of a query point, i.e., the feature vector of the current range. Thus, the sequential search in the domain pool is substituted by multidimensional nearest neighbor searching, which can be run in logarithmic time.

We consider a set of n_D codebook vectors $\mathbf{c}_1, \ldots, \mathbf{c}_{n_D} \in \mathbb{R}^m$ and a range vector $\mathbf{r}_i \in \mathbb{R}^m$. We let $E(\mathbf{c}_j, \mathbf{r}_i)$ denote the norm of the smallest error of an approximation of $\mathbf{r}_i$ by an affine transformation of the codebook vector $\mathbf{c}_j$ as in (6.19), i.e., $E(\mathbf{c}_j, \mathbf{r}_i) = \min_{s,o_1\ldots,o_p \in \mathbb{R}} \|\mathbf{r}_i - (s\mathbf{c}_j + \sum_{k=1}^{p} o_k \mathbf{u}_k)\|_2$. Let $\mathcal{B}$ denote the linear span of the fixed basis vectors $\mathbf{u}_1, \ldots, \mathbf{u}_p$. Then, using (6.27), we get the following important theorem.

Theorem 6.4.1 [80].

Let $m \geq 2$ and $X = \mathbb{R}^m - \mathcal{B}$. Define the function $\Delta : X \times X \to [0, \sqrt{2}]$ by

$$\Delta(\mathbf{c}, \mathbf{r}) = \min\left(\|\phi(\mathbf{r}) + \phi(\mathbf{c})\|_2, \|\phi(\mathbf{r}) - \phi(\mathbf{c})\|_2\right) \tag{6.31}$$

where ϕ is as defined in (6.22). For $\mathbf{c}_j, \mathbf{r}_i \in X$ the error $E(\mathbf{c}_j, \mathbf{r}_i)$ is given by

$$E(\mathbf{c}_j, \mathbf{r}_i) = \langle \mathbf{r}_i, \phi(\mathbf{r}_i) \rangle \, g(\Delta(\mathbf{c}_j, \mathbf{r}_i)) \tag{6.32}$$

where

$$g(\Delta) = \Delta\sqrt{1 - \frac{\Delta^2}{4}} \tag{6.33}$$

The theorem states that the least-squares error $E(\mathbf{c}_j, \mathbf{r}_i)$ is proportional to the simple function g of the Euclidean distance Δ between the projections $\phi(\mathbf{c}_j)$ and $\phi(\mathbf{r}_i)$ (or $-\phi(\mathbf{c}_j)$ and $\phi(\mathbf{r}_i)$). Since $g(\Delta)$ is a monotonically increasing function for $0 \leq \Delta \leq \sqrt{2}$, *the minimization of the errors $E(\mathbf{c}_j, \mathbf{r}_i)$ for $j = 1, \ldots, n_D$ is equivalent to the minimization of the distance expressions $\Delta(\mathbf{c}_j, \mathbf{r}_i)$*. Thus, one may replace the computation and minimization of n_D least-squares errors $E(\mathbf{c}_j, \mathbf{r}_i)$ by the search for the nearest neighbor of $\phi(\mathbf{r}_i) \in \mathbb{R}^m$ in the set of $2n_D$ vectors $\pm\phi(\mathbf{c}_j) \in \mathbb{R}^m$.

The problem of finding N closest neighbors in d-dimensional Euclidean spaces has been thoroughly studied in computer science. For example, a method using kd-trees that runs in expected logarithmic time is presented by Friedman et al. [33]. After a preprocessing step to set up the required kd-tree, which takes $O(N \log N)$ steps, the search for the nearest neighbor of a query point can be completed in expected logarithmic time, $O(\log N)$. However, as the dimension d increases, the performance may suffer. A method that is more efficient in that respect, presented by Arya et al. [1], produces a so-called approximate nearest neighbor. For domain pools that are not large, other methods, which are not based on space-partitioning trees, may be better.

In practice, there is a limit in terms of storage for the feature vectors of domains and ranges. To cope with this difficulty, one may down-filter all ranges and domains to some prescribed dimension of moderate size, e.g., $d = 4 \times 4 = 16$. Moreover, each of the d components of a feature vector may be quantized (8 bits/component suffice). This allows the processing of an increased number of domains and ranges, however, with the implication that the formula of the theorem is no longer exact but only approximate.

The feature vectors can be improved by using an image transformation that concentrates relevant subimage information in its components. For example, in [7,82], the 2D discrete cosine transformation (DCT) of the projected codebook vectors $\pm\phi(\mathbf{c}_j)$ was considered. The distance-preserving property of the unitary transform carries over the result of the above theorem to the frequency domain and nearest neighbors of DCT coefficient vectors will yield the smallest least-squares errors. In practice, one computes the DCT for all domains and ranges. Then, from the resulting coefficients, the DC component is ignored and the next d coefficients are normalized and make up the feature vector.

Because of the downfiltering and the quantization of both the feature vectors and the coefficients $s, o_1, \ldots, o_p$, it can happen that the nearest neighbor in feature vector space is not the codebook vector with the minimum least-squares error using quantized coefficients. Moreover, it could yield a scaling factor s whose magnitude is too large to be allowed. To take that into consideration, one searches the codebook not only for the nearest neighbor of the given query point but also for, say, the next 5 or 10 nearest neighbors (this can still be accomplished in logarithmic time using a priority queue). From this set of neighbors the nonadmissible codebook vectors are discarded and the remaining vectors are compared using the ordinary least-squares approach. This also takes care of the problem that the estimate by the theorem is only approximate. While the domain corresponding to the closest point found may not be the optimal one, there are usually near-optimum alternatives among the other candidates.

Several variations and improvements of the approach by feature vectors have been proposed. In [17], the partial distortion elimination technique, well known from vector quantization, is applied. In [90], the quantization of the scaling parameter is done before the search, which multiplies the size of the domain pool, but errors due to the quantization of scaling factors are avoided. Moreover, the authors propose a range-adaptive choice of parameters for the approximate nearest-neighbor search. In [12], the kd-tree in the search for the nearest neighbor in feature vector space is replaced by an adaptive binary space partition, which significantly speeds up the search.

6.4.2 CLASSIFICATION SCHEMES

In classification methods for fractal image compression, domain blocks are classified into a certain number of classes. For a given range block, only domains, resp. corresponding codebook vectors, from the class that the range belongs to will be searched. This speeds up the search. However, a small loss in reconstruction quality may occur since it is not guaranteed that the optimal domain is found for a given range.

The classification as described here applies to the case $p = 1$, where just one fixed basis vector $\mathbf{u} = (1, \ldots, 1)^T/\sqrt{m}$ is used. However, note that the method extends to the general case allowing $p > 1$, provided that certain modifications are made. Essentially, this amounts to considering the transformed domains $\phi(\mathbf{c}_j)$ instead of the original codebook vectors.

6.4.2.1 Jacquin's Approach

Jacquin [53] used a classification scheme in which domain blocks are classified according to their perceptual geometric features. Three major types of blocks are differentiated: shade blocks, edge blocks, and midrange blocks. In shade blocks, the image intensity varies only very little, while in edge blocks a strong change in intensity occurs, e.g., along a boundary of an object displayed in the image. The class of edge blocks is further subdivided into two subclasses: simple and mixed edge blocks. Midrange blocks have larger intensity variations than shade blocks, but there is no pronounced gradient as in an edge block. Thus, these blocks typically are blocks containing texture.

6.4.2.2 Classification by Intensity and Variance

A more elaborate classification technique was proposed by Boss, Fisher, and Jacobs [30,51]. A square range or domain is subdivided into its four quadrants: upper left, upper right, lower left, and lower right. In the quadrants, the average pixel intensities A_i and the corresponding variances V_i are computed $(i = 1, \ldots, 4)$. It is easy to see that one can always orient (rotate and flip) the range or domain such that the average intensities are ordered in one of the following three ways:

$$\text{Major class 1:} \quad A_1 \geq A_2 \geq A_3 \geq A_4,$$
$$\text{Major class 2:} \quad A_1 \geq A_2 \geq A_4 \geq A_3,$$
$$\text{Major class 3:} \quad A_1 \geq A_4 \geq A_2 \geq A_3.$$

Once the orientation of the range or domain is fixed accordingly, there are 24 different possible orderings of the variances which define 24 subclasses for each major class. If the scaling factor s in the approximation $s\mathbf{c} + o\mathbf{1}$ of the range vector $\mathbf{r}_i$ is negative then the orderings in the classes must be modified accordingly. Thus, for a given range, two subclasses out of 72 need to be searched in order to accommodate positive and negative scaling factors.

Although successful, this approach is not satisfactory in the sense that a notion of neighboring classes is not available. So if the search in one class does not yield a sufficiently strong match for a domain, one cannot extend the search to any neighboring classes. A solution to this problem was given by Caso et al. [13], where the inflexible ordering of variances of an image block was replaced by a vector of variances. These variance vectors are strongly quantized leading to a collection of classes such that each class has a neighborhood of classes which can be searched.

6.4.2.3   Clustering Methods

In clustering methods, domains and ranges are partitioned into clusters. The encoding time for a given range is reduced by searching for domains in the same cluster. An efficient clustering method based on the LBG algorithm was proposed in [60], introducing also a block decimation technique to carry out the clustering and the searching in a low-dimensional space. In [39], the nearest-neighbor approach of Saupe [80] was exploited to yield a distance-based clustering technique. After a set of cluster centers is designed either adaptively from the test image or from a set of training images, the projected codebook vectors $\{\pm\phi(\mathbf{c}_1), \ldots, \pm\phi(\mathbf{c}_{n_D})\}$ are clustered by mapping each vector $\pm\phi(\mathbf{c}_i)$ to its nearest cluster center. A range vector $\mathbf{r}_i$ is encoded in two steps. First, one maps its feature vector $\phi(\mathbf{r}_i)$ to its closest cluster center. Then the range vector $\mathbf{r}_i$ is compared only to the codebook vectors $\mathbf{c}_i$ for which $\phi(\mathbf{c}_i)$ or $-\phi(\mathbf{c}_i)$ is in the cluster of the same center. This corresponds to a one-class search. One can search in more classes by considering the next-nearest cluster centers of $\phi(\mathbf{r}_i)$.

6.4.3   TREE-STRUCTURED METHODS

In the tree-structured search [13], the pool of codebook blocks is recursively organized in a binary tree. Initially, two (parent) blocks are chosen randomly from the pool. Then all codebook blocks are sorted into one of two bins depending on by which of the two parent blocks the given block can be covered best in the least-squares sense. This results in a partitioning of the entire pool into two subsets. The procedure is recursively repeated for each one of them until a prescribed bucket size is reached. Given a range, one can then compare this block with the blocks at the nodes of the binary tree until a bucket is encountered at which point all of the codebook blocks in it are checked. This does not necessarily yield the globally best match. However, the best one (or a good approximate solution) can be obtained by extending the search to some nearby buckets. A numerical test based on the angle criterion is given for that purpose. The procedure is related to the nearest-neighbor approach since the least-squares criterion (minimize $E(\mathbf{c}_j, \mathbf{r}_i)$) is equivalent to the distance criterion (minimize $\Delta(\phi(\mathbf{c}_j), \phi(\mathbf{r}_i))$). Thus, the underlying binary tree can be considered to be a randomized version of the kd-tree structure described above.

6.4.4   MULTIRESOLUTION APPROACHES

Multiresolution approaches for encoder complexity reduction in the context of quadtree partitions were introduced by Dekking [22,23] and by Lin and Venetsanopoulos [64]. The idea is to use the gray value pyramid associated with an image to reduce the cost of the search. The search is first done at a low resolution of the image. If no matches can be found at this resolution, then no matches can be found at a finer resolution. The computational savings are due to the fact that fewer computations of the least-squares are needed at a coarser resolution. More precisely, a gray-valued pyramid of a $2^r \times 2^r$ image f is defined as the sequence of $2^k \times 2^k$ images $f^{(0)}, \ldots, f^{(r)}$, where

$f^{(r)} = f$ and

$$f^{(k)}(i,j) = \frac{1}{4} \sum_{m,l=0}^{1} f^{(k+1)}(2i + m, 2j + l) \tag{6.34}$$

for $k = 0, \ldots, r - 1$ and $0 \leq i, j < 2^k$. Similarly, one can obtain range blocks and domain blocks at resolution k from those at resolution $k + 1$. The basic result of Dekking [23] can be stated as follows:

Theorem 6.4.2.

Let $\mathbf{r}^{(k)}$ be the range vector at resolution k obtained from $\mathbf{r}^{(k+1)}$, a range vector at resolution $k + 1$. Let $\mathbf{c}^{(k)}$ be the codebook vector at resolution k obtained from $\mathbf{c}^{(k+1)}$, the codebook vector at resolution $k + 1$. Then $E(\mathbf{c}^{(k+1)}, \mathbf{r}^{(k+1)}) \geq E(\mathbf{c}^{(k)}, \mathbf{r}^{(k)})$.

However, applying the theorem as stated above will take into consideration only domains of resolution $k + 1$ whose corners are at positions $(2i, 2j)$, since not all domains at resolution $k + 1$ have corresponding domains at resolution k. To circumvent this problem, one may consider a pyramid tree, where every resolution $k + 1$ domain has four resolution k domain children [23]. Note that one cannot discard a $k + 1$ resolution domain simply because its k resolution child has a scaling factor s_k such that $|s_k| > 1$.

6.4.5 FAST SEARCH VIA FAST CONVOLUTION

Most of the techniques discussed above are *lossy* in the sense that they trade in a speedup for some loss in image fidelity. In contrast, with a *lossless* method the codebook vector with the smallest local collage error is obtained rather than an acceptable but suboptimal one. The method presented here takes advantage of the fact that the codebook blocks, taken from the image, are usually overlapping. The fast convolution — based on the convolution theorem and carried out in the frequency domain — is ideally suited to exploit this sort of codebook coherence. The essential part of the basic computation in fractal image compression is a certain convolution [45]. For a range vector $\mathbf{r}$ and codebook vector $\mathbf{c}$ the optimal coefficients are

$$s = \frac{m\langle \mathbf{c}, \mathbf{r} \rangle - \langle \mathbf{c}, \mathbf{1} \rangle \langle \mathbf{r}, \mathbf{1} \rangle}{m\langle \mathbf{c}, \mathbf{c} \rangle - \langle \mathbf{c}, \mathbf{1} \rangle^2}, \qquad o = \frac{1}{m} \left(\langle \mathbf{r}, \mathbf{1} \rangle - s\langle \mathbf{c}, \mathbf{1} \rangle \right) \tag{6.35}$$

where m is the number of pixels in the blocks (compare (6.11) and (6.12)). For any (s, o) the error $E(\mathbf{c}, \mathbf{r})$ can be regarded as a function of $\langle \mathbf{c}, \mathbf{r} \rangle$, $\langle \mathbf{c}, \mathbf{c} \rangle$, $\langle \mathbf{c}, \mathbf{1} \rangle$, $\langle \mathbf{r}, \mathbf{r} \rangle$, and $\langle \mathbf{r}, \mathbf{1} \rangle$. Its evaluation requires 23 floating-point operations. Typically, the computations are organized into two nested loops:

- Global preprocessing: compute $\langle \mathbf{c}, \mathbf{c} \rangle$, $\langle \mathbf{c}, \mathbf{1} \rangle$ for all codebook blocks $\mathbf{c}$.
- For each range $\mathbf{r}$ do:
  - Local preprocessing: compute $\langle \mathbf{r}, \mathbf{r} \rangle$, $\langle \mathbf{r}, \mathbf{1} \rangle$.
  - For all codebook blocks $\mathbf{c}$ do:
    - Compute $\langle \mathbf{c}, \mathbf{r} \rangle$ and $E(\mathbf{c}, \mathbf{r})$.

The calculation of the inner products $\langle \mathbf{c}, \mathbf{r} \rangle$ dominates the computational cost in the encoding. The codebook vectors $\mathbf{c}$ are typically defined by downfiltering the image to half its resolution. Any subblock in the downfiltered image, that has the same shape as the range, is a codebook block for that range. In this setting, the inner products $\langle \mathbf{c}, \mathbf{r} \rangle$ are essentially the finite impulse response (FIR) of the downfiltered image with respect to the range. In other words, the convolution (or, more precisely, the cross-correlation) of the range $\mathbf{r}$ with the downfiltered image is required. This discrete two-dimensional convolution can be carried out more efficiently in the frequency domain when the range block is not too small (convolution theorem). This procedure takes the inner product calculation out of the inner loop and places it into the local preprocessing where the inner products $\langle c, r \rangle$ for *all* codebook vectors c are obtained in one batch by means of fast Fourier transform convolution. Clearly, the method is lossless.

Moreover, the global preprocessing requires a substantial amount of time, but can be accelerated by the same convolution technique. The products $\langle c, 1 \rangle$ are obtained by convolution of the downfiltered image with a range block where all intensities are set equal. The sum of the squares is computed in the same way where all intensities in the downfiltered image are squared before the convolution.

6.4.6 FRACTAL IMAGE COMPRESSION WITHOUT SEARCHING

Complexity reduction methods that are somewhat different in character are based on reducing the domain pool rigorously to a small subset of all possible domains. For example, for each range the codebook block to be used to cover the range is uniquely predetermined to be a specific block that contains the range block [68]. A similar idea was proposed by Hürtgen and Stiller [48]. Here the search area for a domain is restricted to a neighborhood of the current range or additionally a few sparsely spaced domains far from the range are taken into account as an option. Saupe [83] considered a parameterized and nonadaptive version of domain pool reduction by allowing an adjustable number of domains to be excluded (ranging from 0 to almost 100%) and studied the effects on computation time, image fidelity, and compression ratio. He showed that there is no need for keeping domains with low-intensity variance in the pool.

Signes [87] and Kominek [57] pursue similar ideas for domain pool reduction. An adaptive version of spatial search based on optimizing the rate-distortion performance is presented by Barthel [7].

6.5 DECODER COMPLEXITY REDUCTION

6.5.1 FAST DECODING WITH ORTHOGONALIZATION

When Øien's operator (see Section 6.2.1) is used with a quadtree partition and a domain pool where each domain block is a union of range blocks, the decoding has the following nice properties [72]:

- It converges into a finite number of iterations without any constraints on the scaling factors; this number depends only on the domain and range sizes.
- It converges at least as fast as the conventional scheme.
- It can be organized as a pyramid-structured algorithm with a low computational complexity.

6.5.2 HIERARCHICAL DECODING

Baharav et al. [2] proposed a fast decoding algorithm based on the hierarchical properties of the Jacquin–Barnsley operator. The idea is to carry out the iterations at a coarse resolution of the image. Once the fixed point of the operator at the coarse resolution is reached, a deterministic algorithm is used to find the fixed point at any higher resolution. The savings in computation are due to the fact that the iterations are applied to a vector of low dimension. A similar approach was proposed by Fisher and Menlove [31]. Looking at the image pyramid obtained by successive averaging of the original image (see Section 6.4.4), they observe that to each range at the real size (parent range) corresponds a range at a lower resolution (child range). The same observation holds for the domains if they are unions of ranges. Now, it is reasonable to expect that each range at a low resolution can be encoded well by the child of the domain that encodes the parent of the range, together with the same isometry, scaling factor, and offset. Fisher and Menlove [31] first decode the low-resolution image.

This is not computationally expensive because the ranges and domains have a small size. Once the limit point is found, one can upsample it to the real size, getting hopefully a suitable approximation of the real size fixed point. Finally, this approximation is used as a starting vector for the iterations at the real size. Fisher and Menlove report that convergence at the real size was achieved in all cases tested after only two iterations.

6.5.3 CODEBOOK UPDATE

This method works in the spirit of the Gauss–Seidel method. Suppose that the generalized Jacquin–Barnsley operator T is a fractal transform. Then $\mathbf{x}_{f_T}$ can be obtained as the limit of the sequence of iterates $\{\mathbf{x}^{(k)}\}_k$, where $\mathbf{x}^{(0)}$ is any initial vector and $\mathbf{x}^{(k)}$ is defined by

$$\mathbf{x}^{(k+1)} = \mathbf{A}\mathbf{x}^{(k)} + \mathbf{b} \tag{6.36}$$

Here $\mathbf{A}$ is an $MN \times MN$ real matrix and $\mathbf{b}$ is an MN column vector specified by (6.15). Thus, the conventional decoding method (6.36) can be written as

$$x_u^{(k+1)} = \sum_{v=1}^{NM} a_{u,v} x_v^{(k)} + b_u, \quad u = 1, \ldots, NM \tag{6.37}$$

Some studies, e.g., [40,54], proposed to use the new components (pixel intensities) $x_1^{(k+1)}, \ldots, x_{u-1}^{(k+1)}$ instead of the old components $x_1^{(k)}, \ldots, x_{u-1}^{(k)}$ for the computation of $x_u^{(k+1)}$. In this way, we have for $u = 1, \ldots, NM$,

$$x_u^{(k+1)} = \sum_{v \leq u-1} a_{u,v} x_v^{(k+1)} + \sum_{v \geq u} a_{u,v} x_v^{(k)} + b_u \tag{6.38}$$

One straightforward advantage of (6.38) over (6.37) is that it has less storage requirements. Indeed, in (6.38), we do not need the simultaneous storage of $\mathbf{x}^{(k)}$ and $\mathbf{x}^{(k+1)}$ as in conventional decoding.

We now give a matrix form of (6.38), which will be useful in the analysis of the convergence properties. Let the matrix $\mathbf{A}$ be expressed as the matrix sum

$$\mathbf{A} = \mathbf{L} + \mathbf{U} \tag{6.39}$$

where the matrix $\mathbf{L} = (l_{u,v})$ is strictly lower triangular defined by

$$l_{u,v} = \begin{cases} 0, & \text{if } v \geq u \\ a_{u,v}, & \text{otherwise} \end{cases} \tag{6.40}$$

Then (6.38) corresponds to the iterative method

$$\mathbf{x}^{(k+1)} = \mathbf{L}\mathbf{x}^{(k+1)} + \mathbf{U}\mathbf{x}^{(k)} + \mathbf{b} \tag{6.41}$$

or equivalently

$$\mathbf{x}^{(k+1)} = (\mathbf{I} - \mathbf{L})^{-1}\mathbf{U}\mathbf{x}^{(k)} + (\mathbf{I} - \mathbf{L})^{-1}\mathbf{b} \tag{6.42}$$

We now have [40]

Theorem 6.5.1.

If $|s_T(i)| < 1$ for all $1 \leq i \leq n_R$, then the iterative method (6.42) is convergent. Moreover, method (6.42) and the conventional method (6.36) have the same limit vector and $||(\mathbf{I} - \mathbf{L})^{-1}\mathbf{U}||_\infty \leq ||\mathbf{A}||_\infty$.

Furthermore, experiments show that the convergence of decoding (6.42) is faster than that of the conventional method (6.36).

With the conventional method (6.37), changing the order in which the pixel intensities are computed has no effect on the image iterates $\{\mathbf{x}^{(k)}\}_k$. This is not true for (6.38). In [40], it is suggested to apply the method, starting with the pixels in the ranges with highest frequency. The frequency of a subset of $\mathcal{X}$ is defined as follows.

Definition 6.5.2.

Let $D_T(1), \ldots, D_T(n_R)$ be the domains associated to the Jacquin–Barnsley operator T. Let B be a nonempty subset of $\mathcal{X}$. Then the frequency of B is

$$\frac{\sum_{i=1}^{n_R} |B \bigcap D_T(i)|}{|B|} \tag{6.43}$$

The expected gain with the new ordering is twofold. First, this strategy increases the number of pixels that are used in their updated form, yielding, hopefully, a faster convergence. Second, the decoder can identify the ranges that are not covered by the domains, that is, ranges with zero frequency. The pixels' intensities in these ranges need only be computed once during the decoding. Note that the additional step needed for the computation and the sorting of the range frequencies is negligible because it is done once.

6.5.4 OTHER METHODS

The fractal coder of Monro and Dudbridge [68] has a fast noniterative decoder, giving an exact reconstruction of the fixed point. In [27], the dependence graph of the fractal code was exploited to reconstruct some of the range blocks in a noniterative way.

6.6 ATTRACTOR CODING

Since collage coding provides only a suboptimal solution to optimal fractal coding (see (6.7)), it is interesting to ask if better solutions can be found. Using a polynomial reduction from the MAXCUT problem, Ruhl and Hartenstein [78] showed that optimal fractal coding is NP-hard. Thus, it is unlikely that an efficient algorithm can be found for problem (6.7). However, collage coding can be improved [26,28,47,92]. The idea is to start from the solution found by collage coding, fix the domain blocks and the permutations, and optimize the scaling factors and the offsets (considered as continuous variables) by, for example, gradient descent methods, which yield local minima of the reconstruction error. However, after quantization, the PSNR improvement over collage coding is negligible for practical encodings [92]. Moreover, the time complexity of the optimization is too high. In contrast, Barthel and Voyé [8] and Lu [65] suggested to update all code parameters by an iterative procedure. One starts from a solution T_0 found by collage coding. Then at step $n \geq 1$, one modifies the code parameters of all range blocks R_i, $i = 1, \ldots, n_R$, by solving the minimization problem

$$\min_{(D,s,o,\mathbf{P}) \in \mathcal{D} \times \mathcal{S} \times \mathcal{O} \times \mathcal{P}} ||\mathbf{x}_{f^*|R_i} - (s\hat{\mathbf{D}}\mathbf{P}\mathbf{x}_{f_{T_{n-1}}|D} + o\mathbf{1})||_2^2 \tag{6.44}$$

In other words, one does collage coding based on the domain blocks with image intensities from the fixed point of step $n - 1$. This method allows substantial PSNR improvements over collage coding.

However, it has two drawbacks. First, there is no guarantee that the reconstruction error decreases after each step. Second, the procedure is time expensive because every step corresponds to a new encoding of the test image. To accelerate the procedure, Lu proposed to consider at each step only the 10% range blocks R_i for which the ratio between the collage error and the reconstruction error $\|\mathbf{x}_{f^*|R_i} - \mathbf{x}_{f_{T_{n-1}}|R_i}\|_2^2$ is largest. Barthel and Voyé [8] saved time by fixing the range block–domain block association and updating only the scaling factor and the offset. Hamzaoui et al. [41] modified the method in a way to ensure that the reconstruction error is monotonically decreasing. This gives a local search [74] scheme, where the neighborhood of a feasible solution T_n is the set of n_R transforms obtained from T_n by modifying the code parameters of only a single range block according to rule (6.44). The algorithm is as follows.

Local Search Algorithm

1. **Initialization**: Let M be a maximum number of trials. Set $n := 0$, $i := 0$, and $j := 0$. Find an initial feasible solution T_0 by collage coding. Let n_R be the number of range blocks in the partition.
2. Let $r := 1 + (i \bmod n_R)$. Determine for the range block R_r new code parameters by solving the minimization problem

$$\min_{(D,s,o,\mathbf{P})\in\mathcal{D}\times\mathcal{S}\times\mathcal{O}\times\mathcal{P}} \|\mathbf{x}_{f^*|R_r} - (s\hat{\mathbf{D}}\mathbf{P}\mathbf{x}_{f_{T_n}|D} + o\mathbf{1})\|_2^2 \qquad (6.45)$$

   Set $i := i + 1$.
3. Let T_c be the solution obtained from T_n by changing the code parameters of range block R_r according to the result of Step 2. Compute the fixed point of T_c.
4. If $\|f^* - f_{T_c}\|_2 < \|f^* - f_{T_n}\|_2$, set $T_{n+1} := T_c, n := n + 1, j := 0$. Otherwise set $j := j + 1$.
5. If $(i \leq M$ and $j < n_R)$ go to Step 2. Otherwise stop.

When M is set to ∞, the algorithm stops at a local minimum.

The most time-consuming part of the local search algorithm is the computation in Step 3 of the fixed point f_{T_c}. In [43], several techniques for accelerating were proposed. The main idea exploits the fact that T_c and T_n differ only in the code parameters of R_r. Thus, in Step 3 if we start from the current fixed point f_{T_n} and apply T_c once, then only the pixel intensities in R_r have to be updated, which avoids many unnecessary computations. If we now apply T_c to $T_c(f_{T_n})$, then only the range blocks whose domain blocks overlap R_r have to be updated. Note that these range blocks may include R_r. This procedure is repeated until convergence to f_{T_c}.

The iteration scheme $f_{T_n} \rightarrow T_c(f_{T_n}) \rightarrow T_c(T_c(f_{T_n})) \rightarrow \cdots \rightarrow f_{T_c}$ can be implemented as a breadth-first traversal of the dependence graph of T_c, starting from vertex R_r. The first iteration $f_{T_n} \rightarrow T_c(f_{T_n})$ corresponds to visiting the root vertex R_r, and for $k \geq 2$, iteration k corresponds to visiting the children of all vertices visited at iteration $k - 1$. Formally, if we denote by $\{L_k\}_{k \geq 1}$ the sequence of subsets of $\{R_1, \ldots, R_{n_R}\}$ given by $L_1 = \{R_r\}$ and $L_{k+1} = \bigcup_{R_i \in L_k} \{R_j \in V : (R_i, R_j) \in E\}$, then at iteration k, we compute the pixel intensities of only the range blocks in L_k.

The procedure can be accelerated by using the newly computed pixel intensities as soon as they are available. In this situation, the breadth-first traversal of the dependence graph starting from vertex R_r corresponds to the iteration scheme

$$f_{k+1} = T_{c,i_k}(f_k), \ f_0 = f_{T_n} \qquad (6.46)$$

where T_{c,i_k} is a Gauss–Seidel like operator (6.38) such that the pixel intensities of the range blocks in L_k are computed last. It can be proved that iteration (6.46) converges to f_{T_c} [43].

6.7 RATE-DISTORTION CODING

Given an original image f^*, a set of Jacquin–Barnsley operators $\mathcal{T}$, and a bit budget r_0, rate-distortion fractal coding is the discrete constrained optimization problem

$$\min_{T \in \mathcal{T}} \|f^* - T(f^*)\|_2^2 \tag{6.47}$$

subject to

$$|\omega(T)| \le r_0 \tag{6.48}$$

In [86], the generalized BFOS algorithm [16] was applied to the above problem by assuming an HV partition and a fixed codeword length for the domains, the scaling factors, and the offsets.

In [7], it is observed that often best domain candidates are located near the range blocks. Thus, rate-distortion gains can be achieved by allowing domain blocks to have variable-length codewords. For each range block, a codebook of domain blocks is designed such that domain blocks nearer to the range block have shorter codewords. With this setting, optimal domain blocks in the constrained problem are not necessarily those that minimize the collage error. Therefore, the domain blocks are selected based on the approach of Westerink et al. [95]. The encoder starts with a Jacquin–Barnsley operator where all range blocks are encoded with domain blocks having shortest codewords. Then, one tries to reduce the collage error iteratively by allocating at each step more bits to the domain block of one of the range blocks. This range block R_k is one that maximizes the reduction in collage error per additional domain block bit, that is,

$$k = \arg \max_{i=1,\dots,n_R} \lambda_i \tag{6.49}$$

where

$$\lambda_i = \max_m \frac{E(R_i, D) - E(R_i, D_m)}{r_{D_m} - r_D} \tag{6.50}$$

Here D is the domain block that currently encodes R_i, and D_m is any domain block whose codeword length r_{D_m} is larger than r_D, the codeword length of D, and for which the local collage error $E(R_i, D_m)$ is smaller than the local collage error $E(R_i, D)$ associated to R_i and D. The iteration is stopped when no further reduction of the collage error is possible or when the total bit rate becomes larger than the bit budget r_0.

Lu [65] uses an algorithm that handles the constrained problem in the context of quadtree partitions and domain blocks of variable-length codewords. The algorithm, which was previously proposed in [88] for quadtree-based vector quantization, relies on a Lagrange multiplier method and consists of solving the unconstrained optimization problem

$$\min_{T \in \mathcal{T}} \|f^* - T(f^*)\|_2^2 + \lambda |\omega(T)| \tag{6.51}$$

where $\lambda \in [0, \infty)$. Indeed, one can show [29] that for any choice of $\lambda \ge 0$ a solution $T(\lambda)$ to the unconstrained problem (6.51) is a solution to the constrained problem

$$\min_{T \in \mathcal{T}} \|f^* - T(f^*)\|_2^2 \tag{6.52}$$

subject to

$$|\omega(T)| \le |\omega(T(\lambda))| \tag{6.53}$$

In the standard quadtree scheme, a (parent) square block is either accepted in the partition or subdivided into four smaller (child) square blocks. A more adaptive scheme [34] can be obtained by allowing ten other intermediary configurations. The authors use the Lagrange multiplier technique to determine an adaptive partition corresponding to a three-level hierarchy. Finally, the coder of Gharavi–Alkhansari and Huang [36] uses a rate-distortion criterion.

All above algorithms are not guaranteed to find an optimal solution to the original constrained optimization problem (see Box 8 in [73]). However, in practice, there generally exists a bit budget r_0' that is close to r_0 for which these algorithms yield an optimal solution.

6.8 EXTENSIONS

6.8.1 HYBRID METHODS

In this section, we mention hybrid coders in which fractal coding is combined with another method. However, we restrict the discussion to cases where it was shown that the hybrid is better than both constituent methods.

Fractal image compression is similar to mean shape-gain vector quantization [59]. This similarity is exploited in [42], where a mean shape-gain vector quantization codebook is used to accelerate the search for the domain blocks and to improve both the rate-distortion performance and the decoding speed of a fractal coder.

Fractal image compression was combined with the DCT in [6,66,93]. These works transfer the coding in the frequency domain and use rate-distortion-based optimization to optimize the hybrid code. Whereas the first two papers fix the size of the range blocks, the third one uses a quadtree partition, making the rate-distortion optimization more complicated.

Davis [18] analyzed fractal image compression using a wavelet-based framework. He explained the success of fractal coders by their ability to encode zero trees of wavelet coefficients. The most impressive hybrid fractal-wavelet coding results are reported in [56,63].

6.8.2 CHANNEL CODING

Fractal image codes are very sensitive to channel noise because a single error in a code parameter can affect the reconstruction of many parts of the image. This is due to the iterative nature of the decoding and the complex dependency between the range and domain blocks. In [15], a joint source-channel coding system is proposed for fractal image compression. The available total bit rate is allocated between the source code and the channel code using a Lagrange multiplier optimization technique.

6.8.3 PROGRESSIVE CODING

Progressive coding is an important feature of compression schemes. Wavelet coders are well suited for this purpose because the wavelet coefficients can be naturally ordered according to decreasing importance. Progressive fractal coding is straightforward for hybrid fractal-wavelet schemes. In [58], a progressive fractal image coder in the spatial domain is introduced. A Lagrange optimization based on rate-distortion performance estimates determines an optimal ordering of the code bits. The optimality is in the sense that the reconstruction error is monotonically decreasing and minimum at intermediate rates. The decoder recovers this ordering without side information.

6.8.4 POSTPROCESSING

Fractal coding suffers from blocking artifacts, in particular at high compression ratios. Several works [30,65] addressed this problem by using smoothing filters that modify the pixel intensities at range block boundaries with a weighted average of their values. However, due to the recursive nature of the fractal decoder, blocking artifacts also appear inside the range blocks. In [38], a smoothing filter adapted to the partition is applied before each iteration of the decoding process to reduce the blocking artifacts inside the range blocks. After decoding, the reconstructed image is enhanced by applying another adaptive filter along the range block boundaries. This filter is designed to reduce blocking artifacts while maintaining edges and texture of the original image.

6.8.5 COMPRESSION OF COLOR IMAGES

In [11], the G (green) component is encoded individually, and the result is used to predict the encoding of the R (red) and B (blue) components. Fisher [30] transforms the RGB components into YIQ components, which are encoded separately, the I- and Q-channels being encoded at a lower bit rate than the Y-channel. Popescu and Yan [76] transform the RGB components into three approximate principal components and encode each new component separately.

6.8.6 VIDEO CODING

The ideas of fractal coding were also used for the compression of video sequences. Monro and Nichols [69] separately encode each 2-D frame with a fast fractal coder. A uniform partition is used, and the domain block is fixed a priori for a given range block. In [32], the domain blocks are taken from the previously decoded frame, yielding a noniterative decoder. A more sophisticated scheme is proposed in [55]. In [9,61], time is used as a third variable and fractal coding is applied to the 3-D ranges and domains.

6.8.7 APPLICATIONS

Besides image and video compression, the fractal transform was used for other applications in image processing, including watermarking [46], zooming [70], indexing and retrieval [24,94], image segmentation and contour detection [50], recognition of microcalcifications in digital mammograms [62], and denoising [37].

6.9 STATE OF THE ART

In this section, we compare the rate-distortion performance of a state-of-the-art fractal coder in the spatial domain [71] to that of the SPIHT algorithm with arithmetic coding [79].

Figure 6.3 shows the PSNR as a function of the compression ratio for the 8 bits per pixel 512×512 standard images *Lenna* and *Peppers*. Figure 6.4 shows range partitions yielded by the fractal coder. Table 6.1 compares the performance of the two coders for two other standard images.

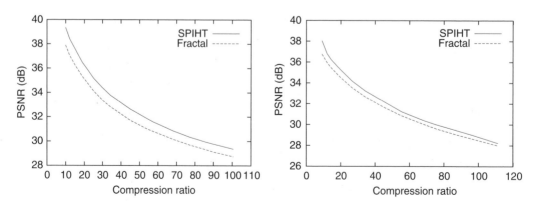

FIGURE 6.3 PSNR as a function of the compression ratio for the *Lenna* image (left) and the *Peppers* image (right).

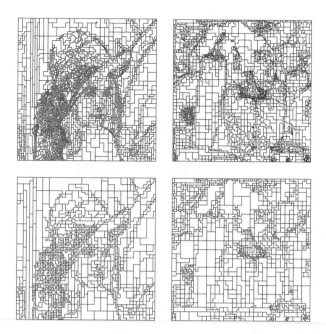

FIGURE 6.4 Range partition for *Lenna* (left) and *Peppers* (right) at compression ratio 32:1 (top) and 70.30:1 (bottom).

TABLE 6.1
PSNR in dB at Various Compression Ratios for the Fractal Coder [71] and SPIHT

Barbara	16:1	32:1	64:1
Fractal	29.09	26.30	24.44
SPIHT	32.11	28.13	25.38
Boat			
Fractal	33.85	30.58	27.82
SPIHT	34.45	30.97	28.16

Note: Results are given for the 512 × 512 Barbara and Boat.

The results show that the PSNR performance of the fractal coder was between 0.3 and 3 dB worse than that of the SPIHT coder. The experiments indicate that the fractal coder is especially less efficient than the wavelet coder at low compression ratios and for textured images. However, as the PSNR is not an absolute measure of quality, the visual quality of the encoded images must also be compared. At low compression ratios, the SPIHT coder has a better visual quality than the fractal coder. When the compression ratio is increased, the comparison becomes difficult as the reconstruction artefacts produced by the two methods differ. Whereas the fractal coder suffers from blocking artefacts, the wavelet-coded images are blurred and show ringing around the edges (see Figure 6.5 and Figure 6.6).

The rate-distortion performance of fractal coding can be improved by combining fractal coding with other coding techniques. For example, the hybrid wavelet-fractal coder of Kim et al. [56] provides slightly better PSNR performance than the SPIHT algorithm for the Lenna image.

FIGURE 6.5 Reconstruction of the original *Lenna* image (top) with the fractal coder of [71] (left) and the SPIHT coder (right) at compression ratio 32:1 (middle) and 70.30:1 (bottom).

6.10 CONCLUSION

We described the basic principles of fractal image compression, reviewed the most important contributions, and provided a comparison to the SPIHT coder. Because of space limitation, it was not possible to mention many other interesting works. An exhaustive online bibliography on fractal image compression can be found at http://www.inf.uni-konstanz.de/cgip/fractal2/html/bibliography.html.

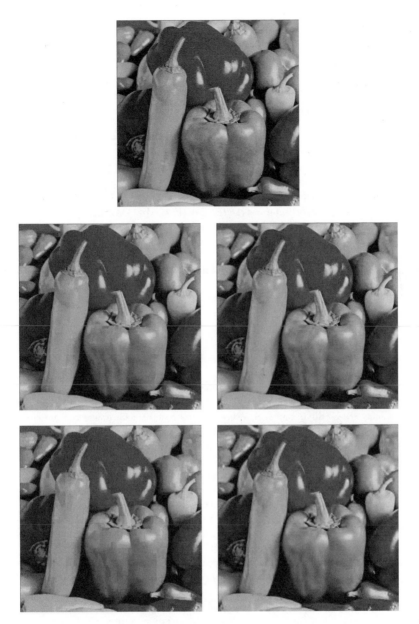

FIGURE 6.6 Reconstruction of the original *Peppers* image (top) with the fractal coder of [71] (left) and the SPIHT coder (right) at compression ratio 32:1 (middle) and 70.30:1 (bottom).

As this chapter shows, a huge effort has gone into fractal coding research, although it is probably fair to say that the attention that wavelet coding received has been much larger. Nonetheless, we feel that fractal image coding has reached its limits; further really significant advances can hardly be expected, perhaps unless one changes the class of image operators considered.

ACKNOWLEDGMENTS

We thank Tilo Ochotta for providing us with the coding results of [71].

REFERENCES

1. S. Arya, D. Mount, N.S. Netanyahu, R. Silverman, and A. Wu, An optimal algorithm for approximate nearest neighbor searching in fixed dimensions, *Proceedings of the 5th Annual ACM-SIAM Symposium on Discrete Algorithms*, 1994, pp. 573–582.
2. Z. Baharav, D. Malah, and E. Karnin, Hierarchical interpretation of fractal image coding and its applications, in *Fractal Image Compression — Theory and Application*, Y. Fisher, Ed., Springer, New York, 1994.
3. M. Barnsley, *Fractals Everywhere*, Academic Press, San Diego, 1988.
4. M. Barnsley and L. Hurd, *Fractal Image Compression*, AK Peters, Wellesley, 1993.
5. M.F. Barnsley and A. Sloan, *A better way to compress images*, BYTE Magazine, 1988, pp. 215–222.
6. K.U. Barthel, Entropy constrained fractal image coding, *Fractals*, 5, April 1997, pp. 17–26.
7. K.U. Barthel, J. Schüttemeyer, T. Voyé, and P. Noll, A new image coding technique unifying fractal and transform coding, *Proceedings of the IEEE ICIP-94*, Vol. 3, Austin, TX, 1994, pp. 112–116.
8. K.U. Barthel and T. Voyé, Adaptive fractal image coding in the frequency domain, *Proceedings of the International Workshop on Image Processing: Theory, Methodology, Systems and Applications*, Budapest, 1994.
9. K.U. Barthel and T. Voyé, Three-dimensional fractal video coding, *Proceedings of the IEEE ICIP-95*, Washington, DC, 1995, Vol. 3, pp. 260–263.
10. T. Bedford, F.M. Dekking, and M.S. Keane, Fractal image coding techniques and contraction operators, *Nieuw Arch. Wisk. (4)*, 10 (1992), 185–218.
11. A. Bogdan, The fractal pyramid with application to image coding, *Proceedings of the IEEE ICASSP-1995*, Detroit, MI, 1995.
12. J. Cardinal, Fast fractal compression of greyscale images, *IEEE Trans. Image Process.*, 10 (1), 2001, 159–164.
13. G. Caso, P. Obrador, and C.-C.J. Kuo, Fast methods for fractal image encoding, *Proceedings of the SPIE VCIP*, Vol. 2501, 1995, pp. 583–594.
14. Y.-C. Chang, B.-K. Shyu, and J.-S. Wang, Region-based fractal image compression with quadtree segmentation, *Proceedings of the IEEE ICASSP'97*, Munich, 1997.
15. Y. Charfi, V. Stankovic, R. Hamzaoui, A. Haouari, and D. Saupe, Joint source-channel fractal image coding with unequal error protection, *Opt. Eng.*, 41 (12), 2002, 3168–3176.
16. P. Chou, T. Lookabaugh, and R. Gray, Optimal pruning with applications to tree-structured source coding and modeling, *IEEE Trans. Inform. Theory*, 35 (2), 1989, 299–315.
17. L. Cieplinski, C. Jedrzejek, and T. Major, Acceleration of fractal image compression by fast nearest-neighbor search, *Fractals*, 5, April 1997, pp. 231–242.
18. G.M. Davis, A wavelet-based analysis of fractal image compression, *IEEE Trans. Image Process.*, 7 (2), 1998, 141–154.
19. F. Davoine, E. Bertin, and J.-M. Chassery, From rigidity to adaptive tessellation for fractal image compression: Comparative studies, *Proceedings of the IEEE 8th Workshop Image and Multidimensional Signal Processing*, Cannes, 1993, pp. 56–57.
20. F. Davoine and J.-M. Chassery, Adaptive delaunay triangulation for attractor image coding, *Proceedings of the 12th International Conference on Pattern Recognition*, Jerusalem, 1994, pp. 801–803.
21. F. Davoine, J. Svensson, and J.-M. Chassery, A mixed triangular and quadrilateral partition for fractal image coding, *Proceedings of the IEEE ICIP-95*, Washington, DC, 1995.
22. F.M. Dekking, Fractal Image Coding: Some Mathematical Remarks on its Limits and its Prospects, Technical Report 95–95, Delft Univeristy of Technology, 1995.
23. F.M. Dekking, An Inequality for Pairs of Martingales and its Applications to Fractal Image Coding, Technical Report 95–10, Delft Univeristy of Technology, 1995.
24. R. Distasi, M. Nappi, and M. Tucci, Fire: fractal indexing with robust extensions for image databases, *IEEE Trans. Image Process.*, 12 (3), 2003, 373–384.
25. R. Distasi, M. Polvere, and M. Nappi, Split decision functions in fractal image coding, *Electron. Lett.*, 34 (8), 1998, 751–753.
26. J. Domaszewicz and V.A. Vaishampayan, Structural limitations of self-affine and partially self-affine fractal compression, *Proceedings of SPIE VCIP*, Vol. 2094, 1993, pp. 1498–1504.

27. J. Domaszewicz and V.A. Vaishampayan, Graph-theoretical analysis of the fractal transform, *Proceedings of the IEEE ICASSP-1995*, Detroit, MI, 1995.

28. F. Dudbridge and Y. Fisher, Attractor optimization in fractal image encoding, *Proceedings of the Conference on Fractals in Engineering*, Arcachon, 1997.

29. H. Everett, Generalized Lagrange multiplier method for solving problems of optimum allocation of resources, *Opera. Res.*, 11 (1963), 399–417.

30. Y. Fisher, *Fractal Image Compression — Theory and Application*, Springer, New York, 1994.

31. Y. Fisher and S. Menlove, Fractal encoding with hv partitions, in *Fractal Image Compression — Theory and Application*, Y. Fisher, Ed., Springer, Berlin, 1994, pp. 119–126.

32. Y. Fisher, D. Rogovin, and T.-P. Shen, Fractal (self-vq) encoding of video sequences, *Proceedings of the SPIE VCIP*, Vol. 2308, Chicago, 1994.

33. J.H. Friedman, J.L. Bentley, and R.A. Finkel, An algorithm for finding best matches in logarithmic expected time, *ACM Trans. Math. Software*, (3), 1977, 209–226.

34. T. Fuchigami, S. Yano, T. Komatsu, and T. Saito, Fractal block coding with cost-based image partitioning, *Proceedings of the International Picture Coding Symposium*, Melbourne, 1996.

35. M. Gharavi-Alkhansari and T.S. Huang, Fractal based techniques for a generalized image coding method, *Proceedings of the IEEE ICIP-94*, Austin, TX, 1994.

36. M. Gharavi-Alkhansari and T.S. Huang, Fractal image coding using rate-distortion optimized matching pursuit, *Proceedings of the SPIE VCIP*, Vol. 2727, Orlando, FL, 1996, pp. 1386–1393.

37. M. Ghazel, G.H. Freeman, and E.R. Vrscay, Fractal image denoising, *IEEE Trans. Image Process.*, (12), 2003, 1560–1578.

38. N.K. Giang and D. Saupe, Adaptive post-processing for fractal image compression, *Proceedings of the IEEE ICIP-2000*, Vancouver, 2000.

39. R. Hamzaoui, Codebook clustering by self-organizing maps for fractal image compression, *Fractals*, 5, April 1997, 27–38.

40. R. Hamzaoui, Fast iterative methods for fractal image compression, *J. Math. Imaging and Vision*, 11 (2), 1999, 147–159.

41. R. Hamzaoui, H. Hartenstein, and D. Saupe, Local iterative improvement of fractal image codes, *Image Vision Compu.*, 18 (6–7), 2000, 565–568.

42. R. Hamzaoui and D. Saupe, Combining fractal image compression and vector quantization, *IEEE Trans. Image Process.*, 9 (2), 2000, 197–208.

43. R. Hamzaoui, D. Saupe, and M. Hiller, Distortion minimization with a fast local search for fractal image compression, *J. Visual Commn. Image Representation*, 12 (4), 2001, 450–468.

44. H. Hartenstein, M. Ruhl, and D. Saupe, Region-based fractal image compression, *IEEE Trans. Image Process.*, 9 (7), 2000, 1171–1184.

45. H. Hartenstein and D. Saupe, Lossless acceleration of fractal image encoding via the fast fourier transform, *Signal Process.: Image Commn.*, 16 (4), 2000, 383–394.

46. M. Haseyama and I. Kondo, Image authentication based on fractal image coding without contamination of original image, *Syst. Comput. Japan*, 34 (9), 2003, 1–9.

47. B. Hürtgen, Performance bounds for fractal coding, *Proceedings of the IEEE ICASSP-1995*, Detroit, MI, 1995.

48. B. Hürtgen and C. Stiller, Fast hierarchical codebook search for fractal coding of still images, *Proceedings of the EOS/SPIE Visual Communications and PACS for Medical Applications'93*, Berlin, 1993.

49. J. Hutchinson, Fractals and self-similarity, *Indiana University J. Math.*, 30 (1981), 713–747.

50. T. Ida and Y. Sambonsugi, Image segmentation and contour detection using fractal coding, *IEEE Trans. Circuits Syst. Video Technol.*, (8), 1998, 968–975.

51. E.W. Jacobs, Y. Fisher, and R.D. Boss, Image compression: a study of the iterated transform method, *Signal Processing*, 29 (3), 1992, 251–263.

52. A.E. Jacquin, A novel fractal block-coding technique for digital images, *Proceedings of the IEEE ICASSP-1990*, Vol. 4, Albuquerque, 1990, pp. 2225–2228.

53. A.E. Jacquin, Image coding based on a fractal theory of iterated contractive image transformations, *IEEE Trans. Image Process.*, 1 (1), 1992, 18–30.

54. H. Kaouri, Fractal coding of still images, *Proceedings of the IEE 6th International Conference on Digital Processing of Signals in Communications*, 1991, pp. 235–239.

55. C.-S. Kim, R.-C. Kim, and S.-U. Lee, Fractal coding of video sequence using circular prediction mapping and noncontractive interframe mapping, *IEEE Trans. Image Process.*, 7 (4), 1998, 601–605.

56. T. Kim, R.V. Dyck, and D.J. Miller, Hybrid fractal zerotree wavelet image coding, *Signal Process.: Image Commn.*, 17 (4), 2002, 347–360.

57. J. Kominek, Convergence of fractal encoded images, *Proceedings of Data Compression Conference*, IEEE Computer Society Press, Snowbird, UT, 1995.

58. I. Kopilovic, D. Saupe, and R. Hamzaoui, Progressive fractal coding, *Proceedings of the IEEE ICIP-01*, Thessaloniki, 2001.

59. S. Lepsøy, P. Carlini, and G. E. Øien, On fractal compression and vector quantization, *Proceedings of NATO ASI Fractal Image Encoding and Analysis*, Trondheim, July 1995, Y. Fisher, Ed., Springer, Berlin, 1998.

60. S. Lepsøy and G.E. Øien, Fast attractor image encoding by adaptive codebook clustering, in *Fractal Image Compression — Theory and Application*, Y. Fisher, Ed., Springer, New York, 1994.

61. H. Li, M. Novak, and R. Forchheimer, Fractal-based image sequence compression scheme, *Opt. Eng.*, 32 (7), 1993, 1588–1595.

62. H. Li, K.J.R. Liu, and S.-C.B. Lo, Fractal modeling and segmentation for the enhancement of microcalcifications in digital mammograms, *IEEE Trans. Med. Imaging*, 16 (6), 1997, 785–798.

63. J. Li and C.-C.J. Kuo, Image compression with a hybrid wavelet-fractal coder, *IEEE Trans. Image Process.*, 8 (6), 1999, 868–873.

64. H. Lin and A.N. Venetsanopoulos, Fast fractal image coding using pyramids, *Proceedings of the 8th International Conference Image Analysis and Processing*, San Remo, 1995.

65. N. Lu, *Fractal Imaging*, Academic Press, New York, 1997.

66. G. Melnikov and A.K. Katsaggelos, A jointly optimal fractal/dct compression scheme, *IEEE Trans. Multimedia*, (4), 2002, 413–422.

67. D.M. Monro, A hybrid fractal transform, *Proceedings of the IEEE ICASSP-1993*, Vol. 5, Minneapolis, 1993, pp. 169–172.

68. D.M. Monro and F. Dudbridge, Fractal approximation of image blocks, *Proceedings of the IEEE ICASSP-1992*, Vol. 3, San Francisco, CA, 1992, pp. 485–488.

69. D.M. Monro and J.A. Nicholls, Low bit rate colour fractal video, *Proceedings of the IEEE ICIP-95*, Washington, DC, 1995.

70. D.M. Monro and P. Wakefield, Zooming with implicit fractals, *Proceedings of the IEEE ICIP-97*, Santa Barbara, CA, 1997.

71. T. Ochotta and D. Saupe, Edge-based partition coding for fractal image compression, *The Arabian J. Sci. Eng.*, *Special Issue on Fractal and Wavelet Methods*, 29, 2C (2004), 63–83.

72. G.E. Øien, L_2 Optimal Attractor Image Coding with Fast Decoder Convergence, Ph.D. thesis, The Norwegian Institute of Technology, Trondheim, Norway, 1993.

73. A. Ortega and K. Ramchandran, Rate-distortion methods for image and video compression, *Signal Process. Mag.* 15,6 (1998), 23–50.

74. C.H. Papadimitriou and K. Steiglitz, *Combinatorial Optimization: Algorithms and Complexity*, Dover, Mineola, New York, 1998.

75. H.-O. Peitgen and D. Saupe, Eds., *The Science of Fractal Images*, Springer, New York, 1988.

76. D. Popescu and H. Yan, Fractal-based method for color image compression, *J. Electron. Imaging*, 4 (1), 1995, 23–30.

77. E. Reusens, Partitioning complexity issue for iterated function systems based image coding, *Proceedings of the VIIth European Signal Processing Conference EUSIPCO'94*, Vol. 1, Edinburgh, 1994, pp. 171–174.

78. M. Ruhl and H. Hartenstein, Optimal fractal coding is np-hard, *Proceedings of the Data Compression Conference*, IEEE Computer Society Press, Snowbird, UT, 1997.

79. A. Said and W. Pearlman, A new fast and efficient image codec based on set partitioning in hierarchical trees, *IEEE Trans. Circuits Syst. Video Technol.*, 6 (3), 1996, 243–250.

80. D. Saupe, From classification to multi-dimensional keys, in *Fractal Image Compression — Theory and Application*, Y. Fisher, Ed., Springer, New York, 1994.

81. D. Saupe, Accelerating fractal image compression by multi-dimensional nearest neighbor search, *Proceedings of Data Compression Conference*, IEEE Computer Society Press, Snowbird, UT, 1995.

82. D. Saupe, Fractal image compression via nearest neighbor search, *Proceedings of NATO ASI Fractal Image Encoding and Analysis*, Trondheim, 1995, Y. Fisher, Ed., Springer, Berlin, 1998.

83. D. Saupe, Lean domain pools for fractal image compression, *J. Electron. Imaging*, 8 (1999), 98–103.

84. D. Saupe and S. Jacob, Variance-based quadtrees in fractal image compression, *Electron. Lett.*, 33 (1), 1997, 46–48.

85. D. Saupe and M. Ruhl, Evolutionary fractal image compression, *Proceedings of the IEEE ICIP-96*, Lausanne, 1996.

86. D. Saupe, M. Ruhl, R. Hamzaoui, L. Grandi, and D. Marini, Optimal hierarchical partitions for fractal image compression, *Proceedings of the IEEE ICIP-98*, Chicago, 1998.

87. J. Signes, Geometrical interpretation of ifs based image coding, *Fractals*, 5, April 1997, 133–143.

88. G.J. Sullivan and R.L. Baker, Efficient quadtree coding of images and video, *IEEE Trans. Image Process.*, (3), 1994, 327–331.

89. L. Thomas and F. Deravi, Region-based fractal image compression using heuristic search, *IEEE Trans. Image Process.*, 4 (6), 1995, 832–838.

90. C.S. Tong and M. Wong, Adaptive approximate nearest neighbor search for fractal image compression, *IEEE Trans. Image Process.*, 11 (6), 2002, 605–615.

91. G. Vines, Orthogonal basis ifs, in *Fractal Image Compression — Theory and Application*, Y. Fisher, Ed., Springer, New York, 1994.

92. E.R. Vrscay and D. Saupe, Can one break the "collage barrier" in fractal image coding, in *Fractals: Theory and Applications in Engineering*, M. Dekking, J.L. Vehel, E. Lutton, and C. Tricot, Eds., Springer, London, 1999, pp. 307–323.

93. P. Wakefield, D. Bethel, and D. Monro, Hybrid image compression with implicit fractal terms, *Proceedings of the IEEE ICASSP-1997*, Vol. 4, Munich, 1997, pp. 2933–2936.

94. Z. Wang, Z. Chi, and D. Feng, Content-based image retrieval using block-constrained fractal coding and nona-tree decomposition, *IEE Proc. Vision Image Signal Process.*, 147 (1), 2000, pp. 9–15.

95. P.H. Westerink, J. Biemond, and D.E. Boekee, An optimal bit allocation algorithm for sub-band coding, *Proceedings of the IEEE ICASSP'88*, New York, 1988, pp. 757–760.

Part II

New Directions

7 Beyond Wavelets: New Image Representation Paradigms

Hartmut Führ, Laurent Demaret, and Felix Friedrich

CONTENTS

7.1 Introduction 179
7.2 The Problem and Some Proposed Solutions 180
    7.2.1 Wedgelets 184
    7.2.2 Curvelets 187
    7.2.3 Alternative Approaches 192
7.3 Digital Wedgelets 193
    7.3.1 Rapid Summation on Wedge Domains: Discrete Green's Theorem 194
    7.3.2 Implementation 195
7.4 Digital Curvelets: Contourlets 196
7.5 Application to Image Compression 199
    7.5.1 Experimental Approximation Properties 199
    7.5.2 Coding Schemes 202
7.6 Tentative Conclusions and Suggestions for Further Reading 204
Acknowledgment 205
References 205

7.1 INTRODUCTION

Despite the huge success of wavelets in the domain of image compression, the failure of two-dimensional multiresolution wavelets when dealing with images of the cartoon class, i.e., images consisting of domains of smoothly varying gray values, separated by smooth boundaries, has been noted repeatedly. This failure can be discerned visually, in the form of characteristic compression artifacts, but it can also be derived theoretically. In this chapter, we review some of the constructions that were proposed as a remedy to this problem. We focus on two constructions: wedgelets [16] and curvelets [6]. Both systems stand for larger classes of image representation schemes; let us just mention ridgelets [4], contourlets [12,14], beamlets [17], platelets [35], and surflets [7] as close relatives.

The chapter starts with a discussion of the shortcomings of wavelet orthonormal bases. The reason for expecting good approximation rates for cartoon-like images is the observation that here the information is basically contained in the edges. Thus, ideally, one expects that smoothness of the boundaries should have a beneficial effect on approximation rates. However, the tensor product wavelets usually employed in image compression do not adapt to smooth boundaries, because of the isotropic scaling underlying the multiresolution scheme. The wedgelet scheme tries to overcome this by combining adaptive geometric partitioning of the image domain with local regression on the image segments. A wedgelet approximation is obtained by minimizing a functional that weighs model complexity (in the simplest possible case: the number of segments) against approximation error. By contrast, the curvelet approach can be understood as a directional filterbank, designed and sampled so as to ensure that the system adapts well to smooth edges (the key feature here turns out to be *hyperbolic scaling*) while at the same time providing a *frame*, i.e., a system of building blocks with properties similar to an orthonormal bases. Here nonlinear approximation is achieved by a simple truncation of the frame coefficients.

After a presentation of these constructions for the continuous setting, we then proceed with a description of methods for their digital implementation. We sketch a recently developed, particularly efficient implementation of the wedgelet scheme, as well as the contourlet approach to curvelet implementation, as proposed by Do and Vetterli [12–14].

In the last section, we present some numerical experiments to compare the nonlinear approximation behavior of the different schemes, and contrast the theoretical approximation results to the experiments. We close by commenting on the potential of wedgelets and curvelets for image coding. Clearly, the nonlinear approximation behavior of a scheme can only be used as a first indicator of its potential for image coding. The good approximation behavior of the new methods for *small* numbers of coefficients reflects their ability to pick out the salient geometric features of an image rather well, which could be a very useful property for hybrid approaches.

7.2   THE PROBLEM AND SOME PROPOSED SOLUTIONS

Besides the existence of fast decomposition and reconstruction algorithms, the key feature that paved the way for wavelets is given by their ability to effectively represent discontinuities, at least for one-dimensional signals. However, it has been observed that the tensor product construction is not flexible enough to reproduce this behavior in two dimensions. Before we give a more detailed analysis of this failure, let us give a heuristic argument based on the wavelet coefficients displayed in Figure 7.1. Illustrations like this are traditionally used to demonstrate how wavelets pick salient (edge) information out of images. However, it has been observed previously (e.g., [13]) that Figure 7.1 in fact reveals a weakness of wavelets rather than a strength, showing that wavelets detect isolated *edge points* rather than *edges*. The fact that the edge is smooth is not reflected adequately; at each scale j the number of significant coefficients is proportional to 2^j times the length of the boundary, regardless of its smoothness.

A more quantitative description of this phenomenon can be given in terms of the *nonlinear approximation error*. All theoretical considerations below refer to a Hilbert space of (one- or two-dimensional) signals, and the approximation performance is measured by the Hilbert space norm. In

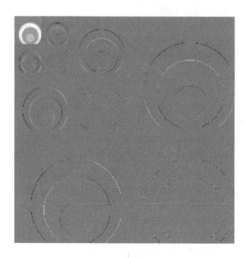

FIGURE 7.1 Wavelet coefficients of an image with smooth edges. The detail images are renormalized for better visibility.

the continuous domain setting, the Hilbert space will be $L^2(\mathbb{R}^d)$, the space of functions f on $\mathbb{R}^d$ that are square-integrable with respect to the usual Lebesgue measure, with scalar product

$$\langle f, g \rangle = \int_{\mathbb{R}^d} f(x)\overline{g(x)} \, dx \tag{7.1}$$

and norm $\|f\|_2 = \langle f, f \rangle^{1/2}$. In the discrete domain setting, the signals under consideration are members of $\ell^2(\mathbb{Z}^d)$, the square-summable sequences indexed by $\mathbb{Z}^d$. To keep the notation in the following arguments simple, we let C denote a constant that may change from line to line.

Now suppose we are given an orthonormal basis $(\psi_\lambda)_{\lambda \in \Lambda}$ of the signal Hilbert space. For a signal f and $N \geq 0$, we let $\epsilon_N(f)$ denote the smallest possible squared error that can be achieved by approximating f by a linear approximation of (at most) N basis elements, i.e.,

$$\epsilon_N(f) = \inf \left\{ \left\| f - \sum_{\lambda \in \Lambda'} \alpha_\lambda \psi_\lambda \right\|^2 : \Lambda' \subset \Lambda \text{ with } |\Lambda'| = N, (\alpha_\lambda) \in \mathbb{C}^\Lambda \right\}$$

The study of the nonlinear approximation error can be seen as a precursor to rate-distortion analysis; for more details on nonlinear approximation and its connections to compression we refer to the survey given in [29], or to Chapter IX in Mallat's book [23]. Since we started with an orthonormal basis, the approximation error is easily computed from the expansion coefficients $(\langle f, \psi_\lambda \rangle)_{\lambda \in \Lambda}$, by the following procedure that relates the asymptotic behavior of the approximation error to the decay behavior of the coefficients: reindex the coefficients to obtain a sequence $(\theta_m)_{m \in \mathbb{N}}$ of numbers with decreasing modulus. Then the Parseval relation associated to the orthonormal basis yields

$$\epsilon_N(f) = \sum_{m=N+1}^{\infty} |\theta_m|^2 \tag{7.2}$$

Let us now compare the approximation behavior of one- and two-dimensional wavelet systems. We only give a short sketch of the argument, which has the purpose to give a closer description of the dilemma surrounding two-dimensional wavelets, and to motivate the constructions designed as remedies. Generally speaking, the mathematics in this chapter will be held on an informal level.

Let $f : [0,1] \to \mathbb{R}$ be a bounded function that is piecewise n-times continuously differentiable, say outside a finite set $S \subset [0,1]$ of singularities; we will call these functions *piecewise C^n*. Let $(\psi_{j,k})_{j \in \mathbb{Z}, k \in \mathbb{Z}}$ be a wavelet orthonormal basis consisting of compactly supported functions with n vanishing moments. Recall, e.g. from [23], that the wavelet orthonormal basis is obtained from a suitable mother wavelet ψ via

$$\psi_{j,k}(x) = 2^{j/2} \psi(2^j x - k) \tag{7.3}$$

If ψ is supported in the interval $[-r, r]$, then $\psi_{j,k}$ is supported in $[2^{-j}(k - r), 2^{-j}(k + r)]$.

The structure of the following argument will be encountered several times in this chapter. Basically, we differentiate between two cases, depending whether the support of the wavelet $\psi_{j,k}$ contains a singularity or not. If it does contain a singularity, we employ fairly rough estimates. In the other case, the smoothness of the signal and the oscillatory behavior of the wavelet combine to yield the appropriate decay behavior. The overall estimate of the decay behavior then crucially relies on the fact that the first case does not occur too often.

In the one-dimensional setting, on each dyadic level $j \geq 0$, corresponding to scale 2^{-j} the number of positions k such that the support of $\psi_{j,k}$ contains a singularity of f is fixed (independent of j), and for these coefficients we can estimate

$$|\langle f, \psi_{j,k}\rangle| = \left| \int_{\mathbb{R}} f(x)\overline{\psi_{j,k}(x)}\,\mathrm{d}x \right|$$

$$\leq \sup_{x\in\mathbb{R}}|f(x)| \int_{\mathbb{R}} 2^{j/2}|\psi(2^j x - k)|\,\mathrm{d}x$$

$$= \|f\|_\infty 2^{-j/2} \int_{\mathbb{R}} |\psi(x)|\,\mathrm{d}x$$

$$= \|f\|_\infty \|\psi\|_1 2^{-j/2}$$

Likewise, for $j < 0$, the number of k such that the support of $\psi_{j,k}$ has nonempty intersection with $[0, 1]$ is fixed, and for each such k the roles of ψ and f in the previous argument can be exchanged to yield the estimate

$$|\langle f, \psi_{j,k}\rangle| \leq \|f\|_1 \|\psi\|_\infty 2^{j/2} \tag{7.4}$$

Thus, the two cases considered so far yield for each $\pm j = m \geq 0$ a constant number of coefficients of size $\leq C 2^{-m/2}$. Sorting these coefficients by size yields a decay behavior of order $O(2^{-m/2})$.

We next consider the remaining coefficients, i.e., $j \geq 0$ and $\mathrm{supp}(\psi_{j,k})$ does not contain a singularity. Here we use the vanishing moment condition, which requires for $0 \leq i < n$ that

$$\int_{\mathbb{R}} \psi(x)x^i\,\mathrm{d}x = 0 \tag{7.5}$$

Note that this property is inherited by $\psi_{j,k}$. Next we approximate f by its n-term Taylor polynomial of f at $x_0 = 2^{-j}k$, yielding

$$f(x) = \sum_{j=0}^{n-1} \frac{f^{(j)}(x_0)}{j!}(x - x_0)^j + R(x) \tag{7.6}$$

with $|R(x)| \leq C|x - x_0|^n \leq C 2^{-jn}$, for a suitable constant C independent of k. Now (7.5) implies that

$$|\langle f, \psi_{j,k}\rangle| = |\langle R, \psi_{j,k}\rangle|$$

$$= \left| \int_{\mathbb{R}} R(x)\overline{\psi_{j,k}(x)}\,\mathrm{d}x \right|$$

$$\leq C 2^{-jn}\|\psi_{j,k}\|_1 = C 2^{-j(n+1/2)}$$

Observe that there are $O(2^j)$ such coefficients. Hence, sorting these coefficients we arrive at a decay behavior of $O(m^{-n-1/2})$. Thus, if we arrange the coefficients corresponding to both the cases into a single sequence $(\theta_m)_{m\in\mathbb{N}}$ of decreasing modulus, we again obtain a decay of $O(m^{-n-1/2})$, which can be substituted into (7.2) to yield $\epsilon_N(f) \leq CN^{-2n}$.

The constructions presented in this chapter are to a large extent motivated by the desire to achieve a similar behavior in two dimensions. First, however, we need to define the analog of piecewise C^n. Our image domain is the square $[0, 1[^2$. We call an image $f : [0, 1[^2 \to \mathbb{R}$ *piecewise smooth* if it is of the form

$$f(x) = f_1(x) + \mathbf{1}_\Omega(x)f_2(x) \tag{7.7}$$

see Figure 7.2 for an illustration. Here $\mathbf{1}_\Omega$ is the indicator function of a compact subset $\Omega \subset [0, 1[^2$ with a boundary $\partial\Omega$ that is C^2, by which we mean that there is a twice continuously differentiable

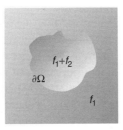

FIGURE 7.2 Example of a piecewise smooth function of the type (7.7).

parametrization of $\partial\Omega$. The functions f_1 and f_2 belong to suitable classes of smooth functions, that may depend on the setting.

For the case that both f_1 and f_2 are C^2 as well, there exist theoretical estimates which yield that generally the optimal approximation rate will be of $O(N^{-2})$ [8]. We are going to show that wavelet bases fall short of this. For the following discussion, it suffices to assume that f_1 and f_2 are in fact constant. Observe that the estimates given below can be verified directly for the two-dimensional Haar wavelet basis and the special case that $f_1 = 1$, $f_2 = 0$, and Ω is the subset of $[0, 1]$ below the diagonal $x = y$. This is a particularly simple example, where the pieces f_1 and f_2 are C^∞, the boundary is a straight line (hence C^∞) and not particularly ill-adapted to the tensor-product setting, *and yet* wavelet bases show poor nonlinear approximation rates. As we will see below, the reason is that as in one dimension the wavelets situated at the boundary still contribute most to the signal energy, but this time their number is no longer under control.

We fix a two-dimensional wavelet basis, constructed in the usual way from a one-dimensional multiresolution analysis; see [23] or Chapter 1 of this volume. Our aim is to describe the approximation error of f in this basis. The argument follows the same pattern as for the one-dimensional case. Whenever a wavelet does not meet the boundary of Ω, the smoothness of the functions f_1, f_2 entails that the wavelet coefficients can be estimated properly. The problems arise when we consider those wavelets that meet the boundary. As before, for each wavelet of scale 2^{-j} meeting the boundary of Ω,

$$|\langle f, \psi_{j,k,l} \rangle| \sim 2^{-j} \tag{7.8}$$

Note that the tilde notation is a sharpened formulation, meaning that there exist upper *and* lower estimates between the two sides, at least for sufficiently many coefficients. (This is easily seen for the example involving Haar wavelets and diagonal boundary.) Hence, even though the scale-dependent decay behavior for the coefficients corresponding to singularities is better than in one dimension, it holds for a crucially *larger* number of coefficients, which spoils the overall performance of the wavelet system. More precisely, as the supports of the wavelets are (roughly) squares of size $\sim 2^{-j}$ shifted along the grid $2^{-j}\mathbb{Z}^2$, the number of wavelets at scale 2^{-j} meeting the boundary is of $O(2^j)$. Thus we obtain $|\theta_m| \sim m^{-1}$, and this results in $\epsilon_N(f) \sim N^{-1}$.

A few observations are in order here: First, note that the arguments we present are indifferent to the smoothness of the boundary; for any smooth boundary of finite length that is not strictly horizontal or vertical we would obtain a similar behavior. This is the blindness of wavelet tensor products to edge smoothness, which we already alluded to above. By construction, wavelets are only designed to represent discontinuities in the horizontal or vertical directions, and cannot be expected to detect connections between neighboring edge points. It should also be noted that the problem cannot be helped by increasing the number of vanishing moments of the wavelet system. (Again, this is readily verified for the diagonal boundary case.)

In the following subsections, we describe recently developed schemes that were designed to improve on this, at least for the continuous setting. The digitization of these techniques will be the subject of Sections 7.3 and 7.4. The following remark contains a disclaimer that we feel to be necessary in connection with the transferal of notions and results from the continuous to the discrete domain.

Remark 7.1

In this chapter, we describe schemes that were originally designed for continuous image domains, together with certain techniques for digitization of these notions. In this respect, this chapter reflects the current state of discussion. The continuous domain setting is in many ways more accessible to mathematical analysis, as witnessed by the smoothness properties that were at the basis of our discussion. The continuous domain viewpoint is advocated, e.g., in the survey paper [15]; we specifically refer to [15, Section IV] for a discussion of the relevance of continuous domain arguments to coding.

It is however not at all trivial to decide how results concerning asymptotic behavior for continuous image domains actually apply to the analysis and design of image approximation schemes for the discrete setting. Observe that all nonlinear approximation results describing the asymptotic behavior for images with bounded domain necessarily deal with *small-scale* limits; for pixelized images, this limit is clearly irrelevant. Also, as we will encounter below, in particular in connection with the notion of angular resolution, the discussion of the continuous setting may lead to heuristics that hardly make sense for digital images.

Note that also in connection with coding applications, the relevance of asymptotic results is not altogether clear: These results describe the right end of the nonlinear approximation curves. Thus they describe how effectively the approximation scheme adds finer and finer details, for numbers of coefficients that are already large, which in compression language means *high-bit-rate coding*. By contrast, from the point of view of compression the left end of the nonlinear approximation curve is by far more interesting. As the approximation results in Section 7.5.1 show, this is also where the new schemes show improved performance, somewhat contrary to the asymptotic results developed for the continuous setting.

7.2.1 WEDGELETS

Wedgelets were proposed by Donoho [16] as a means of approximating piecewise constant images with smooth boundaries. The wedgelet dictionary by definition is given by the characteristic functions of wedge-shaped sets obtained by splitting dyadic squares along straight lines. It is highly redundant, and thus the problem of choosing a suitable representation or approximation of a signal arises. However, wedgelet approximation is not performed by pursuit algorithms or similar techniques typically encountered in connection with redundant dictionaries, but rather driven by a certain functional that depends on a regularization parameter. As we will see below, this approach results in fast approximation algorithms.

For the description of wedgelets, let us first define the set of dyadic squares of size 2^{-j},

$$Q_j = \left\{ [2^{-j}k : 2^{-j}(k+1)[\times [2^{-j}\ell : 2^{-j}(\ell+1)[: 0 \le k, \ell < 2^j \right\} \tag{7.9}$$

and $Q = \bigcup_{j=0}^{\infty} Q_j$. A *dyadic partition* of the image domain is given by any partition (tiling) Q of $[0, 1[^2$ into *disjoint* dyadic squares, not necessarily of constant size. A *wedgelet partition* is obtained by splitting each element $q \in Q$ of a dyadic partition Q into (at most) two *wedges*, $q = w_1 \cup w_2$, along a suitable straight line. The admissible lines used for splitting elements of Q_j are restricted to belong to certain prescribed sets L_j; we will comment on the choice of these sets below. A *wedgelet segmentation* is a pair (g, W) consisting of a wedge partition W and a function g that is constant on all $w \in W$. See Figure 7.3 for an example.

A *wedgelet approximation* of an image f is now given as the minimizer of the functional

$$H_{\lambda, f}(g, W) = \|f - g\|_2^2 + \lambda |W| \tag{7.10}$$

over all admissible wedgelet segmentations (g, W). Here λ acts as a *regularization* or *scale* parameter: For $\lambda = 0$, the minimization algorithm will return the data f, whereas $\lambda \to \infty$ will eventually produce

FIGURE 7.3 (See color insert following page 336) Image IBB North: (a) original image, (b) wedge reconstruction $\lambda = 0.012$, and (c) with corresponding wedge grid superimposed.

a constant image as minimizer. We denote the minimizer of $H_{\lambda,f}$ as $(\widehat{g}_\lambda, \widehat{W}_\lambda)$. The following remark collects the key properties that motivate the choice of wedgelets and the associated functional.

Remark 7.2

(1) Given the optimal partition $\widehat{W}_\lambda$, the optimal $\widehat{g}_\lambda$ is found by a simple projection procedure: For each $w \in \widehat{W}_\lambda$, $\widehat{g}_\lambda|_w$ is simply the mean value of g over w. Hence, finding the optimal wedgelet segmentation is equivalent to finding the optimal partition.

(2) Dyadic partitions are naturally related to quadtrees. More precisely, given a dyadic partition W, consider the set V of all dyadic squares q such that there exists $p \in W$ with $p \subset q$. The inclusion relation induces a *quadtree structure* on V, i.e., V is a directed graph with the property that each node has either zero or four descendants. In this tree, W is just the set of *leaves*, i.e., the set of nodes without descendants.

The quadtree structure is the basis for a fast algorithm for the computation of the optimal wedgelet segmentation W_λ, by recursive application of the following principle: Let $[0, 1[^2 = q_1 \cup q_2 \cup q_3 \cup q_4$ be the decomposition into the four smaller dyadic squares. Then, for a fixed parameter λ, three cases may occur:

(1) $\widehat{W}_\lambda = \{[0, 1[^2\}$;
(2) $\widehat{W}_\lambda$ is obtained by a wedgesplit applied to $[0, 1[^2$;
(3) $\widehat{W}_\lambda = \bigcup_{i=1}^{4} \widehat{V}_\lambda^i$, where each $\widehat{V}_\lambda^i$ is the optimal wedgelet segmentation of q_i associated to the restriction of f to q_i, and to the regularization parameter λ.

Note that for a fixed λ with minimizer $(\widehat{g}_\lambda, \widehat{W}_\lambda)$, $\widehat{g}_\lambda$ is the minimizer of the norm distance $\|f - g\|_2^2$ among all admissible wedgelet segmentations (g, W) with at most $N = |\widehat{W}_\lambda|$ wedges. This observation is used as the basis for the computation of nonlinear approximation rates below.

Let us next consider the nonlinear approximation behavior of the scheme. The following technical lemma counts the dyadic squares meeting the boundary $\partial\Omega$. Somewhat surprisingly, the induced dyadic partition grows at a comparable speed.

Lemma 7.3

Let f be piecewise constant, with C^2 boundary $\partial\Omega$. Let $Q_j(f)$ denote the set of dyadic square $q \in Q_j$ meeting $\partial\Omega$. Then, there exists a constant C such that $|Q_j(f)| \leq 2^j C$ holds for all $j \geq 1$. Moreover, for each j there exists a dyadic partition W_j of $[0, 1]^2$ containing $Q_j(f)$, with $|W_j| \leq 3C2^j$.

Proof

The statement concerning $Q_j(f)$ is straightforward from a Taylor approximation of the boundary. The dyadic partition W_j is constructed inductively: W_{j+1} is obtained by replacing each dyadic square in $Q_j(f)$ by the four dyadic squares of the next scale. Hence

$$|W_{j+1}| = |W_j| - |Q_j(f)| + 4|Q_j(f)| = |W_j| + 3|Q_j(f)| \tag{7.11}$$

Now an easy induction shows the claim on $|W_j|$. □

We obtain the following approximation result. The statement is in spirit quite close to the results in [16], except that we use a different notion of resolution for the wedgelets, which is closer to our treatment of the digital case later on.

Theorem 7.4

Let f be piecewise constant with C^2 boundary. Assume that the set L_j consists of all lines taking the angles $\{-\pi/2 + 2^{-j}\ell\pi : 0 \leq \ell < 2^j\}$. Then the nonlinear wedgelet approximation rate for f is $O(N^{-2})$, meaning that for $N \in \mathbb{N}$ there exists a wedgelet segmentation (g, W) with $|W| \leq N$ and $\|f - g\|_2^2 \leq CN^{-2}$.

Proof

For $N = 2^j$, the previous lemma provides a dyadic partition W_j into $O(N)$ dyadic squares, such that $\partial\Omega$ is covered by the elements of $Q_j \cap W_j$. Observe that only those dyadic squares contribute to the squared approximation error. In each such square, a Taylor approximation argument shows that the boundary can be approximated by a straight line in $O(2^{-2j})$ precision. The required angular resolution allows to approximate the optimal straight line by a line from L_j up to the same order of precision. Now the incurred squared L^2-error is of order $O(2^{-3j})$; the additional $O(2^{-j})$ factor is the diameter of the dyadic squares. Summing over the $O(2^j)$ squares yields the result. □

We note that the theorem requires that the number of angles increases as the scale goes to zero; the *angular resolution of L_j* scales linearly with 2^j. Observe that this requirement does not make much sense as we move on to digital images. In fact, this is the first instance where we encounter the phenomenon that intuitions from the continuous model prove to be misleading in the discrete domain.

7.2.2 CURVELETS

Curvelets are in many ways conceptually closer to wavelets than wedgelets. While wedgelet approximation may be read as the approximation of the image using building blocks that are specifically chosen for the image at hand, curvelet approximation is based on the decomposition of the image into a *fixed* system of components, prescribed without prior knowledge of the image. The *curvelet system* is a family of functions $\gamma_{j,l,k}$ indexed by a scale parameter j, an orientation parameter l, and a position parameter $k \in \mathbb{R}^2$, yielding a *normalized tight frame* of the image space. The latter property amounts to postulating for all $f \in L^2(\mathbb{R}^2)$,

$$\|f\|_2^2 = \sum_{j,k,l} |\langle f, \gamma_{j,k,l} \rangle|^2 \tag{7.12}$$

or equivalently, that each f can be expanded as

$$f = \sum_{j,k,l} \langle f, \gamma_{j,k,l} \rangle \gamma_{j,k,l} \tag{7.13}$$

see [23, Chapter V]. This expansion is reminiscent of orthonormal bases; however, the elements of a tight frame need not be pairwise-orthogonal; in fact they need not even be linearly independent.

In the curvelet setting, image approximation is performed by expanding the input in the curvelet frame and quantizing the coefficients, just as in the wavelet setting. However, the effectiveness of the approximation scheme critically depends on the *type of scaling*, and the sampling of the various parameters. Unlike the classical, group theoretical construction of 2D wavelets, as introduced by Antoine and Murenzi [1,2], which also incorporates scale, orientation, and position parameters, the scaling used in the construction of curvelets is *anisotropic*, resulting in atoms that are increasingly more needlelike in shape as the scale decreases.

Let us know delve into the definition of curvelets. The following construction is taken from [6], which describes the most recent generation of curvelets. A precursor was constructed in [5] ("curvelets 99" in the terminology of [6]), which has a more complicated structure, relying on additional windowing and the ridgelet transform. A comparison of the two types of curvelets is contained in [6]. Both constructions are different realizations of a core idea, which may be summarized by the catchphrase that the curvelet system corresponds to a *critically sampled, multiscale directional filterbank, with angular resolution behaving like* $1/\sqrt{\text{scale}}$.

As the filterbank view suggests, curvelets are most conveniently constructed on the frequency side. The basic idea is to cut the frequency plane into subsets that are cylinders in polar coordinates. The cutting needs to be done in a smooth way, however, in order to ensure that the resulting curvelets are rapidly decreasing.

For this purpose, we fix two window functions:

$$v : [-\pi, \pi] \to \mathbb{C}, \qquad w : \mathbb{R}^+ \to \mathbb{C} \tag{7.14}$$

Both v and w are assumed smooth; in addition, for v we require that its 2π-periodic extension v_{per} is smooth as well. In addition, we pick w to be compactly supported. v acts as angular window in order to guarantee that the functions constructed from v are even. Moreover, we impose that v fulfills

(for almost every ϑ)

$$|v_{\text{per}}(\theta)|^2 + |v_{\text{per}}(\theta + \pi)|^2 = 1 \tag{7.15}$$

which guarantees that the design of curvelets later on covers the full range of angles. Equation (7.15) allows one to construct partitions of unity of the angular domain into a dyadic scale of elements: Defining $v_{j,l} = v(2^j\theta - \pi l)$, for $l = 0, \ldots, 2^j - 1$ on $[-\pi, \pi]$, and extended periodically, it is easily verified that

$$\sum_{l=0}^{2^j - 1} |v_{j,l}(\theta)|^2 + |v_{j,l}(\theta + \pi)|^2 = 1 \tag{7.16}$$

A similar decomposition is needed for the scale variable. Here we make the additional assumption that

$$|w_0(s)|^2 + \sum_{j=1}^{\infty} |w(2^j s)|^2 = 1 \tag{7.17}$$

for a suitable compactly supported C^∞-function w_0. w should be thought of as a bump function concentrated in the interval $[1, 2]$.

In the following, we will frequently appeal to polar coordinates, i.e., we will identify $\xi \in \mathbb{R}^2 \backslash \{0\}$ with $(|\xi|, \theta_\xi) \in \mathbb{R}^+ \times]-\pi, \pi]$. The scale and angle windows allow a convenient control on the design of the curvelet system. The missing steps are now to exert this control to achieve the correct scale-dependent angular resolution, and to find the correct sampling grids (which shall depend on scale and orientation). For the first part, filtering with the scale windows $(w_j)_{j \geq 0}$ splits a given signal into its frequency components corresponding to the annuli $A_j = \{\xi \in \mathbb{R}^2 : 2^j \leq |\xi| < 2^{j+1}\}$. In each such annulus, the number of angles should be of order $2^{j/2}$, following the slogan that the angular resolution should scale as $1/\sqrt{\text{scale}}$. Thus we define the scale-angle window function $\eta_{j,l}$, for $j \geq 1$ and $0 \leq l \leq 2^{\lfloor j/2 \rfloor} - 1$, on the Fourier side by

$$\widehat{\eta}_{j,l}(\xi) = w_j(|\xi|)(v_{2^{\lfloor j/2 \rfloor},l}(\theta_\xi) + v_{2^{\lfloor j/2 \rfloor},l}(\theta_\xi + \pi)) \tag{7.18}$$

In addition, we let $\widehat{\eta}_{0,0}(\xi) = w_0(|\xi|)$, which is responsible for collecting the low-frequency part of the image information. Up to normalization and introduction of the translation parameter, the family $(\eta_{j,l})_{j,l}$ is the curvelet system. By construction, $\widehat{\eta}_{j,l}$ is a function that is concentrated in the two opposite wedges of frequencies

$$W_{j,l} = \{\xi \in \mathbb{R}^2 : 2^j \leq |\xi| \leq 2^{j+1}, \theta_\xi \quad \text{or} \quad \theta_\xi + \pi \in [2^{-\lfloor j/2 \rfloor} l, 2^{-\lfloor j/2 \rfloor}(l+1)]\} \tag{7.19}$$

as illustrated in Figure 7.4.

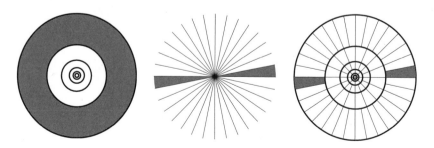

FIGURE 7.4 Idealized frequency response of the curvelet system. Filtering using the scale window w_j, followed by filtering with the angular window $v_{j,l}$, yields the frequency localization inside the wedge $W_{j,l}$. Observe the scaling of the angular resolution, which doubles at every other dyadic step.

Now (7.17) and (7.16) imply for almost every $\xi \in \mathbb{R}^2$ that

$$\sum_{j,l} |\widehat{\eta}_{j,l}(\xi)|^2 = 1 \tag{7.20}$$

and standard arguments allow to conclude from this that convolution with the family $(\eta_{j,l})_{j,l}$ conserves the L^2-norm of the image, i.e.,

$$\|f\|_2^2 = \sum_{j,l} \|f * \eta_{j,l}\|_2^2 \tag{7.21}$$

The definition of the *curvelet frame* is now obtained by critically sampling the isometric operator $f \mapsto (f * \eta_{j,l})_{j,l}$. The following theorem states the properties of the resulting system; see [6] for a proof.

Theorem 7.5

There exist frequency and scale windows v, w, normalization constants $c_{j,l} > 0$, and sampling grids $\Gamma_{j,l} \subset \mathbb{R}^2$ (for $j \geq 0, l = 0, \ldots, 2^{\lfloor j/2 \rfloor} - 1$) with the following properties: Define the index set

$$\Lambda = \{(j,l,k) : j \geq 0, l = 0, \ldots, 2^{\lfloor j/2 \rfloor}, k \in \Gamma_{j,l}\}$$

Then the family $(\gamma_\lambda)_{\lambda \in \Lambda}$, defined by

$$\gamma_{j,l,k}(x) = c_{j,l}\eta_{j,l}(x - k)$$

is a normalized tight frame of $L^2(\mathbb{R})$, yielding an expansion

$$f = \sum_{\lambda \in \Lambda} \langle f, \gamma_\lambda \rangle \gamma_\lambda, \qquad \forall f \in L^2(\mathbb{R}^2) \tag{7.22}$$

The following list collects some of the geometric features of the curvelet system:

1. The shift in the rotation parameter implies that $\gamma_{j,l}(x) = \gamma_{j,0}(R_{\theta_{j,l}}x)$, where $\theta_{j,l} = \pi l 2^{-\lfloor j/2 \rfloor}$, and R_θ denotes the rotation matrix

$$R_\theta = \begin{pmatrix} \cos\theta & \sin\theta \\ -\sin\theta & \cos\theta \end{pmatrix} \tag{7.23}$$

2. $\widehat{\gamma}_{j,0,0}$ is essentially supported in a union of two rectangles of dimensions $O(2^j \times 2^{\lfloor j/2 \rfloor})$, and the associated sampling lattice can be chosen as $\Gamma_{j,0} = \delta_{1,j}\mathbb{Z} \times \delta_{2,j}\mathbb{Z}$, with $\delta_{1,j} \sim 2^{-j}$ and $\delta_{2,j} \sim 2^{-\lfloor j/2 \rfloor}$. As could be expected from the previous observation, $\Gamma_{j,l} = R_{\theta_{j,l}}\Gamma_{j,0}$.
3. By the previous observation, the change of sampling lattices from j to $j+2$ follows an *anisotropic scaling law*,

$$\Gamma_{j+2,0} \approx D^{-1}\Gamma_{j,0}, \quad \text{where } D = \begin{pmatrix} 4 & 0 \\ 0 & 2 \end{pmatrix} \tag{7.24}$$

The discussion in [6] suggests that, at least conceptually, it is useful to think of all curvelets to be descended from the two basic curvelets $\gamma_{1,0}$ and $\gamma_{2,0}$, by iterating the relation $\gamma_{j+2,0}(x) \approx \det(D)^{1/2}\gamma_{j,0}(Dx)$.

4. Summarizing, the $\gamma_{j,l}$ are a system of rapidly decreasing functions that oscillate at speed of order $2^{\lfloor j/2 \rfloor}$, primarily in the $(\cos\theta_{j,l}, \sin\theta_{j,l})$ direction. As $j \to \infty$, the essential support of $\gamma_{j,l,0}$ scales like a rectangle of size $2^{-j} \times 2^{-\lfloor j/2 \rfloor}$, when viewed in the appropriate coordinate system.

The following theorem shows that up to a logarithmic factor the curvelet system yields the desired nonlinear approximation behavior for piecewise C^2 functions. One of the remarkable features of the theorem is that the approximation rate is already achieved by simple nonlinear truncation of Equation (7.22). Observe that this is identical with best N-term approximation only for orthogonal bases; however, the curvelet system is only a tight frame, and cannot be expected to be an orthogonal basis.

Theorem 7.6 ([6, Theorem 1.3])

Let f be a piecewise C^2 function with C^2 boundary. For $N \in \mathbb{N}$, let $\Lambda_N(f) \subset \Lambda$ denote the indices of the N largest coefficients. Then there exists a constant $C > 0$ such that

$$\left\| f - \sum_{\lambda \in \Lambda_N} \langle f, \gamma_\lambda \rangle \gamma_\lambda \right\|_2^2 \le CN^{-2}(\log N)^3 \tag{7.25}$$

For a detailed proof of the theorem we refer to [6]. In the following we present a shortened version of the heuristics given in [6]. They contrast nicely to the wavelet case discussed above, and motivate in particular the role of anisotropic scaling for the success of the curvelet scheme. A graphically intuitive way of understanding the proof is to first observe that the locally smooth parts of the image are captured largely by the translates of the low-pass filter $\gamma_{0,0,0}$. The remaining image information is therefore located in the edge, and the whole point of the curvelet construction is to provide "brushstrokes" that are increasingly precise as their scale goes to zero, both in the possible orientations and in their breadth (due to the anisotropic scaling).

Suppose we are given a piecewise smooth image f, and a fixed scale index j. We start the argument by a geometric observation that motivates the use of anisotropic scaling and rotation. Recall from the wavelet case that $O(2^j)$ dyadic squares of size 2^{-j} are needed to cover the boundary $\partial\Omega$. This time we consider a covering by rectangles of size $2^{-\lfloor j/2 \rfloor} \times 2^{-j}$, which may be arbitrarily rotated. Then a Taylor approximation of the boundary shows that this can be done by $O(2^{j/2})$ such rectangles, which is a vital improvement over the wavelet case.

Next we want to obtain estimates for the scalar products $\langle f, \gamma_{j,l,k} \rangle$, depending on the position of the curvelet relative to the boundary. Recall that $\gamma_{j,l,k}$ is a function that has elongated essential support of size $2^{-\lfloor j/2 \rfloor} \times 2^{-j}$, in the appropriately rotated coordinate system, and oscillates in the "short" direction.

Then there are basically three cases to consider, sketched in Figure 7.5:

(1) *Tangential:* The essential support of $\gamma_{j,l,k}$ is close in position and orientation to one of the covering boxes, i.e., it is tangent to the boundary.
(2) *Transversal:* The essential support is close in position to one of the covering boxes, but not in orientation. Put differently, the support intersects the boundary at a significant angle.
(3) *Disjoint:* The essential support does not intersect the boundary.

The argument rests on the intuition that only the tangential case yields significant coefficients. One readily expects that the disjoint case leads to negligible coefficients: the image is smooth away from the boundary, hence the oscillatory behavior of the curvelet will cause the scalar product to be small. By looking more closely at the *direction* of the oscillation, we can furthermore convince ourselves that the transversal case produces negligibly small coefficients as well: The predominant

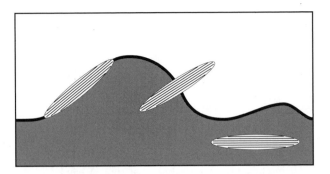

FIGURE 7.5 An illustration of the three types of curvelet coefficients. The essential supports of the curvelets are shown as ellipses, with indicated oscillatory behavior. From left to right: tangential, transversal, and disjoint case.

part of the essential support is contained in regions where f is smooth, and the oscillations across the short direction imply that this part contributes very little to the scalar product.

Thus, we have successfully convinced ourselves that only the tangential case contributes significant coefficients. Here we apply the same type of estimate that we already used in the wavelet cases, i.e.,

$$|\langle f, \gamma_{j,l,k}\rangle| \le \|f\|_\infty \|\gamma_{j,l,k}\|_1 \le C2^{-3j/4} \tag{7.26}$$

because of the choice of normalization coefficients $c_{j,l}$. Since there are $O(2^{j/2})$ boxes, the sampling of the position and angle parameter in the curvelet system implies also $O(2^{j/2})$ coefficients belonging to the tangential case. Ignoring the other coefficients, we therefore have produced — rather informal — evidence for the statement that the sorted coefficients obey the estimate

$$|\theta_m| \le Cm^{-3/2} \tag{7.27}$$

As we have already mentioned above, the normalized tight frame property of the curvelet system is equivalent to the statement that the *coefficient operator* assigning each function f its curvelet coefficients $(\langle f, \gamma_{j,l,k}\rangle_{(j,l,k)\in\Lambda})$ is an isometric mapping from $L^2(\mathbb{R}^2)$ to $\ell^2(\Lambda)$, the latter being the Hilbert space of square-summable coefficient sequences. The adjoint of this operator is the *reconstruction operator* mapping each coefficient sequence $(\alpha_\lambda)_{\lambda\in\Lambda} \in \ell^2(\Lambda)$ to the sum $\sum_{\lambda\in\Lambda} \alpha_\lambda \gamma_\lambda$; as the adjoint of an isometry, this operator is normdecreasing. This allows to finish the argument by the estimate

$$\left\| f - \sum_{\lambda\in\Lambda_N} \langle f, \gamma_\lambda\rangle\gamma_\lambda \right\|_2^2 = \left\| \sum_{\lambda\in\Lambda\backslash\Lambda_N} \langle f, \gamma_\lambda\rangle\gamma_\lambda \right\|_2^2 \le \sum_{\lambda\in\Lambda_N} |\langle f, \gamma_\lambda\rangle|^2$$

$$= \sum_{m=N+1}^\infty |\theta_m|^2 \le CN^{-2}$$

Observe that the logarithmic factor in the statement of Theorem 7.6 has disappeared in the course of the argument. This is just an indicator of the degree of oversimplification of the presentation.

Remark 7.7

We close the section by citing another observation from [6], which allows a neat classification of wavelet, curvelet, and ridgelet schemes by means of their angular resolution: Wavelets have a constant angular resolution, for curvelets the angular resolution behaves like $1/\sqrt{\text{scale}}$, and for ridgelets like $1/\text{scale}$.

7.2.3 ALTERNATIVE APPROACHES

Besides the two methods described in the previous subsections, various recent models for the information content of natural images were developed from different heuristic principles. It is outside the scope of this chapter to describe all of them in detail; in the following we briefly sketch some of the more prominent approaches.

Another interesting recently developed method that combines wavelet theory with geometric partitioning of the image domain are bandelets [20,21], which make use of redundancies in the geometric flow, corresponding to local directions of the image gray levels considered as a planar one-dimensional field. The geometry of the image is summarized with local clustering of similar geometric vectors, the homogeneous areas being taken from a quadtree structure. A bandelet basis can be viewed as an adaptive wavelet basis, warped according to the locally selected direction. Bandelet decomposition achieves optimal approximation rates for C^α functions. This method presents similarities with optimal wedgelet decompositions in that it uses geometric partitioning of the image domain, according to the minimization of a certain complexity-distortion functional. For instance, bandelets decomposition combined with a rate-distortion method leads to a quite competitive compression scheme. The geometric part can be used to incorporate prior knowledge, allowing to adapt the scheme to very specific classes of images, such as ID photos; see the web site [22] for the description of a bandelet coder of ID images.

Another approach that uses oriented wavelet bases has been introduced under the name "directionlets" [30]. Like wavelets, directionlets rely on separable filters; however, the filtering is performed along digital lines that are not necessarily horizontal or vertical, resulting in improved approximation rates for certain classes of primarily geometric images.

A further approximation scheme is based on the use of triangulations, which corresponds to a quite different philosophy. By their flexibility, adaptive irregular triangulations have very good approximation behavior. It can be shown that the optimal rates of approximation can be attained (see [21]) when we require that every conform triangulation is allowed for the representation. The main problem encountered by these methods is the sheer number of possible triangulations. In practice, especially for the purpose of implementation, one is forced to consider highly reduced triangulations classes, while still trying to obtain nearly optimal results.

To mention an example of such an approach, the method proposed in [10] uses a greedy removal of pixels, minimizing at each step the error among the possible triangulations. The class of triangulations under consideration is reduced to the set of Delaunay triangulations of a finite set of pixel positions, which allows a simplified parameterization, only using the point coordinates, without any connectivity information about the according triangulation. This fact is employed in a suited scattered data compression scheme. For natural images, the rate-distortion performances achieved are comparable with those obtained by wavelet methods, leading to very different kind of artifacts. In particular, it avoids ringing artifacts, but smooths textured areas.

An approach which is in a sense dual to the majority of the schemes described here are *brushlets*, introduced by Meyer and Coifman [24]. While most of the approaches we mentioned so far involve some form of spatial adaptation to the image content, brushlet approximations are based on the adaptive tiling of the frequency plane. As might be expected from this description, the experiments in [24] show that brushlets are quite well adapted to the representation of periodic textures, which shows that the brushlet approach is in a sense complementary to geometric approaches such as wedgelets. By construction, brushlets have trouble dealing with piecewise smooth images, which constitute the chief class of benchmark signals in this chapter. It is well known that the Fourier transform is particularly ill-suited to dealing with piecewise smooth data. Hence, any scheme that uses the Fourier transform of the image as primary source of information will encounter similar problems; for brushlets this effect can also be examined in the examples displayed in [24].

Finally, let us mention dictionary-based methods, usually employed in connection with pursuit algorithms. As most of the approaches described in this chapter are based more or less explicitly on redundant systems of building blocks, there are necessarily some similarities to dictionary-based methods. The use of highly redundant dictionaries for image representations is the subject of a separate chapter in this volume, to which we refer the interested reader.

7.3 DIGITAL WEDGELETS

Let us now turn to discrete images and algorithms. In this section, we describe a digital implementation of Donoho's wedgelet algorithm. For notational convenience, we suppose the image domain to be $\Omega = \{0, \ldots, 2^J - 1\} \times \{0, \ldots, 2^J - 1\}$. In this setting, dyadic squares are sets of the type $\{2^j k, \ldots, 2^j(k+1) - 1\} \times \{2^j \ell, 2^j(\ell+1) - 1\}$, with $0 \leq k, \ell < 2^{J-j}$. Our goal is to describe an efficient algorithm that for a given image $f \in \mathbb{R}^\Omega$ computes a minimizer of

$$H_{\lambda,f}(g, W) = \|f - g\|_2^2 + \lambda |W| \tag{7.28}$$

where W is a wedge partition of Ω and g is constant on each element of W. As in the continuous setting, wedge partitions are obtained from dyadic partitions by splitting dyadic squares along straight lines. It turns out that there are several options of defining these notions; already the digitization of lines is not as straightforward an issue as one might expect.

In the following, we use the definitions underlying a recent implementation, described in more detail in [11,19]. Other digitizations of wedgelets can be used for the design of wedgelet algorithms, and to some extent, the following definitions are just included for the sake of concreteness. However, as we explain below, they also provide particularly fast algorithms.

We fix a finite set $\Theta \subset]-\pi/2, \pi/2[$ of *admissible angles*. The admissible discrete wedgelets are then obtained by splitting dyadic squares along lines meeting the x-axis at an angle $\theta \in \Theta$.

Definition 7.8

For $\theta \in]-\pi/2, \pi/2]$ let $v_\theta^\perp = (-\sin\theta, \cos\theta)$. Moreover, define

$$\delta = \max\{|\sin\theta|/2, |\cos\theta|/2\} \tag{7.29}$$

The **digital line through the origin in direction** v_θ is then defined as

$$L_{0,\theta} = \{p \in \mathbb{Z}^2 : -\delta < \langle p, v_\theta^\perp \rangle \leq \delta\} \tag{7.30}$$

Moreover, we define $L_{n,\theta}$, for $n \in \mathbb{Z}$, as

$$L_{n,\theta} = \begin{cases} \{p + (n, 0) : p \in L_{0,\theta}\}, & |\theta| > \pi/4 \\ \{p + (0, n) : p \in L_{0,\theta}\}, & |\theta| \leq \pi/4 \end{cases} \tag{7.31}$$

In other words, $L_{n,\theta}$ is obtained by shifting $L_{0,\theta}$ by integer values in the vertical direction for *flat lines*, and by shifts in the horizontal direction for *steep lines*. In [11] we prove that the set $(L_{n,\theta})_{n \in \mathbb{Z}}$ partitions $\mathbb{Z}^2$, i.e., $\mathbb{Z}^2 = \bigcup_{n \in \mathbb{Z}}^\bullet L_{n,\theta}$; see also [19, Section 3.2.2].

Now we define the *discrete wedge splitting* of a square.

Definition 7.9

Let $q \subset \mathbb{Z}^2$ be a square, and $(n, \theta) \in \mathbb{Z} \times [-\pi/2, \pi/2[$. The *wedge split induced by* $L_{n,\theta}$ is the partition of q into the sets $\{w_{n,\theta}^1(q), w_{n,\theta}^2(q)\}$ defined by

$$\begin{cases} w_{n,\theta}^1(q) = \bigcup_{k \leq n} L_{k,\theta} \cap q \\ w_{n,\theta}^2(q) = \bigcup_{k > n} L_{k,\theta} \cap q \end{cases} \tag{7.32}$$

Our description of discrete wedgelets is somewhat nonstandard owing to the fact that we use a globally defined set of angles and lines. The advantage of our definition is that it allows the efficient solution of the key problem arising in rapid wedgelet approximation, namely the efficient computation of image mean values over wedges of varying shapes and sizes. In a sense, this latter problem is the only serious challenge that is left after we have translated the observations made for the continuous setting in Remark 7.2 to the discrete case. Again the minimization problem is reduced to finding the best wedgelet segmentation, and the recursive minimization procedure is fast, provided that for every dyadic interval the optimal wedgesplit is already known. The latter problem requires the computation of mean values for large numbers of wedges, and here our definitions pay off.

In the following two sections, we give algorithms dealing with a somewhat more general model, replacing locally constant by locally polynomial approximation. In other words, we consider minimization (7.28) for functions g that are given by polynomials of fixed degree r on the elements of the segmentation. Thus, the following also applies to the platelets introduced in [35]. The more general problem requires the computation of higher degree image moments over wedge domains, but is otherwise structurally quite similar to the original wedgelet problem. There exists a freely available implementation of models up to order 2, which can be downloaded from [34].

Section 7.3.1 sketches the key technique for moment computation. It relies on precomputed lookup tables containing cumulative sums over certain image domains. The number of lookup tables grows linearly with the number of angles in Θ and the number of required moments. This way, the *angular resolution* of the discrete wedges can be prescribed in a direct and convenient way, and at linear cost, both computational and in terms of memory requirements. Section 7.3.2 contains a summary of the algorithm for the minimization of (7.28). For more details refer to [11,19].

7.3.1 RAPID SUMMATION ON WEDGE DOMAINS: DISCRETE GREEN'S THEOREM

As explained above, efficient wedgelet approximation requires the fast evaluation integrals of the form $\int_w f(x,y)\,\mathrm{d}x\,\mathrm{d}y$, over all admissible wedges w. For higher order models, image moments of the form $\int_w f(x,y)x^i y^j\,\mathrm{d}x\,\mathrm{d}y$ need to be computed; for the discrete setting, the integral needs to be replaced by a sum. The following is a sketch of techniques developed to provide a fast solution to this problem.

For exposition purposes, let us first consider the continuous setup. We let Q_+ denote the positive quadrant, $Q_+ = \mathbb{R}_0^+ \times \mathbb{R}_0^+$. Given $z \in Q_+$, and $\theta \in\,] -\pi/2, \pi/2]$, let $S_\theta(z) = z + \mathbb{R}^-(\cos\theta, \sin\theta) \cap Q^+$. Moreover, denote by $\Omega_\theta(z) \subset Q^+$ the domain that is bounded by the coordinate axes, the vertical line through z, and $S_\theta(z)$; see Figure 7.6. Define the auxiliary function $K_\theta : Q^+ \to \mathbb{R}$ as

$$K_\theta(z) = \int_{\Omega_\theta(z)} f(x,y)\,\mathrm{d}x\,\mathrm{d}y \tag{7.33}$$

note that this implies $K_{\pi/2} = 0$, as the integral over a set of measure zero.

Let us now consider a wedge of fixed shape, say a trapezoid w, with corners z_1, z_2, z_3, z_4, as shown in the right-hand part of Figure 7.6. Then Equation (7.33) implies that

$$\int_w f(x,y)\,\mathrm{d}x\,\mathrm{d}y = K_\theta(z_4) - K_\theta(z_3) - K_0(z_1) + K_0(z_2) \tag{7.34}$$

In order to see this, observe that $w = \Omega_\theta(z_4) \setminus (\Omega_\theta(z_3) \cup \Omega_0(z_1))$. Hence the integral over w is obtained by subtracting from $K_\theta(z_4)$ the integrals over $\Omega_\theta(z_3)$ and $\Omega_0(z_1)$, and then adding the part that is subtracted twice, i.e., the integral over $\Omega_\theta(z_3) \cap \Omega_0(z_1) = \Omega_0(z_2)$.

Note that the evaluation on the right-hand side of (7.34) involves only four operations, supposing that K_θ and K_0 are known. Similar results can then be obtained for the different kind of wedge domains arising in the general scheme, and more generally for all polygonal domains with boundary

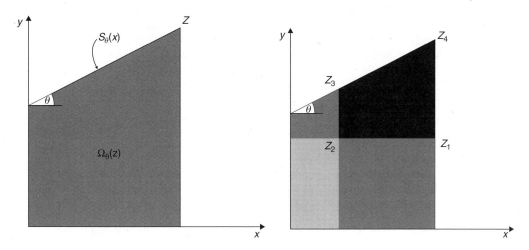

FIGURE 7.6 Left: The sets $S_\theta(z)$ and $\Omega_\theta(z)$. Right: An illustration of the argument proving (7.34).

segments belonging to angles in Θ. As a side remark, these considerations in fact just describe a special case of Green's theorem; see [11] for a more complete discussion of this connection, and for a description of the discrete implementation of this formula.

The discrete analogs K_θ^d of the auxiliary functions can be stored in matrices of the same size as the image, and they are efficiently computable in linear time by cumulative summation first in the vertical direction, and then along the lines $L_{n,\theta}$. As the main result of this discussion, we record the following theorem. For the rest of this section, $N = 2^{2J}$ denotes the number of pixels.

Theorem 7.10

For any angle $\theta \in]-\pi/2, \pi/2]$, the auxiliary matrix K_θ^d is computable in $O(N)$. After computing K_θ^d and K_0^d, the sum $\sum_{(x,y)\in w} f(x,y)$ is obtainable using at most six operations, for every wedge domain w obtained by splitting a dyadic square along a line with angle θ.

7.3.2 IMPLEMENTATION

Now, combining Donoho's observations from Remark 7.2 with the techniques outlined in the previous section, we obtain the following algorithm for the minimization of (7.28).

(1) Compute the auxiliary matrices $K_{\theta,i,j}^d$, for all $\theta \in \Theta$, which are necessary for the computation of the moment of index i,j to be used in the next steps. Local regression of order r requires $(r+1)(r+2)/2$ such moment matrices. By the considerations in the previous subsection, the overall memory and time requirements for this computation step is therefore $(r+1)$ $(r+2)/2 \times N \times |\Theta|$.

(2) For each dyadic square q, we need to select a best local wedge-regression model among the possible wedge splitting of this square. For each digital line l, we compute the $(r+1)(r+2)/2$ moments in fixed time, using the appropriate version of Equation (7.34). This allows to solve the corresponding regression problems over w_l^1 and w_l^2, which requires $O(r^3)$ flops. Finally, we compute the according discrete l^2-error. This procedure applies to the $|\Theta|2^{j+1}$ admissible discrete lines passing through q.

For each q, we need then to store the line $\hat{l}_{n_q,\theta_q}$ which corresponds to the minimal error, the associated two sets of optimal coefficients of the local regression models, and the incurred squared error E_q.

The whole procedure needs to be carried out for all $(2N - 1)/3$ dyadic squares.

(3)   Once Step 2 is finished, we are in a position to determine the wedgelet partition $\widehat{W}_\lambda(f)$, which minimizes (7.28) for a given parameter λ, using the algorithm sketched in Remark 7.2. The algorithm runs through all dyadic squares, starting from the smallest ones, i.e., single pixels.

Hence, if we consider a dyadic square q, its children q_i, $i = 1, \ldots, 4$, have already been treated, and we know an optimal partition for each q_i, denoted by $\widehat{W}_\lambda(q_i)$, and also the associated error $E_{q_i,\lambda}$ and penalization $\lambda |(\widehat{W}_\lambda(q_i))|$.

The optimal partition of A is then the result of the comparison of two partitions, $\widehat{W}(q)$ and $\cup_{i=1}^4 \widehat{W}_\lambda(q_i)$. The associated error is given by

$$E_{q,\lambda} = \min \left\{ E_q + \lambda |\widehat{W}(q)|, \sum_{i=1}^4 E_{q_i,\lambda} \right\} \tag{7.35}$$

and according to the result of the comparison, we store the corresponding optimal wedge partition $\widehat{W}_{q,\lambda}$. The process stops at the top level, yielding the minimizer $(\widehat{g}_\lambda, \widehat{W}_\lambda)$.

We summarize the results concerning the computational costs in the following proposition.

Proposition 7.11

Step 1 requires $O(aNr^3)$ flops and a memory storage of $O(aN^2r^3)$. Step 2 also requires $O(aNr^3)$ flops and a memory storage in $O(rN)$. Step 3 requires $O(N)$ flops.

The following observations are useful for fine-tuning the algorithm performance:

Remark 7.12

(a) In actual implementation, allocation for the auxiliary matrices storing K_θ^d turns out to be a crucial factor. A closer inspection shows that in the Steps 1 and 2 the angles in Θ can be treated *consecutively*, thus reducing memory requirement to $O(Nr^3)$. This results in a considerable speedup.

(b) The use of a fixed set Θ of angles for splitting dyadic squares of varying size is not very efficient. For small dyadic squares, a small difference in angles yields identical wedgesplits. Roughly speaking, a dyadic square of size 2^j can resolve $O(2^j)$ angles. It is possible to adapt the algorithm to this scaling. Note that this scale-dependent angular resolution is precisely the inverse of what is prescribed by Theorem 7.4. Numerical experiments, documented in [19, Section 6.3], show that this scale-dependent angular resolution leads to the same approximation rates as the use of a full set of 2^J angles, valid for all scales.

(c) A further highly interesting property of the algorithm is the fact that only the last step uses the regularization parameter. Thus the results of the previous steps can be recycled, allowing fast access to $(\widehat{g}_\lambda, \widehat{W}_\lambda)$ for *arbitrary* parameters λ.

7.4   DIGITAL CURVELETS: CONTOURLETS

The curvelet construction relies on features that are hard to transfer to the discrete setting, such as polar coordinates and rotation. Several approaches to digital implementation have been developed since the first inception of curvelets; see, e.g., [12,13,28]. In the following we present the approach introduced by Do and Vetterli [13], which to us seems to be the most promising among the currently available implementations, for several reasons. It is based on fast filterbank algorithms with perfect reconstruction; i.e., the tight frame property of curvelets is fully retained, in an algorithmically efficient manner. Moreover, the redundancy of the transform is 1.333, which is by far better than the factor $16J + 1$ (J is the number of dyadic scales in the decomposition) reported in [28]. It is clear that from a coding perspective, redundancy is a critical issue.

The starting point for the construction of contourlets is the observation that computing curvelet coefficients can be broken down into the following three steps (compare Figure 7.4):

(1) Bandpass filtering using the scale windows w_j.
(2) Directional filtering using the angular windows $v_{j,l}$.
(3) Subsampling using the grids $\Gamma_{j,l}$, resulting in a tight frame expansion with associated inversion formula.

The discrete implementation follows an analogous structure (see Figure 7.7 and Figure 7.8):

(1) The image is passed through a pyramid filterbank, yielding a sequence of bandpassed and subsampled images.
(2) Directional filterbanks [3,12] are applied to the difference images in the pyramid, yielding directionally filtered and critically subsampled difference images. The angular resolution is controlled in such a way as to approximate the scaling of the angular resolution prescribed for the curvelet system.
(3) The directional filterbanks have an inherent subsampling scheme, which makes them orthogonal when employed with perfect reconstruction filters. Combining this with a perfect reconstruction pyramid filterbank, the whole system becomes a perfect reconstruction filterbank with a redundancy factor of 1.333 inherited from the pyramid filter.

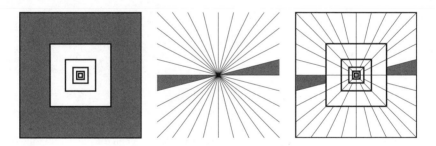

FIGURE 7.7 Idealized frequency response of the contourlet system. The scaling of the angular resolution is controlled by employing a suitable directional filterbank.

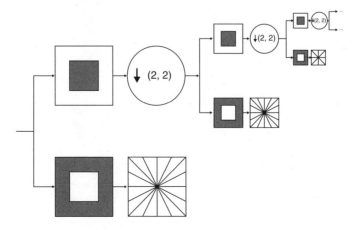

FIGURE 7.8 Structure of the contourlet decomposition.

The filterbank uses time-domain filtering, leading to linear complexity decomposition and reconstruction algorithms. The effect of combined bandpass and directional filtering can be inspected in the sample decomposition of a geometric test image in Figure 7.9. The filterbank implementation computes the coefficients of the input image with respect to a family of discrete curvelets or *contourlets*. A small sample of this family is depicted in Figure 7.10, showing that the anisotropic scaling properties of the continuous domain curvelet system are approximately preserved.

In connection with angular resolution, we note a similar phenomenon as for the discrete wedgelet case. Recall that for wedgelets, the continuous domain theory prescribed an angular resolution that increases as the scale of the dyadic squares decreases, and that this requirement made little sense for digital images. In the curvelet/contourlet case, the anisotropic scaling amounts to increasing the angular resolution for large frequencies, which cannot be carried out indefinitely for the discrete domain. As we will see in Section 7.5.1, contourlets outperform wavelets in the low-bit-coding area, showing an improved ability to pick out salient image structures, including large-scale structures. It seems reasonable to expect that this ability is further improved if we allow the curvelets more orientations in the large scales, which is rather the opposite heuristic to the one proposed for the continuous domain.

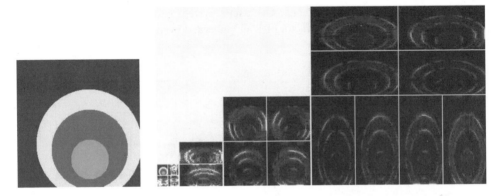

FIGURE 7.9 Sample image circles, decomposed by subsequent bandpass and directional filtering.

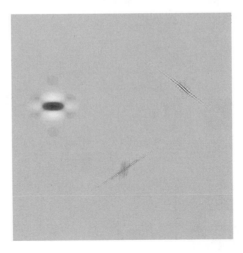

FIGURE 7.10 A sample of three contourlets of different scales and orientations; the gray-scale is manipulated to improve visibility of the different contourlets. Observe the change in aspect ratios as the scale decreases.

7.5 APPLICATION TO IMAGE COMPRESSION

The novel image representations induced by the methods surveyed in this chapter present potential alternatives to the standard image compression schemes, mainly based on wavelet transform. First, in Section 7.5.1, we perform some experiments about the practical approximation properties of the representation model. Then in Section 7.5.2, we focus on the first attempts in the direction of effective coding–decoding schemes of the underlying information.

7.5.1 EXPERIMENTAL APPROXIMATION PROPERTIES

We have conducted tests with real and artificial images to compare the different approximation schemes in terms of coefficients used vs. distortion for the reconstructed images. We used standard test images; see Figure 7.11. The wavelet approximations were obtained by the standard matlab implementation, using the db4 filters. The contourlet approximations were obtained using the matlab contourlet toolbox [9], developed by Do and collaborators. For wedgelets we used our implementation available in [34]. The resulting rate-distortion curves are displayed in Figure 7.12.

The capacity to achieve high theoretical rates of approximation is an important indicator for the potential of a geometrical method in the field of image compression. In the case of two-dimensional locally smooth functions with regular boundaries, we already remarked that the rates obtained with wedgelets or curvelets are of higher order than those induced by the classical decomposition framework (Fourier decompositions, wavelet frames [25,26]).

A naive transferral of the approximation results obtained in Section 7.1 would suggest that the new schemes outperform wavelets in the high-bit-rate area, i.e., as the number of coefficients per pixel approaches 1. However, for all images, wavelets have superior approximation rates in these areas. By contrast, contourlets and wedgelets perform consistently better in the low-bit-rate area. Given the fact that the contourlet system has the handicap of a redundancy by a factor of 1.333, and the fact that the approximation is obtained by simple thresholding, the consistently good approximation behavior of contourlets for extremely small numbers of coefficients is remarkable. Wedgelets, on the other

FIGURE 7.11 Test images; the usual suspects: *Barbara* (512 × 512), *Peppers* (512 × 512), *Cameraman* (256 × 256), *Baboon* (512 × 512), and *Circles* (256 × 256).

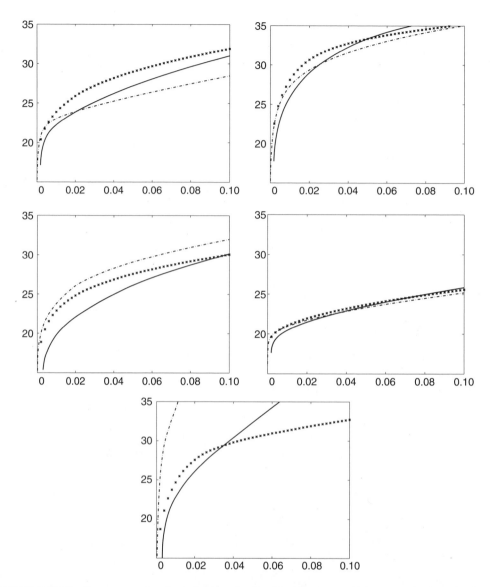

FIGURE 7.12 Nonlinear approximation behavior, visualized by plotting coefficients per pixels against PSNR (db). We compare wavelets (solid), contourlets (crosses), and wedgelets (dashed), corresponding to the images. From top left to bottom: *Barbara* (512 × 512), *Peppers* (512 × 512), *Cameraman* (256 × 256), *Baboon* (512 × 512), and *Circles* (256 × 256).

hand, perform best when dealing with images that are of a predominantly geometric nature, such as the cameraman or the circles. This of course was to be expected. Similarly, the trouble of wedgelets in dealing with textured regions could be predicted beforehand. By contrast, contourlets also manage to represent textured regions to some accuracy, as can be seen in the nonlinear approximation plot for the Barbara image, and in the reconstructions of Baboon and Barbara in Figure 7.13 and Figure 7.14.

Clearly, PSNR is not the only indicator of visual quality. Figure 7.13 and Figure 7.14 present reconstructions of our sample images using 0.01 coefficients per pixel. We already remarked that contourlets are superior to wedgelets when it comes to approximating textured regions (cf. Baboon, Barbara). On the other hand, contourlets produce wavelet-like ringing artifacts around sharp edges

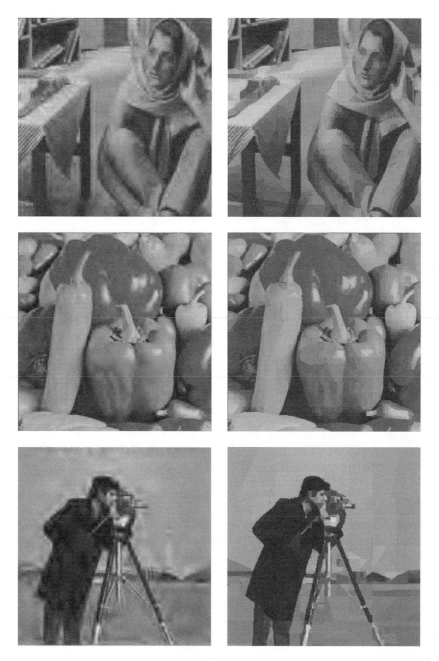

FIGURE 7.13 Sample reconstructions using 0.01 coefficients per pixel, for contourlets (left) and wedgelets (right). Top row: *Barbara*, with contourlets: 23.89 db, wedgelets: 23.02 db; middle row: *Peppers*, with contourlets: 28.12, wedgelets: 27.84 db; bottom row: *Cameraman*, with contourlets: 22.90 db, wedgelets: 23.82 db.

(cf. Cameraman, Circles). Here the wedgelets produce superior results, both visually and in terms of PSNR. As a general rule, the artifacts due to contourlet truncation are visually quite similar to wavelet artifacts. On the other hand, typical wedgelet artifacts come in the form of clearly discernible edges or quantization effects in the representation of color gradients. To some extent, these effects can be ameliorated by employing a higher-order system, such as platelets.

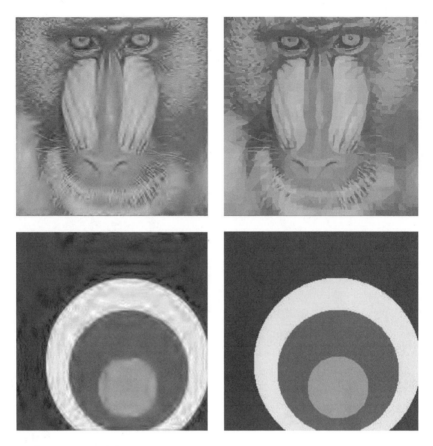

FIGURE 7.14 Sample reconstructions using 0.01 coefficients per pixel, for contourlets (left) and wedgelets (right). Top row: *Baboon*, with contourlets: 21.05 db, wedgelets: 20.89 db; bottom row: *Circles*, with contourlets: 26.56 db, wedgelets: 34.12 db.

Summarizing the discussion, the results suggest that contourlets and wedgelets show improved approximation behavior in low-bit-rate areas. Here the improvement is consistent, and somewhat contradictory to the theoretical results which motivated the design of these systems in the first place. Both contourlets and wedgelets are able to well represent the coarse-scale geometric structures inherent in the image. As more and more details are required, wedgelets fail to efficiently approximate textured regions, while in the contourlet case the overall redundancy of the contourlet system increasingly deteriorates the performance.

7.5.2 CODING SCHEMES

As we have already remarked on various occasions, it remains an open question to decide whether these approximation rates constitute an adequate framework for the case of compression of discrete images represented with a discrete set of values. The experiments in Section 7.5.1 confirm that owing to the discretization effects, the theoretical approximation rates are not observed in practice, even for reasonably big sizes of images. On the other hand, for representations by very few coefficients, where the discretization effect is negligible, the asymptotical rates do not bring a very relevant information. It is also obvious that the choice of the L^2-error for measuring the distortion also leads to some undesired artifacts. For instance, this kind of measure also incorporates some noise inherent to natural images [8], and is thus poorly adapted to the human visual systems.

Now let us turn to the other main ingredient required for a complete effective compression scheme, i.e., a suited coding method which captures the remaining structural redundancies in the representation model. In the case of wavelets, tree-based methods [27,26] allow an efficient coding of coefficients. Apparently, analogs for contourlets have not been developed so far. Recall that a contourlet-based coding scheme would also have to make up for the redundancy factor of 1.333 induced by the contourlet decomposition. In any case, we are not aware of an existing implementation of a compression scheme based on contourlets.

Thus, we restrict the following discussion to the application of wedgelets to image compression, which is mainly due to the work of Wakin [31,33]. The first attempts are based on a model mixing cartoon and texture coding [32]. More than for its efficiency, this method is interesting for the sake of understanding the difficulties occurring in the coding of wedge representations of images. The heuristic behind this method consists in considering a natural image as the combination of two independent components, one containing the textures, the other corresponding to a simple-edge cartoon model, containing only the sharp-edge information. Then, a separated coding of each component is performed. The cartoon component is treated with the help of wedgelet approximation, whereas the residual error image inferred from this first approximation is coded with wavelets in a classical way. On the following, we focus on the coding of the tree-based wedgelet representation of the cartoon component. The decoder needs to know the structure of the tree (a node of the tree can be either a square leaf, a wedge leaf, or subdivided), the wedge parameters for the wedge leaves, and the corresponding quantized, optimized constant values for the selected wedges and squares.

Such a *naive coding* of a wedgelet decomposition avoids most ringing artifacts around the edges, but still suffers of a bad punctual details and texture rendering. In terms of PSNR, it remains inferior to wavelet coding (like JPEG2000). Note that this coding scheme is suboptimal, mainly because it does not model the dependencies between neighboring coefficients, and also because of redundancies between wavelet and wedgelet representations, inherent to this scheme.

For the problem of dependencies between different edges, a possible modeling is the MGM (multiscale geometry model). It relies on a kind of multiscale Markov model for the structural dependencies between wedge orientations and positions; indeed they make use of the probabilities of an angle in a child dyadic square to be selected, conditionally to the optimal angle selected in the parent dyadic square. Note that this model only takes into account the Hausdorff distance between the parents lines and the children lines. Contrarily to the contextual coding used in JPEG2000 [27], where the Markov models are adaptively updated, this model does not adapt to the image contents, but is rather based on an *ad hoc* assumption concerning the correlation between geometrical structure of parents and children dyadic squares. This joint coding of the wedge parameters allows significant coding gains when compared to an independent coding.

Deriving an efficient compression scheme depends also on the possibility to prune the tree adequately. In (7.28), the penalization used for the pruning of the tree corresponds to a balance between the distortion in the reconstruction of the image and the complexity of the model measured by the number of pieces retained for the representation of the image. In the compression context, it is interesting to consider a modified penalization, which takes into account the coding cost. The problem reduces then to the minimization of a functional of the form

$$H_{\lambda,f}(g, W) = \|f - g\|_2 2 + \lambda[- \log(P(W))] \tag{7.36}$$

where P is an entropy measure depending on some probabilistic assumptions on the data. This measure is used for evaluating the amount of bits required for the coding of the wedge tree and parameters' information.

The most interesting method is the W-SFQ (wedgelet-space frequency quantization) compression scheme proposed in [33] and based on the use of wedgeprints. The main difference with the previous scheme consists in acting directly in the wavelet coefficients domain, instead of the image domain.

The method is mainly an enhancement of the SFQ scheme, which was originally proposed in [36]. SFQ is a zerotree coding where the coefficients clustering are optimized according to a rate-distortion criterion. It is a very efficient method outperforming the standard SPIHT (set partitioning into hierarchical trees [26]) in many cases, especially for low bit rates. It still suffers, however, from the limits of zerotree wavelet coders for the coding of significant wavelet coefficients along edges.

Whereas SFQ considers two possibility for a coded node, either being a zerotree (all its descendants being considered unsignificant) or a significant, coded coefficient; W-SFQ introduces a third possibility, a node can be a wedge, the wavelet coefficients descendants of this node being evaluated from the optimal wedge function. In other words, W-SFQ is a zerotree where the clustering of coefficients is more suited to geometry, with the help of wedge functions. This clustering together with the associated function is also called wedgeprint. Despite the high coding cost of the wedges, the coherence is ensured by the rate-distortion optimization: a wedge is only chosen when its cost remains low toward the gain in distortion. In [33], an enhanced version of this coding scheme is proposed with some additional ingredients:

- The use of the MGM model which is an efficient tool to code efficiently deeper wedge tilings; a more accurate description of the geometry of a contour than with a single wedgeprint is possible.
- A smoothing parameter for the edges, which takes into account blurring artifacts in original image, due to pixelization.
- A specific coding for textures.

This enhanced version model allows to code the information for less larger wedgeprints. The finer wedgeprints are then coded at low cost, thanks to the MGM. Encouraging results were obtained for some natural images. For instance, W-SFQ outperforms SFQ at very low bit rate, for some natural images [33], with poor texture contents, dominated by geometrical structures (and hence closer to the cartoon model).

Note that the MGM model in this context remains relatively rudimentary. Unlike many classical coding schemes, it is a nonadaptive model. Furthermore, it only takes into account correlations between neighboring scales. One could expect improved compression rates with the help of a modeling of the spatial correlations between wedge parameters of neighboring wedgeprints.

7.6   TENTATIVE CONCLUSIONS AND SUGGESTIONS FOR FURTHER READING

The construction and efficient implementation of new multiscale methods for the representation of relevant image structures is an ongoing endeavor. It seems that while the existing constructions already manage to outperform wavelets, at least in certain respects, both theoretically and in practice, a consistent improvement over wavelets, in particular in the domain of image compression, has not yet been achieved. In this respect one should acknowledge the fact that wavelet compression, as it is incorporated into JPEG2000, is the product of at least ten years of activity that took place between the recognition of the potential of wavelets for image coding, and their eventual inclusion into an image compression standard. As both the diversity of the approaches and their computational complexity — at least in comparison to wavelets — seem to indicate, a further improvement will require considerable effort, and it is hard to predict which method will prove to be the most effective. In this chapter, we have tried to describe some of the approaches that have been developed so far, and for further reading, we refer to our main sources, which were [6,11,13,14,16]. A closer description of curvelets and ridgelets can also be found in the book [2]. Promising alternatives that we mentioned in Section 7.2.3 but did not treat in more detail are bandelets [20, 21] and directionlets [30]. An approach that allows the combination of different approaches, e.g., wavelets and wedgelets, are footprints (see the discussion of wedgeprints in the previous subsection). A introduction to footprints can be found in [18].

ACKNOWLEDGMENT

HF and FF acknowledge additional funding through the European Research Training Network HASSIP, under the contract HPRN-CT-2002-00285.

REFERENCES

1. J.P. Antoine and R. Murenzi, Two-dimensional directional wavelets and the scale-angle representation, *IEEE Signal Process.*, 52, 259–281, 1996.
2. J.P. Antoine, R. Murenzi, P. Vandergheynst, and S.T. Ali, *Two-dimensional Wavelets and Their Relatives*, Cambridge University Press, London; New York, 2004.
3. R. Bamberger and M. Smith, A filter bank for the directional decomposition of images: Theory and design, *IEEE Trans. Signal Proc.*, 40, 882–893, 1992.
4. E. Candes and D. Donoho, Ridgelets: a key to higher-dimensional intermittency?, *R. Soc. Lond. Philos. Trans. Ser. A Math. Phys. Eng. Sci.*, 357, 2495–2509, 1999.
5. E. Candes and D. Donoho, in *Curvelets — A Surprisingly Effective Nonadaptive Representation for Objects with Edges, Curves and Surfaces,* L. Schumaker et al., Eds. Vanderbilt University Press, Nashville, TN, 1999.
6. E. Candes and D. Donoho, New tight frames of curvelets and optimal representations of objects with C^2 singularities, *Commun. Pure Appl. Math.*, 57, 219–266, 2004.
7. V. Chandrasekaran, M. Wakin, D. Baron, and R. Baraniuk, Surflets: a sparse representation for multidimensional functions containing smooth discontinuities, *IEEE Symposium on Information Theory*, Chicago, IL, in press, June 2004.
8. A. Cohen, W. Dahmen, I. Daubechies, and R. DeVore, Tree approximation and optimal encoding, *J. Appl. Comp. Harm. Anal.*, 11, 192–226, 2001.
9. http://www.ifp.uiuc.edu/~minhdo/software/, containing a downloadable contourlet toolbox.
10. L. Demaret, N. Dyn, and A. Iske, *Image Compression by Linear Splines over Adaptive Triangulations*, University of Leicester, 7 January 2005 [to appear].
11. F. Friedrich, L. Demaret, H. Führ, and K. Wicker, *Efficient Moment Computation Over Polygonal Domains with an Application to Rapid Wedgelet Approximation*, Submitted 2005. Available at http://ibb.gsf.de/preprints.php.
12. M. Do, Directional Multiresolution Image Representation, Thesis, Swiss Federal Institute of Technology, 2001.
13. M. Do and M. Vetterli, The contourlet transform: an efficient directional multiresolution image representation, *IEEE Trans. Image Process*, Vol. 14, no. 12, Dec. 2005, pp. 2091–2106.
14. M. Do and M. Vetterli, Contourlets, in *Beyond Wavelets*, G.V. Welland, Ed., Academic Press, New York, 2003.
15. D. Donoho, M. Vetterli, R. DeVore, and I. Daubechies, Data compression and harmonic analysis, *IEEE Trans. Inform Theory*, 44, 2435–2476, 1998.
16. D. Donoho, Wedgelets: nearly minimax estimation of edges, *Ann. Statist.*, 27, 859–897, 1999.
17. D. Donoho and X. Huo, Beamlets and multiscale image analysis, in T.J. Barth, et al., Ed., *Multiscale and Multiresolution Methods. Theory and Applications*, Lecture Notes in Computer Science and Engineering, Vol. 20, Springer, Berlin, 2002, pp. 149–196.
18. P.L. Dragotti and M. Vetterli, Wavelet footprints: theory, algorithms and applications, *IEEE Trans. Signal Proc.*, 51, 1307–1323, 2003.
19. F. Friedrich, Complexity Penalized Segmentations in 2D, PhD Thesis, Technische Universität München, 2005.
20. E. LePennec and S. Mallat, Bandelet representations for image compression, *IEEE Proceedings of the International Conference on Image Processing*, Vol. 1, 2001, pp. 12–14.

21. E. LePennec and S. Mallat, Sparse geometric image representations with bandelets, *IEEE Trans. Image Process.*, Vol. 14, no. 4, April 2005, pp. 423–438.

22. LetItWave, http://www.letitwave.fr/

23. S. Mallat, *A Wavelet Tour of Signal Processing*, Academic Press, New York, 1998.

24. F. Meyer and R. Coifman, Brushlets: a tool for directional image analysis and image compression, *Appl. Comp. Harm. Anal.*, 4, 147–187, 1997.

25. M. Shapiro, An embedded hierarchical image coder using zerotrees of wavelet coefficients, *IEEE Trans. Signal Process.*, 41, 3445–3462, 1993.

26. A. Said and W.A. Pearlman, A new, fast, and efficient image codec based on set partitioning in hierarchical trees, *IEEE Trans. Circuits Syst. Video Technol.*, 6, 243–250, 1996.

27. D. Taubman, High performance scalable image compression with EBCOT, *IEEE Trans. Image Process.*, 1158–1170, 2000.

28. J. Starck, E. Candes, and D. Donoho, The curvelet transform for image denoising, *IEEE Trans. Image Process.*, 11, 670–684, 2002.

29. M. Vetterli, Wavelets, approximation and compression, *IEEE Signal Proc. Mag.*, 18, 59–73, 2001.

30. V. Velisavljević, B. Beferull-Lozano, M. Vetterli, and P.L. Dragotti, Directionlets: anisotropic multi-directional representation with separable filtering, *IEEE Trans. Image Proc.* [to appear].

31. M.B. Wakin, Image Compression using Multiscale Geometric Edge Models, MSc. thesis, Rice University, Houston, 2002.

32. M.B. Wakin, J.K. Romberg, H. Choi, and R.G. Baraniuk, Image compression using an efficient edge cartoon + texture model, *Proceedings, IEEE Data Compression Conference — DCC'02*, Snowbird, April 2002, pp. 43–52.

33. M. Wakin, J. Romberg, H. Choi, and R. Baraniuk, Wavelet-domain approximation and compression of piecewise smooth images, *IEEE Trans. Image Process.*, May 2006 [to appear].

34. www.wedgelets.de.

35. R.M. Willett and R.D. Nowak, Platelets: A multiscale approach for recovering edges and surfaces in photon-limited medical imaging, *IEEE Trans. Medical Imaging*, 22, 332–350, 2003.

36. Z. Xiong, K. Ramchandran, and M.T. Orchard, Space-frequency quantization for wavelet image coding, *IEEE Trans. Image Process.*, 6, 677–693, 1997.

8 Image Coding Using Redundant Dictionaries

Pierre Vandergheynst and Pascal Frossard

CONTENTS

8.1 Introduction 207
    8.1.1 A Quick Glance at Digital Image Compression 208
    8.1.2 Limits of Current Image Representation Methods 209
8.2 Redundant Expansions 209
    8.2.1 Benefits of Redundant Transforms 209
    8.2.2 Nonlinear Algorithms 210
        8.2.2.1 A Wealth of Algorithms 210
        8.2.2.2 Highly Nonlinear Approximations 210
        8.2.2.3 Greedy Algorithms: Matching Pursuit 212
    8.2.3 A Scalable Image Encoder 214
        8.2.3.1 Overview 214
        8.2.3.2 Matching Pursuit Search 215
        8.2.3.3 Generating Functions of the Dictionary 215
        8.2.3.4 Anisotropy and Orientation 216
        8.2.3.5 Dictionary 217
        8.2.3.6 Coding Stage 218
        8.2.3.7 Coefficient Quantization 218
        8.2.3.8 Rate Control 220
    8.2.4 Experimental Results 220
        8.2.4.1 Benefits of Anisotropy 220
        8.2.4.2 Coding Performance 222
    8.2.5 Extension to Color Images 223
    8.2.6 High Adaptivity 227
        8.2.6.1 Importance of Adaptivity 227
        8.2.6.2 Spatial Adaptivity 227
        8.2.6.3 Rate Scalability 229
8.3 Discussions and Conclusions 230
    8.3.1 Discussions 230
    8.3.2 Extensions and Future Work 231
Acknowledgments 231
References 232

8.1 INTRODUCTION

Image compression has been key in enabling what already seems to be two of the major success stories of the digital era: rich media experience over the internet and digital photography. What are the technologies lying behind such industry flagships and, more importantly, what is the future of these technologies are some of the central questions of this chapter.

We begin with a quick review of the state of the art. Then, identifying some weaknesses of the actual systems together with new requirements from applications, we depict how novel algorithms based on redundant libraries could lead to new breakthroughs.

8.1.1   A Quick Glance at Digital Image Compression

Modern image compression algorithms, most notably JPEG, have been designed following the transform coding paradigm. Data are considered as the output symbols x_n of a random source X, which can have a complicated probability density function. In order to reduce redundancy between symbols, one seeks a new representation of the source by applying a suitable linear transformation T. The new symbols $Y = T \cdot X$ will then be quantized and entropy-coded. Very often a scalar quantizer will be applied to the transform coefficients y_n. It is a standard result in data compression that, in order to maximize the performance of such a system, the transform T should be chosen so that it yields uncorrelated coefficients. In this regard, the optimal transform is thus the Karhunen–Loeve transform (KLT). It is one of the beauty of transform coding that such a simple and complete analysis is possible. It also leads us to a few important comments about the whole methodology. First, the KLT is a data-dependent and complex transform. Using it in practice is at the least difficult, but usually impossible, as it would require to send the basis that represents T to the decoder for each source. Very often, one seeks a linear transform that performs close to the KLT and this is one of the reasons why the direct cosine transform (DCT) was chosen in JPEG. Second, the optimality of the transform coding principle (KLT plus scalar quantizer) can only be ensured for simple models (e.g., Gaussian cyclostationary). In practice, for natural data, this kind of modeling is far from truth. Finally, the role of the transform in this chain is relegated to its role of providing uncorrelated coefficients for feeding the scalar quantizer. Nothing about the main structures of the signal and the suitability of the transform to catch them is ever used.

Based on these observations the research community started to consider other alternatives

- Replacing scalar quantization by vector quantization (VQ), which can be seen as a way to overcome the limits of transform coding while also putting more emphasis on the content of the signal
- Searching and studying new transforms, better suited to represent the content of the signal
- Completely replacing transform coding by other techniques

Out of the many interesting techniques that have emerged based on these interrogations, wavelet-based techniques have had the largest impact. Indeed, in these last few years, image compression has been largely dominated by the use of wavelet-based transform coding techniques. Many popular compression algorithms use wavelets at their core (SPIHT and EBCOT) and the overall success of this methodology resulted in the actual JPEG-2000 standard for image compression [38]. As it was quickly realized, there is more to wavelets than their simple use as a decorrelating transform. On the conceptual point of view, we see three main reasons for their success: (1) fast algorithms based on filter banks or on the lifting scheme, (2) nice mathematical properties, and (3) smart adaptive coding of the coefficients.

Efficient algorithms are, of course, of paramount importance when putting a novel technique to practice, but the overall power of wavelets for image compression really lies in the second and third items. The mathematical properties of wavelets have been well studied in the fields of computational harmonic analysis (CHA) and nonlinear approximation theory. Generally, the central question that both theories try to answer (at least in connection with data compression) is: given a signal, how many wavelet coefficients do I need to represent it up to a given approximation error? There is a wealth of mathematical results that precisely relate the decay of the approximation error with the smoothness of the original signal, when N coefficients are used. By modeling a signal as a piecewise smooth function, it can be shown that wavelets offer the best rate of nonlinear approximation. By this we mean that approximating functions that are locally Hölder α with discontinuities, by their N

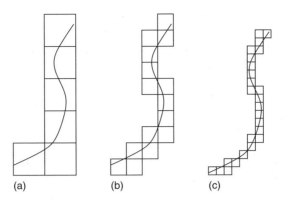

FIGURE 8.1 Inadequacy of isotropic refinement for representing contours in images. The number of wavelets intersecting the singularity is roughly doubled when the resolution increases: (a) 6 coefficients; (b) 15 coefficients; and (c) 25 coefficients.

biggest wavelet coefficients, one obtains an approximation error in the order of $N^{-\alpha}$ and that this is an optimal result (see [9,15] and references therein). The key to this result is that wavelet bases yield very sparse representations of such signals, mainly because their vanishing moments *kill* polynomial parts, while their multiresolution behavior allows to localize discontinuities with few non-negligible elements. Now, practically speaking, the real question should be formulated in terms of bits: how many bits do I need to represent my data up to a given distortion? The link between both questions is not really trivial: it has to take into account both quantization and coding strategies. But very efficient wavelet coding schemes exist, and many of them actually use the structure of nonnegligible wavelet coefficients across subbands.

8.1.2 LIMITS OF CURRENT IMAGE REPRESENTATION METHODS

While the situation described above prevails in one dimension, it gets much more problematic for signals with two or more dimensions, mainly because of the importance of geometry. Indeed, an image can still be modeled as a piecewise smooth 2-D signal with singularities, but the latter are not point-like anymore. Multidimensional singularities may be highly organized along embedded submanifolds and this is exactly what happens at image contours, for example. Figure 8.1 shows that wavelets are inefficient at representing contours because they cannot deal with the geometrical regularity of the contours themselves. This is mainly due to the isotropic refinement implemented by wavelet basis: the dyadic scaling factor is applied in all directions, where clearly it should be fine along the direction of the local gradient and coarse in the orthogonal direction in order to efficiently *localize* the singularity in a sparse way. This is the reason why other types of signal representation, like redundant transforms, certainly represent the core of new breakthroughs in image coding, beyond the performance of orthogonal wavelet transforms.

8.2 REDUNDANT EXPANSIONS

8.2.1 BENEFITS OF REDUNDANT TRANSFORMS

In order to efficiently represent contours, beyond the performance of wavelet decompositions, anisotropy is clearly desirable in the coding scheme. Several authors have explored the rate–distortion characteristics of anisotropic systems for representing edge-dominated images [11,20]. These preliminary studies show that for images that are smooth away from a smooth edge (typically a C^2

rectifiable curve), a rate–distortion (R–D) behavior of the form

$$D(R) \asymp \frac{\log R}{R^2} \tag{8.1}$$

can be reached. Comparing this with the associated wavelet R–D behavior, i.e., $D(R) \asymp R^{-1}$, one clearly sees how the use of a geometry-adapted system of representation can boost coding expectations. It is important to realize here that it is really the anisotropic scaling of the basis functions that allows for such performances. Simply using an anisotropic basis with multiple orientations but a fixed isotropic scaling law would not provide such results (though it may improve visual quality for instance).

Candes and Donoho [4] have recently proposed a construction called the *curvelet transform*, which aims at solving the lack of flexibility of wavelets in higher dimensions. Basically, curvelets satisfy an anisotropic scaling law that is adapted to representing smooth curves in images. Curvelet tight frames have been shown to achieve a much better nonlinear approximation rate than wavelets for images that are smooth away from a $\mathcal{C}^2$ edge. Very interesting results have been reported for statistical estimation and denoising [37] and efficient filter bank implementations have been designed [12]. On the coding side, curvelets satisfy the localization properties that lead to (8.1) and there is thus hope to find efficient compression schemes based on the curvelet transform, even though such results have not yet been reported.

8.2.2 NONLINEAR ALGORITHMS

8.2.2.1 A Wealth of Algorithms

Clearly, another way to tackle the problem of higher dimensional data representation would be to turn to nonlinear algorithms. The interested reader searching a way through the literature might feel as if he/she had suddenly opened Pandora's box! Various algorithms exist, and they all differ in philosophy. Before moving to the particular case of interest in this chapter, we thus provide a basic road map through some of the most successful techniques.

- *Wavelet footprints* [16]: for piecewise polynomial signals, the significant wavelets of predefined singularities are grouped together into a footprint dictionary. The algorithm locates singularities and then selects the best footprints in the dictionary. In 1-D, it reaches the near-optimal bound. In 2-D, the situation gets complicated by the problem of chaining these footprints together along contours.
- *Wedgeprints* [43]: weds wavelets and wedgelets [13] by grouping wavelet coefficients into wedgeprints in the wavelet domain. One advantage is that all is computed based on the wavelet coefficients: they are sorted in a tree-like manner according to their behavior as smooth, wedgeprints, or textures. Markov trees help ensuring that particular grouping of coefficients do make sense (i.e., they represent smooth edges). It reaches the near-optimal bound in 2-D.
- *Bandelets* [27]: the image is processed so as to find its edges. The wavelet transform is then warped along the geometry in order to provide a sparse expansion. It reaches the near-optimal bound for all smooth edges.

8.2.2.2 Highly Nonlinear Approximations

Another interesting way of achieving sparsity for low bit-rate image coding is to turn to very redundant systems. In particular, we will now focus on the use of highly nonlinear approximations in redundant dictionaries of functions.

Highly nonlinear approximation theory is mainly concerned with the following question: given a collection $\mathcal{D}$ of elements of norm 1 in a Banach space[1] $\mathcal{H}$, find an exact N-sparse representation of any signal s:

$$s = \sum_{k=0}^{N-1} c_k g_k \tag{8.2}$$

The equality in (8.2) may not need to be reached, in which case a N-term approximant $\tilde{s}_N$ is found:

$$\tilde{s}_N = \sum_{k=0}^{N-1} c_k g_k, \quad \|s - \tilde{s}_N\| \leq \epsilon(N) \tag{8.3}$$

for some approximation error ϵ. Such an approximant is sometimes called (ϵ, N)-sparse.

The collection $\mathcal{D}$ is often called a dictionary and its elements are called atoms. There is no particular requirement concerning the dictionary, except that it should span $\mathcal{H}$, and there is no prescription on how to compute the coefficients c_k in Eq. (8.2). The main advantage of this class of techniques is the complete freedom in designing the dictionary, which can then be efficiently tailored to closely match signal structures.

Our ultimate goal would be to find the best, that is the sparsest possible representation of the signal. In other words, we would like to solve the following problem:

$$\text{minimize } \|c\|_0 \quad \text{subject to } s = \sum_{k=0}^{K-1} c_k g_{\gamma_k}$$

where $\|c\|_0$ is the number of nonzero entries in the sequence $\{c_k\}$. If the dictionary is well adapted to the signal, there are high hopes that this kind of representation exists, and would actually be sparser than a nonlinear wavelet-based approximation. The problem of finding a sparse expansion of a signal in a generic dictionary $\mathcal{D}$ leads to a daunting NP-hard combinatorial optimization problem. This is however not true anymore for particular classes of dictionaries. Recently, constructive results were obtained by considering incoherent dictionaries [14,18,23], i.e., collections of vectors that are not too far from an orthogonal basis. These results impose very strict constraints on the dictionary, but yield a striking improvement: they allow to solve the original NP-hard combinatorial problem by linear programming. As we will now see, this rigidity can be relaxed when we turn to the problem of finding sufficiently good N-term approximants, instead of exact solutions to Eq. (8.2).

In order to overcome this limitation, Chen et al. [5] proposed to solve the following slightly different problem:

$$\text{minimize } \|c\|_1 \quad \text{subject to } s = \sum_{k=0}^{K-1} c_k g_{\gamma_k}$$

Minimizing the ℓ_1 norm helps to find a sparse approximation, because it prevents diffusing the energy of the signal over a lot of coefficients. While keeping the essential property of the original problem, this subtle modification leads to a tremendous change in the very nature of the optimization challenge. Indeed, this ℓ_1 problem, called *basis pursuit* (BP), is a much simpler problem, that can be efficiently solved by linear programming using, for example, interior point methods.

---

[1] A Banach space is a complete vector space B with a norm $\|v\|$; for more information refer to [34].

Constructive approximation results for redundant dictionaries however do not abound, contrary to the wavelet case. Nevertheless, recent efforts pave the way toward efficient and provably good nonlinear algorithms that could lead to potential breakthroughs in multidimensional data compression. For illustration purposes, let us briefly comment on the state of the art.

Recently, many authors focused on incoherent dictionaries, or, equivalently, dictionaries whose coherence μ is smaller than a sufficiently small constant C (i.e., $\mu < C$), whereas the coherence of a dictionary $\mathcal{D}$ is defined as

$$\mu = \sup_{\substack{i,j \\ i \neq j}} |\langle g_i, g_j \rangle| \tag{8.4}$$

Coherence is another possible measure of the redundancy of the dictionary and Eq. (8.4) shows that $\mathcal{D}$ is not too far from an orthogonal basis when its coherence is sufficiently small (although it may be highly overcomplete). Let us first concentrate on a dictionary $\mathcal{D}$ that is given by the union of two orthogonal bases in $\mathbb{R}^N$, i.e., $\mathcal{D} = \{\psi_i\} \cup \{\phi_j\}$, $1 \leq i, j \leq N$. Building on early results of Donoho and Huo [14], Elad and Bruckstein [18] have shown a particularly striking and promising result: if $\mathcal{D}$ is the concatenated dictionary described above with coherence μ and $s \in \mathbb{R}^N$ is any signal with a sufficiently sparse representation,

$$s = \sum_i c_i g_i \quad \text{with } \|c\|_0 < \frac{\sqrt{2} - 0.5}{\mu} \tag{8.5}$$

then this representation is the unique sparsest expansion of s in $\mathcal{D}$ and can be exactly recovered by BP. In other words, we can replace the original NP-hard combinatorial optimization problem of finding the sparsest representation of s by the much simpler ℓ_1 problem. These results have been extended to arbitrary dictionaries by Gribonval and Nielsen [23], who showed that the bound in Eq. (8.5) can be refined to

$$\|c\|_0 < \frac{1}{2}\left(1 + \frac{1}{\mu}\right) \tag{8.6}$$

So far the results obtained are not constructive. They essentially tell us that, if a sufficiently sparse solution exists in a sufficiently incoherent dictionary, it can be found by solving the ℓ_1 optimization problem. Practically, given a signal, one does not know whether such a solution can be found and the only possibility at hand would be to run BP and check *a posteriori* that the algorithm finds a sufficiently sparse solution. These results also impose very strict constraints on the dictionary, i.e., sufficient incoherence. But this has to be understood as a mathematical artifice to tackle a difficult problem: managing dependencies between atoms in order to prove exact recovery of a unique sparsest approximation. When instead one wants to find sufficiently good N-term approximants, such a rigidity may be relaxed as shown in practice by the class of greedy algorithms described in the next section.

8.2.2.3   Greedy Algorithms: Matching Pursuit

Greedy algorithms iteratively construct an approximant by selecting the element of the dictionary that best matches the signal at each iteration. The pure greedy algorithm is known as *Matching Pursuit* (MP) [30]. Assuming that all atoms in $\mathcal{D}$ have norm 1, we initialize the algorithm by setting $R_0 = s$ and we first decompose the signal as

$$R_0 = \langle g_{\gamma_0}, R_0 \rangle g_{\gamma_0} + R_1 \tag{8.7}$$

Clearly, g_{γ_0} is orthogonal to R_1 and we thus have

$$\|R_0\|^2 = |\langle g_{\gamma_0}, R_0 \rangle|^2 + \|R_1\|^2 \tag{8.8}$$

If we want to minimize the energy of the residual R_1 we must maximize the projection $|\langle g_{\gamma_0}, R_0 \rangle|$. In the next step, we simply apply the same procedure to R_1, which yields

$$R_1 = \langle g_{\gamma_1}, R_1 \rangle g_{\gamma_1} + R_2 \tag{8.9}$$

where g_{γ_1} maximizes $|\langle g_{\gamma_1}, R_1 \rangle|$. Iterating this procedure, we thus obtain an approximant after M steps:

$$s = \sum_{m=0}^{M-1} \langle g_{\gamma_m}, R_m \rangle g_{\gamma_m} + R_M \tag{8.10}$$

where the norm of the residual (approximation error) satisfies

$$\|R_M\|^2 = \|s\|^2 - \sum_{m=0}^{M-1} |\langle g_{\gamma_m}, R_m \rangle|^2 \tag{8.11}$$

Some variations around this algorithm are possible. An example is given by the weak greedy algorithm [10], which consists of modifying the atom selection rule by allowing to choose a slightly suboptimal candidate:

$$|\langle R_m, g_{\gamma_m} \rangle| \geq t_m \sup_{g \in \mathcal{D}} |\langle R_m, g \rangle|, \quad t_m \leq 1 \tag{8.12}$$

It is sometimes convenient to rephrase MP in a more general way, as a two-step algorithm. The first step is a selection procedure that, given the residual R_m at iteration m, will select the appropriate element of $\mathcal{D}$:

$$g_{\gamma_m} = \mathcal{S}(R_m, \mathcal{D}) \tag{8.13}$$

where $\mathcal{S}$ is a particular selection operator. The second step simply updates the residual:

$$R_{m+1} = \mathcal{U}(R_m, g_{\gamma_m}) \tag{8.14}$$

One can easily show that MP converges [25] and even converges exponentially in the strong topology in finite dimension (see [30] for a proof). Unfortunately, this is not true in general in infinite dimension, even though this property holds for particular dictionaries [42]. However, De Vore and Temlyakov [10] constructed a dictionary for which even a good signal, i.e., a sum of two dictionary elements, has a very bad rate of approximation: $\|s - s_M\| \geq CM^{-1/2}$. In this case, a very sparse representation of the signal exists, but the algorithm dramatically fails to recover it! Notice though, that this again does in no way rule out the existence of particular classes of very good dictionaries.

A clear drawback of the pure greedy algorithm is that the expansion of s on the linear span of the selected atoms is not the best possible one, since it is not an orthogonal projection. Orthogonal MP [7,33] solves this problem by recursively orthogonalizing the set of selected atoms using a Gram–Schmidt procedure. The best M-term approximation on the set of selected atoms is thus computed and the algorithm can be shown to converge in a finite number of steps, but at the expense of a much bigger computational complexity.

At the same time, greedy algorithms offer constructive procedures for computing highly nonlinear N-term approximations. Although the mathematical analysis of their approximation properties is complicated by their nonlinear nature, interesting results are emerging (see, for example, [22,24,40,41]). Let us briefly illustrate one of them.

Theorem 8.2.1

Let $\mathcal{D}$ be a dictionary in a finite/infinite-dimensional Hilbert space and let μ: $\max_{k\neq l}|\langle g_k, g_l\rangle|$ be its coherence. *For any finite index set I of size $card(I) = m < (1 + 1/\mu)/2$ and any $s = \sum_{k\in I} c_k g_k \in span(g_k, k \in I)$, MP:*

    1. picks up only "correct" atoms at each step ($\forall n$, $k_n \in I$);
    2. converges exponentially

$$\|f_n - f\|^2 \leq ((1 - 1/m)(1 + \mu))^n \|f\|^2.$$

The meaning of this theorem is the following. Take a dictionary for which interactions among atoms are small enough (low coherence) and a signal that is a superposition of atoms from a subset $\{g_k, k \in I\}$ of the dictionary. In this case, MP will only select those correct atoms and no other. The algorithm thus exactly identifies the elements of the signal. Moreover, since MP is looping in a finite dimensional subset, it will converge exponentially to f. The interested reader will find in [23,40] similar results for the case when the signal is not an *exact* superposition of atoms, but when it can be well *approximated* by such a superposition. In this case again, MP can identify those correct atoms and produce N-term approximants that are close to the optimal approximation.

The choice of a particular algorithm generally consists in trading off complexity and optimality, or more generally efficiency. The image compression scheme presented in this chapter proposes to use MP as a suboptimal algorithm to obtain a sparse signal expansion, yet an efficient way to produce a progressive low bit-rate image representation with a controlled complexity. Matching Pursuit, as already stressed before, iteratively chooses the best matching terms in a dictionary. Despite its possible numerical complexity in the signal representation, it is very easy to implement. Moreover, since there is almost no constraint on the dictionary itself, MP clearly stands as a natural candidate to implement an efficient coding scheme based on anisotropic refinement, and such a construction is detailed in the next section.

8.2.3 A Scalable Image Encoder

8.2.3.1 Overview

The benefits of redundant expansions in terms of approximation rate have been discussed in the first part of this chapter. The second part now describes an algorithm that builds on the previous results and integrates nonlinear expansions over an anisotropically refined dictionary, in a scalable MP image encoder. The advantages offered by both the greedy expansion and the structured dictionary are used to provide flexibility in image representation.

The encoder can be represented as in Figure 8.2. The input image is compared to a redundant library of functions using a MP algorithm. Iteratively, the index of the function that best matches the (residual) signal is sent to an entropy coding stage. The corresponding coefficient is quantized, and eventually entropy-coded. The output of the entropy coder block forms the compressed image bitstream. The decoder performs inverse entropy coding, inverse quantization, and finally reconstructs the compressed image by summing the dictionary functions, multiplied by their respective coefficients.

Clearly, the transform only represents one single stage in the compression chain. In order to take the benefit of the improved approximation rate offered by redundant signal expansions, the quantization and entropy coding stage have also to be carefully designed. All the blocks of the compression algorithm have to be adapted to the specific characteristics of MP expansions. It is important to note that the real benefits of redundant transforms in image compression, can only be appreciated when all the blocks of the image encoder are fully optimized.

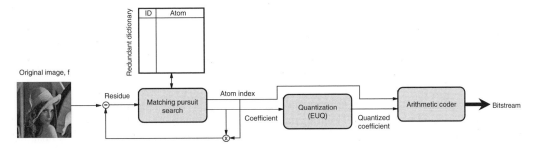

FIGURE 8.2 Block diagram of the MP image encoder.

Alternative image representation methods based on MP, have been proposed in the literature. One of the first papers that proposed to use MP for representing images is [2]. This first work does however not propose a coder implementation, and the dictionary is different than the one proposed in this chapter. MP has been used for coding the motion estimation error in video sequences [32], in a block-based implementation. This coder, contrarily to the one proposed below, makes use of subblocks, which, in a sense, limits the efficiency of the expansion. At the same time, it has been designed to code the residual error of motion estimation, which presents very different characteristics than edge-dominated natural images. The coder presented in the remainder takes benefit of the properties of both redundant expansions and anisotropic functions to offer efficient and flexible compression of natural images.

8.2.3.2 Matching Pursuit Search

One of the well-known drawbacks of MP is the complexity of the search algorithm. The computations to find the best atom in the dictionary have to be repeated at each iteration. The complexity problem can be alleviated in replacing full-search methods, by optimization techniques, such as implementations based on Tree Pursuit [26]. Although such methods greatly speed up the search, they often sacrifice in the quality of the approximation. They sometimes get trapped in local minima, and may choose suboptimal atoms, which do not truly maximize the projection coefficient $|\langle g_\gamma | \mathcal{R}f \rangle|$. Other solutions can be found in efficient implementations of the MP algorithm, in taking benefit from the structure of the signal and the dictionary. The dictionary can, for example, be decomposed in incoherent blocks, and the search can thus be performed independently in each incoherent block, without penalty.

The actual implementation of the MP image encoder described here still performs a full search over the complete dictionary, but computes all the projections in the Fourier domain [19]. This tremendously reduces the number of computations, in the particular case of our dictionary built on anisotropic refinement of rotated atoms. The number of multiplications in this case only depends on the number of scales and rotations in the dictionary, and does not depend any more on the number of atom translations. The MP search in the Fourier domain allows to decrease the number of computations, possibly however at the expense of an increase in memory resources.

8.2.3.3 Generating Functions of the Dictionary

As presented in the previous section, a structured dictionary is built by applying geometric transformations to a generating mother function g. The dictionary is built by varying the parameters of a basis function, in order to generate an overcomplete set of functions spanning the input image space. The choice of the generating function, g, is driven by the idea of efficiently approximating contour-like singularities in 2-D. To achieve this goal, the atom is a smooth low-resolution function in the direction of the contour, and behaves like a wavelet in the orthogonal (singular) direction.

In other words, the dictionary is composed of atoms that are built on Gaussian functions along one direction and on second derivative of Gaussian functions in the orthogonal direction, that is

$$g(\vec{p}) = \frac{2}{\sqrt{3\pi}}(4x^2 - 2)\exp(-(x^2 + y^2)) \tag{8.15}$$

where $\vec{p} = [x, y]$ is the vector of the image coordinates and $||g|| = 1$. The choice of the Gaussian envelope is motivated by the optimal joint spatial and frequency localization of this kernel. The second derivative occurring in the oscillatory component is a trade-off between the number of vanishing moments used to filter out smooth polynomial parts and ringing-like artifacts that may occur after strong quantization. It is also motivated by the presence of second derivative-like filtering in the early stages of the human visual system [31].

The generating function described above is however not able to efficiently represent the low-frequency characteristics of the image at low rates. There are two main options to capture these features: (1) to perform a low-pass filter of the image and send a quantized and downsampled image or (2) to use an additional dictionary capable of representing the low-frequency components. This second approach also has the advantage of introducing more natural artifacts at very low bit rate, since it tends to naturally distribute the available bits between the low and high frequencies of the image. A second subpart of the proposed dictionary is therefore formed by Gaussian functions, in order to keep the optimal joint space–frequency localization. The second generating function of our dictionary can be written as

$$g(\vec{p}) = \frac{1}{\sqrt{\pi}}\exp(-(x^2 + y^2)) \tag{8.16}$$

where the Gaussian has been multiplied by a constant in order to have $||g(\vec{p})|| = 1$.

8.2.3.4 Anisotropy and Orientation

Anisotropic refinement and orientation is eventually obtained by applying meaningful geometric transformations to the generating functions of unit L^2 norm, g, described earlier. These transformations can be represented by a family of unitary operators $U(\gamma)$, and the dictionary is thus expressed as

$$\mathcal{D} = \{U(\gamma)g, \ \gamma \in \Gamma\} \tag{8.17}$$

for a given set of indexes Γ. Basically, this set must contain three types of operations:

- Translations $\vec{b}$, to move the atom all over the image
- Rotations θ, to locally orient the atom along contours
- Anisotropic scaling $\vec{a} = (a_1, a_2)$, to adapt to contour smoothness

A possible action of $U(\gamma)$ on the generating atom g is thus given by

$$U(\gamma)g = \mathcal{U}(\vec{b}, \theta)D(a_1, a_2)g \tag{8.18}$$

where $\mathcal{U}$ is a representation of the Euclidean group,

$$\mathcal{U}(\vec{b}, \theta)g(\vec{p}) = g(r_{-\theta}(\vec{p} - \vec{b})) \tag{8.19}$$

r_θ is a rotation matrix, and D acts as an anisotropic dilation operator:

$$D(a_1, a_2)g(\vec{p}) = \frac{1}{\sqrt{a_1 a_2}}g\left(\frac{x}{a_1}, \frac{y}{a_2}\right) \tag{8.20}$$

It is easy to prove that such a dictionary is overcomplete using the fact that, under the restrictive condition $a_1 = a_2$, one gets 2-D continuous wavelets as defined in [1]. It is also worth stressing that, avoiding rotations, the parameter space is a group studied by Bernier and Taylor [3]. The advantage of such a parametrization is that the full dictionary is invariant under translation and rotation. Most importantly, it is also invariant under isotropic scaling, e.g., $a_1 = a_2$. These properties will be exploited for spatial transcoding in the next sections.

8.2.3.5   Dictionary

Since the structured dictionary is built by applying geometric transformations to a generating mother function g, the atoms are therefore indexed by a string γ composed of five parameters: translation $\vec{b}$, anisotropic scaling $\vec{a}$, and rotation θ. Any atom in our dictionary can finally be expressed in the following form:

$$g_\gamma = \frac{2}{\sqrt{3\pi}}(4g_1^2 - 2)\exp(-(g_1^2 + g_2^2)) \qquad (8.21)$$

with

$$g_1 = \frac{\cos(\theta)(x - b_1) + \sin(\theta)(y - b_2)}{a_1} \qquad (8.22)$$

and

$$g_2 = \frac{\cos(\theta)(y - b_2) - \sin(\theta)(x - b_1)}{a_2} \qquad (8.23)$$

For practical implementations, all parameters in the dictionary must be discretized. For the anisotropic refinement (AR) atoms subdictionary, the translation parameters can take any positive integer value smaller than the image dimensions. The rotation parameter varies by increments of $\frac{\pi}{18}$, to ensure the overcompleteness of the dictionary. The scaling parameters are uniformly distributed on a logarithmic scale from one up to an eighth of the size of the image, with a resolution of one third of octave. The maximum scale has been chosen so that at least 99% of the atom energy lies within the signal space when it is centered in the image. Experimentally, it has been found that this scale and rotation discretization choice represents a good compromise between the size of the dictionary, and the efficiency of the representation. One can choose a finer resolution for scale and rotation, getting generally more accuracy in the initial approximations. There is however a price to pay in terms of atom coding and search complexity. Finally, atoms are chosen to be always smaller along the second derivative of the Gaussian function than along the Gaussian itself, thus maximizing the similarity of the dictionary elements with edges in images. This limitation allows to limit the size of the dictionary.

For the Gaussian (low-frequency) subdictionary, the translation parameters vary exactly in the same way as for the AR atoms, but the scaling is isotropic and varies from $\frac{1}{32}\min(W, H)$ to $\frac{1}{4}\min(W, H)$ on a logarithmic scale with a resolution of one third of octave (W and H are image width and height, respectively). The minimum scale of these atoms has been chosen to have a controlled overlap with the AR functions, i.e., large enough to ensure a good coverage of the signal space, but small enough to avoid destructive interactions between the low-pass and the band-pass dictionary. This overlap has been designed so that <50% of the energy of the Gaussians lies in the frequency band taken by the AR functions. The biggest scale for these Gaussian atoms has been chosen so that at least 50% of the atom energy lies within the signal space when centered in the image. Lastly, owing to isotropy, rotations are obviously useless for this kind of atoms. Sample atoms are shown in Figure 8.3.

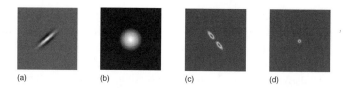

FIGURE 8.3 Sample anisotropic atoms with a rotations of $\frac{5}{18}\pi$ radians and scales of 4 and 8 (a), sample Gaussian function (b), and their respective transforms (c, d).

8.2.3.6 Coding Stage

Compact signal representations also necessitate an efficient entropy coding stage, to remove statistical redundancy left in the signal representation. This stage is crucial in overcomplete signal expansions, since the dictionary is inherently more redundant than in the case of common orthogonal transforms. Optimal coding in redundant expansions is however still an open research problem, which is made nontrivial by the large number of parameters in the case of image coding.

Efficient coding of MP parameters has been proposed in [32], for example, with a smart scanning of atom positions within image blocks. The coder presented in this section aims at producing fully scalable image streams. Such a requirement truly limits the options in the entropy coding stage, since the atom order is given by the magnitude of their coefficients, as discussed in the previous paragraph. The scalable encoder therefore implements an adaptive arithmetic coding, with independent contexts for position, scale, rotation, and coefficient parameters. The core of the arithmetic coder is based on [44], with the probability update method from [17]. As the distribution of the atom parameters (e.g., positions or scales) is dependent on the image to be coded, the entropy coder first initializes the symbol probabilities to a uniform distribution. The encoded parameters are then sent in their natural order, which results in a progressive stream, that can eventually be cut at any point to generate rate-scalable streams.

Finally, recall that flexibility is the main motivation for choosing this kind of arithmetic coder. It can be imagined that more efficient coders could, for example, try to estimate the parameter's distribution in order to optimally distribute the bits. Alternatively, grouping atoms according to their position parameters might also increase the compression ratio when combined with differential coding. Similar methods could be applied to rotation or scale indexes. However, the generated stream would not be progressive anymore, and scalability would only be attained in this case by stream manipulations, and more generally transcoding.

8.2.3.7 Coefficient Quantization

One of the crucial points in the MP encoder is the coefficient quantization stage. Since coefficients computed by the MP search take real values, quantization is a mandatory operation in order to limit the coding rate. Redundant signal expansions present the advantage that quantization error on one coefficient may be mitigated by later MP iterations, when the quantization is performed in the loop [8]. The encoder presented in this section however uses a different approach, which performs quantization *a posteriori*. In this case, the signal expansion does not depend on the quantization, and hence the coding rate. *A posteriori* quantization and coding allow for one single expansion to be encoded at different target rates. This is particularly interesting in scalable applications, which represent the main target for the image coder under consideration here. Since the distortion penalty incurred by *a posteriori* quantization is moreover generally negligible [21], this design choice is justified by an increased flexibility in image representation.

The proposed coder uses a quantization method specifically adapted to the MP expansion characteristics, the *a posteriori* rate-optimized exponential quantization. It takes benefit from the fact that the MP coefficient energy is upper-bounded by an exponential curve, decaying with the coefficient order. The quantization algorithm strongly relies on this property, and the exponential upper-bound directly determines the quantization range of the coefficient magnitude, while the coefficient sign is reported on a separate bit. The number of quantization steps is then computed as the solution of a rate–distortion optimization problem [21].

Recall that the coefficient c_{γ_n} represents the scalar product $\langle g_{\gamma_n}, \mathcal{R}^n f \rangle$. It can be shown that its norm is upper-bounded by an exponential function [29], which can be written as

$$|c_{\gamma_n}| \leq (1 - \alpha^2 \beta^2)^{n/2} \|f\| \tag{8.24}$$

where $\|f\|$ is the energy of the signal to code, β a constant depending on the construction of the dictionary, and α a suboptimality factor depending on the MP implementation (for a full-search algorithm as the one used in this paper, $\alpha = 1$). The coefficient upper-bound thus depends on both the energy of the input function and the construction of the dictionary. Since the coefficients cannot obviously bring more energy than the residual function, the norm of the coefficient is strongly related to the residual energy decay curve.

Choosing the exponential upper-bound from Eq. (8.24) as the limit of the quantization range, the number of bits to be spent on each coefficient remains to be determined. The rate–distortion optimization problem shows that the number of quantization levels have also to follow a decaying exponential law given by

$$n_j = \sqrt{\frac{\|f\|^2 (1 - \beta^2)^j \log 2}{6\lambda}} \tag{8.25}$$

where n_j is the number of quantization levels for coefficient c_j, and λ the Lagrangian multiplier that drives the size of the bitstream [21].

In practice, the exponential upper-bound and the optimal bit distribution given by Eq. (8.25) are often difficult to compute, particularly in the practical case of large dictionaries. To overcome these limitations, the quantizer uses a suboptimal but very efficient algorithm based on the previous optimal results. The key idea lies in a dynamic computation of the redundancy factor β from the quantized data. Since this information is also available at the decoder, this one is able to perform the inverse quantization without any additional side information.

In summary, the quantization stage of the coefficients is implemented as follows. The coefficients are first reordered, and sorted in the decreasing order of their magnitude (this operation might be necessary since the MP algorithm does not guarantee a strict decay of the coefficient energy). Then let $Q[c_k]$, $k = 1, \ldots, j - 1$, denote the quantized counterparts of the $j - 1$ first coefficients. Owing to the rapid decay of the magnitude, coefficient c_j is very likely to be smaller than $Q[c_{j-1}]$. It can thus be quantized in the range $[0, Q[c_{j-1}]]$. The number of quantization levels at step j is theoretically driven by the redundancy factor as given by Eq. (8.25). The adaptive quantization uses an estimate of the redundancy factor to compute the number of quantization levels as

$$n_j = (1 - \tilde{\beta}_{j-1}^2)^{1/2} n_{j-1} \tag{8.26}$$

The estimate of the redundancy factor $\tilde{\nu}$ is recursively updated as

$$\tilde{\beta}_j = \sqrt{1 - \left(\frac{Q[c_j]}{\|f\|}\right)^{2/j}} \tag{8.27}$$

Finally, the quantization range is given by the quantized coefficient $Q[c_j]$.

8.2.3.8 Rate Control

The quantization algorithm presented above is completely determined by the choice of n_0, the number of bits for the first coefficient, and a positive value of N, the number of atoms in the signal expansion. When the bitstream has to conform to a given bit budget, the quantization scheme parameters n_0 and N can be computed as follows. First, β is estimated with Eq. (8.27) by training the dictionary on a large set of signals (e.g., images), encoded with the adaptive quantization algorithm. The estimation quite rapidly tends to the asymptotic value of the redundancy factor. The estimation of β is then used to compute λ as a function of the given bit budget R_b which has to satisfy

$$R_b = \sum_{j=0}^{N-1} \log_2 n_j + \sum_{j=0}^{N-1} a_j$$

$$= \sum_{j=0}^{N-1} \log_2(1 - \beta^2)^{1/2} + N \log_2 n_0 + N A \qquad (8.28)$$

where a_j represents the number of bits necessary to code the parameters of atom g_{γ_j} (i.e., positions, scales, and rotation indexes), and $A = E[a_j]$ represents the average index size. From Eq. (8.25), the value of λ determines the number of bits of the first coefficient n_0. Under the reasonable condition that the encoder does not code atoms whose coefficients are not quantized (i.e., $n_j < 2$), the number of atoms to be coded, N, is finally determined by the condition $(1 - \beta^2)^{(N-1)/2} n_0 \le 2$. The adaptive quantization algorithm is then completely determined, and generally yields bit rates very close to the bit budget.

8.2.4 Experimental Results

8.2.4.1 Benefits of Anisotropy

Anisotropy and rotation represent the core of the design of our coder. To show the benefits of anisotropic refinement, our dictionary has been compared to four different dictionaries, in terms of the quality of the MP expansion. The first dictionary uses the real part of oriented Gabor atoms generated by translation $\vec{b}$, rotation θ, and isotropic scaling a of a modulated Gaussian function

$$U(a, \theta, \vec{b})g(\vec{x}) = \frac{1}{a} g(a^{-1} r_{-\theta}(\vec{x} - \vec{b})) \qquad (8.29)$$

with

$$g(\vec{x}) = e^{i\vec{\omega}_0 \cdot \vec{x}} e^{-\|\vec{x}\|^2/2} \qquad (8.30)$$

The next dictionary is an affine Weyl–Heisenberg dictionary [6] built by translation, dilation, and modulation of the Gabor–generating atom of Eq. (8.30):

$$U(a, \vec{\omega}, \vec{b})g(\vec{x}) = \frac{1}{a} e^{i\vec{\omega} \cdot (\vec{x} - \vec{b})} g(a^{-1}(\vec{x} - \vec{b})) \qquad (8.31)$$

where again, as we are dealing with real signals, only the real part is used. The other two dictionaries are simply built on orthogonal wavelet bases. Figure 8.4 shows the reconstructed quality as a function of the number of iterations in the MP expansion using different types of dictionaries. In this figure, the comparison is performed with respect to the number of terms in the expansion, in order to emphasize the approximation properties (the behavior of the coding rate is discussed below). Clearly, overcompleteness and anisotropic refinement allow to outperform the other dictionaries, in terms of approximation rate, which corresponds to the results presented in [11,20]. As expected, the orthogonal

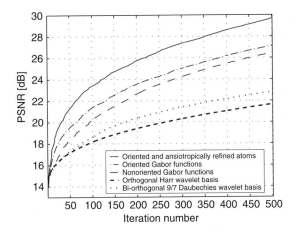

FIGURE 8.4 Comparison of the MP approximation rate for *Lena* (128 × 128 pixels), using five different dictionaries (anisotropic scaling, Gabor wavelets, Weyl–Heisenberg dictionary, an orthogonal Haar wavelet basis, and a biorthogonal Daubechies 9/7 basis, with five levels of decomposition).

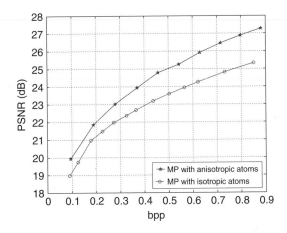

FIGURE 8.5 Comparison of the rate–distortion characteristic of a decomposition using a dictionary built on anisotropic refinement and a dictionary without anisotropic refinement. The basis functions are the same for isotropic and anisotropic functions, with the same angle discretization (allowing 18 different angles) and with spatial translation resolution of 1 pixel.

bases offer the lowest approximation rates due to the fact that these kinds of bases cannot deal with the smoothness of edges. We can thus deduce that redundancy in a carefully designed dictionary provides sparser signal representations. This comparison shows, as well, that the use of rotation is also of interest since the oriented Gabor dictionary gives better results than the modulated one. It is worth noticing that rotation and anisotropic scaling are true 2-D transformations: the use of nonseparable dictionaries is clearly beneficial to efficiently approximate 2-D objects. Separable transforms, although they may enable faster implementations, are unable to cope with the geometry of edges.

It is interesting now to analyze the penalty of anisotropy on the coding rate. In our coder, the addition of anisotropy induces the cost of coding an additional scaling parameter for each atom. To highlight the coding penalty due to anisotropic refinement, the image has also been coded with the same dictionary, built on isotropic atoms, all other parameters staying identical to the proposed scheme. Figure 8.5 illustrates the quality of the MP encoding of *Lena*, as a function of the coding rate,

(a)   (b)

FIGURE 8.6 *Lena* (512 × 512) encoded at 0.16 bpp: (a) MP, 31.0610 dB; (b) JPEG-2000, 31.9285 dB.

with both dictionaries. To perform the comparison, the isotropic and the anisotropic dictionaries are generated with the same generating function and with the same discretization of the parameters (three scales per octave and an angle resolution of 10°). The anisotropy however implies the coding of one additional scale parameter. It is shown that the dictionary based on anisotropic refinement provides superior coding performance, even with longer atom indexes. The penalty due to the coding cost of one additional scale parameter, is largely compensated by a better approximation rate. Anisotropic refinement is thus clearly an advantage in MP image coding.

8.2.4.2 Coding Performance

The objective of this section is to emphasize the potential of redundant expansion's low-rate compression of natural images, even though the MP encoder is not fully optimized yet, as it has been discussed in Section 8.2.3.

Figure 8.6 presents a comparison between images compressed with MP and JPEG-2000.[2] It can be seen that the PSNR rating is in favor of JPEG-2000, which is not completely surprising since a lot of research efforts are being put in optimizing the encoding in JPEG-2000-like schemes. Interestingly, however, the image encoded with MP is visually more pleasant than the JPEG-2000 version. The coding artifacts are quite different, and the degradations due to MP are less annoying to the human visual system than the ringing due to wavelet coding at low rate. The detailed view of the hat, as illustrated in Figure 8.7, confirms this impression. It can be seen that the JPEG-2000 encoder introduces quite a lot of ringing, while the MP encoder concentrates its effort on providing a good approximation of the geometrical components of the hat structure. JPEG-2000 has difficulties to approximate the 2-D-oriented contours, which are generally the most predominant components of natural images. And this is clearly one of the most important advantages of the MP coder built on anisotropic refinement, which is really efficient to code edge-like features.

To be complete, Figure 8.8 shows the rate–distortion performance of the MP encoder for common test images, at low to medium bit rates. It can be seen that MP provides better PSNR rating than JPEG-2000 at low coding rates. However, the gap between both coding schemes decreases rapidly when the bit rate increases, as expected. MP and overcomplete expansions are especially efficient for low bit-rate coding. They rapidly capture the most important components of the image, but MP then suffers from its greedy characteristic when the rate increases. It has to be noted also that the bitstream header penalizes JPEG-2000 compared to MP, where the syntactic information is truly minimal (at most a few bits). This penalty becomes particularly important at a very low bit rate.

---

[2] All results have been generated with the Java implementation available at http://jj2000.epfl.ch/ with default settings.

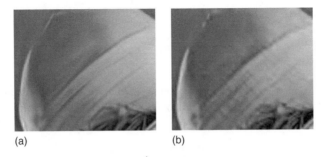

(a)   (b)

FIGURE 8.7 Detailed view of *Lena* (512 × 512) encoded at 0.16 bpp: (a) MP; (b) JPEG-2000.

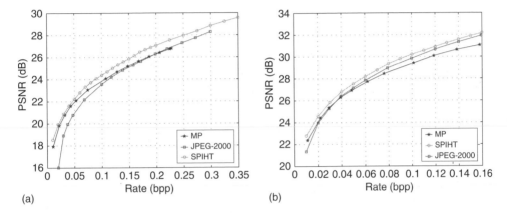

(a)   (b)

FIGURE 8.8 (See color insert following page 336) Distortion–rate performance for JPEG-2000, SPIHT, and the proposed MP coder for common test images. (a) *Cameraman* (256 × 256); (b) *Lena* (512 × 512).

The performance of the proposed coder is also compared to the SPIHT encoder [35], which introduces a minimal syntactic overhead. SPIHT almost always outperforms the proposed coder on the complete range of coding rate, and tends to perform similarly to JPEG-2000 for high rates. However, the stream generated by the SPIHT encoder in general does not provide scalability, while MP and JPEG-2000 offer increased flexibility for stream adaptation.

Finally, the proposed encoder performs reasonably well in terms of rate–distortion performance, especially at low rates. The distortion is in general visually less annoying in the MP coding algorithm. The artifacts introduced by MP (basically a simplification or refinable sketch of the image) are indeed less annoying for the human observer than the ringing introduced by the wavelets in JPEG-2000. When the rate increases, the saturation of the quality can be explained by the limitations of redundant transforms for high rate approximations. Hybrid coding schemes could provide helpful solutions for high rate coding.

8.2.5 EXTENSION TO COLOR IMAGES

The MP encoder presented in the previous section can be extended to code color images, using a similar principle. Instead of performing independent iterations in each color channel, a vector search algorithm can be implemented in a color image encoder. This is equivalent to using a dictionary of P vector atoms of the form $\{\vec{g_\gamma} = [g_\gamma, g_\gamma, g_\gamma]\}_{\gamma \in \Gamma}$. In practice though, each channel is evaluated with one single component of the vector atom, whose global energy is given by adding together its respective contribution in each channel. MP then naturally chooses the vector atom, or equivalently the vector component g_γ with the highest energy. Hence, the component of the dictionary chosen at

each MP iteration satisfies

$$\max_{\gamma_n} \sqrt{\langle R^n f^1, g_{\gamma_n}\rangle^2 + \langle R^n f^2, g_{\gamma_n}\rangle^2 + \langle R^n f^3, g_{\gamma_n}\rangle^2} \qquad (8.32)$$

where $R^n f^i$, $i = 1, 2, 3$, represents the signal residual in each of the color channels. Note that this is slightly different than the algorithm introduced in [28], where the supnorm of all projections is maximized:

$$\max_{i} \sup_{\mathcal{D}} |\langle R^n f^i, g_{\gamma_n}\rangle| \qquad (8.33)$$

All signal components f^i are then jointly approximated through an expansion of the form

$$f^i = \sum_{n=0}^{+\infty} \langle R^n f^i, g_{\gamma_n}\rangle g_{\gamma_n}, \quad \forall i = 1, 2, 3 \qquad (8.34)$$

Note that channel energy is conserved, and that the following Parseval-like equality is verified:

$$\|f^i\|^2 = \sum_{n=0}^{+\infty} |\langle R^n f^i, g_{\gamma_n}\rangle|^2, \quad \forall i = 1, 2, 3 \qquad (8.35)$$

The search for the atom with the highest global energy necessitates the computation of the three scalar products $c_n^i = \langle R^n f^i, g_{\gamma_n}\rangle$, $i = 1, 2, 3$, for each atom g_{γ_n}, and for each iteration of the MP expansion. The number of scalar products can be reduced by first identifying the color channel with the highest residual energy, and then performing the atom search in this channel only. Once the best atom has been identified, its contribution in the other two channels is also computed and encoded. The reduced complexity algorithm obviously performs in a suboptimal way compared to the maximization of the global energy, but in most of the cases the quality of the approximation only suffers a minimal penalty. (Figure 8.9 is an example of an MP performed in the most energetic channel.)

An important parameter of the color encoder is the choice of color space. Interestingly, experiments show that the MP coder tends to prefer highly correlated channels. This can be explained by the fact that atom indexes carry higher coding costs than coefficients. Using correlated channels basically means that the same structures are found, and thus the loss of using only one index for all channels is minimized. The choice of the RGB color space thus seems very natural. This can also be highlighted by the following experiments. The coefficients $[c_n^1, c_n^2, c_n^3]$ of the MP decomposition can

(a)   (b)

FIGURE 8.9 (See color insert following page 336) Japanese woman coded with 1500 MP atoms, using the most energetic channel search strategy in YUV color space (a) and in RGB color space (b).

be represented in a cube, where the three axes correspond to the red, green, and blue components, respectively (see Figure 8.10a). It can be seen that the MP coefficients are interestingly distributed along the diagonal of the color cube, or equivalently that the contribution of MP atoms is very similar in the three color channels. This very nice property is a real advantage in overcomplete expansions, where the coding cost is mainly due to the atom indexes. On the contrary, the distribution of MP coefficients, resulting from the image decomposition in the YUV color space, does not seem to present any obvious structure (see Figure 8.10b). In addition, the YUV color space has been shown to give quite annoying color distortions for some particular images (see Figure 8.9, for example).

Owing to the structure of the coefficient distribution, centered around the diagonal of the RGB cube, efficient color quantization does not anymore consist of coding the raw values of the R, G, and B components, but instead of coding the following parameters: the projection of the coefficients on the diagonal, the distance of the coefficients to the diagonal, and the direction where it is located. This is equivalent to coding the MP coefficients in an HSV color space, where V (Value) becomes the projection of RGB coefficients on the diagonal of the cube, S (Saturation) the distance of the coefficient to the diagonal, and H (Hue) the direction perpendicular to the diagonal, where the RGB coefficient is located. The HSV values of the MP coefficients present the following characteristic distributions. The value distribution is Laplacian, centered at zero (see Figure 8.11c), saturation presents an exponential distribution (see Figure 8.11b), and a Laplacian-like distribution with two peaks can be observed for hue values (Figure 8.11a). Finally, once the HSV coefficients have been calculated from the available RGB coefficients, the quantization of the parameters is performed as follows:

- The value is exponentially quantized with the quantizer explained before (see [21]). The number that will be given as input to the arithmetic coder will be $N_j(l)$ — Quant(V), where $N_j(l)$ is the number of quantization levels that are used for coefficient l.
- Hue and saturation are uniformly quantized.

Finally coefficients and indexes are entropy-coded, by the same technique used earlier for gray-scale images. Compression performances of this algorithm are illustrated in Figure 8.12, where a comparison with JPEG-2000 is also provided. It can be seen that MP advantageously compares to JPEG-2000, and even performs better at low bit rates. This can be explained by the property of MP to immediately capture most of the signal features in a very few iterations and across channels. Note that the PSNR values have been computed in the Lab color space in order to match the human visual system perception.

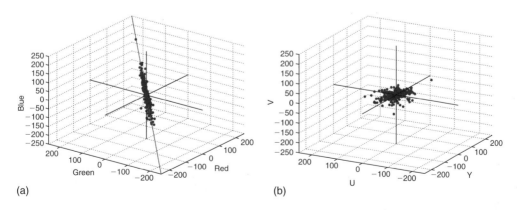

(a)   (b)

FIGURE 8.10 Distribution of the MP coefficients when MP is performed in the RGB or YUV color space: (a) RGB space; (b) YUV space.

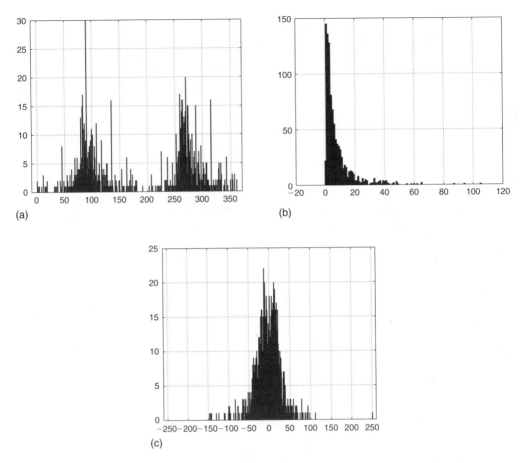

FIGURE 8.11 Histograms of the MP coefficients when represented in HSV coordinates. (a) Hue; (b) saturation; and (c) value.

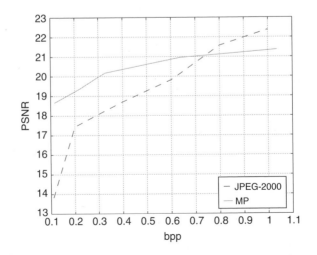

FIGURE 8.12 PSNR comparison between JPEG-2000 and MP. The PSNR has been computed in the CIELAB color space.

8.2.6 High Adaptivity

8.2.6.1 Importance of Adaptivity

As outlined in the previous section, one of the main advantages of the MP coder is to provide highly flexible streams at no additional cost. This is very interesting in present-day visual applications involving transmission and storage, like database browsing or pervasive image and video communications. We call adaptivity the possibility for partial decoding of a stream, to fulfill decoding constraints given in terms of rate, spatial resolution, or complexity. The challenge in scalable coding is to build a stream decodable at different resolutions without any significant loss in quality by comparison to nonadaptive streams. In other words, adaptive coding is efficient if the stream does not contain data redundant to any of the target resolutions.

In image coding, adaptivity generally comprises rate (or SNR) adaptivity and spatial adaptivity. On the one hand, the most efficient rate adaptivity is attained with progressive or embedded bitstreams, which ensure that the most important part of the information is available, independently of the number of bits used by the decoder [36,39]. In order to enable easy rate adaptation, the most important components of the signals should be placed near the beginning of the stream. The encoding format also has to guarantee that the bitstream can be decoded, even when truncated. On the other hand, efficient adaptive coding schemes, like JPEG-2000 or the coder proposed in [45] are generally based on subband decompositions, which provide intrinsic multiresolution representations. However, spatial adaptivity is generally limited to octave-based representations, and different resolutions can only be obtained after nontrivial transcoding operations.

Multidimensional and geometry-based coding methods can advantageously provide high flexibility in the stream representation and manipulation. In this section, we will emphasize the intrinsic spatial and rate adaptivity of the bitstreams created with our MP image coder. First, owing to the geometrical structure of the proposed dictionary, the stream can easily and efficiently be decoded at any spatial resolution. Second, the embedded bitstream generated by the MP coder can be adapted to any rate constraints, while the receiver is guaranteed to always get the most energetic components of the MP representation. Most importantly, MP streams offer the advantage of decoupling spatial and rate adaptivity, which can be performed independently. Adaptive decoding is now discussed in more detail in the remainder of the section.

8.2.6.2 Spatial Adaptivity

Owing to the structured nature of our dictionary, the MP stream provides inherent spatial adaptivity. The group law of the similitude group of $\mathbb{R}^2$ indeed applies [1] and allows for invariance with respect to *isotropic* scaling of α, rotation of Θ, and translation of $\vec{\beta}$. Let us remind the reader that the dictionary is built by acting on a mother function with a set of operators realizing various geometric transformations (see Eqs. (8.17)–(8.18)). When considering only isotropic dilations, i.e., $a_1 = a_2$ in (8.18), this set forms a group: the similitude group of the 2-D plane. Therefore, when the compressed image $\hat{f}$ is submitted to any combination of these transforms (denoted here by the group element η), the indexes of the MP stream can simply be transformed with help of the group law

$$\mathcal{U}(\eta)\hat{f} = \sum_{n=0}^{N-1} \langle g_{\gamma_n} | \mathcal{R}^n f \rangle \, \mathcal{U}(\eta) g_{\gamma_n} = \sum_{n=0}^{N-1} \langle g_{\gamma_n} | \mathcal{R}^n f \rangle \, \mathcal{U}(\eta \circ \gamma_n) g \qquad (8.36)$$

In the above expression, $\gamma_n = (\vec{a}_n, \theta_n, \vec{b}_n)$ represents the parameter strings of the atom encoded at iteration n, with scaling $\vec{a}_n$, rotation θ_n, and translation $\vec{b}_n$, and $\eta = (\alpha, \Theta, \vec{\beta})$ represents the geometric transformation that is applied to the set of atoms. The decoder can apply the transformations to the encoded bitstream simply by modifying the parameter strings of the unit-norm atoms, according to

the group law of similitude, where

$$(\vec{a}, \theta, \vec{b}) \circ (\alpha, \Theta, \vec{\beta}) = (\alpha \, \vec{a}, \theta + \Theta, \vec{b} + \alpha \, r_{\Theta} \, \vec{\beta}) \tag{8.37}$$

In other words, if $\eta_{\alpha} = (\alpha, 0, 0)$ denotes the isotropic scaling by a factor α, the bitstream of an image of size $W \times H$, after entropy decoding, can be used to build an image at any resolution $\alpha W \times \alpha H$ simply by multiplying positions and scales by the scaling factor α (from Eqs. (8.37) and (8.18)). The coefficients also have to be scaled with the same factor to preserve the energy of the different components. The quantization error on the coefficient will therefore also vary proportionally to the scaling factor, but the absolute error on pixel values will remain almost unchanged, since the atom support also varies. Finally, the scaled image is obtained by

$$\mathcal{U}(\eta_{\alpha})\hat{f} = \alpha \sum_{n=0}^{N-1} c_{\gamma_n} g_{\eta_{\alpha} \circ \gamma_n} \tag{8.38}$$

The modified atoms $g_{\eta_{\alpha} \circ \gamma_n}$ are simply given by Eqs. (8.21) to (8.23), where $\vec{b}$ and $\vec{a}$ are respectively replaced by $\alpha \, \vec{b}$ and $\alpha \, \vec{a}$. It is worth noting that the scaling factor α can take any positive real value as long as the scaling is isotropic. Atoms that become too small after transcoding are discarded. This allows for further bit-rate reduction, and avoids aliasing effects when $\alpha < 1$. The smallest atoms generally represent high-frequency details in the image, and are located toward the end of the stream. The MP encoder initially sorts atoms along their decreasing order of magnitude, and scaling does not change this original arrangement.

Finally, scaling operations are quite close to image editing applications. The main difference is in the use of the scaling property. Scaling will be used at a server, within intermediate network nodes, or directly at the client in transcoding operations, while it could be used in the authoring tool for editing. Even in editing, the geometry-based expansion provides an important advantage over conventional downsampling or interpolation functions, since there is no need for designing efficient filters. Other image-editing manipulations, such as rotation of the image, or zooming in a region of interest, can easily be implemented following the same principles.

The simple spatial adaption procedure is illustrated in Figure 8.13, where the encoded image of size 256×256 has been rescaled with irrational factors $\sqrt{\frac{1}{2}}$ and $\sqrt{2}$. The smallest atoms have been discarded in the downscaled image without impairing the reconstruction quality. The up-scaled image provides quite a good quality, even if very high-frequency characteristics are obviously missing since they are absent from the initial (compressed) bitstream. Table 8.1 shows rate–distortion performance for spatial resizing of the 256×256 *Lena* image compressed at 0.3 bpp with the proposed MP coder

FIGURE 8.13 *Lena* image of size 256×256 encoded with MP at 0.3 bpp (center) and decoded with scaling factors of $\sqrt{\frac{1}{2}}$ (left) and $\sqrt{2}$ (right).

TABLE 8.1

Comparison of Spatial Adaptivity of the MP Encoder and JPEG-2000. PSNR Values are Compared to Quality Obtained Without Transcoding (w/o tr.)

Encoder		128 × 128	256 × 256	512 × 512
Matching pursuit	PSNR	27.34	30.26	27.5
	rate (bpp)	0.8	0.3	0.08
	PSNR (w/o tr.)	27.4	30.26	27.89
JPEG-2000	PSNR	27.18	29.99	—
	rate (bpp)	1.03	0.3	—
	PSNR (w/o tr.)	33.75	29.99	—

and JPEG-2000. It presents the PSNR values of the resized image as well as the rate after transcoding. It also shows the PSNR values for encoding directly at the target spatial resolutions for equivalent rates. The PSNR values have been computed with reference to the original 512×512 pixel *Lena* image, successively downsampled to 256×256 and 128×128 pixel resolutions. This is only one possibility for computing the low-resolution reference images and other more complex techniques, involving, for example, filtering and interpolation, could be adopted. The choice of such a low-resolution reference image was done in order not to favor one algorithm or the other. If a Daubechies 9/7 filter had been chosen, JPEG would have given better results. On the contrary, if a Gaussian filter had been chosen, MP would have given better results. Note that the transcoding operations for JPEG-2000 are kept very simple for the sake of fairness; the high-frequency subbands are simply discarded to get the lowest resolution images.

Table 8.1 clearly shows that our scheme offers results competitive with respect to state-of-the-art coders like JPEG-2000 for octave-based downsizing. In addition it allows for nondyadic spatial resizing as well as easy upscaling. The quality of the down-scaled images are quite similar, but the JPEG-2000-transcoded image rate is largely superior to the MP stream one. The scaling operation does not significantly affect the quality of the image reconstruction from MP streams. Even in the upscaling scenario, the transcoded image provides a very good approximation of the encoding at the target (higher) resolution. In the JPEG-2000 scenario however, the adaptation of the bitstream has a quite big impact on the quality of the reconstruction, compared to an encoding at the target resolution. Note, however, that the PSNR value is highly dependent on the choice of the reference images, which in this case are simply downsampled from the original version.

8.2.6.3 Rate Scalability

MP offers an intrinsic multiresolution advantage, which can be efficiently exploited for rate adaptivity. The coefficients are by nature exponentially decreasing so that the stream can simply be truncated at any point to provide a SNR-adaptive bitstream, while ensuring that the most energetic atoms are kept. The simplest possible rate adaption algorithm that uses the progressive nature of the MP stream works as follows. Assume an image has been encoded at a high target bit rate R_b, using the rate controller described in Section 8.2.3. The encoded stream is then restricted to lower bit budgets r_k, $k = 0, \ldots, K$, by simply dropping the bits $r_k + 1$ to R_b. This simple rate adaption, or filtering operation, is equivalent to dropping the last iterations in the MP expansion, focusing on the highest energy atoms.

Figure 8.14 illustrates the rate adaptivity performance of the MP encoder. Images have been encoded with MP at a rate of 0.17 bpp and truncated to lower rates r_k. For comparison, the bitstream has also been encoded directly at different target rates r_k, as described in Section 8.2.3. It can be seen

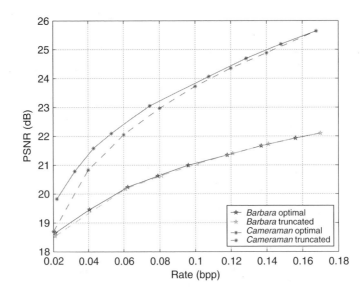

FIGURE 8.14 Rate–distortion characteristics for MP encoding of the 256×256 *Barbara* and *Cameraman* images at 0.17 bpp, and truncation/decoding at different (smaller) bit rates.

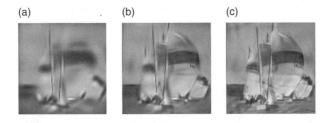

FIGURE 8.15 **(See color insert following page 336)** Matching pursuit bitstream of *sail* image decoded after 50, 150, and 500 coefficients. (a) 50 coefficients; (b) 150 coefficients; and (c) 500 coefficients.

that there is a very small loss in PSNR with respect to the optimal MP stream at the same rate. This loss is due to the fact that the rate truncation simply results in dropping iterations, without using the optimal quantizer settings imposed by rates r_k as proposed in Section 8.2.3. The quantization parameters are not optimal anymore with respect to the truncation rate, but the penalty is quite low away from very low coding rates. The loss in performance is larger for images that are easier to code, since the decay of the coefficients is faster. Nevertheless, both optimal and truncated rate–distortion curves are quite close, which shows that a simple rate adaption method, though quite basic, is very efficient.

Finally, rate scalability is also almost automatic for the color image stream. Figure 8.15 shows the effects of truncating the MP expansion with different numbers of coefficients. It can be observed again that the MP algorithm will first describe the main objects in a sketchy way (keeping the colors) and then it will refine the details.

8.3 DISCUSSIONS AND CONCLUSIONS

8.3.1 DISCUSSIONS

The results presented in this chapter show that redundant expansions over dictionaries of nonseparable functions may represent the core of new breakthroughs in image compression. Anisotropic refinement and orientation of dictionary functions allow for very good approximation rates, due to their ability to capture two-dimensional patterns, especially edges, in natural images. In addition, multidimensional representations may generate less annoying artifacts than wavelet or DCTs, that introduce ringing or blocking artifacts at low rates.

Matching pursuit is however just one (suboptimal) method that allows to solve the NP-hard problem of finding the best signal expansion in an overcomplete dictionary. It provides a computationally tractable solution with a very simple decoder implementation, and has the advantage to generate a scalable and progressive bitstream. Owing to its greedy nature, MP however presents some limitations at high rate. An hybrid scheme could help in high bit-rate coding, and provide a simple solution to the limitations of MP.

Finally, the image encoder presented in this chapter mainly aims at illustrating the potential of redundant expansions in image compression. It has been designed in order to provide scalable streams, and this requirement limits the encoding options that could be proposed, and thus possibly the compression efficiency. It is clear that the proposed MP encoder is not fully optimized, and that numerous research problems remain to be solved before one can really judge the benefit of redundant transforms in image compression. The approximation rate has been proven to be better than the rate offered in the orthogonal transform case but the statistics of coefficients in subband coding, for example, present a large advantage in terms of compression. It is thus too early to claim that MP image coding is the next breakthrough in terms of compression, but it already presents a very interesting alternative with competitive quality performance and increased flexibility. The current coding scheme of the MP coefficients is not optimal, contrarily to the very efficient coding of wavelet coefficients in JPEG-2000. The advantage of the multidimensional decomposition in terms of approximation rate, is thus significantly reduced under a rate–distortion viewpoint. The design of better coding scheme, carefully adapted to the characteristics of the MP representation, however represents a challenging research problem.

8.3.2 EXTENSIONS AND FUTURE WORK

One of the striking advantages of using a library of parameterized atoms is that the reconstructed image becomes parameterized itself: it is described by a list of geometrical features, together with coefficients indicating the "strength" of each term. This list can be manipulated as explained in Section 8.2.6. But these features can be used to perform many other different tasks. For example, these features can be thought of as a *description* of the image and can thus be used for recognition or classification. The description could also be easily manipulated or altered to encrypt the image or to insert an invisible signature.

The ideas of using redundant expansions to code visual information could be further extended to video signals. In this case, the coder can follow two main design strategies; one based on motion estimation, and the other based on temporal transform. In the first case, a MP encoder can be used to code the residue from the motion estimation stage. The characteristics of this residue are however quite different than the features present in natural images. The dictionary need to be adapted to this mainly high-frequency motion noise. In the scenario where the coding is based on a temporal transform, MP could work with a dictionary of three-dimensional atoms, where the third dimension represents the temporal component. In this case, atoms live in a block of frames, and the encoder works very similarly to the image encoder.

ACKNOWLEDGMENTS

The authors thank Rosa M. Figueras i Ventura for her help in designing the image coder, and the production of the experimental results presented in this chapter.

REFERENCES

1. J.-P. Antoine, R. Murenzi, and P. Vandergheynst, Directional wavelets revisited: Cauchy wavelets and symmetry detection in patterns, *Appl. Comput. Harmonic Anal.*, 6, 314–345, 1999.

2. F. Bergeaud and S. Mallat, Matching pursuit of images, *Proceedings of IEEE International Conference on Image Processing*, Washington DC, October 1995, vol. I, pp. 53–56.

3. D. Bernier and K. Taylor, Wavelets from square-integrable representations, *SIAM J. Math. Anal.*, 27, 594–608, 1996.

4. E.J. Candès and D.L. Donoho, Curvelets — a surprisingly effective nonadaptive representation for objects with edges, in A. Cohen, C. Rabut, and L.L. Schumaker, Eds., *Curves and Surfaces*, Vanderbilt University Press, Nashville, TN, 2000, pp. 105–120.

5. S. Chen, D.L. Donoho, and M.A. Saunders, Atomic decomposition by basis pursuit, *SIAM J. Sci. Comp.*, 20, 33–61, 1999.

6. Z. Cvetkovic and M. Vetterli, Tight Weyl-Heisenberg Frames in $l^2(Z)$, *IEEE Trans. Signal Process.*, 46, 1256–1259, 1998.

7. G.M. Davis, S. Mallat, and Z. Zhang, Adaptive time-frequency decompositions, *SPIE J. Opt. Eng.*, 33, 2183–2191, 1994.

8. C. De Vleeschouwer and A. Zakhor, In-loop atom modulus quantization for matching pursuit and its application to video coding, *IEEE Trans. Image Process.*, 12, 1226–1242, 2003.

9. R.A. De Vore, Nonlinear approximation, *Acta Numerica*, vol. 7, pp. 51–150, 1998.

10. R.A. De Vore and V.N. Temlyakov, Some remarks on greedy algorithms, *Adv. Comp. Math.*, 5, 173–187, 1996.

11. M.N. Do, P.L. Dragotti, R. Shukla, and M. Vetterli, On the compression of two dimensional piecewise smooth functions, *IEEE International Conference on Image Processing (ICIP)*, 2001.

12. M.N. Do and M. Vetterli, Contourlets: a directional multiresolution image representation, *Proceedings of the IEEE International Conference on Image Processing*, Rochester, September 2002, vol. 1, pp. 357–360.

13. D.L. Donoho, Wedgelets: nearly-minimax estimation of edges, *Ann. Stat.*, 27, 859–897, 1999.

14. D.L. Donoho and X. Huo, Uncertainty principles and ideal atomic decompositions, *IEEE Trans. Info. Theory*, 47, 2845–2862, 2001.

15. D.L. Donoho, M. Vetterli, R.A. De Vore, and I. Daubechies, Data compression and harmonic analysis, *IEEE Trans. Info. Theory*, 44, 391–432, 1998.

16. P.L. Dragotti and M. Vetterli, Wavelet footprints: theory, algorithms and applications, *IEEE Trans. Signal Process.*, 51, 1306–1323, 2003.

17. D.L. Duttweiler and C. Chamzas, Probability estimation in arithmetic and adaptive-huffman entropy coders, *IEEE Trans. Image Process.*, 4, 237–246, 1995.

18. M. Elad and A.M. Bruckstein, A generalized uncertainty principle and sparse representations in pairs of bases, *IEEE Trans. Info. Theory*, 48, 2558–2567, 2002.

19. R.M. Figueras i Ventura, O. Divorra Escoda and P. Vandergheynst, Matching Pursuit Full Search Algorithm, Technical Report, F-Group (LTS-2), ITS, EPFL, 2003.

20. R.M. Figueras i Ventura, L. Granai, and P. Vandergheynst, R-D analysis of adaptive edge representations, *Proceedings of IEEE International Workshop on Multimedia Signal Processing (MMSP02)*, St Thomas, US Virgin Islands, 2002.

21. P. Frossard, P. Vandergheynst, R.M. Figueras i Ventura, and M. Kunt, A posteriori quantization of progressive matching pursuit streams, *IEEE Trans. Signal Process.*, 52, 525–535, 2004.

22. A.C. Gilbert, S. Muthukrishnan, and M.J. Strauss, Approximation of functions over redundant dictionaries using coherence, *Proceedings of the 14th Annual ACM-SIAM Symposium on Discrete Algorithms*, 2003.

23. R. Gribonval and M. Nielsen, Sparse Representations in Unions of Bases, Technical Report 1499, IRISA, Rennes, France, 2003.

24. R. Gribonval and P. Vandergheynst, On the Exponential Convergence of Matching Pursuits in Quasi-Incoherent Dictionaries, Technical Report 1619, IRISA, Rennes, France, 2003.

25. L. Jones, On a conjecture of huber concerning the convergence of projection pursuit regression, *Ann. Stat.*, 15, 880–882, 1987.

26. P. Jost, P. Vandergheynst, and P. Frossard, Tree-Based Pursuit, Technical Report 2004.13, EPFL-Signal Processing Institute, July 2004.

27. E. Le Pennec and S.G. Mallat, Geometrical image compression with bandelets, in *Visual Communications and Image Processing 2003*, T. Ebrahimi and T. Sikora, Eds., SPIE, Lugano, Switzerland, 2003, pp. 1273–1286.

28. A. Lutoborski and V.N. Temlyakov, Vector greedy algorithms, *J. Complexity*, 19, 458–473, 2003.

29. S. Mallat, *A Wavelet Tour of Signal Processing*, Academic Press, New York 2nd. ed., 1999.

30. S.G. Mallat and Z. Zhang, Matching pursuits with time-frequency dictionaries, *IEEE Trans. Signal Process.*, 41, 3397–3415, 1993.

31. D. Marr, *Vision*, Freeman, San Francisco, 1982.

32. R. Neff and A. Zakhor, Very low bit-rate video coding based on matching pursuits, *IEEE Trans. Circuits Syst. Video Technol.*, 7, 158–171, 1997.

33. Y.C. Pati, R. Rezaifar, and P.S. Krishnaprasad, Orthogonal matching pursuit: recursive function approximation with applications to wavelet decompositions, *Proceedings of the 27th Asilomar Conference on Signals, Systems and Computers*, 1993.

34. W. Rudin, *Real and Complex Analysis*, McGraw-Hill, New York 1987.

35. A. Said and W.A. Pearlman, Reversible image compression via multiresolution representation and predictive coding, *Proceedings of the SPIE — Visual Communication and Image Processing*, 1993, vol. 2094, pp. 664–674, http://www.cipr.rpi.edu/research/SPIHT/.

36. A. Said and W.A. Pearlman, A new, fast, and efficient image codec based on set partitioning in hierarchical trees, *IEEE Trans. Circuits Syst. Video Technol.*, 6, 243–250, 1996.

37. J.-L. Starck, E.J. Candès, and D.L. Donoho, The curvelet transform for image denoising, *IEEE Trans. Image Process.*, 11, 670–684, 2002.

38. D. Taubman and M. Marcellin, *JPEG2000: Image Compression Fundamentals, Standards and Practice*, Kluwer Academic Publishers, Boston, 2001.

39. D. Taubman and A. Zakhor, Multirate 3-D subband coding of video, *IEEE Trans. Image Process.*, 3, 572–588, 1994.

40. J.A. Tropp, Greed is Good: Algorithmic Results for Sparse Approximation, Technical Report, ICES, University of Texas at Austin, Austin, USA, 2003.

41. J.A. Tropp, Just Relax: Convex programming methods for Subset Selection and Sparse Approximation, Technical Report, ICES, University of Texas at Austin, Austin, USA, 2004.

42. L.F. Villemoes, Nonlinear approximation with walsh atoms, in A. Le Méhanté, C. Rabut, and L.L. Schumaker, Eds., *Surface Fitting and Multiresolution Methods*, Chamonix, July 1996, Vanderbilt University Press, Nashville, TN, 1997, pp. 329–336.

43. M. Wakin, J. Romberg, H. Choi, and R. Baraniuk, Geometric methods for wavelet-based image compression, *International Symposium on Optical Science and Technology*, San Diego, CA, August 2003.

44. I.H. Witten, R.M. Neal, and J.G. Cleary, Arithmetic coding for data compression, *Commun. ACM*, 30, 520–540, 1987.

45. J.W. Woods and G. Lilienfield, A resolution and frame-rate scalable subband/wavelet video coder, *IEEE Trans. Circuits Syst. Video Technol.* 11, 1035–1044, 2001.

9 Distributed Compression of Field Snapshots in Sensor Networks

Sergio D. Servetto

CONTENTS

9.1 Introduction 235
    9.1.1 Distributed Image Coding and Wireless Sensor Networks 236
    9.1.2 The Sensor Broadcast Problem 236
        9.1.2.1 Problem Formulation 236
        9.1.2.2 Relevance of the Problem 237
    9.1.3 Data Compression Structures 238
        9.1.3.1 Data Compression Using Independent Encoders 239
        9.1.3.2 Exploiting Correlations in the Source 239
    9.1.4 Organization of the Chapter 239
9.2 Distributed Compression of Sensor Measurements 240
    9.2.1 Information-Theoretic Bounds for Bandlimited Images 240
    9.2.2 Distributed Computation of Decorrelating Transforms 241
    9.2.3 Images Constrained by Physical Laws 241
9.3 Transforms for Distributed Decorrelation of Bandlimited Images 243
    9.3.1 The "Drop-Data" Transform 243
    9.3.2 Linear Signal Expansions with Bounded Communication 244
        9.3.2.1 Wavelets and Sensor Broadcast 244
        9.3.2.2 Definition of the Coding Strategy 244
        9.3.2.3 Differentiating between Local and Global Communication
                 Requirements 246
        9.3.2.4 Communication within a Coherence Region 246
        9.3.2.5 Global Communication 246
        9.3.2.6 In Summary 247
9.4 Physically Constrained Nonbandlimited Images 248
    9.4.1 Wave Field Models 248
    9.4.2 Sampling and Interpolation 249
    9.4.3 Compression 250
9.5 Literature Review 250
9.6 Conclusion 251
References 251

9.1 INTRODUCTION

Digital cameras use solid-state devices as image sensors, formed by large arrays of photosensitive diodes. During the short period of time in which the shutter is open, each diode records the intensity of the light that falls on it by accumulating a charge. This brightness measurement is then digitized and stored in memory, forming a picture. Imagine now that we had one particular digital camera with some rather unusual characteristics:

- Besides measuring brightness, diodes can also compute/communicate.
- No central storage, diodes store an encoding of the whole image.
- Communication among diodes happens over a wireless channel.

Such a "digital camera" is not as contrived an example as it may seem: this camera is actually a *wireless sensor network*, in which nodes observe portions of an image, and then need to cooperate so that a copy of the entire image can be stored at all sensors. This chapter takes some initial steps toward understanding the problem of distributed in-network compression of data fields captured by large sensor arrays.

9.1.1 DISTRIBUTED IMAGE CODING AND WIRELESS SENSOR NETWORKS

The view of a sensor network as a digital camera with communication constraints is not conventional, but we believe it is a very valid one. A vision has been laid out some years ago, according to which it would be possible to observe physical phenomena at unprecedented high resolutions, and to control such phenomena, by means of very large collections of very small devices equipped with sensing, communication, and actuation capabilities. According to this vision, *sensor networks* would be deeply embedded in the environment, to provide an interface between the Internet and the physical world [18]. Since then, a great deal of activity in the area of sensor networking has taken place, dealing with a variety of questions in networking, communications, and signal processing. However, by thinking of a sensor network as a *network*, or a *communications* device, as opposed to thinking of it as a *data acquisition* device (such as a camera), a stronger emphasis is implicitly placed on the communications aspects, rather than on the very important problems of distributed sampling, interpolation, and compression that such networks give rise to. Yet very challenging problems in signal processing arise in sensor networks: by introducing communication constraints as well as distributed processing constraints, we are confronted with a new set of technical challenges that completely transform the nature of the data acquisition problem. It is the goal of this chapter, therefore, to present an overview on a number of distributed signal processing problems related to this new form of image data compression.

9.1.2 THE SENSOR BROADCAST PROBLEM

9.1.2.1 Problem Formulation

To make our arguments precise, consider the following setup. A sensor network consists of n nodes v_i, randomly placed on a square grid of unit area, at locations $(x_i, y_i) \in [0, 1] \times [0, 1]$ (for $i = 1, \ldots, n$). Each v_i observes a sample S_i of a stochastic process $S(x, y)$ defined over the field (e.g., light intensity, also referred to as an image or as a network snapshot). In the sensor broadcast problem, each v_i wants to form a rate-constrained estimate $\hat{S}(x, y)$ of the image $S(x, y)$ defined over $[0, 1] \times [0, 1]$. For this purpose, each v_i can communicate with other nodes, only sending messages to and receiving messages from nodes within some transmission range d_n (which in general will depend on the total number of nodes in the network), and these links have some finite capacity L. This setup is illustrated in Figure 9.1.

Given an estimate $\hat{S}(x, y)$, we measure the distortion between S and $\hat{S}$ using the mean-squared error metric:

$$\Delta(S, \hat{S}) = \int_0^1 \int_0^1 |S(x, y) - \hat{S}(x, y)|^2 \, dx \, dy \qquad (9.1)$$

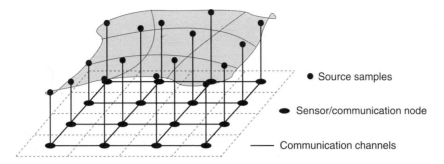

FIGURE 9.1 Setup for the sensor broadcast problem. By having all nodes broadcasting an encoding of their observations to every other node in the network, all nodes are able to form an estimate of the entire image.

for the continuous field, and

$$D(S, \hat{S}) = \sum_{i=1}^{n} |S(x_i, y_i) - \hat{S}(x_i, y_i)|^2 \tag{9.2}$$

for the sampled field. We denote by $R_S(\Delta)$ the rate/distortion function of the continuous time process, and $R_S(D, n)$ the rate/distortion function of the sampled process (for n samples). Note that for $n \to \infty$ we have that $D/n \to \Delta$; therefore, we will be particularly interested in studying the behavior of $R_S(D, n)$ in the regime where n grows asymptotically large, but $\Delta \approx D/n$ remains constant.

A *coding strategy* for the sensor broadcast problem consists of an algorithm to distribute the measurements collected by each sensor to every other sensor. There are two parameters of particular interest to us in assessing the performance of a coding strategy: its rate-distortion performance (i.e., the total number of bits needed to compress the continuous field with an average distortion per unit area of not more than Δ), and its communication complexity (i.e., the number of bits that need be communicated among nodes to accomplish this).

Our goal in this work is to begin a study of coding strategies for broadcast. As a first attempt on the problem, we consider a setup in which $S(x, y)$ is either a continuous-space bandlimited Gaussian process (Section 9.2.1), or the solution of a partial differential equation (Section 9.2.3), and we focus on the asymptotic behavior of coding strategies in the regime of high density of nodes.

9.1.2.2   Relevance of the Problem

There are many reasons that make broadcast a very important problem in the context of sensor networks. From a purely theoretical point of view, as formulated above, this problem is of great interest because of its extremal properties. In our setup we work under a network model first analyzed by Gupta and Kumar [26], for which they showed that as the number of nodes in the network grows to infinity, the per-node throughput of the network tends to be 0. Then, our sensor nodes generate Gaussian (i.e., maximum entropy) images, which they want to communicate to every other node in the network. That is, we consider a highly throughput-constrained network, in which nodes generate what is arguably the largest possible amount of traffic — broadcast of maximum entropy images.

Our interest in the broadcast problem is not motivated only by theory questions. In terms of applications, a most compelling one is the use of a broadcast protocol in the construction of a distributed transmission array for the sensor reachback problem [4]. In this problem, the goal is to move the field of observations picked up by all sensors to a far receiver. What makes the reachback problem interesting is the fact that typically, each individual sensor does not have enough resources to generate a strong information bearing signal that can be detected reliably at the far receiver. A

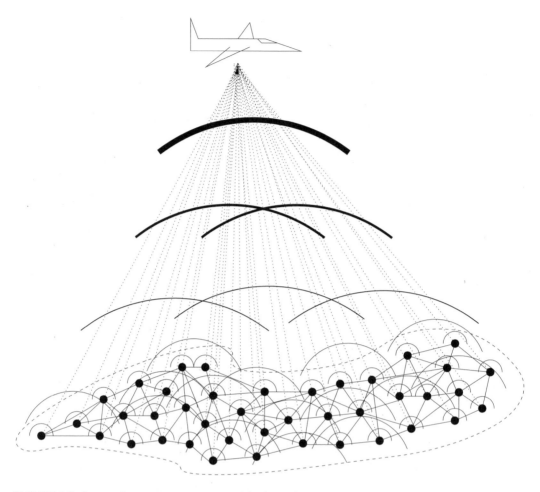

FIGURE 9.2 Cooperation among sensors to reach back to a far receiver. A good analogy to describe the role of a sensor broadcast protocol in the context of reachback is that of a *symphonic orchestra*. When all instruments in the orchestra play independently, all we hear is noise; but when they all play according to a common script, the music from all instruments is combined into a coherent play. The bits distributed during sensor broadcast play the role of that common script for a later cooperative transmission step.

sensor broadcast protocol, however, allows all nodes to agree on a common stream of bits to send, and then all these nodes can synchronize their transmissions so as to generate a strong signal at the far point, a signal that results from coherently superimposing all the weak signals generated by each of the sensors [29,30]. This setup is illustrated in Figure 9.2.

9.1.3 DATA COMPRESSION STRUCTURES

In standard communication networks that support the transmission of analog sources (e.g., voice, images, video), every information source independently encodes and compresses the data at the edge of the network, which is then routed through a high-speed backbone to destination. Networking and physical layer aspects are naturally kept separate in that traditional network structure. It is natural therefore to ask how would the amount of traffic generated by the sensors scale with network size, if the sensors operate as independent encoders, following the traditional model.

9.1.3.1 Data Compression Using Independent Encoders

For illustration purposes we consider a simple example: suppose that S_i is uniform in the range $[0, 1]$, that each node uses a scalar quantizer with B bits of resolution (i.e., the quantization step is 2^{-B}), and that the distortion is measured in the mean-square sense (i.e., $d(S, \hat{S}) = E(\|S - \hat{S}\|^2)$). On this particular source the average distortion achieved by such a quantizer is $\delta(B) = \frac{1}{12} 2^{-2B}$ [20]. Hence, solving for B in $D/n = \frac{1}{12} 2^{-2B}$, we find that to maintain an average per-node distortion over the entire network of D/n each sample requires $B = \lceil \frac{1}{2} \log_2(n/12D) \rceil$ bits. As a result, keeping D/n fixed at any constant value, the total amount of traffic generated by the whole network scales *linearly* in network size. Interestingly, even using optimal vector quantizers at each node, if the compression of the node samples is performed without taking into consideration the statistics of the entire field, the amount of traffic generated by the entire network still scales linearly in network size: in other words, one could certainly reduce the number of bits generated for a fixed distortion level D/n, but this reduction would only affect constants hidden by the big-oh notation, and the $O(n)$ scaling behavior of network traffic would remain unchanged.

Once we have determined how much data our particular coding strategy (independent quantizers at each node) produces, we need to know if the network has enough capacity to transport all that data. And for independent encoders, the answer is *no*, since the total amount of traffic that the network can carry is at most $O(\sqrt{n})$ [26], well below the $O(n)$ number of bits generated by the network.

9.1.3.2 Exploiting Correlations in the Source

As pointed out in [46], the scaling analysis of Gupta and Kumar [26] for independent encoders is not well suited for sensor networks: sensor data is increasingly dependent as the density of nodes increases. And indeed, if the data is so highly correlated that all sensors observe essentially the same value, at least intuitively it seems clear that almost no exchange of information at all is needed for each node to know all other values: knowledge of the local sample and of the global statistics already provide a fair amount of information about remote samples. This naturally raises a number of issues about data compression under the distribution constraints imposed by a sensor network:

1. What are suitable bounds for the performance of *distributed* algorithms to encode an over-sampled field of sensor measurements?
2. What are efficient signal processing architectures for this new form of massively distributed data compression, capable of closely approximating the performance bounds above?

As we will see below, neither of these questions admit simple answers.

9.1.4 Organization of the Chapter

As we hope it is clear from this introduction, distributed compression of images is not a straightforward extension of classical image coding problems. When we put communication constraints among sensing devices, new technical challenges arise that simply do not exist in the classical problem. The goal of this chapter is to present a survey of known results, and to identify problems on which further work is necessary.

The rest of this chapter is organized as follows. In Section 9.2 we present a discussion on various aspects of our data compression problem that illustrate fundamental differences with classical coding problems. In Section 9.3 we consider the classical decorrelation problem, now in the distributed setup. In Section 9.4 we go a step further, and consider a situation in which the physical process that leads to the formation of an image could potentially have a significant impact on the design of compression algorithms and on their performance. The chapter concludes with a literature survey in Section 9.5, and final remarks in Section 9.6.

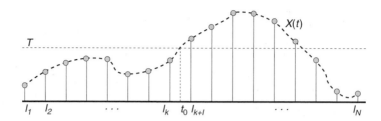

FIGURE 9.3 Choose a threshold value T that the sensor field, here illustrated in 1D only for simplicity, crosses with probability 1 — this value could be thought of as a boundary between quantization cells. Then, the location t_0 of the *first* time the process hits that value can be estimated with arbitrary accuracy as $n \to \infty$. Now, since t_0 is a function of the whole quantized field, by the data processing inequality of information theory, its entropy must be a lower bound for the joint entropy of all quantized values. But since t_0 is continuous-valued, its discrete entropy is infinite.

9.2 DISTRIBUTED COMPRESSION OF SENSOR MEASUREMENTS

9.2.1 INFORMATION-THEORETIC BOUNDS FOR BANDLIMITED IMAGES

The natural first place where to start an information theoretic analysis for a data compression problem is to study the rate–distortion function of the data field to be compressed, as this function provides the ultimate bounds on the compressibility of this data.

To within quantities that differ at most by constants independent of network size (but depend on specifics of the codes and protocols used, and on the bandwidth of the process being sensed), in [45], it is established that the rate–distortion function of any bandlimited and spatially homogeneous sensor field has an asymptotic rate of growth of $\Theta(\log(n/D))$ (where D/n is the average distortion in the reconstruction of a sample in the sensor field, and n the number of nodes in the network. So from this observation, and from the fact that by keeping D/n constant we have that $\Theta(\log(n/D))$ is also a constant (independent of n), one can conclude that there exist algorithms for compressing such images down to a constant number of bits, independent of n. However, even though algorithms for compressing oversampled fields down to a constant number of bits do exist, there is a legitimate question on whether those algorithms are suitable for implementation in a sensor array. This is because the high-dimensional vector quantizers used in the proof of achievability of the rate–distortion bound may or may not be directly applicable in the context of sensor networks, due to the fact that such nonseparable structures may or may not be well suited for *distributed* processing.

To further illustrate the point that this coding problem is significantly more elaborate than it might appear at first sight, consider the following example (adapted from Marco et al. [38]). Take S to be bandlimited with bandwidth W in both spatial dimensions, and define a matrix $P_{ij} = S(i/\sqrt{n}, j/\sqrt{n})$, for integer $1 \leq i, j \leq \sqrt{n}$ — then, under the assumption of n being asymptotically large, we will have $\sqrt{n} \gg W$, and hence the field S is clearly oversampled. For simplicity, we assume there is a sensor at each location $(i/\sqrt{n}, j/\sqrt{n})$, and *all* sensors pick up a sample of the continuous field S. The sample P_{ij} is first scalar quantized with a quantizer q inducing an average distortion D/n, and then entropy-coded with a Slepian-Wolf code such that the total number of bits spent on encoding the entire field is the joint entropy $\mathcal{H}(q(P_{ij}) : 1 \leq i, j \leq \sqrt{n})$. Then, each sensor floods the network with their encodings generated as above. The rate requirements for this construction were established in [38], where it is shown that as $n \to \infty$, $\mathcal{H}(q(P_{ij}) : 1 \leq i, j \leq \sqrt{n}) \to \infty$ as well. A formal proof of this fact could be fairly long, but the basic intuition was outlined in Marco et al. [38] and is captured by Figure 9.3.

The basic result of Marco et al. [38] then is that, for any one fixed scalar quantizer used at all nodes in the network, the gap between the rate–distortion function of the continuous field ($R_S(\Delta)$), and the joint entropy of all the quantized symbols ($\mathcal{H}(q(P_{ij}){:}1 \leq i,j \leq \sqrt{n})$), grows unbound. Thus, the use of a single scalar quantizer followed by an entropy coding step *is not* a viable strategy for our distributed image coding problem.

From these arguments we also get a hint on the nature of the technical challenges that must be faced in the distributed setup, not present in the classical image coding problem: what kind of coding strategies admit distributed implementations, and what is the best performance achievable under such decentralization constraints?

9.2.2 DISTRIBUTED COMPUTATION OF DECORRELATING TRANSFORMS

So, unlike in classical (not distributed) coding problems, for distributed compression of highly correlated images, quantization followed by entropy coding leads to significant performance degradation, relative to the best possible achievable. Now, whereas the result of Marco et al. [38] certainly paints a very discouraging picture (it hints an apparent impossibility of solving the problem of data aggregation in dense sensor networks), we argue that perhaps transform coding methods might hold the key to solving this problem.

If we inspect the proof of the $\Theta(\log(n/D))$ bound in [45], we recognize that a key step in that proof was the fact that almost all eigenvalues of the correlation matrix Σ vanish for large n. This means, there exists a linear function $f : \mathbb{R}^n \to \mathbb{R}^n$ of the entire set of samples $\{S_i\}_{i=1,\dots,n}$, such that the transform field $\{S_i'\}_{i=1,\dots,n} = f(\{S_i\}_{i=1,\dots,n})$ is such that $S_i' \sim \mathcal{N}(0, \lambda_i^{(n)})$. This f is the standard Karhunen–Loève transform: we have that $\Sigma = \mathbf{U}\Lambda\mathbf{U}^\mathrm{T}$, where $\Lambda = \mathrm{diag}(\lambda_1^{(n)}, \dots, \lambda_n^{(n)})$ is a matrix that has the eigenvalues of Σ in its main diagonal (and has zero entries everywhere else), and $\mathbf{U} = [\phi_1^{(n)}, \dots, \phi_n^{(n)}]$ is an orthonormal basis of eigenvectors of Σ — then, $\{S_i'\}_{i=1,\dots,n} = \mathbf{U}\{S_i\}_{i=1,\dots,n}$. And therefore, if a genie replaced the field $\{S_i\}_{i=1,\dots,n} \sim \mathcal{N}(0, \Sigma)$ with the field $\{S_i'\}_{i=1,\dots,n} \sim \mathcal{N}(0, \Lambda)$, then using an argument entirely analogous to that used in the proof of $R_S^{\mathrm{Crit}}(D,n) = \Theta(\log(n/D))$ later in Section 9.3.1, we could show that scalar quantization plus entropy coding of this transform data again produces an encoding of the continuous field of length $\Theta(\log(n/D))$ and average distortion $\Delta \approx D/n$.

Unfortunately, although scalar processing of the coefficients of a KLT of the sensor readings would indeed provide a finite length encoding for the continuous field, the KLT fails the test of admitting a distributed implementation. This is because the model assumed for S involves a spatial homogeneity assumption, captured in the form of assuming Σ is a doubly-Toeplitz matrix. Then it follows from Szegö's theorem that as we let network size $n \to \infty$, the eigenvectors ϕ_k become two-dimensional complex exponentials [24]. And this illustrates the nature of the difficulties we have to decorrelate sensor data: since these eigenvectors are in general not compactly supported, to be able to compute each individual sample of the transform vector Y we require access to *all* samples of S. So, our goal is to decorrelate sensor data because in that way we will be able to provide efficient algorithms to solve the sensor broadcast problem ... but we find that to decorrelate, a straightforward implementation of the equations defining the optimal transform requires us to solve a sensor broadcast problem first. A most important question then, the one that holds the key to being able to realize practical implementations of sensor broadcast protocols, is if it is at all possible to decorrelate sensor data using transforms with limited communication requirements.

9.2.3 IMAGES CONSTRAINED BY PHYSICAL LAWS —

Oftentimes, the physical process leading to the formation of an image $S(x, y)$ results in images that cannot be modeled as bandlimited processes, as in the examples considered above. In this case, the meaning of spatial sampling and reconstruction operations is not even clear, and much less clear is how to compress these images.

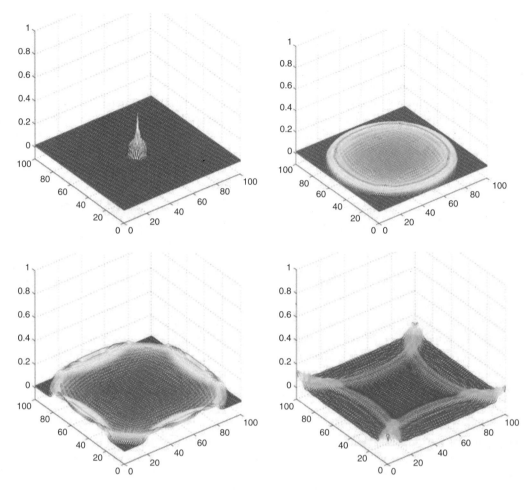

FIGURE 9.4 A 2-D wave field, in which an excitation is applied at the center location. The two horizontal axes represent space in a 2-D membrane, and the vertical axis represents pressure intensity at each location. An excitation is applied at the center location (top left), setting up a traveling wave of pressures on this membrane. Once the waves hit the boundary of the membrane, they are reflected back. These images show snapshots (at four different times) of the solution (in space) of the wave equation. Complete animations (MPEG files) can be downloaded from http://cn.ece.cornell.edu/research/#ddspcc.

As an illustrative example, consider the case when measurements correspond to pressure intensities at different locations in space of a *wave field*. Wave field models are mathematical abstractions for physical process that exhibit wave-like behavior: acoustics, electromagnetics, and seismic activity are some well-known examples. Perhaps the simplest incarnation of a wave field model is given by the following setup: a rectangular membrane in two dimensions (particles in the membrane only oscillate on a plane), no energy dissipation within the membrane, a perfectly reflecting boundary (no transmission of energy outside of the membrane), and an external disturbance $s(t)$ at a given location (x, y) in the membrane. The source $s(t)$ creates pressure waves in the membrane, and pressures must satisfy a differential equation known as the *wave* equation. A numerical solution of the wave equation for this setup is illustrated in Figure 9.4.

Performing *digital* signal processing operations on wave fields is not a straightforward task. To start with, by considering wave fields defined over finite membranes, we have that such fields cannot be bandlimited in space, and thus standard sampling results do not apply. Furthermore, the fact that the signal acquisition process is performed by a *distributed* device means that standard techniques for dealing with analog signals with too much bandwidth (such as, using an analog prefilter prior to sampling) are not feasible in this context, since no component of the system has access to the full signal field.

For the sake of argument, however, assume that the sampling and reconstruction problems are dealt with somehow. Then there are many other problems we still need to deal with:

- *Distributed compression.* How do we efficiently compress the samples of a wave field?
- *Distributed filtering.* How do we design filters that transform a sampled field into another desired one? And how do we implement these filters in a distributed manner?
- *Distributed signal detection/parameter estimation.* Given a set of samples encoding a wave field, how do we detect the presence of particular signals? And how do we estimate parameters of these signals?

Note that these are all classical tasks in digital signal processing. However, there are unique new challenges to deal with here. Some of them are "pure DSP" questions: with new sampling and reconstruction kernels, we just cannot blindly apply classical signal processing algorithms, we need to take into account how transformations of the samples affect the reconstructed continuous field. Other challenges however arise when we consider the "network factor": these new algorithms must admit distributed implementations, must operate under severe communication constraints, and must scale up to very large numbers of sensor/actor nodes.

9.3 TRANSFORMS FOR DISTRIBUTED DECORRELATION OF BANDLIMITED IMAGES

As discussed in Sections 9.2.1 and 9.2.2, distributed decorrelating transforms may play a crucial role in finding solutions to the sensor broadcast problem, in the case of bandlimited images. In this section we review various possible transforms.

9.3.1 THE "DROP-DATA" TRANSFORM

Perhaps the simplest-minded strategy one could conceive for implementing a distributed decorrelating transform consists of just *dropping* data: if the image is oversampled, keep a number of samples as determined by the Nyquist rate, and drop the rest [45].

Define a matrix $P_{ij} = S(i/W, j/W)$, for integer $1 \le i, j \le W$, and assume there is a sensor at each location $(i/W, j/W)$. The sample P_{ij} is first scalar-quantized with a quantizer inducing an average distortion D/n, and then entropy-coded with a Huffman code designed for $f_{S(i/W, j/W)}(s)$, the marginal distribution for that sample. Then, sensors at locations $(i/W, j/W)$ flood the network with their encodings generated as above. For this coding strategy, we need to determine the total number of bits generated, and the average distortion with which S is reconstructed.

Once the flood is complete, all nodes have access to an approximation of each sample P_{ij} that resulted from quantizing the original data with distortion D/n. But it is a well-known result from the rate–distortion theory that the reconstruction of the continuous source S with distortion Δ per unit of area or less is equivalent to the reconstruction of the samples P_{ij} with per-sample distortion $\Delta \approx D/n$ or less [6, Section 4.6.3]. Therefore, the distortion constraint is verified.

Let $R_S^{\mathrm{Crit}}(D, n)$ denote the total number of bits generated by the network when encoding a critically sampled version of the field S with total distortion D and n samples, and let $R_{P_{ij}}(D/n)$ denote the number of bits needed to code the sample P_{ij} with distortion D/n as described above. Then,

$$R_S^{\text{Crit}}(D,n) \stackrel{(a)}{=} \sum_{i=1}^{W} \sum_{j=1}^{W} R_{P_{ij}}(D/n)$$

$$\stackrel{(b)}{\leq} \sum_{i=1}^{W} \sum_{j=1}^{W} \left[\frac{1}{2} \log\left(\frac{\sigma^2}{D/n} \right) + 0.255 \right]$$

$$= W^2 (\log(n/D) + \log(\sigma^2) + 0.255)$$

$$\stackrel{(c)}{=} c_1 \log(n/D) + c_2$$

$$= \Theta(\log(n/D)),$$

where (a) follows from the fact that samples are encoded independently of each other at the sensors; (b) is a standard result from information theory which states that, at high resolutions, the performance gap between a scalar quantizer followed by an entropy code and the rate–distortion bound for that source is at most 0.255 bits/sample [23, pp. 2333], and σ^2 is the variance of each sample (obtained from the correlation matrix of S); and (c) follows from defining the constants $c_1 = W^2$ and $c_2 = W^2(\log(\sigma^2) + 0.255)$.

From the preceding paragraphs, we see that the "drop-data" transform does indeed allow us to generate finite-length encodings with bounded distortion, in the regime of high densities. However, there is a clearly unsatisfactory aspect to this solution: if the only way to exploit correlations due to oversampling consists of discarding data, we could have started the process by sampling at the critical rate. But there are strong reasons why we might need to oversample: if we did not know the spatial bandwidth of the process being sensed (for example, because that bandwidth changes over time), a perfectly feasible solution to cope with this problem consists of deploying a dense sensor array, and have the array compress its oversampled measurements. To compress this information by first decorrelating using the "drop-data" transform, we would require knowledge of the spatial bandwidth parameters at each point in time when an image is collected — but this information might not be available to us in general. Thus, there is practical interest in transforms capable of compressing *all* information efficiently, not just a carefully selected subset of samples.

9.3.2 LINEAR SIGNAL EXPANSIONS WITH BOUNDED COMMUNICATION

9.3.2.1 Wavelets and Sensor Broadcast

Wavelets with compact support naturally provide a mechanism for trading off communication complexity for decorrelation performance. The basic idea based on which we develop practical algorithms in this paper is that compactly supported wavelets, while still having good decorrelation properties, do not require access to the entire field of measurements to compute its coefficients. To illustrate this, consider the simple graph of dependencies illustrated in Figure 9.5.

Essentially, orthogonal wavelets with compact support are obtained from a single filter h having a finite impulse response [12]. As a result, a coefficient at scale k depends only on samples that span a range of $O(2^k)$ sensors (with constants that depend only on the length of the filters) — that is, this range is independent of the number of nodes in the network, leading to a localized communication pattern in which nodes only require access to data from other nearby nodes.

9.3.2.2 Definition of the Coding Strategy

Define a matrix $P_{ij} = S(i/\sqrt{n}, j/\sqrt{n})$, for integer $1 \leq i, j \leq \sqrt{n}$, and again we have that under the assumption of n asymptotically large and all sensor nodes turned on, $\sqrt{n} \gg W$, resulting in an

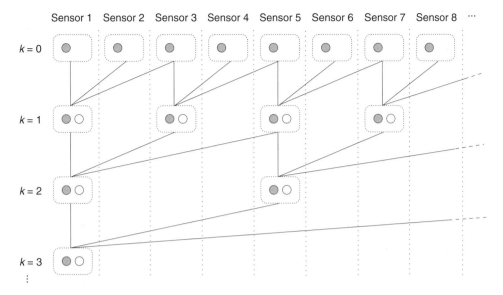

FIGURE 9.5 To illustrate the usefulness of compactly supported wavelets for the sensor broadcast problem, a sensor is represented by a dashed box at the top row, containing a gray circle that represents its measurement — a sample of the projection of the signal f at the finest scale. Dashed boxes in lower rows correspond to computations performed by particular sensors: gray circles represent coefficients of projections onto coarse spaces, white circles represent coefficients of projections onto detail spaces. Lines connecting boxes denote dependencies between coefficients: in this case, three lines correspond to filters with three taps. An equivalent representation of the sequence of gray circles at the top level is given by the sequence of gray circles at the bottom level, and all white circles.

oversampled representation of S. Then each node performs the following steps:

1. Quantize P_{ij} at high (but finite) rates to obtain a $\tilde{P}_{ij} \approx P_{ij}$. Define $y_{ij}^{(0),LL} = \tilde{P}_{ij}$.
2. For integer $1 \leq m \leq \log_2(\sqrt{n})$, repeat:
   2.1. Partition the network area $[0,1] \times [0,1]$ into squares $\Box_{ij}^m$ of size $2^m/\sqrt{n}$, centered at locations $(i2^m/\sqrt{n}, j2^m/\sqrt{n})$ of the form $0 \leq i,j \leq \sqrt{n}/2^m$ (for integer i,j).
   2.2. For all boxes $\Box_{ij}^m$:
      2.2.1. Choose one node within the box to carry out computations, all others remain silent.
      2.2.2. Wait until all lowpass coefficients from neighboring boxes at scale $m-1$ arrive.
      2.2.3. Compute the coefficients at scale m. Let $y_{ij}^{(m),LL}$ denote the lowpass coefficient, $d_{ij}^{(m),LH}/d_{ij}^{(m),HL}/d_{ij}^{(m),HH}$ the highpass coefficients.
      2.2.4. Estimate the variance of $d^{(m),LH}/d^{(m),HL}/d^{(m),HH}$.
      2.2.5. If these variances are above a threshold T, re-quantize $y_{ij}^{(m),LL}/d_{ij}^{(m),LH}/d_{ij}^{(m),HL}/d_{ij}^{(m),HH}$, with a scalar quantizer inducing distortion D/n, and flood the network with these new quantized values; from this point on, only relay flood messages for other nodes.
      2.2.6. If not, relay $y_{ij}^{(m),LL}$ to $\Box_{ij}^{m+1}$, so that some node at scale $m+1$ in charge of computing the next set of lowpass and highpass coefficients has access to the data needed for that purpose.

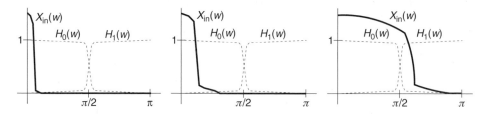

FIGURE 9.6 An oversampled signal is characterized by having a discrete-time Fourier transform with a vanishingly small support. As this signal is repeatedly filtered and subsampled, the net result is to widen the support of the input spectrum, until eventually that spectrum crosses the $\pi/2$ threshold: when that happens, a nonnegligible variance will be detected in line 2(b).iv of the protocol, and the node will flood and switch to relay mode only. At this point, after repeated downsampling, the flooded signal is sampled at most at twice the Nyquist rate.

    2.3.   Done with this iteration. If the box to which a node belongs did not decide to flood their coarsely quantized samples, the node goes on to the next iteration. Else, stop and get out of this loop.

  3.   Wait until floods are received covering the whole network. Use the received data to interpolate the field $\hat{S}$.

9.3.2.3   Differentiating between Local and Global Communication Requirements

A key property of the coding strategy defined above is the fact that, with high probability, all nodes in a box $\square_{ij}^m$ will choose to stay within the loop at step 2(c) if $2^m/\sqrt{n} < 1/W$, and will choose to exit the loop if $2^m/\sqrt{n} \geq 1/W$. The reason for this behavior is explained in Figure 9.6.

Based on this property of the protocol, we are naturally led to distinguish between two phases in its operation:

- A phase during which no floods of information over the entire network occur, but nodes cooperate to exchange some high-rate messages among them within "small" boxes — small relative to the spatial bandwidth of the process.
- A phase during which boxes stop growing in larger in size and smaller in number, but during which the information generated by a box floods the entire network.

To evaluate the rate–distortion performance for this coding strategy, we need to treat these two different phases of the coding strategy in different ways.

9.3.2.4   Communication within a Coherence Region

Let η denote the number of nodes in a box of size $1/W \times 1/W$ (i.e., $\eta = n/W^2$, so η is the *oversampling factor*). Then, the total number of bits exchanged within a box of size $1/W \times 1/W$ is $\Theta(\log(\eta))$, as explained in Figure 9.7.

9.3.2.5   Global Communication

From the argument developed in Figure 9.6, we see that with high probability, the number of boxes that participate in a flood is W^2. Therefore, this problem is reduced to the case of the drop-data transform, and we have that the total amount of rate generated by flood operations is $\Theta(\log(n/D))$.

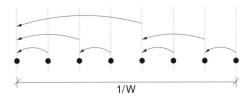

FIGURE 9.7 Without loss of generality, assume the node in charge of carrying out computations within a box $\square_{ij}^m$ is located in the upper-left corner, and for simplicity consider a 1D figure. In this case we have 8 nodes per coherence region, and the node with the heaviest load is the one in the corner: this one must receive a number of messages from other nodes within its box which, due to the dyadic subsampling mechanism of the filter bank, is logarithmic in the number of nodes in the box.

9.3.2.6   In Summary

A few remarks are in order about this proposed protocol.

First of all, it should be noted that its main goal is to process the (oversampled) entire field of samples to eliminate redundant information, in an attempt to flood the network with an amount of data that is proportional to what would have been generated if the continuous field had been critically sampled. Therefore:

- If the statistics of the field are known beforehand, from a pure data compression point of view, there is no reason whatsoever to oversample the field — using this protocol, we are able to obtain at best the same performance as we would obtain by critical sampling, at significantly higher complexity.
- If the statistics of the field are *not* known in advance, then this protocol becomes attractive. The reason is that the penalty in rate for oversampling is logarithmic in the oversampling ratio, and *only* within a region of space that is of size inversely proportional to the spatial bandwidth of the field of measurements.[1] As a result, it seems perfectly within the realm of what one could reasonably argue to be practical to, if needed, oversample by several orders of magnitude, as a means to provide robustness against inaccurate knowledge of the field statistics.

Also, about the filtering mechanism. One should note that, although wavelets with compact support did provide the original motivation to consider transforms with bounded communication requirements, beyond the fact that they are built on FIR filters, there is very little about wavelets that is actually used in the protocol above — specifically, we discard a significant amount of information, so there is no apparent reason for requiring filter banks with a perfect reconstruction property.

Finally, about the assumption that the computation of coefficients $y^{(m),LL}$ within small boxes (prior to the flooding step) can be carried out with high enough precision: it is *not* enough to assume high rate quantization in this context to justify the approximation — with high densities and distributed processing, also comes the need to iterate the filter bank many times, and hence to quantize at intermediate steps of the computation. Analyzing this issue exactly involves modeling quantization noise, studying the effects of finite precision on filter implementations [41, Chapter 6], the development of integer-to-integer transforms [22], and the lifting scheme for implementing wavelets [13].

---

[1] In the context of a different problem (Distributed Classification), D'Costa and Sayeed [15] observed a similar behavior, whereby nodes need to communicate intensively within local coherence regions, but only need to exchange low-rate feature vectors on a global scale.

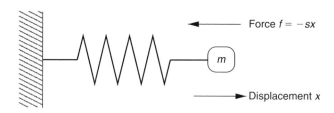

FIGURE 9.8 Spring–mass model for an elementary (small) oscillation. A small mass is tied to one end of a spring, which at the other end is fixed. Motion is one-dimensional. The mass is displaced from its rest position by a distance x, and the spring exerts a force on that mass proportional to this distance.

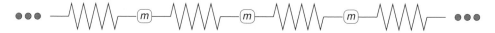

FIGURE 9.9 Multiple springs and masses. In this setup, displacing one of the elemental masses from its rest position, besides setting that mass in motion as in the simpler model of Figure 9.8, also has the effect of inducing motion on neighboring masses. Therefore, the oscillations of individual masses are clearly not independent, but coupled through the springs that bind them.

9.4   PHYSICALLY CONSTRAINED NONBANDLIMITED IMAGES

There are many types of images for which the bandlimited model assumed in the previous section does not apply, and one example on which we focus in this section is the case of wave fields, as argued in Section 9.2.3. In this section we consider the problem of distributed compression of wave field signals.

9.4.1   WAVE FIELD MODELS

What is a wave field? Perhaps the simplest explanation can be given in terms of the spring–mass model for an oscillator [34, Chapter 1]. This model is illustrated in Figure 9.8.

In the simple physical system of Figure 9.8, when energy is supplied in the form of displacing the elemental mass from its rest position, the spring produces another force that attempts to bring the mass back to rest. This force sets the mass in motion, and results in a certain pressure on the mass. Clearly, the motion will be an oscillation in the horizontal direction: the spring pulls the mass back, then pushes it away, then pulls it back again, and so on, until all the energy supplied by the initial displacement is dissipated (or forever, if there is no dissipation at all).

The next step is to extend this model to a one-dimensional vibration. And for this purpose, consider a chain of masses all interconnected by springs, as illustrated in Figure 9.9.

In the simple system of Figure 9.9, motion is still constrained to be one-dimensional. But from here, it is not hard to visualize two- and three-dimensional arrangements of elementary masses coupled by springs, and therefore we have all we need to provide some intuition as to what is a wave field: *the scalar field of pressures acting on each individual mass, resulting from a given excitation.* Many physical phenomena can be modeled using the simple construction of very small masses coupled by springs. The physics of acoustic wave propagation [34],[2] and the physics of seismic wave

---

[2] Only in the regime of small vibrations. For sharp discontinuities in pressure fields such as would arise, for example, in the shock wave of an explosion, the model is not applicable.

propagation [7], are well-known examples in which this model offers very accurate description of such phenomena.

At least intuitively, from the spring–mass model, it seems clear that pressure fields are highly constrained both in time and in space:

- Pressures (forces) acting on individual mass elements are determined by pressures (forces) acting on neighboring mass elements, and by forces that offer resistance to their motion.
- Some mass elements undergo forced oscillations provided by an external excitation: this is where waves draw their energy from.
- The motion of mass elements at edge locations in the membrane is constrained by some boundary conditions (such as having smaller, possibly zero velocity) different from those at locations away from this boundary.

All these constraints are captured in the form of a differential equation that pressure fields must satisfy at every point of space and time, known as the *wave equation*. An example of a two-dimensional wave field obtained by numerically solving one particular form of the wave equation was illustrated in Figure 9.4.

9.4.2 SAMPLING AND INTERPOLATION

For illustration purposes, consider the setup shown in Figure 9.4: a 2D membrane with no dissipation of energy in its interior and perfectly reflecting boundaries. If we apply an excitation s on this membrane, the resulting field of pressures p is constrained by the *forced* form of the wave equation

$$\nabla^2 \hat{p} + \frac{\omega^2}{c^2} \hat{p} + \hat{s} = 0 \tag{9.3}$$

where ∇^2 is the Laplacian operator, $\hat{s}$ the spectrum of the source pressure excitation, $\hat{p}$ the spectrum of the resulting pressure field, ω the frequency variable, and c the speed of propagation in the membrane (this equation is given in the frequency domain) [7]. Now, this equation holds for a source s arbitrarily distributed in space, and for arbitrary boundary conditions of the membrane. If we now specialize this equation to the case of a point source at the center, rectangular and perfectly reflecting boundary, then this equation can be solved analytically, and we have that

$$\hat{p}(x, y, \omega) = \hat{s}(\omega) \cdot \hat{h}_{(x,y)}(\omega) \tag{9.4}$$

where $\hat{h}_{(x,y)}$ is defined by

$$\hat{h}_{(x,y)}(\omega) = \frac{4}{L_x L_y} \sum_{m,n} \frac{\cos(k_n x_0)\cos(l_m y_0)\cos(k_n x)\cos(l_m y)}{k_n^2 + l_m^2 - \omega^2/c^2} \tag{9.5}$$

Equation (9.4) provides much insight into the structure of wave fields generated by point sources:

- The pressure signal observed at an arbitrary location (x, y) can be expressed as the convolution of the source signal $s(t)$ with a space-dependent filter $h_{(x,y)}(t)$. For obvious reasons, we refer this function as the *impulse response of the membrane*, and its Fourier transform as the *frequency response*.
- The filter $h_{(x,y)}(t)$ depends on a number of parameters: the location (x, y) of the field measurement, the location (x_0, y_0) of the source signal, the membrane dimensions L_x and L_y, and the speed c of wave propagation.

- $\hat{h}_{(x,y)}(\omega)$ has an asymptote at frequencies of the form

$$c\sqrt{\left(\frac{n\pi}{L_x}\right)^2 + \left(\frac{m\pi}{L_y}\right)^2} \tag{9.6}$$

for all $m, n \in \mathbb{Z}$. These special frequencies are denoted by ω_{nm}, and are referred to as the *modes* of the membrane [34].

With these results, now we can pose and answer the following sampling question: how many pressure sensors must be placed on this membrane, so that based on samples in time collected by each sensor, and in space collected by the different sensors, we gather enough information to reproduce the field at an arbitrary location? And the surprising answer is that, under these very special assumptions, *only one sensor is enough*. This is easy to see: in general, the pressure signals at arbitrary locations (x, y) and (x', y') are related by $\hat{p}(x', y', \omega) = \hat{p}(x, y, \omega)\hat{h}_{(x',y')}(\omega)/\hat{h}_{(x,y)}(\omega)$. This is immediate since, from Equation (9.4), $\hat{p}(x, y, \omega)/\hat{h}_{(x,y)}(\omega) = \hat{s}(\omega)$. Therefore, provided we neither put our sensor, nor try to obtain the pressure signal, at locations for which the frequency response $\hat{h}$ has a zero (and we can show this happens over a set of locations of measure zero), the temporal samples at a single location provide enough information to reconstruct the pressure measurement at any other arbitrary location. Thus, in this case, the measurements of a single sensor are enough to interpolate the field at an arbitrary location [37].

We should highlight that not requiring a prefilter not only is a key difference with previous work on sampling for nonbandlimited signals, but it is also one that makes our interpolation kernel particularly appropriate for use in a sensor network. This is because, whereas classical A/D devices can certainly include a prefilter in their circuitry, when a signal is acquired by a *distributed* device in which none of its components have access to the full signal field (such as the sensor network), filtering before sampling becomes a much harder, perhaps impossible task.

9.4.3 COMPRESSION

For the highly restrictive model set up above, we see that the broadcast problem admits a trivial solution: *each sensor can reconstruct the entire image based only on its own measurements in time, and on knowledge of the membrane's impulse response $h_{(x,y)}(t)$ at all locations (x, y)*. Of course, in reality waves *will* dissipate energy as they travel in a medium, boundaries *will* reflect only a fraction of the energy of a wave back, there *will* be inhomogeneities both in the medium and at the boundaries, boundaries *will not* be perfect rectangles, and measurements *will* be corrupted by noise. So, it is very clear that in a realistic setting, one *should not* expect a single sensor to provide enough information to reconstruct an entire wave field. However, and this was the whole point of this example, one *should* expect that the structure of this data field (in this case, the PDE model that describes the constraints imposed by the physics of wave propagation) will play a significant role in our ability to solve broadcast problems.

9.5 LITERATURE REVIEW

Wave propagation problems have been studied for a long time in physics and engineering, and they are reasonably well understood. Some textbooks we found particularly useful are [7,32,34]. The standard mathematical tool for dealing with these problems are differential equations with boundary value constraints, and Hilbert space methods for their solution. And in this area, textbooks we relied on heavily are [9,10,27].

In the context of distributed signal processing and communications problems, a fair amount of work has been done on the topics of distributed data compression [1,5,19,21,31,38,42,45,46,48], distributed transmission and beamforming [3,8,29,39,44,54], time synchronization [17,28,33,53], and distributed inference [14,25,40]. Related distributed control problems have been considered in [11,36].

Issues related to data models appropriate for sensor readings were discussed in [5,47]. Related parameter estimation problems were considered in [43].

In the context of classical (meaning, not distributed) signal processing problems, questions involving sound fields and light fields have received some attention in the past. A number of references on data compression for light fields is available from http://www.stanford.edu/˜chuoling/lightfield.html. In the context of sound fields, the impulse response $h_{(x,y)}(t)$ has been referred to as the *plenacoustic* function [35]. Spatial sampling and reconstruction of this function has been considered in [2], and spectral properties of this function were studied in [16], under a far-field assumption. A closely related signal processing problem to those considered in this paper is the problem of wave field *synthesis*: in this case, the goal is to generate a prespecified wave field from a finite number of fixed pointsources. Various groups, primarily in Germany and in the Netherlands, have studied this problem. A number of references on various aspects of this problem are available from http://www.lnt.de/LMS/.

Sampling theory has a long and rich history, nicely surveyed in [49]. Some recent contributions to this problem can be found in [50–52].

9.6 CONCLUSION

In this chapter we have considered a new form of image data compression problems, in which we introduce communication constraints among image capture elements. The area is new, and there is plenty of opportunity for studying interesting problems at the intersection of traditional disciplines such as signal processing, communications, control, information theory, and distributed algorithms. Our goal in this presentation was to highlight the fundamental differences between classical and distributed image coding problems, as well as promising research directions.

REFERENCES

1. J. Aćimović, R. Cristescu, and B. Beferull-Lozano, Efficient distributed multiresolution processing for data gathering in sensor networks, *Proceedings of the IEEE International Conference on Acoustics, Speech, and Signal Processing (ICASSP)*, Philadelphia, PA, 2005.
2. T. Ajdler and M. Vetterli, The plenacoustic function: sampling and reconstruction, *Proceedings of the IEEE International Conference on Acoustics, Speech, and Signal Processing (ICASSP)*, 2003.
3. B. Ananthasubramaniam and U. Madhow, Virtual radar imaging for sensor networks, *Proceedings of the International Workshop on Information Processing Sensor Networks (IPSN)*, 2004.
4. J. Barros and S.D. Servetto, Coding theorems for the sensor reachback problem with partially cooperating nodes, *Discrete Mathematics and Theoretical Computer Science (DIMACS) Series on Network Information Theory*, Piscataway, NJ, 2003.
5. B. Beferull-Lozano, R.L. Konsbruck, and M. Vetterli. Rate/distortion problem for physics based distributed sensing. *Proceedings of the IEEE International Conference on Acoustics, Speech, and Signal Processing (ICASSP)*, 2004.
6. T. Berger, *Rate Distortion Theory: A Mathematical Basis for Data Compression*, Prentice-Hall, Englewood Cliff, NJ, 1971.
7. A.J. Berkhout, *Applied Seismic Wave Theory*, Elsevier, Amsterdam, 1987.
8. S.H. Breheny, R. D'Andrea, and J.C. Miller, Using airborne vehicle-based antenna arrays to improve communications with UAV clusters, *Proceedings of the IEEE Conference Decision Control*, 2003.
9. G.F. Carrier and C.E. Pearson, *Partial Differential Equations, Theory and Technique*, Academic Press, New York, 1976.
10. R.V. Churchill, *Fourier Series and Boundary Value Problems*, McGraw-Hill, New York, 1969.

11. R. D'Andrea and G.E. Dullerud, Distributed control design for spatially interconnected systems, *IEEE Trans. Autom. Control*, 48, 1478–1495, 2003.

12. I. Daubechies, *Ten Lectures on Wavelets*, SIAM, Philadelphia, PA, 1992.

13. I. Daubechies and W. Sweldens, Factoring wavelet transforms into lifting steps, *J. Fourier Anal. Appl.*, 4, 245–267, 1998.

14. A. D'Costa, V. Ramachandran, and A. Sayeed, Distributed classification of Gaussian space-time sources in wireless sensor networks, *IEEE J. Select. Areas Commun.*, 22, 1026–1036, 2004.

15. A. D'Costa and A.M. Sayeed, Collaborative signal processing for distributed classification in sensor networks, *Proceedings of the 2nd International Workshop on Information Processing in Sensor Networks (IPSN)*, Palo Alto, CA, 2003.

16. M. Do, Toward sound-based synthesis: the far-field case, *Proceedings of the IEEE International Conference on Acoustics, Speech, and Signal Processing (ICASSP)*, 2004.

17. J. Elson, L. Girod, and D. Estrin, fine-grained network time synchronization using reference broadcasts, *Proceedings of the 5th Symposium on Optical System Design Implementation (OSDI)*, Boston, MA, 2002.

18. D. Estrin, *Embedding the Internet*, Talk presented at the MIT LCS 35th Anniversary Colloquium, April 1999. Available from http://www.isi.edu/scadds/papers/lcsapril99c.pdf.

19. M. Gastpar, P.L. Dragotti, and M. Vetterli, The Distributed Karhunen-Loève Transform, Technical Report IC/2002/36, Ecole Polytechnique Fédérale de Lausanne, 2003. Available from http://www.eecs.berkeley.edu/~gastpar/ .

20. A. Gersho and R.M. Gray, *Vector Quantization and Signal Compression*, Kluwer, Dordrecht, 1992.

21. B. Girod, A. Aaron, S. Rane, and D. Rebollo-Monedero, Distributed video coding, *Proc. IEEE*, 93, 71–83, 2005.

22. V.K. Goyal, Transform coding with integer-to-integer transforms, *IEEE Trans. Inform. Theory*, 46, 465–473, 2000.

23. R.M. Gray and D.L. Neuhoff, Quantization, *IEEE Trans. Inform. Theory*, 44, 2325–2383, 1998.

24. U. Grenander and G. Szegö, *Toeplitz Forms and their Applications*, University of California Press, Berkeley, 1958.

25. D. Guo and X. Wang, Dynamic sensor collaboration via sequential Monte Carlo, *IEEE J. Select. Areas Commun.*, 22, 1037–1047, 2004.

26. P. Gupta and P. R. Kumar, The capacity of wireless networks, *IEEE Trans. Inform. Theory*, 46, 388–404, 2000.

27. K.E. Gustafson, *Introduction to Partial Differential Equations and Hilbert Space Methods*, 2nd ed., Wiley, New York, 1987.

28. A. Hu and S.D. Servetto, *Algorithmic Aspects of the Time Synchronization Problem in Large-Scale Sensor Networks*, ACM/Kluwer Mobile Networks and Applications. Special issue with selected papers from ACM WSNA 2003. To appear. Preprint available from http://cn.ece.cornell.edu/.

29. A. Hu and S.D. Servetto, DFSK: Distributed frequency shift keying modulation in dense sensor networks, *Proceedings of the IEEE International Conference on Communication (ICC)*, Paris, France, 2004.

30. A. Hu and S.D. Servetto, Distributed signal processing algorithms for the physical layer of a large-scale sensor networks, in *Wireless Sensor Networks*, I. Stojmenovic, Ed., Wiley, New York, 2005. To appear.

31. C. Intanagonwiwat, R. Govindan, D. Estrin, J. Heidemann, and F. Silva, Directed diffusion for wireless sensor networking, *IEEE/ACM Trans. Networking*, 11(1), 2–16, 2003.

32. J.B. Jones and R.E. Dugan, *Engineering Thermodynamics*, Prentice-Hall, Englewood Cliff, NJ, 1996.

33. R. Karp, J. Elson, D. Estrin, and S. Shenker, Optimal and Global Time Synchronization in Sensornets, Technical Report 9, Center for Embedded Network Sensing, UCLA, 2003. Available from http://lecs.cs.ucla.edu/.

34. L.E. Kinsler, A.R. Frey, A.B. Coppens, and J.V. Sanders, *Fundamentals of Acoustics*, 4th ed., Wiley, New York, 2000.

35. M. Kubovy and D. van Valkenburg, Auditory and visual objects, *Cognition*, 80(1–2), 97–126, 2001.

36. M.D. Lemmon, Q. Ling, and Y. Sun, Overload management in sensor-actuator networks used for spatially distributed control systems, *Proceedings of the ACM Sensory system*, 2003.

37. G.N. Lilis, M. Zhao, and S.D. Servetto, Distributed sensing and actuation on wave fields, *Proceedings of the 2nd Sensor and Actor Networks Protocols and Applications* (*SANPA*), Boston, MA, 2004. Available from http://cn.ece.cornell.edu/.

38. D. Marco, E.J. Duarte-Melo, M. Liu, and D.L. Neuhoff, On the many-to-one transport capacity of a dense wireless network and the compressibility of its data, *Proceedings of the International workshop on Information Processing in Sensor Networks* (*IPSN*), 2003.

39. I. Maric and R. Yates, Efficient multihop broadcast for wideband systems, *Discrete Mathematics and Theoretical Computer Science* (*DIMACS*) *Series on Signal Processing for Wireless Transmission*, Piscataway, 2002.

40. R. Nowak, Distributed EM algorithms for density estimation and clustering in sensor networks, *IEEE Trans. Signal Proc.*, 51(8), 2245–2253, 2003.

41. A.V. Oppenheim and R.W. Schafer, *Discrete-Time Signal Processing*, 2nd ed., Prentice-Hall, Englewood Cliff, NJ, 1999.

42. S. Pattem, B. Krishnamachari, and R. Govindan, The impact of spatial correlation on routing with compression in wireless sensor networks, *Proceedings of the International workshop on Information Processing in Sensor Networks* (*IPSN*), 2004.

43. L.A. Rossi, B. Krishnamachari, and C.-C.J. Kuo, Distributed parameter estimation for monitoring diffusion phenomena using physical models, *Proceedings of the IEEE SECON*, 2004.

44. A. Scaglione and Y.W. Hong, Opportunistic large arrays, *IEEE International Symposium on Advanced Wireless Communication* (*ISWC02*), Victoria, BC, 2002.

45. A. Scaglione and S.D. Servetto, *On the Interdependence of Routing and Data Compression in Multi-Hop Sensor Networks*, 2003. ACM/Kluwer Mobile Networks and Applications. Special issue with selected papers from ACM MobiCom 2002. To appear. Available from http://cn.ece.cornell.edu/.

46. S.D. Servetto, Lattice quantization with side information: codes, asymptotics, and applications in sensor networks, *IEEE Trans. Inform. Theory*, March 2004. In Press. Preprint available from http://cn.ece.cornell.edu/.

47. S.D. Servetto, On the feasibility of large-scale wireless sensor networks, *Proceedings of the 40th Allerton Conference on Communication, Control and Computing*, Urbana, IL, 2002.

48. S.D. Servetto, Sensing Lena — massively distributed compression of sensor images, *Proceedings of the IEEE International Conference on Image Processing* (*ICIP*), 2003. Invited paper.

49. M. Unser, Sampling – 50 years after shannon, *Proc. IEEE*, 88, 569–587, 2000.

50. R. Venkataramani and Y. Bresler, Perfect reconstruction formulas and bounds on aliasing error in sub-nyquist nonuniform sampling of multiband signals, *IEEE Trans. Inform. Theory*, 46, 2173–2183, 2000.

51. R. Venkataramani and Y. Bresler, Multiple-input multiple-output sampling: necessary density conditions, *IEEE Trans. Inform. Theory*, 50(8), 1754–1768, 2004.

52. M. Vetterli, P. Marziliano, and T. Blu, Sampling signals with finite rate of innovation, *IEEE Trans. Signal Proc.*, 50, 1417–1428, 2002.

53. H. Wang, L. Yip, D. Maniezzo, J.C. Chen, R.E. Hudson, J. Elson, and K. Yao, A wireless time-synchronized COTS sensor platform: applications to beamforming, *Proceedings of the IEEE CAS Workshop on Wireless Communication Networking*, Pasadena, CA, 2002.

54. K. Yao, R.E. Hudson, C.W. Reed, D. Chen, and F. Lorenzelli, Blind beamforming on a randomly distributed sensor array system, *IEEE J. Select Areas Commun.*, 16(8), 1555–1567, 1998.

10 Data Hiding for Image and Video Coding

Patrizio Campisi and Alessandro Piva

CONTENTS

10.1 Introduction 255
10.2 Data Hiding for Image and Video Compression 256
    10.2.1 Data in Image 257
    10.2.2 Data in Video 264
10.3 Data Hiding for Error Concealment 268
    10.3.1 Error Concealment for Resynchronization 269
    10.3.2 Error Concealment for the Recovery of MV Values in Lost Blocks 270
    10.3.3 Error Concealment for the Recovery of Pixel Values in Lost Blocks 271
    10.3.4 Recovery of Pixel Values in Lost Blocks through Self-Embedding Methods 275
10.4 Final Comments 277
10.5 Further Readings 278
Acknowledgments 278
References 278

10.1 INTRODUCTION

Data hiding is the general process by which a discrete information stream, the *mark*, is merged within media content by imposing imperceptible changes on the original *host* signal. Other requirements, other than transparency, can be needed according to the specific application that is taken into account.

In the recent years, there has been an always growing interest in data hiding within multimedia owing to its potential for improved multimedia transmission, signal captioning, transaction tracking in media commerce, copy control, authentication, time stamps, copy protection through the development of digital watermarking technology, and so forth. In particular, the increasing diffusion of digital image and video transmission applications, which, owing to the limited transmission rates currently available, require the use of compression techniques, has driven toward the application of data hiding for improved image and video compression as well as for error concealment in images and videos. In fact, data-hiding principles can be used to discard the perceptually redundant information and then to properly embed part of the information to transmit in the host data, thus obtaining improved data-coding schemes that lead to significant bit saving. Possible applications are image-in-image, video-in-video, and audio-in-video embedding. On the other hand, data-hiding algorithms can be used to embed redundant information in a coded bit stream in order to improve its resilience with respect to channel noise. In fact, compressed digital video streaming, based on predictive coding, is very susceptible to channel noise, since errors can propagate through the decoded sequence generating at the receiver display a content with annoying visual artifacts. However, by embedding redundant information about the transmitted data into coded bit stream, the effect of the potential channel errors can be counteracted by the decoder by using the side information.

The chapter is organized as follows. Some data-hiding techniques for improved image and video compression are described in Section 10.2. How data hiding can be used to improve error concealment performance in image and video coding is described in Section 10.3. Conclusions are drawn in Section 10.4, and eventually in Section 10.5 further readings are suggested.

10.2 DATA HIDING FOR IMAGE AND VIDEO COMPRESSION

The goals of data hiding and perceptual coding can be viewed as being somewhat contradictory.

Perceptual coding refers to the lossy compression of multimedia signals using perceptual models. These approaches discard the information that cannot be perceived by the human perceptual system. These modifications are imposed on the signal in such a way as to reduce the number of information bits required for transmission and storage of the content.

On the other hand, any data-hiding method aims at embedding information into a host in a transparent way. In order to obtain imperceptibility, a data-hiding method must exploit the characteristics of the human perceptual system to properly select the media features where to embed the mark in such a way that a human observer does not have an altered perception of the marked media with respect to the original one.

Therefore, imperceptible data hiding and perceptual coding hinder each other: the former uses the irrelevant information to mask the presence of the hidden data, whereas the latter attempts to remove redundant and irrelevant information from a signal, thus decreasing the overall possible compression ratio.

When using an ideal perceptual lossless compression algorithm, all the information that cannot be perceived by a human observer is removed. In this situation, no room for imperceptible data hiding is available, thus making transparent data hiding incompatible with ideal perceptual coding.

However, actual compression algorithms are far from being ideal. This implies that data hiding can be performed in those parts of the host data that, although imperceptible, have not been eliminated. Several methods, in which it is assumed that each process hinders, not helps, the objective of the other, have been developed to find an appropriate compromise between data hiding and compression.

Some authors have taken a different perspective, trying to identify how data hiding can be used to design more powerful compression schemes. They rely on the nonideality of perceptual coding schemes in order to perform the embedding of large volume of data. Besides the bit saving in the transmission/storage process, this use of data hiding can give other side benefits. For example, when dealing with a color video, its chrominance components can be embedded into the luminance ones in order to both reduce the amount of bits to transmit and to provide the opportunity to view the video in a progressive way from a monochrome to a color version. Moreover, the hiding of an audio stream into the corresponding video stream facilitates the task of maintaining synchronization between the two data streams under different attacks. In video streaming applications, a video, at spatial resolution smaller than that of the host video, can be embedded into the host, thus obtaining an enhanced picture-in-picture (PiP) system that uses a unique data stream instead of the two used in conventional PiP systems.

When using data hiding for enhanced image/video coding, the capacity requirement, that is, the capability of the host data to mask an amount of data as large as possible, is crucial to obtain a significant bit saving. Robustness may not be of primary concern given that data embedding is performed in the already coded bit stream, as well as simultaneously to compression. On the contrary, whether or not data hiding is performed before coding, the hidden data must survive to compression.

In Section 10.2.1, we will describe some approaches that have been used for hiding image in image [3,4,8,14,31–33,48]; and in Section 10.2.2, the algorithms that have been employed for embedding data in video [22,23,37,49].

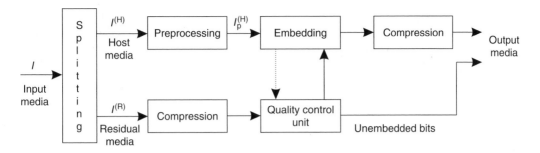

FIGURE 10.1 Block diagram of the data-hiding-based improved compression scheme described in [37,48,49].

10.2.1 DATA IN IMAGE

In [48] a data-hiding algorithm for improved image coding, based on projection, quantization, and perturbation in the discrete cosine transform (DCT) domain, is presented. The proposed coding scheme is depicted in Figure 10.1. Specifically, an image I is split into two parts having the same size: a host image $I^{(H)}$ and a residual image $I^{(R)}$. The residual image is first compressed, in order to reduce the payload, and then embedded into a preprocessed version of the host image. However, in the embedding stage, it is necessary to compromise between the payload size and the host image capacity. This role is played by the "quality control unit" that controls the amount of the bits that can be embedded according to the host image capacity and that gives as output the bits that cannot be embedded in the host part. The marked image is then JPEG compressed. More in detail, the implemented splitting method employs a one-dimensional wavelet transform applied to the input image. The so-obtained low-pass image is used as the host image $I^{(H)}$ and the high-pass image constitutes the residual image $I^{(R)}$. In order to increase the data-hiding capability of the approach, the residual image is coded using a modified embedded zerotree wavelet coder [30] before being embedded. Moreover, the overall data-hiding method has been made robust to JPEG compression, at a designed quality level, by introducing a preprocessing stage where the host image is JPEG coded and decoded with the assigned quality level. The so-obtained image $I_p^{(H)}$ is then used as host image for the data embedding. The first step of the embedding procedure consists of partitioning $I_p^{(H)}$ into $n \times n$ pixel blocks, which are then projected onto the DCT domain. Then, given a generic transformed block k, its coefficients are arranged into a vector $\mathbf{X}_k^{(H)}$ which is projected along a random direction $\mathbf{Z}_k$ generated according to a user-defined random key, thus obtaining

$$p_k = \langle \mathbf{X}_k^{(H)}, \mathbf{Z}_k \rangle \tag{10.1}$$

where $\langle \cdot, \cdot \rangle$ denotes the inner product operator. The projection p_k is then quantized using a frequency-masking-based threshold T_k. The quantized value $\bar{p}_k$ is then perturbed as follows:

$$\tilde{p}_k = \begin{cases} \bar{p}_k + T_k/4 & \text{for } b_k = 1 \\ \bar{p}_k - T_k/4 & \text{for } b_k = 0 \end{cases} \tag{10.2}$$

having indicated with b_k the bit to be embedded in the kth block. The new vector with the embedded data can be obtained by requiring that the perturbation is along the direction $\mathbf{Z}_k$, that is

$$\overline{\mathbf{X}}_k^{(H)} = \mathbf{X}_k^{(H)} + (\tilde{p}_k - \bar{p}_k)\mathbf{Z}_k \tag{10.3}$$

The inverse DCT is then performed on $\overline{\mathbf{X}}_k^{(H)}$, thus obtaining the marked block coefficients, which, thanks to the performed frequency masking, appear indistinguishable from the original ones although carrying the requested information. These operations are iterated for each block of the image. The host image with the embedded data is finally JPEG-compressed at the desired quality. The inserted bits are recovered by performing inverse operations with respect to the ones made in the embedding stage. Let $\hat{\mathbf{X}}_k^{(H)}$ be the vector containing the DCT coefficients of the kth block of the received embedded image. The extracted bit $\hat{b}_k$ is obtained by projecting $\hat{\mathbf{X}}_k^{(H)}$ onto the random direction $\mathbf{Z}_k$ as follows:

$$\hat{b}_k = \begin{cases} 1 & \text{if } (\langle \hat{\mathbf{X}}_k^{(H)}, \mathbf{Z}_k \rangle / T_k - \lfloor \langle \hat{\mathbf{X}}_k^{(H)}, \mathbf{Z}_k \rangle / T_k \rfloor) > 0 \\ 0 & \text{otherwise} \end{cases} \tag{10.4}$$

where $\lfloor \cdot \rfloor$ indicates the operator that gives the largest integer smaller than or equal to a given number. It is worth noting that more than one bit can be embedded into an image block. In fact, for example, it is possible to embed two bits by perturbing along two orthogonal directions $\mathbf{Z}_k^{(1)}$ and $\mathbf{Z}_k^{(2)}$. Since the projections are orthogonal, the perturbations along the two directions do not affect each other. Therefore, given that we are considering an image of size $N \times N$, segmented into blocks of size $n \times n$, and that m bit are embedded per block, the hidden bit rate is m/n^2 bits/pixel. In [48] experiments conducted on the image *Lena* of size 256×256, 8 bits/pixel are shown. The splitting has been performed using the 9-7 biorthogonal wavelet filters. The preprocessing consists in JPEG compression with quality factor 75% on the host image and subsequent decoding. The data-hiding compression method allows obtaining an image representation using 57,657 bits, thus leading to a bit saving of 33.8% with respect to plain JPEG with quality factor 75%, which is represented by 87,155 bits, still maintaining approximately the same perceptual appearance. The method has been also applied by the same authors to hiding data in video as described in Section 10.2.2.

In [3,4] a progressive data-hiding-based compression scheme, properly designed in order to trade off between the goals of data hiding and perceptual coding, is proposed. Specifically, the compression scheme is designed in order to embed the image color information in the image gray-scale component, without impairing the perceptual appearance of the embedded image. This method improves the compression efficiency with respect to JPEG and SPIHT [30] compression methods. Moreover, it gives the opportunity, at the receiving side, of viewing the image progressively from a monochrome version, at different detail levels, to a color one according to the user's needs. In Figure 10.2, a block diagram representation of the approach is provided. First, the given image X is represented in the YIQ color coordinate system, where Y is the luminance and I and Q are the chrominance components, which jointly represent hue and saturation of the color. Then, the chrominances are represented by their subsampled versions and hidden into the luminance component in such a way that the perceptual appearance of the host is not impaired. The embedding is performed in the wavelet domain because

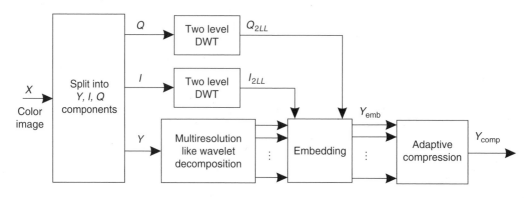

FIGURE 10.2 Compressive data-hiding scheme [3,4] for improved compression of color images.

of the spatial and frequency localization it provides, which well suits the behavior of the human visual system (HVS).

The embedding regions are determined by projecting the luminance image onto the wavelet domain and by choosing those subbands which can transparently hide the chrominance components. After embedding the color information, an adaptive scheme is used to achieve further compression. More in detail, the choice of the YIQ color space has been done since it nearly provides as much energy compaction as a theoretically optimal decomposition performed using the Karhunen–Loeve transform. The bandwidths of the chrominance components are much smaller than that of the luminance. This implies that I and Q can be represented by their subsampled versions without any perceptual quality loss of the reconstructed image. Specifically, a two-level pyramidal discrete wavelet transform (DWT) is performed on both I and Q, thus obtaining the subbands I_{2LL} and Q_{2LL}, that is, the low-pass chrominance replicas at the coarsest resolution, which are used to represent the color information. Then, as shown in Figure 10.3, an unconventional two-level multiresolution-like wavelet decomposition is performed on the luminance component. The first level of the multiresolution decomposition is obtained by performing a DWT onto Y. In particular, an eight-tap Daubachies filter is used. This leads to the subbands Y_{LL} and Y_{LH}, Y_{HL}, and Y_{HH}, which take into account the image at coarser resolution and the "horizontal," "vertical," and "diagonal" details of the image also at coarser resolution, respectively. The subbands Y_{LH} and Y_{HL} are chosen to host the chrominance information. The rational behind this choice relies on the observation that, in order to obtain a good trade-off between robustness and transparency, many watermarking techniques use "middle frequency" coefficients. This makes the subbands Y_{LH} and Y_{HL} suitable to host the data, whereas the subband Y_{HH} is not. Then the subbands Y_{LH} and Y_{HL} are further wavelet-decomposed, thus leading to the subbands $Y_{\alpha,\beta}$ with $\alpha \in \{ll, hl, lh, hh\}$ and $\beta \in \{HL, LH\}$. This represents the unconventional part of the scheme because usually the Y_{LL} band is further decomposed instead of its details.

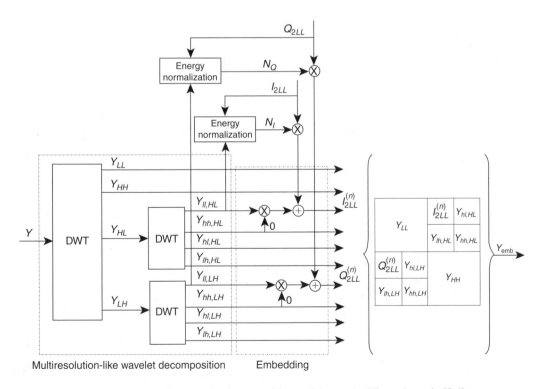

FIGURE 10.3 Multiresolution-like wavelet decomposition and data-embedding scheme in [3,4].

In particular, $Y_{ll,HL}$ and $Y_{ll,LH}$ represent the low-pass subbands, at coarser resolution, obtained from the high-frequency subbands Y_{HL} and Y_{LH}, respectively. It is expected that their energy contribution is relatively small compared to the energy of the remaining subbands of the set $Y_{\alpha,\beta}$. Therefore, they can be zeroed and the subsampled chrominance components I_{2LL} and Q_{2LL} can be embedded in place of the subbands $Y_{ll,HL}$ and $Y_{ll,LH}$ (see Figure 10.3). However, before the embedding, the energies of I_{2LL} and Q_{2LL} have to be normalized to the values of the corresponding host subbands as not to impair the perceptual appearance of the reconstructed image, thus obtaining $I_{2LL}^{(n)}$ and $Q_{2LL}^{(n)}$, respectively. It should be noted that the normalization values have to be transmitted to the decoder since they are necessary to properly reconstruct the color information. To this end, they are embedded in the bit stream's header. Let $Y_{HL}^{(e)}$ and $Y_{LH}^{(e)}$ be the subbands which host I_{2LL} and Q_{2LL}, obtained by calculating the inverse discrete wavelet transform (IDWT) on the subbands specified by Equations (10.5) and (10.6):

$$\left\{ I_{2LL}, Y_{hh,HL}, Y_{hl,HL}, Y_{lh,HL} \right\} \overset{\text{IDWT}}{\to} Y_{HL}^{(e)} \tag{10.5}$$

$$\left\{ Q_{2LL}, Y_{hh,LH}, Y_{hl,LH}, Y_{lh,LH} \right\} \overset{\text{IDWT}}{\to} Y_{LH}^{(e)} \tag{10.6}$$

Therefore, the luminance and the embedded chrominance components are represented, in a perceptually lossless manner, by the subbands

$$Y_{\text{emb}} = \left\{ Y_{LL}, Y_{HH}, Y_{HL}^{(e)}, Y_{LH}^{(e)} \right\} \tag{10.7}$$

The embedding strategy leads to no perceptual degradation of the image luminance component and to a very low mean square error as experimentally verified in [4]. The final step of the algorithm consists in the *adaptive compression stage* that codes each of the Y_{emb} subbands separately in order to guarantee the desired bit rate, related to the maximum distortion allowed, for each subband. Given the desired global bit rate b_{tot}, and the bit rates $b_{LL}, b_{HH}, b_{HL}, b_{LH}$ for the subbands $Y_{LL}, Y_{HH}, Y_{HL}^{(e)}, Y_{LH}^{(e)}$, respectively, the following relation must hold:

$$b_{\text{tot}} = b_{LL} + b_{HH} + b_{HL} + b_{LH} \tag{10.8}$$

Moreover, the bit rate b_{LL} is the first to be designed, since it plays a significant role in the decoded image appearance. The remaining bit rates b_{HH}, b_{HL}, and b_{LH} are automatically assigned by the coder in such a way that a higher bit rate is assured to the subbands having higher energy, according to the following formulas:

$$b_{LH} = \frac{\mathcal{E}_{LH}}{\mathcal{E}_{HH}} b_{HH} \qquad b_{HL} = \frac{\mathcal{E}_{HL}}{\mathcal{E}_{HH}} b_{HH} \tag{10.9}$$

where $\mathcal{E}_\gamma$ ($\gamma \in \{LH, HL, HH\}$) are the energies of the different subbands. After having chosen b_{tot} and b_{LL}, according to the user's needs, the bit rates for each subband are then obtained through Equations (10.8) and (10.9). Finally, each subband is compressed, at the rates previously evaluated, using the SPIHT coder. At the receiving side, the color image is reconstructed, from the compressed bit stream, performing dual operations of the ones done during the coding stage. First, the single subbands are decoded, thus obtaining an estimation (denoted by $\widehat{\ }$) $\widehat{Y}_{\text{emb}}$ of Y_{emb}. Then, $\widehat{I}_{2LL}$ and $\widehat{Q}_{2LL}$ are extracted from $\widehat{Y}_{HL}^{(e)}$ and $\widehat{Y}_{LH}^{(e)}$ (see Equations (10.5) and (10.6)) and the components $\widehat{I}$ and $\widehat{Q}$ are obtained by applying a two-level inverse DWT. After having zeroed $\widehat{Y}_{ll,HL}$ and $\widehat{Y}_{ll,LH}$, the subbands $\widehat{Y}_{HL}$ and $\widehat{Y}_{LH}$ are reconstructed by performing one-level inverse DWT. Finally, an estimation of the luminance $\widehat{Y}$ is achieved by applying a one-level inverse DWT to the subbands $\widehat{Y}_{LL}, \widehat{Y}_{HH}, \widehat{Y}_{HL}$, and $\widehat{Y}_{LH}$.

FIGURE 10.4 For each row from left to right: original image (24 bits/pixel), compressed image using the proposed approach, compressed image using SPIHT, compressed image using the JPEG method. *First row*: *Lena*; compression at 0.15 bits/pixel (for JPEG the maximum allowed compression rate is 0.25 bits/pixel). *Second row*: *Baboon*; compression at 0.30 bits/pixel. *Third row*: *GoldHill*; compression at 0.30 bits/pixel.

Experimental results obtained using the compressive data-hiding approach for the compression of 24-bit color images at 0.15, 0.30, and 0.45 bpp are shown in Figure 10.4. A more comprehensive simulation results set is given in [4]. For comparison purposes, the same test images are compressed, at the same ratios, using JPEG and SPIHT. The experimental results point out that the compressive data-hiding approach performs better than the other methods, especially at low bit rates. The performance comparison has been performed perceptually as well as using the normalized color distance (NCD) metric [26]. As a further characteristic, the compressive data-hiding approach gives the opportunity of viewing the image progressively from a monochrome version at different details, to a full detailed color image.

In [14,31,33], the authors present some methods allowing to embed high volumes of data in images still satisfying the imperceptibility constraint. The embedding methods are based on the selection, driven by perceptual criteria, of the regions where to embed the mark. To accomplish this task, two approaches are proposed: entropy thresholding (ET) and selectively embedding in coefficients (SEC). The embedding is performed by choosing a scalar quantizer in a set as described in [9]. The methods are detailed in [31] for uncoded and in [14] for coded hidden data. Moreover, they are presented in a unique framework in [33]. Specifically, the ET method consists in partitioning the image into blocks of 8×8 pixels, which are then DCT transformed. Let c_{ij}, $i, j \in \{0, 1, \ldots, 7\}$, be the DCT transform coefficients of the generic block. For each block, the energy is computed as follows:

$$E = \sum_{i=0}^{7} \sum_{j=0}^{7} |c_{ij}|^2 \qquad (10.10)$$

with $(i,j) \neq (0,0)$. The block under analysis is selected for the embedding if its energy E given by Equation (10.10) is greater than a threshold value t. In order to make the method robust against JPEG compression up to a predefined quality factor QF, the coefficients of the selected blocks are divided by the corresponding values of the quantization matrix M_{ij}^{QF}, with $i,j \in \{0,1,\ldots,7\}$, thus obtaining $\tilde{c}_{ij} = c_{ij}/M_{ij}^{\text{QF}}$. The so-obtained coefficients $\tilde{c}_{ij}$ are then scanned according to a zig-zag strategy, collected in an one-dimensional vector with elements $\tilde{\mathbf{c}} = \{\tilde{c}_k\}$. Then the first n coefficients of the vector $\tilde{\mathbf{c}}$, with the exception of the DC coefficient, are used to host the data. The embedding is performed by quantizing $\tilde{c}_k$ by means of a quantizer $Q_b(\cdot)$ chosen between two, $Q_0(\cdot)$ and $Q_1(\cdot)$, according to the bit $b \in \{0,1\}$ to embed [9]. Thus, the embedded coefficient $\tilde{d}_k$ is obtained as follows:

$$\tilde{d}_k = \begin{cases} Q_b(\tilde{c}_k) & \text{if } 1 \leq k \leq n \\ \tilde{c}_k & \text{otherwise} \end{cases} \qquad (10.11)$$

For each block, the embedded coefficients are then scanned to form an 8×8 matrix, multiplied by the used quantization matrix and inverse DCT transformed. It is worth pointing out that JPEG compression can lower the global block energy; therefore, it is necessary to verify that the block energy exceeds the threshold both before and after the embedding in order to avoid errors at the receiving side. If the current block does not pass the test no embedding is performed in it, and the next block is examined for potential data insertion. When the SEC method is employed, the embedding regions selection is performed at a coefficient level rather than at a block level as for ET. After having obtained the vector $\tilde{\mathbf{c}}$ as for the ET approach, the first n coefficients and the DC coefficients are selected. Then, they are quantized to the nearest integer value, whose magnitude is indicated with r_k. The embedding is then performed using the proper scalar quantizer according to the bit to embed as follows:

$$\tilde{d}_k = \begin{cases} Q_b(\tilde{c}_k) & \text{if } 1 \leq k \leq n \text{ and } r_k > t \\ r_k & \text{if } r_k = t \\ \tilde{c}_k & \text{otherwise} \end{cases} \qquad (10.12)$$

Since the SEC method relies on a pixel-by-pixel analysis, it introduces smaller perceptual degradation than the ET approach for the same amount of data to embed. The uncoded methods have been designed to be robust against JPEG compression attack. Although limiting the embedding capacity, to face other attacks, some coding strategies can be implemented. In [14,33], the use of Reed Solomon (RS) codes [41] for the ET method and of repeat-accumulate (RA) [11] coding for the SEC scheme are introduced, thus making the embedding method robust against JPEG attacks, additive white Gaussian noise (AWGN) attacks, wavelet compression attacks, and image resizing. At the decoder side, the location of the embedded data is estimated and hard-decision decoding is performed for both the ET and the SEC method when JPEG attacks are carried out. Soft decision provides better performances with respect to hard decision when the attack statistics are known. It can be effectively employed for RA-coded SEC scheme under AWGN attacks. Specifically, the RA decoder uses the sum–product algorithm [15]. For those applications where the embedding of large volume of data is of main concern, the capacity of the host image is first evaluated according to the aforementioned criteria, then the image to be embedded is JPEG compressed at a bit rate such that it fits in the host image. The experimental results reported in [33], some of which are reported in Table 10.1, show that both the methods allow hiding a large volume of data without perceptually impairing the host image.

They refer to the bit amount that can be hidden into the 512×512 *Lena* image using the uncoded embedding methods described previously (ET, SEC) at different quality factors QF and threshold values t. From the experimental results it is possible to verify that, for a defined value of the threshold, higher QFs correspond to higher embedding capacity. In fact, higher compression factors (lower QFs) correspond to a more severe quantization. This implies that, considering for example the

TABLE 10.1
Experimental Results for the Image _Lena_

| | QF = 25 | | | | | QF = 50 | | QF = 75 | |
| | t = 0 | | t = 1 | t = 2 | | t = 0 | | t = 0 | |
	ET	SEC	ET	SEC	ET	SEC	ET	SEC
Bits	6240	11044	4913	2595	15652	18786	34880	31306

Source: From K. Solanki, N. Jacobsen, U. Madhow, B.S. Manjunath, and S. Chandrasekaran, _IEEE Trans. Image Process_, 13, 1627–1639, 2004. With Permission.

ET method, the global block energy is likely to decrease after quantization and can assume values lower than the selected threshold. As a result, the block cannot be used in the embedding stage thus reducing the global image capability to host data. The same motivations can be used to explain the similar behavior when the SEC method, where the embedding is performed at a coefficient level rather than at a block level, is employed. Moreover, for a predefined QF, when the threshold value increases the embedding capacity decreases. In [33] broader results sets, which refer to the coded version of the methods, are presented. However, uncoded schemes are used when the embedding capacity is of primary concern and robustness only against the JPEG compression attack is required. It is worth pointing out that the system, in its uncoded version, is designed to be robust against a predefined level of JPEG attack. This implies that even if the attack is milder the quality of the recovered embedded image is given by the foreseen level of attack.

An attempt to overcome this problem is presented in [32], where an image-in-image hiding method that leads to perceptual as well as mean square error improvements, when the level of JPEG attack decreases, is proposed. The first step of the embedding strategy consists in splitting the image to embed W into a digital part W_D and an analog part W_A. Specifically, W_D is obtained by JPEG compressing W, whereas the analog part W_A is the residual error between W and W_D. The host image is segmented into 8×8 blocks, which are then DCT-transformed. For each host block, few low-frequency coefficients, with the exclusion of the DC coefficient, are collected in the vector $\tilde{\mathbf{c}}^A = \{\tilde{c}_k^A\}$ and reserved for W_A embedding. The other low- and middle-frequency coefficients, collected in the vector $\tilde{\mathbf{c}}^D = \{\tilde{c}_k^D\}$, are employed to host W_D. The embedding of the digital part is performed using the RA-coded SEC approach described in [33]. The embedding of the analog part is performed by first quantizing the host coefficient, $\tilde{c}_k^A$, with a uniform quantizer of step Δ. Then, the analog residual m (belonging to W_A), which has been scaled to lie in the interval $(0, \Delta)$, is embedded into the host coefficient to produce the embedded coefficient $\tilde{d}_k$ as follows:

$$\tilde{d}_k = \begin{cases} \Delta(\lfloor \tilde{c}_k^A/\Delta \rfloor) + m & \text{if } \lfloor \tilde{c}_k^A/\Delta \rfloor \text{ is even} \\ \Delta(\lfloor \tilde{c}_k^A/\Delta \rfloor + 1) - m & \text{if } \lfloor \tilde{c}_k^A/\Delta \rfloor \text{ is odd} \end{cases} \qquad (10.13)$$

Then, for each block, the embedded coefficients are scanned to form an 8×8 matrix and inverse DCT transformed. The decoding of the analog and of the digital parts is performed separately. Specifically, the digital part is obtained using the iteratively sum–product algorithm for RA decoding. The analog part is estimated by means of an MMSE decoder, obtained in the hypothesis of a uniform quantization attack, that the residual data have a uniform distribution probability, and that the reconstruction points of the attacker are known to the decoder but not to the encoder. Experimentations have been conducted using 512×512 host images and both 128×128 and 256×256 images to embed. JPEG attacks at different quality factors have been performed. The experimental results highlight an improvement of the perceived quality of the extracted image as well as of its mean square error as the JPEG attacks became milder.

10.2.2 DATA IN VIDEO

The data-embedding method sketched by the block diagram in Figure 10.1, specialized to data embedding in image in Section 10.2.1, can be straightforwardly applied to video sequences [37,49] as input media. The host and the residual media can be chosen according to the specific application. For example, when dealing with MPEG-2-coded video sequences, the intra-frames (I-frames) can represent the host media, whereas the bidirectionally predicted frames (B-frames) represent the residual media. The predicted frames (P-frames) can be used either as the host part or the residual part depending on the compression scheme. In [37,49], experiments on raw videos are reported. As an example, a gray-scale video of 311 frames having size 120×160 is embedded into a video having the same length and having size 240×360. Both the video have a temporal resolution of 30 frames/sec. With reference to the general scheme of Figure 10.1, the embedding is done as described in Section 10.2.1. Specifically, the compression stage of Figure 10.1, acting on the residual media, performs MPEG-2 compression at a bit rate of 294 bytes per frame. The single frames of the host video are segmented into blocks of size 8×8 and 2 bits per block are embedded. After the embedding, no visible degradations of the host video sequence are reported by the authors. The same approach has been used to embed speech streams in a video. The video that has been used in the experiments consists of 250 frames of size 360×240 and four speech streams have been considered as residual media. The speech streams are sampled at 8 kHz and represented with 8 bits per sample. A code excited linear prediction (CELP) voice compression method has been used to perform compression on the residual media at 2400 bits per second. The host media, after the speech streams embedding, has been compressed using motion JPEG: at compression ratio 3:1 the bit error rate (BER) was 0.1%, and at compression ratio 7:1 the BER was 1%. When independent Gaussian noise at PSNR of 30 dB is added to the host media with the embedded residual media a BER equal to 0.2% was obtained, whereas at PSNR of 28.1 dB the BER was 1.2%.

In [23], a source and channel coding for hiding large volume of data, video and speech, in video is presented. Although in [23], both the cases were considered, where the host data was not available to the receiver and the case where it was taken into account, in this summary only the first situation is considered. The method can be briefly summarized as follows. The host data is first orthogonally transformed, and then a set of transform coefficients is properly chosen. A prediction is then performed on these coefficients, which are then grouped into vectors and perturbed in order to host the data to embed. A source coding by vector quantization (VQ) [13] as well as channel-optimized vector quantization (COVQ) [12], which better performs on noisy channels, are carried out on the data to embed before the perturbation of the transform coefficients. In order to make the embedding imperceptible, the perturbations are constrained by the maximum mean square error that can be introduced into the host.

In more detail, with reference to Figure 10.5, the data to embed is source-coded by VQ or COVQ, thus obtaining a sequence of symbols belonging to a Q-ary alphabet $\{s_1, s_2, \ldots, s_Q\}$ having dimension k. Following the notation introduced in [23], let $X = (x_1, x_2, \ldots, x_N)$ be the host data sequence, $C = (c_1, c_2, \ldots, c_N)$ the transform coefficients obtained by means of an orthogonal transform, $\tilde{C} = (\tilde{c}_1, \tilde{c}_2, \ldots, \tilde{c}_N)$ the perturbed transform coefficients, and $\tilde{X} = (\tilde{x}_1, \tilde{x}_2, \ldots, \tilde{x}_N)$ the inverse transform coefficients. The transparency constraint is expressed in terms of the mean square error as follows:

$$\frac{1}{N} \sum_{i=1}^{N} |x_i - \tilde{x}_i|^2 = \frac{1}{N} \sum_{i=1}^{N} |c_i - \tilde{c}_i|^2 < P \qquad (10.14)$$

where the identity holds since an orthogonal transform is used and P specifies the maximum mean square error for the considered applications. However, according to the amount of data to hide, a smaller set of $M < N$ host coefficients can be chosen for data embedding. The coefficients where to embed the data can be selected *a priori* in the transform domain in order to guarantee

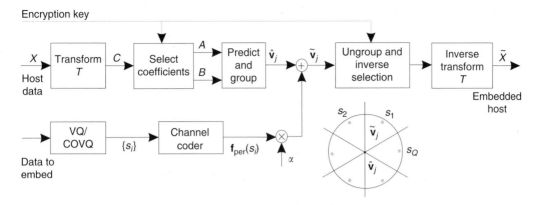

FIGURE 10.5 Data-hiding scheme [23] for embedding large volume of data in video.

the imperceptibility constraint. Also, they can be selected according to a pseudo-random-generated encryption key, which introduces an additional level of security other than those introduced by source and channel coding. Moreover, the use of an encryption key counteracts the burst errors that can occur when, for example, attacks like compression are performed. Once the host coefficients are chosen, they are collected in the ensemble $A = \{a_1, a_2, \ldots, a_M\}$, whereas all the remaining coefficients are put in the ensemble $B = \{b_1, b_2, \ldots, b_{N-M}\}$. For each element a_i of the ensemble A, a deterministic predicted version $\hat{a}_i$ is obtained from the elements belonging to the ensemble B:

$$\hat{a}_i = f_{a_i}(B), \quad i = 1, 2, \ldots, M \tag{10.15}$$

where f_{a_i} is the predictor of the coefficient a_i on the base of the coefficients in B. The coefficients $\hat{a}_i$ are grouped into vectors $\hat{\mathbf{v}}_j, j = 1, 2, \ldots, M/k$, of dimension k, which are then perturbed as follows:

$$\tilde{\mathbf{v}}_j = \hat{\mathbf{v}}_j + \alpha \mathbf{f}_{per}(s_i) \tag{10.16}$$

with α being the parameter that determines the embedding strength. The ensemble of vectors $\mathbf{f}_{per}(s_i)$, $i = 1, 2, \ldots, Q$, constitutes a noise-resilient channel codebook having size Q and dimension k. The perturbation is made in such a way that the perturbed host vector coincides with one of the symbols $\{s_1, s_2, \ldots, s_Q\}$. The transparency constraint (10.14) is then expressed for each of the perturbed vectors as follows:

$$\frac{1}{k}||\mathbf{v}_j - \tilde{\mathbf{v}}_j||^2 < P_c = \frac{N}{M}P, \quad j = 1, 2, \ldots, M/k \tag{10.17}$$

where $||\cdot||$ represents the Euclidean norm, which implies that the perturbed vectors all lie within a sphere of radius $\sqrt{kP_c}$. Eventually, the elements of the vectors $\tilde{\mathbf{v}}_j, j = 1, 2, \ldots, M/k$, are ungrouped and inserted back into their original locations in the transform domain. The inverse transform allows obtaining the embedded data $\tilde{X}$. At the receiving side, the generic received perturbed vector can be expressed as $\mathbf{w}_j = \tilde{\mathbf{v}}_j + \mathbf{n}_j$, with $\mathbf{n}_j$ being the additive noise, which takes into account the alterations introduced by the transmission–compression stages. The decoding procedure is carried out by first performing an estimation $\hat{\mathbf{v}}'_j$ of $\hat{\mathbf{v}}_j$ using the received coefficients belonging to the set B and contaminated by noise. Then, the k-dimensional space, where $\hat{\mathbf{v}}'_j$ is partitioned into Q different regions, each representing one of the symbols $\{s_1, s_2, \ldots, s_Q\}$, by centering the decision boundaries around $\hat{\mathbf{v}}'_j$ (see Figure 10.5). If the perturbed vector $\tilde{\mathbf{v}}_j$, which represent the symbol s_i, is received as $\mathbf{w}_j$ owing to the noise, the decoding will be correct if $\mathbf{w}_j$ is in the decision region pertaining to s_i. It is worth pointing out that, from a methodological point of view, the method presented in [23] embraces the one described in [37] based on quantization of the host data and subsequent perturbation, and it allows obtaining a higher embedding capacity than that can be obtained by using [37]. The procedure presented in [23] and summarized here can be used to hide large volume of data in video. Specifically, each frame of the video is transformed using a two-level wavelet decomposition performed using orthogonal Daubechies filters of length 6. Middle subbands are chosen for data embedding in order to compromise between robustness to compression and imperceptibility. The used prediction function is the function zero, i.e., the chosen coefficients are zeroed out. In fact, since reasonable wavelet domain coefficients predictors are not available, the zeroing out method seems to be the only feasible one. Then the embedding procedure follows as already specified. Experimental results pertaining the embedding of video-in-video using the aforementioned approach have been presented in [23]. Specifically, a video at spatial resolution 88×72 and temporal resolution 7.5 frames/sec is embedded into 176×144 QCIF video at 30 frames/sec. The chrominance components of both the videos are subsampled by a factor of 2 along the vertical as well as the horizontal direction. The video to embed is partitioned into blocks of 2×2 pixels, and the four pixels from the luminance component together with the corresponding two pixels of the two chrominance components are collected in a vector of six elements. For each frame of the video to embed 1584 vectors are thus considered. In one implementation proposed in [23], only the luminance component of the host video is used for the embedding. The subband $2HH$ is composed by 1584 coefficients, which are grouped into 396 vectors of dimension 4. Since each frame to embed is composed by 1584 vectors, if all the coefficients of the chosen host subband are used, four frames of the host video are necessary to embed the data. The source codebook is of size 256. Vector quantization and COVQ at three different noise levels are considered. The channel codebook is also of size 256 and it has been derived from the D_4 lattice [6]. The experimentations have been performed using as host video "Mother and Daughter" and as video to embed "Hall Monitor". After the embedding, the host video is compressed using H.263 at 30 frames/sec. The PSNRs for the extracted video vs. the host video bit rate after H.263 compression at 30 frames/sec for the described scenario, reported in [23], are given in Figure 10.6. The experimentations highlight that, in the presence of noise, COVQ gives better performance than VQ. In another implementation, besides the luminance, even the chrominances are used to host the data. Specifically, referring to the host video, four luminance wavelet coefficients from the $2HH$ subband and two other coefficients from the chrominance subbands are collected into vectors of dimension 6. As in the other application, four frames of the host video are necessary to host a frame of the video to embed. The experiments have been performed embedding the video "Coastguard" into the video "Akiyo". The source codebook is of size 72.

Both standard VQ as well as COVQ have been implemented. The channel codebook is also of size 72 and it has been derived from the E_6 lattice [6]. The experimental results, extracted from [23], are given in Figure 10.7. As in the other application, the experimentations point out that in presence of noise, COVQ gives better performance than VQ. The aforementioned embedding procedure has also been employed to hide speech in video [22,23]. Specifically, several scenarios for hiding speech, sampled at 8 kHz and represented by 16 bits/sample, in a QCIF video having temporal resolution of 30 frames/sec are presented. In order to improve robustness to video coding, temporal redundancy is introduced by embedding the same information into successive frames. The embedded video is then H.263-coded. The speech embedding in video has been performed both using only the luminance $2HH$ subband and the luminance and chrominances $2HH$ subbands. Vector quantization using codebooks of different size and channel codes extracted from the lattices $D_4, E_6, K_{12}, E_8, \Lambda_{16}, G_{24}$ [6] have been used. A performance comparison among these implementations is provided in [22,23]. In summary, the experimentations have pointed out that the extracted video and speech from the host data are of good perceived quality even for high compression ratios. As a further application of the methodology described in [23], we can mention the embedding of color signature image into host image [8].

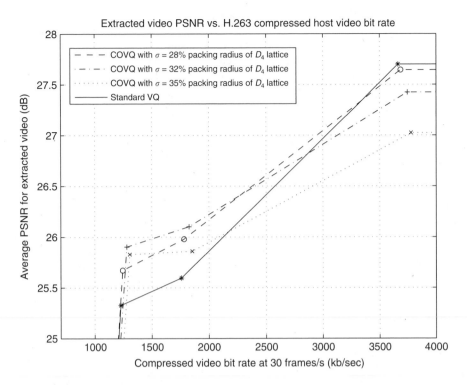

FIGURE 10.6 Average PSNR of the extracted *Hall Monitor* (QCIF) sequence vs. the bit rate for H.263 compressed host *Mother and Daughter* bit stream at 30 frames/sec, for 256 standard size VQ and COVQ.

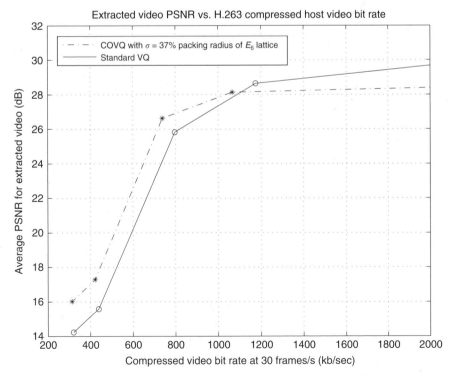

FIGURE 10.7 Average PSNR of the extracted *Coastguard* (QCIF) sequence vs. the bit rate for H.263 compressed host *Akiyo* bit stream at 30 frames/sec, for 72 size standard VQ and COVQ.

10.3 DATA HIDING FOR ERROR CONCEALMENT

In the past few years, the development of error-resilient image and video-coding algorithms, which counteract the effect of noisy channels, has received an increasing attention [38].

Error concealment makes use of inherent characteristics of the visual data, like spatial or temporal correlation, and attempts to obtain a close approximation of the original signal that is least objectionable to human perception. An error concealment system has to cope with three problems: first of all, error detection, which is the localization of errors due to damaged/lost data blocks at the decoder, has to be accomplished; when errors are detected, a resynchronization step is required in order to prevent the propagation of errors to other blocks; finally, the recovery or reconstruction of damaged/lost blocks allows to finalize the error concealment. The recovery at the decoder generally consists of two steps: first, the decoder estimates some features of lost information (like the edge orientation to help spatial interpolation, or the motion vectors for temporal interpolation). Second, the decoder reconstructs lost information from the estimated features and the features of neighboring blocks using a spatial, transform-domain, or temporal interpolation.

Recently, data hiding has been proposed to improve the performance of error concealment algorithms by conveying to the decoder some side information that can be exploited for one of more of the three steps of the error concealment process, i.e., error detection, resynchronization, and data recovery. The improvement is possible since, thanks to data hiding, the decoder will have the availability of a higher amount of information about the lost data, or a more precise information, since performing features extraction at the encoder is more effective than doing it at the decoder. Another advantage is the fact that if the hidden information concerns the estimate of some features of lost data, the computational burden of this estimation is shifted from the decoder to the encoder, that usually has more resources and often can perform the task off-line.

Data hiding allows an image or video-coding system to remain fully compliant: in fact, a decoder with a data-hiding detector can extract and exploit the supplementary information, whereas a standard decoder will simply ignore that information. A drawback of data hiding is that, in general, it could introduce an overhead to the bit rate or a visual quality degradation: so, the design of the algorithm has to take care of it to avoid the potential negative impact on the bit rate or video quality.

In the literature, several data-hiding algorithms for error detection and concealment in image and video transmission have been proposed. It is not the scope of this section to make a complete review of them; on the contrary, the aim is to give an introductory description of how data hiding can be exploited to provide to the decoder some side information useful for error concealment.

A data-hiding algorithm for error concealment purposes can be described according to a set of characteristics: a first classification can be done according to the type of content data hiding is applied to, that is an image or a video bit stream. Moreover, for the video data, we have to distinguish data hiding into intraframes or into interframes. In the literature, it is possible to find algorithms working on images in JPEG or in JPEG2000 format, and video content coded in MPEG2, MPEG4, H.263, H.263+, or H.264 standard.

A second classification can be done according to the type of embedding rule used to conceal the side information to be sent to the decoder and the host features where embedding is accomplished. Concerning the embedding rule, in this application case issues of data hiding like security and robustness to attacks are not of concern, whereas the speed of the algorithm is important. These reasons pushed to the choice of simple embedding rules, like the even–odd embedding: the host feature is modified in such a way that it takes an odd value if the bit to be embedded is a "1", and an even value if the bit is a 0. An exception to this general choice can be found in [34,35], where, as will be described later in more detail, a modification of half-pixel motion estimation scheme is proposed to embed two bits. Concerning the host features where the side information is hidden, in most of the schemes proposed so far, the choice has been to modify the quantized DCT coefficients, since most of the image and video-coding schemes rely on this mathematical transformation; otherwise, the motion vectors (MV) are modified, or some other transform coefficients (like the integer transform in the case of H.264/AVC standard).

A third classification can be done according to the data to be concealed, and their use at the decoder. As already described, an error concealment scheme has to cope with the issues of error detection, resynchronization, and reconstruction of damaged/lost blocks. In the first works on data hiding for error concealment, proposed data-hiding algorithms were only able to detect the errors in the received bit stream. Next, the application of data hiding was extended to cope with the following problems: the resynchronization of the coded stream, the recovery of the lost motion vectors, and the reconstructions of pixels belonging to the lost blocks. This classification is the one chosen for the description in more detail of the algorithms presented in literature, which is carried out in the following subsections. For lack of space, the analysis does not include the description of those algorithms working only with the problem of error detection.

10.3.1   ERROR CONCEALMENT FOR RESYNCHRONIZATION

When errors modify the image or video stream, a resynchronization step is required in order to prevent the propagation of errors to other blocks; also in this case, data-hiding methods can carry to the decoder some synchronization bits that are used at the decoder to understand where the error has happened. In the following, some proposed methods based on this approach are analyzed.

Robie and Mersereau [27–29] propose a data-hiding algorithm against errors in MPEG-2 video standard, applied to I-frames as well as to P- and B-frames. The authors noted that the biggest cause of errors in MPEG-2 decoder is the loss of synchronization, and that, after resynchronization is obtained, the loss of differentially encoded DCT-DC values (I-frames) and MV (P-/B-frames) creates very noticeable artifacts. According to this analysis, the authors decided that the most important information to be protected are: the number of bits for each macroblock (MB); final DCT-DC coefficients for I-frames; final motion vectors for P- and B-frames. The encoder, thus, for a given slice, collects into a string of bits I_S the following data: the number of bits for each MB; the final DCT-DC coefficients (in the case the slice belongs to an I-frame), or the final motion vectors (in the case the slice belongs to a B- or a P-frame), and the number of byte alignment bits at the conclusion of the slice. This information is embedded into the next slice, by properly modifying the least significant bit (LSB) of quantized DCT-AC coefficients. The embedding process is the following: given a DCT 8×8 block (image data for I-frame, error information for P- or B-frames), its coefficients are quantized; next, the quantized coefficients $X(k)$, provided that they are higher than a given threshold, are modified into X', according to this embedding rule:

$$|X'(k)| = \begin{cases} |X(k)| & \text{if } \mathrm{LSB}(X(k)) == I_S(k) \\ |X(k)| + 1 & \text{if } \mathrm{LSB}(X(k))! = I_S(k) \end{cases}$$

Finally, the coefficients are Huffman and run-length encoded, followed by their packetization into a standard MPEG-2 video stream. At the decoder, if an error is recorded in a slice, the data remaining in the slice up to the next start of slice header are stored until the entire frame is decoded. Next the data hidden into the next slide are recovered and used: the size of the MBs can locate the beginning of the next MB to be decoded; DCT-DC or motion vector values are computed using the final values sent and the differential values working "backwards". If, because of multiple errors, the data are not available, then the decoder will attempt to conceal errors by using an early resynchronization (ER) scheme. A set of test has been carried out to compare the proposed method (Stegano) against a simple temporal error concealment (Temp), where the information from the previous frame is taken without correction for motion, and against an ER scheme (Resynch), with the presence of uniform distributed error with $Pe = 10^{-4}$. The PSNR of three different video sequences was computed: for the sequence *Flower*, Stegano scored a PSNR of 24.99 dB (10.28 for Temp, and 22.27 for Resynch); for *Cheer* a PSNR of 18.98 dB (13.34 for Temp, and 18.82 for Resynch); and for *Tennis* 28.07 dB (15.19 for Temp, and 24.13 for Resynch) were obtained.

10.3.2 ERROR CONCEALMENT FOR THE RECOVERY OF MV VALUES IN LOST BLOCKS

The reconstruction of the motion vectors belonging to the lost blocks can be improved through data-hiding methods carrying to the decoder some parity bits, which are used with the correctly received MVs of spatially and temporally neighboring blocks. In the following, some proposed methods based on this approach are analyzed.

Song and Liu [34,35], propose a data-embedding technique for fractional-pixel-based video coding algorithms such as H.263 and MPEG-2. The authors modify the standard motion estimation procedure at fractional-pel precision, in such a way that two bits can be embedded in a motion vector for a inter-mode MB. These bits are used to embed a side information, which is used to protect MVs and coding modes of MBs in one frame, into the MVs of the next frame. In an H.263 encoder, for every intermode-coded MB, an integer-pixel MV is found. Then half-pixel-based motion estimation refinement is obtained by looking for the MV with minimal sum of absolute difference (SAD) among eight half-pixel locations around the integer position previously determined, and the integer position itself. The standard scheme is modified to embed two bits data as follows: after the integer-pel motion estimation, the eight possible MV candidated half-pixel positions are classified into four sets, where each set corresponds to a couple of bits. Now, the new MV position is not looked from the all eight half-pel locations and integer-pel, but from the positions belonging to the set specified by the two bits to be embedded. Thanks to this embedding method, for a QCIF video 176×144 at most 198 bits per frame can be hidden (there are 99 MBs 16×16 in one frame), even if we have to consider that some MBs in a frame are not coded or are INTRA-mode coded. Moreover, data embedding increases the variance of motion estimation error by four times so that the same error as integer-pixel resolution without data hiding is obtained. At the decoder, to recover the information hidden into an intermode-coded MB, its MV are decoded first; the two embedded bits are extracted by looking at the couple of bits corresponding to the position set that this MV belongs to. The authors decided that important information to be protected in video coding are the MVs and coding modes (Inter, Skip, Intra) for each MB. In an interframe k, the differentially Huffman-coded MV (DHCMV) for each MB are collected. A one bit prefix is added to each DHCMV to indicate whether or not this MB has MV. If an MB does not have MV, one more bit is used to specify its coding mode (Skip or Intra-mode). The DHCMVs of the 11 MBs composing each Group of Blocks (GOB) are concatenated to a bit-string, and the nine bit-strings (for a QCIF video there are nine GOB) are arranged row by row. Because the lengths of these bit-strings are different, shorter rows are padded with 0's. Then the rows of bit-strings are modulo-2 addition coded, obtaining a parity bit-sequence that will be embedded in the MVs of inter-mode coded MBs in the following frame $k + 1$, with the method described. Let us note that if the number of bits to be hidden is more than the available number, the information bits are dropped because of consistency requirement. At the decoder side, if one GOB in frame k is corrupted, the parity code of DHCMVs in frame k can be extracted from MVs in frame $k + 1$ after it is decoded. Then the DHCMVs of the corrupted GOB can be recovered by modulo-2 addition using the extracted parity code and the correctly received DHCMVs of the other GOBs of the frame k. However, the residual data of the corrupted GOB cannot be recovered, so that the video decoding process will try to estimate the other missing data. It has to be noted that if more than one GOBs are corrupted in one frame, the proposed MV recovery using embedding cannot be employed because of the limited embedding capacity. Experimental tests have been carried out using the base-mode H.263 on QCIF sequences Car phone, Foreman, and Miss America, coded at 10 frames/sec and bit-rate 48 kb/sec and a memoryless lossy channel where each GOB has lossy probability p_{GOB} 0.01, 0.05, and 0.10. In most situations, all the parity bits were totally embedded in the next frame. For video coding without data embedding, the MVs of the corrupted GOB are simply replaced by the MVs of MBs above the damaged GOB for error concealment. Experimental result demonstrate that the proposed scheme is more effective than conventional concealment schemes when the probability of GOB error is lower than 10%. The modifications introduced by the embedding algorithm slightly degrade the video quality: the PSNR of the video sequence without the embedding algorithm is about 1 dB higher than the video with embedding.

Yin et al. [46] propose to use data-hiding for improving the performance of error concealment for the video stream of MPEG sequences over Internet, by protecting the MV in P-frames. The authors note that error-resilient tools available in MPEG (such as resynchronization, data partitioning, reversible VLC coding, intra-refresh) are designed to combat random bit errors, but are not effective to packet loss that can happen during transmission over Internet. Data hiding can thus help to improve the performance of error resilience in such a conditions. The algorithm has been designed to work with MPEG-1 coded video, transmitted over Internet where packet losses can happen, each packet containing one row of MBs with a resynchronization mark. Error correction capability are achieved transmitting to the decoder through a data-hiding algorithm some parity bits as in [34,35]; however, Song et al. insert parity bits across GOB in the same frame, limiting the effectiveness of data hiding only to the cases where one GOB in a frame is lost, making it not suitable for Internet transmission. Here, frame-wise parity bits for MV across a pre-set group of interframes (all the P-frames in the same Group of Pictures [GOP]) are computed, and embedded to the successive intraframe or interframe through a method named protect motion vector by data embedding (PMVDE). The bits of the differentially Huffman-coded MVs of each P-frame are arranged row by row. If in an MB there are not MVs, a bit "0" is assigned to it; moreover, one bit is used to indicate an MB as intra or inter, shorter rows are then padded with zeros. Then the modulo-2 sum of the rows of coded motion vectors is generated. These parity bits are embedded in some quantized DCT coefficients of the following I-frame, chosen according to an unspecified human visual model. Also the embedding rule has not been specified into the paper. At the decoder, the embedded parity bits can be extracted, so that one parity bit will be used to correct one bit error along each bit plane. More in detail, the error concealment method works as described: first of all, the position of lost packet will be detected by checking the packet numbers at the decoder. For I-frame, a multidirectional interpolation (MDI) is performed. For P-frames, of all slices in the same position in each P-frame in a GOP, if only one slice is lost, the MVs of that slice can be recovered exactly from the quantized DCT coefficients of the corresponding slice in the following I-frame (or P-frame), assuming it is received correctly. Otherwise, median filter(MF) can be used to interpolate lost motion vectors from neighboring motion vectors. So, this method will work if no more than one slice in the same position in the GOP is lost. Experiments have been carried out on video sequences MPEG-1 coded, size QSIF (176×112) at 10 frames/sec, and GOP of type IPPPP. The tests demonstrate that the proposed method PMVDE increases negligibly the bit rate, and may slightly worsen the video quality for I-frame and produce drift errors for P-frames. Experiments show that the method is effective for packet loss up to 25–30%.

10.3.3 ERROR CONCEALMENT FOR THE RECOVERY OF PIXEL VALUES IN LOST BLOCKS

The reconstruction of intensity values of pixels belonging to lost blocks is usually carried out through temporal and spatial interpolation. The aim of data hiding, in this case, is to provide some side information helping to improve the quality of the interpolation process. In the following, some proposed methods based on this approach are analyzed.

In [44,45], two data-hiding methods for error concealment of images are proposed. The authors assume that damaged image areas can be correctly detected and that corrupted region is block-based. Both the proposed methods work by dividing the host image into 8×8 blocks, to be compliant with standard codecs. Each block A has associated a companion block Ac into which the features of block

A is embedded. At the decoder, if block A is lost, the features in companion block and the neighboring blocks of the lost one are used to recover A. In [44], the side information transmitted to the decoder concerning a block A is its so-called Q-signature, constituted by some quantized DWT coefficients of A. In particular, two bits are used for each of the two quantized low-frequency DWT coefficients, and six bits for the position of each of the selected midfrequency DWT coefficient. In this case, the side information gives the position, whereas the value is estimated by the neighborhood blocks. In [45], for error concealment at the decoder, a known multidirectional interpolation (MDI) method is exploited, attempting to derive local geometric information of lost or damaged data from surrounding pixels, and use this information for content-dependent interpolation. The MDI method is divided into two steps: first, local geometrical structure, such as edge directions, of damaged block are estimated from surrounding correctly received blocks; second, damaged block along edges by surrounding correctly received blocks is interpolated. In the proposed method, the first step is modified, in such a way that the encoder itself extracts edge directions from the original content block instead of estimating them at the decoder from the surrounding blocks. The feature extraction is the following: first, a content block is denoted as a smooth block, if there is no edge point in it, or as an edge block: one bit is used to code the type of block (smooth/edge). For an edge block, the main edge orientation is quantized to one of m equally spaced directions in the range 0 to 180, and $b = \log_2 m$ bits will be used to denote the edge direction index. Thus one bit is required to indicate a smooth block, and $1 + b$ for an edge block. This information is embedded into the quantized DCT coefficients of its companion block, by using an odd–even embedding rule (a coefficient value is forced to be even/odd to embed a 0/1). Only low-frequency DCT coefficients are modified, skipping the DC coefficient and the first two lowest AC coefficients. At decoder, if a lost block is detected, its block type and edge features are extracted from the companion block, then bilinear interpolation is performed to reconstruct the lost block. Some tests, carried out to the image *Lenna*, compressed JPEG with quality factor 75% are presented. Experimental results show that the proposed algorithm slightly outperforms the original error concealment method, estimating direction from neighboring blocks: PSNR improves of 0.4 dB. The main advantage consists in a lower computational burden at the decoder (about -30%). The limit of the method consists of the fact that if a lost block has a complex geometry, results are not good: in fact, due to the limited payload, only one edge direction is transmitted at the decoder through data hiding, even if the block has more than one edge.

Zeng [47] proposes to use data hiding to transmit some high-level side information, as opposed to simply some redundant information, to the decoder to improve the performances of an adaptive error concealment system. The adaptive system aims to exploit the advantages of different EC strategies by choosing between a spatial and a temporal error concealment scheme. However, it is difficult for the decoder to decide which strategy to use for a specific case. On the contrary, the performance of different EC strategies can be easily evaluated by the encoder. Thus, the proposed algorithm determines the EC mode information (spatial or temporal EC) at the encoder side, which is then concealed into the compressed bit stream and can be later extracted by the decoder to help the EC operation. Other advantages of this approach are the fact that the side information computed by the encoder is more accurate than what can be determined at the decoder side, since usually more information is available at the encoder side. The system works according to this procedure: the encoder computes the error concealed MBs using both spatial and temporal error concealment separately. Using the original video frame as reference, it is identified as the method reconstructing a better MB. This one bit information, referred to as EC mode, is then embedded into the next GOB. At the decoder, when an MB or a GOB is lost, this EC mode information can be extracted from the neighboring and correctly received GOBs, so that the appropriate EC algorithm (spatial or temporal) will be applied. The embedding rule is an even–odd one, in the sense that one bit is hidden into a 8×8 block by forcing the sum of the quantized AC levels to be odd or even according to the value of the bit. The EC mode bits of the even GOBs are embedded into the coded 8×8 blocks of the odd GOBs, and vice versa. In the case that there are less coded 8×8 blocks in a frame than the required bits, the side information is concealed sequentially until all coded 8×8 blocks have been used, and temporal EC will be used for the lost MBs unprotected by the data-hiding scheme. Some tests have been done considering a worse-case scenario where a packet loss rate of 50% is measured. The bit overhead and the visual quality degradation introduced by the data-hiding process were negligible. It is seen that the side information, even if consisting of just one bit, helps to significantly improve the EC performance, when compared with the scheme estimating the type of EC to be used directly at the decoder: the PSNR gain is up to 4 to 5 dB for some frames.

In [25], a data-hiding-based error concealment (DH-EC) algorithm for H.264/AVC video transmission is proposed. H.264/AVC [42] is the newest video coding standard of the ITU-T Video Coding Experts Group and the ISO/IEC Moving Picture Experts Group, achieving a significant improvement in rate-distortion efficiency with respect to existing standards. In the standardization of H.264/AVC for the development of the reference software, it has been assumed that bit errors are detected by the lower layer entities, and bit-erroneous packets have been considered as being discarded by the receiver, so that any transmission error results in a packet loss. Therefore, the error resiliency of the H.264 test model decoder consists in addressing the reaction of the decoder to slice losses. Error concealment is a nonnormative feature in the H.264 test model. Two well-known concealment algorithms, weighted pixel value averaging for intrapictures and boundary-matching-based motion vector recovery for interpictures, were tailored for H.264 as described in [40]. Weighted pixel value averaging (defined in the following test model error concealment [TM-EC]) operates by concealing a lost block from the pixel values of spatially adjacent MBs. Each pixel value in an MB to be concealed is formed as a weighted sum of the closest boundary pixels of the selected adjacent MBs. The weight, which is associated with each boundary pixel, is proportional to the inverse of the distance between the pixel to be concealed and the boundary pixel. In [25], an error concealment technique based on the use of data hiding is proposed to protect the intraframes of an H.264 video stream; the method, named Data-Hiding-based Error Concealment (DH-EC), has been compared against TM-EC. First of all, the authors analyzed where the TM-EC fails. As already described, TM-EC is based on a spatial interpolation method applied to the MB of the lost slice, whose size is 16 pixel high by $16N$ wide (where N is the number of the MBs inside each slice). When an inner pixel in the slice is reconstructed by means of interpolation, owing to the excessive spatial distance between the pixel itself and the neighboring MBs, an unsatisfying result will be obtained. This fact drove the authors to the idea to use data hiding for carrying to the decoder some additional information useful to increase the quality of interpolation for the inner pixels of the MB: in particular, it was decided to make available at the decoder the values of some inner pixels to be used together with the pixels of neighboring MBs to better reconstruct all the lost pixels through a bilinear interpolation process. In particular, six inner pixels in every MB of the slice to be reconstructed were chosen (see Figure 10.8). The transmitted side information allows to improve the result of the interpolation process, since the distance between interpolating pixels has been reduced from 16 (i.e., the height of a slice) in the case of TM-EC to five in the case of the DH-EC method, resulting in a better visual quality of the reconstructed frame. To reduce the amount of side information to be transmitted, for each of the six chosen pixels, the value is converted in an 8-bit binary format and only the first five most significant bits (MSB) are saved in a buffer. This process is repeated for every MB composing the intraframe. At the end, the buffer will contain the five MSB of each inner pixel in a sequential order, so that it will be possible at the decoder side to identify the position of each pixel value in the image according to the order into the buffer. The content of this buffer is then hidden into the coded video stream allowing to supply the decoder with a raw estimation of the original values of these particular inner pixels. Concerning the embedding rule, the side information concerning the slice n is hidden in the following slide $n + 1$. In this way, the proposed method can protect all the slices of a frame except the last one. The particular features chosen for embedding the bits stored in the buffer are some quantized AC coefficients of the integer transform of the 16 blocks of size 4×4 composing the MB of the slice $n + 1$. The AC coefficients to be modified are the first seven in the double-scan order used in H.264 standard. The

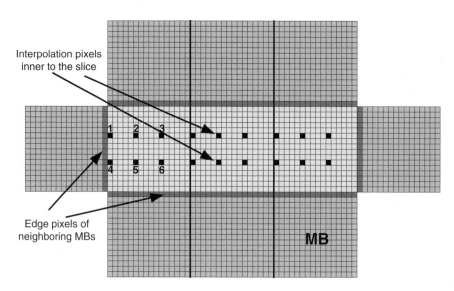

Interpolation pixels
inner to the slice

Edge pixels of
neighboring MBs

MB

FIGURE 10.8 The pixels of the slice that are transmitted to the decoder through the Data-Hiding method.

information bits are embedded by modifying the value of the chosen coefficients (provided that the value is not zero) according to an even/odd rule, so that if the bit b to be embedded is "0" the AC coefficient value is set to be even whereas if the bit is "1" the AC coefficient is set to be odd. This is achieved in the following way: if b is "0" and the AC coefficient is even, then no modification will be applied; if b is "0" and AC coefficient is odd, then its value will be modified, by adding or subtracting 1. If b is "1" and the AC coefficient is odd, then no modification will be applied, whereas if b is "1" and the AC coefficient is even, then its value will be modified, again by adding or subtracting 1. At the decoder, the embedded information has to be recovered from the bit stream to be used in the bilinear interpolation method allowing to reconstruct the damaged slice. The first step in decoding phase is the entropy decoding of the bit stream, which provides the transformed and quantized AC coefficients. To recover the values of the interpolating pixels, the inverse method previously explained will be applied. Each AC coefficient, modified during embedding, is scanned with the double-scan procedure. If the AC coefficient value is even then the embedded bit b will be "0", whereas if the AC coefficient value is odd b will be "1". Every nonzero AC coefficient in the selected positions is checked, to recover the whole binary buffer. At the end of this process, the buffer will contain the binary values of the pixels to be used in the bilinear interpolation process. The position of the bits in the buffer gives us also the spatial position of these pixels inside the MB. However, six pixels are not enough to apply the bilinear interpolation algorithm to reconstruct with good results the lost MB. For a 16×16 MB, 16 pixels are required: as illustrated in Figure 10.9, six pixels are the one recovered with the data-hiding method (highlighted with the letter B in the figure); eight edge pixels are obtained by duplicating the neighboring pixels belonging to the MB above and to the MB below the current one (highlighted with the letter A in the figure); finally, two edge pixels on the right side are the duplicate of the two left-side hidden pixels belonging to the MB on the right of the current one. In this way, the maximum distance between two interpolating pixels is reduced to five.

Experimental results concern video sequences coded according to the H.264 standard at a bit rate of about 100 kb/sec. In Figure 10.10, a frame of the *Foreman* sequence is shown; in particular, the frame on the upper left position is a received frame with two lost slices (each of 11 MB); in the upper center position, there is the same frame after the conventional TM-EC method has been applied. Finally, on the upper right position the frame after the proposed DH-EC algorithm has been

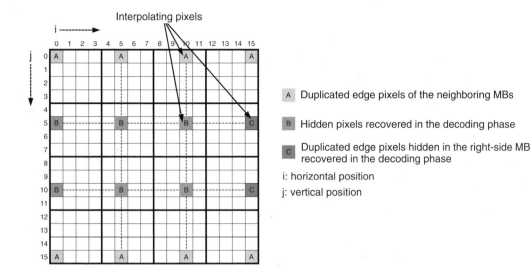

FIGURE 10.9 The 16 interpolating pixels of an MB used in the bilinear interpolation method.

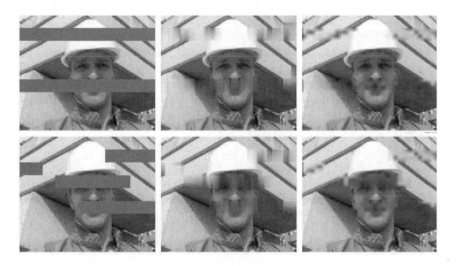

FIGURE 10.10 A frame of the *Foreman* sequence: the frame with some lost slices (left); the same frame after the TM-EC method has been applied (center); the frame after the DH-EC algorithm has been applied (right).

applied is presented. In the lower row, the same frames are shown, in the case where three slices, each composed by 6 MB, were lost. As it appears, the visual quality of the frames where the proposed Error Concealment algorithm is used is clearly increased.

10.3.4 RECOVERY OF PIXEL VALUES IN LOST BLOCKS THROUGH SELF-EMBEDDING METHODS

One particular class of methods for the reconstruction of lost blocks is given by the self-embedding methods. Basically, in all these algorithms a low-quality version of the frame or image to be protected is embedded into itself by means of a data-hiding method. In decoding, if some parts of the image/frame have been lost, the concealed low-quality version can be extracted and used for error concealment. In general, these algorithms exhibit as a critical problem the capacity, since to reconstruct a coarse version of the original image useful for the reconstruction of lost blocks, a large number of bits has to be concealed. Moreover, most of these methods are also computationally expensive, so that their use appears very difficult for video applications. Notwithstanding these drawbacks, it is worth to describe the trends of research on these algorithms.

In [43], a method is proposed to embed a low-quality version of the image, given by a subset of DCT coefficients, into the LSB of its blockwise DCT. The steps composing the algorithm is summarized here. First, the bits for embedding are generated: the image is 8×8 DCT transformed, and all the blocks are classified as smooth, busy, or very busy blocks depending on their activity. The blocks are then quantized with a quantization step depending on their type, and some DCT coefficients, whose number also depends on the block type, are selected for embedding. Next, the host image is processed to select the features where the embedding bits will then be casted: the image is again 8×8 DCT transformed, and quantized with a fine step; the blocks are again classified as smooth, busy, and very busy blocks depending on their activity. Bits that represent the low version of a given block are embedded into a host block of the same type, by modifying the LSB of some DCT coefficients (once more, their number and position depends on the block type). Finally, some side information to be used for reconstruction is generated; however, it is not clear how this information will be made available to the decoder.

In [1,5], a low version of the frame is self-embedded by using a modified version of Cox's [7] method into the DCT of the frame itself. First, a one-level DWT is computed; the low-pass band is dithered to obtain a binary image with size 1/4 of the original size, named final marker. Each pixel of the marker is repeated four times, and the result is multiplied with a pseudo-noise image to obtain the watermark. The watermark is then added to the midfrequencies DCT coefficients of the original frame. At the decoder, the marked DCT coefficients are extracted and multiplied by the pseudo-noise image. The groups of four pixels are averaged to obtain a binary image, which is finally low-pass filtered to obtain a gray image, representing a low version of the original frame. This image can be used to reconstruct lost blocks of the received frame.

Liu and Li [16] propose to embed the DCT-DC coefficient of each 8×8 block into one of its neighboring blocks using a multidimensional lattice encoding method. At the decoder, the extracted DC coefficients are used to reconstruct the lost data. In particular, it can be used as the intensity value of all the pixels of the lost block, or it can help to improve a spatial interpolation method to reconstruct a lost block through the adjacent ones. In this last case, the performance improves by about 0.3 dB with respect to the interpolation without the use of the side information.

Lu [17] proposes a hiding technique where an authentication information is used for error detection, then a recovery information, which is an approximate version of its original data, is used for error concealment. The authentication information is a semi-fragile watermark, created as a set of relationships between the magnitude of some parent–child pairs of DWT coefficients. The recovery information is an image reconstructed from the frequency subbands at the third level of a DWT of the image. In particular, for each DWT coefficient to be concealed only five bits are hidden. The information is embedded by properly modifying the relationships between some parent–child pairs of DWT coefficients in the first and second level of decomposition and some sibling relationships, that is between two DWT coefficients at the same location and scale, but at different orientations.

In [39], an error concealment method through data hiding for JPEG2000 coded images is proposed. JPEG 2000 can separate an image into a region of interest (ROI) and a region of background (ROB). The method conceals some information about the ROI into the ROB, and embeds a fragile watermark into the ROI for error detection. At the decoder, if an error is detected into the ROI, hidden data are extracted from the ROB and used for the reconstruction. In particular, ROI and ROB are transformed into the wavelet domain through a shape-adaptive wavelet transform; DWT coefficients of ROI are coded with a rate-distortion optimized coder, and this bit stream, truncated at a fixed point, is embedded into the DWT coefficients of ROB by using a proper quantization rule. At the decoder, the concealed bit stream can be extracted and used to reconstruct the lost part of ROI.

Munadi et al. [24] propose to use a JPEG2000 codec to perform a multilayer self-embedding: a JPEG2000 bit stream packs the image data in layers, where the most significant layer (MSL) contains the bits pertaining to the most significant contribution to the image quality, and the least significant layer (LSL) contains the least important bits. Given a GOP consisting of I-P-P-P-P-frames, the JPEG200 codec is applied to each frame. The MSL of the intraframe is extracted; these bits are embedded into the LSL of the prediction error of current frame. Next, the MSL of the first P-frame is hidden in the LSL of the second P-frame and so on. Error concealment is achieved by recovering the lost information of a frame using the data embedded into the LSL of the following frame.

10.4   FINAL COMMENTS

The increasing diffusion of digital image and video transmission applications, which, due to the limited transmission rates currently available, require the use of compression techniques, has driven toward the application of data hiding for improved image and video compression as well as for error concealment in images and videos.

In this chapter, some data-hiding algorithms designed to embed large volume of data in the host media while satisfying the transparency constraint have been described. The embedding of image-in-image, based on quantization and projection in the DCT domain, robust to JPEG attack has been presented in [48]. Its extension for video-in-video applications has been presented in [37,49]. In [3,4], a progressive data-hiding-based compression scheme that performs the embedding, in the DWT domain, of the chrominance components of a color image into its luminance using perceptual criteria has been detailed. In [14,31,33], image-in-image embedding schemes based on the choice of scalar quantizers and making use of channel coding to obtain better resilience to attacks, such as JPEG, JPEG2000, AWGN, and resizing, have been detailed. These approaches, although limiting the perceived distortion while hiding large amount of data, do not allow recovering the hidden data if the attack is more severe than that the embedding scheme was designed for. Moreover, if the attack is less severe, the quality of the recovered data corresponds to the one obtained when considering the foreseen (more severe) level of attack. These drawbacks have been removed in the approach presented in [32] that heavily relies on [33], but that leads to recovered data whose quality varies accordingly to the level of the attack. In [22,23], a prediction–estimation-based scheme, operating in the DWT domain and making use of source and channel coding for hiding video and speech in video, has been proposed. The method has been tested to be robust against H.263 compression at different video bit rates. It is worth pointing out that this approach can be thought to somehow include the method presented in [37] as a special case and it provides higher capacity than [37]. Moreover, the use of an encryption-key-based shuffling and grouping of coefficients makes very difficult to the unauthorized retrieval of the data. On the other side, obtaining good predictors across the wavelet subbands is a very difficult problem to cope with. The general principle presented in [22,23] has also been employed in [8] for the embedding of color image in larger host images.

Data hiding has also been proposed to improve the performance of error concealment algorithms by conveying to the decoder some side information that can be exploited for the error concealment process, with particular reference to error detection, resynchronization, and data recovery, while allowing the image or video coding system to remain fully compliant. Here, data-hiding algorithms for error concealment purposes have been classified according to a set of characteristics: the type of content data hiding is applied to; the type of embedding rule used to conceal the side information, and the host features where embedding is accomplished; the data to be concealed, and their use at the decoder. In particular, the most important algorithms proposed so far have been analyzed, grouping them according to the last of the above characteristics, i.e., the use of concealed data at the decoder.

The analysis carried out demonstrate that data hiding can be useful for error concealment, since it improves the available information to the decoder. However, it must be taken into account that the payload at disposal is rather low, so that an interested researcher will have to carefully design the data-hiding method according to the particular aspect of error concealment it is interested in, e.g., a higher protection of motion vectors, or a better resynchronization step, or a better interpolation process.

10.5 FURTHER READINGS

In Section 10.2, some algorithms designed to hide huge quantities of data into a host signal have been discussed without addressing the watermarking capacity estimation. However, the analytical evaluation of the watermarking capacity is an important issue because it provides theoretical bounds to the amount of bits that can be reliably embedded into host data. For example, in [2], a procedure to evaluate the image capacity for nonadditive embedding technique and non-Gaussian host data is provided. In this section, far from being exhaustive, we give some hints about some interesting readings published up to date.

In the last few years, information theoretical models for data hiding have been developed. These models can be used to provide estimations of the maximum-achievable host data-embedding capacity, under some assumptions about the host statistics, the information available to the encoder, decoder, attacker, and the distortion level for the data hider and the attacker. Data hiding was first formulated in [18] as a "game" between two cooperative players, the hider and the decoder, and an opponent, the attacker. Then, extension of this work has been provided in [20], where closed formula for the capacity are expressed in case of host data having both Gaussian and Bernoulli distribution. For non-Gaussian distributions, upper bounds are also given. In [21], a source that emits a sequence of independent and identically distributed Gaussian random vectors is considered. The capacity evaluation for this source, which can be represented by means of parallel Gaussian channels, has been derived. Relying on the theoretical issues developed in [20,21], in [19], closed-form expressions of the capacity, obtained making some assumptions about the source statistics, are given for compressed and uncompressed host images. In [10], the coding capacity of the watermarking game for Gaussian covertext and almost sure squared-error distortion constraint has been evaluated. Some assumptions made in [20] have been dropped in [36], thus obtaining a more general game model.

ACKNOWLEDGMENTS

P. Campisi thanks Dr. D. Mukherjee for having provided Figure 10.6 and Figure 10.7.

REFERENCES

1. C.B. Adsumilli, M.C.Q. de Farias, M. Carli, and S.K. Mitra, A hybrid constrained unequal error protection and data hiding scheme for packet video transmission, *Proceedings of IEEE International Conference on Acoustics, Speech, and Signal Processing (ICASSP '03)*, Vol. 5, Hong Kong, April 2003, pp. 680–683.
2. M. Barni, F. Bartolini, A. De Rosa, and A. Piva, Capacity of full frame DCT image watermarks, *IEEE Trans. Image Process.*, 9, 1450–1455, 2000.
3. P. Campisi, D. Kundur, D. Hatzinakos, and A. Neri, Hiding-based compression for improved color image coding, in *Proc. SPIE, Security and Watermarking of Multimedia Contents IV*, Vol. 4675, Edward J. Delp III, Ping W. Wong, Eds, April 2002, pp. 230–239.

4. P. Campisi, D. Kundur, D. Hatzinakos, and A. Neri, Compressive data hiding: an unconventional approach for improved color image coding, *EURASIP J. Appl. Signal Process.*, (2), Vol. 2002, February 2002, pp. 152–163.

5. M. Carli, D. Bailey, M. Farias, and S.K. Mitra, Error control and concealment for video transmission using data hidings, *Proceedings of the 5th International Symposium on Wireless Personal Multimedia Communications*, Vol. 2, Honolulu, Hawaii, USA, October 2002, pp. 812–815.

6. J.H. Conway and N.J.A. Sloane, *Sphere Packings, Lattices and Groups*, 2nd ed., Springer, New York, 1993.

7. I.J. Cox, J. Kilian, T. Leighton, and T. Shamoon, Secure spread spectrum watermarking for multimedia, *IEEE Trans. Image Process.*, 6, 1673–1687, 1997.

8. J.J. Chae, D. Mukherjee, and B.S. Manjunath, Color image embedding using lattice structures, *Proceedings of IEEE International Conference on Image Processing, ICIP'98*, Vol. 1, Chicago, IL, October 1998, pp. 460–464.

9. B. Chen and G.W. Wornell, Quantization index modulation: a class of provably good methods for digital watermarking and information embedding, *IEEE Trans. Inf. Theory*, 5, 47, 1423–1443, 2001.

10. A.S. Cohen and A. Lapidoth, The Gaussian watermarking game, *IEEE Trans. Inf. Theory*, 48, pp. 1639–1667, 2002.

11. D. Divsalar, H. Jin, and R.J. McEliece, Coding theorems for turbo-like codes, *36th Allerton Conference on Communications, Control, and Computing*, Sept. 1998, pp. 201–210.

12. N. Farvardin and V. Vaishampayan, On the performance and complexity of channel-optimized vector quantizers, *IEEE Trans. Inf. Theory*, 37, 155–160, 1991.

13. A. Gersho and R.M. Gray, *Vector Quantization and Signal Compression*, Kluwer, Norwell, MA, 1992.

14. N. Jacobsen, K. Solanki, U. Madhow, B.S. Manjunath, and S. Chandrasekaran, Image adaptive high volume data hiding based on scalar quantization, *Proceedings of Military Communication Conference, MILCOM 2002*, Vol. 1, Anaheim, CA, USA, October 2002, pp. 411–415.

15. F.R. Kschischang, B.J. Frey, and H.-A. Loeliger, Factor graphs and the sum-product algorithm, *IEEE Trans. Inf. Theory*, 2, 47, 498–519, 2001.

16. Y. Liu and Y. Li, Error concealment for digital imagers using Data hiding, *Proceedings of Ninth DSP (DSP 2000) Workshop*, Hunt, TX, USA, October 15–18, 2000.

17. C.-S. Lu, Wireless multimedia error resilience via a data hiding technique, *Proceedings of IEEE Workshop on Multimedia Signal Processing*, St. Thomas, US Virgin Islands, December 2002, pp. 316–319.

18. J.A. O'Sullivan, P. Moulin, and J.M. Ettinger, Information theoretic analysis of steganography, *Proceedings of IEEE International Symposium on Information Theory*, Cambridge, MA, USA, Aug. 1998, p. 297.

19. P. Moulin and M.K. Mihçak, A framework for evaluating the data-hiding capacity of image sources, *IEEE Trans. Image Process.*, 11, pp. 1029–1042, 2002.

20. P. Moulin and J.A. O'Sullivan, Information theoretic analysis of information hiding, *IEEE Trans. Inf. Theory*, 49, 563–593, 2003.

21. P. Moulin and M.K. Mihçak, The parallel-Gaussian watermarking game, *IEEE Trans. Inf. Theory*, 50, 272–289, 2004.

22. D. Mukherjee, J.J. Chae, and S.K. Mitra, A source and channel coding approach to data hiding with application to hiding speech in video, *Proceedings of IEEE International Conference on Image Processing, ICIP'98*, Vol. 1, Chicago, IL, USA, October 1998, pp. 348–352.

23. D. Mukherjee, J.J. Chae, S.K. Mitra, and B.S. Manjunath, A source and channel-coding framework for vector-based data hiding in video, *IEEE Trans. Circuits Syst. Video Technol.*, 10, 630–645, 2000.

24. K. Munadi, M. Kurosaki, and H. Kiya, Error concealment using a digital watermarking technique for interframe video coding, *Proceedings of ITC-CSCC, The 2002 International Technical Conference On Circuits/systems, Computers and Communications*, Puket, Thailand, July 2002.

25. A. Piva, R. Caldelli, and F. Filippini, Data hiding for error concealment in H.264/AVC, *Proceedings of 2004 Workshop on Multimedia Signal Processing*, Siena, Italy, Sept.–Oct., 2004.

26. K.N. Plataniotis and A.N. Venetsanopoulos, *Color Image Processing and Applications*, Springer, Berlin, 2000.

27. D.L. Robie and R. Mersereau, Video error correction using steganography, *Proceedings of IEEE 2001 International Conference on Image Processing (ICIP 2001)*, Vol. I, Thessaloniki, Greece, October 2001, pp. 930–933.

28. D.L. Robie and R. Mersereau, Video error correction using data hiding techniques, *Proceedings of IEEE 2001 Fourth Workshop on Multimedia Signal Processing*, Vol. I, Cannes, France, October 2001, pp. 59–64.

29. D.L. Robie and R. Mersereau, Video error correction using steganography, *EURASIP J. Appl. Signal Process.*, 2, 164–173, 2002.

30. A. Said and W.A. Pearlman, A new fast and efficient image codec based on set partitioning in hierarchical trees, *IEEE Trans. Circuits Syst. Video Technol.*, 6, 243–250, 1996.

31. K. Solanki, N. Jacobsen, S. Chandrasekaran, U. Madhow, and B.S. Manjunath, High-volume data hiding in images: introducing perceptual criteria into quantization based embedding, *Proceedings of IEEE International Conference on Acoustics, Speech, and Signal Processing, ICASSP 2002*, Vol. 4, Orlando, Fl, USA, May 2002, pp. 3485–3488.

32. K. Solanki, O. Dabeer, B.S. Manjunath, U. Madhow, and S. Chandrasekaran, A joint source-channel coding scheme for image-in-image data hiding, *Proceedings IEEE International Conference on Image Processing, ICIP'03*, Vol. 2, Barcelona, Spain, September 2003, pp. 743–746.

33. K. Solanki, N. Jacobsen, U. Madhow, B.S. Manjunath, and S. Chandrasekaran, Robust image-adaptive data hiding using erasure and error correction, *IEEE Trans. Image Process.*, 13, 1627–1639, 2004.

34. J. Song and K.J.R. Liu, A data embedding scheme for h.263 compatible video coding, *Proceedings of IEEE International Symposium on Circuits and Systems (ISCAS'99)*, Vol. I, Orlando, FL, USA, May–June 1999, pp. 390–393.

35. J. Song and K.J.R. Liu, A data embedded video coding scheme for error-prone channels, *IEEE Transactions on Multimedia*, 3, 415–423, 2001.

36. A. Somekh-Baruch and N. Merhav, On the capacity game of public watermarking systems, *IEEE Trans. Inf. Theory*, 50, 511–524, 2004.

37. M.D. Swanson, B. Zhu, and A.H. Tewfik, Data hiding for video-in-video, *Proceedings of IEEE International Conference on Image Processing, ICIP'97*, Vol. 2, Santa Barbara, CA, USA, October 1997, pp. 676–679.

38. Y. Wang and Q.-F. Zhu, Error control and concealment for video communication: a review, *Proc. IEEE*, 86, 974–997, 1998.

39. J. Wang and L. Ji, A region and data hiding based error concealment scheme for images, *IEEE Trans. Consumer Electron.*, 47, 257–262, 2001.

40. Y.-K. Wang, M.M. Hannuksela, V. Varsa, A. Hourunranta, and M. Gabbouj, The error concealment feature in the H.26l test model, *Proceedings of IEEE International Conference on Image Processing (ICIP 2002)*, Vol. 2, Rochester, NY, USA, September 2002, pp. 729–732.

41. S.B. Wicker and V.K. Bhargava, *Reed-Solomon Codes and their Applications*, IEEE Press, New York, 1994.

42. T. Wiegand, G. Sullivan, G. Bjntegaard, and A. Luthra, Overview of the H.264/AVC video coding standard, *IEEE Trans. Circuits Syst. Video Technol.*, 13, 560–576, 2003.

43. G. Wu, Y. Shao, L. Zhang, and X. Lin, Reconstruction of missing blocks in image transmission by using self-embedding, *Proceedings of 2001 International Symposium on Intelligent Multimedia, Video and Speech Processing*, Hong Kong, May 2001, pp. 535–538.

44. P. Yin and H.H. Yu, Multimedia data recovery using information hiding, *Proceedings of Global Telecommunications Conference (GLOBECOM '00)*, Vol. 3, San Francisco, CA, USA, Nov.–Dec., 2000, pp. 1344–1348.

45. P. Yin, B. Liu, and H.H. Yu, Error concealment using data hiding, *Proceedings of 2001 IEEE International Conference on Acoustics, Speech, and Signal Processing*, Vol. 3, Salt Lake City, UT, USA, May 2001, pp. 1453–1456.

46. P. Yin, M. Wu, and B. Liu, A robust error resilient approach for mpeg video transmission over internet, in *Visual Communication and Image Processing, Proceedings of SPIE*, Vol. 4671, C.-C.J. Kuo, Ed., San Jose, CA, USA, January 2002, pp. 103–111.

47. W. Zeng, Spatial-temporal error concealment with side information for standard video codecs, *Proceedings of IEEE International Conference on Multimedia and Expo*, Vol. 2, Baltimore, (MA), USA, July 2003, pp. 113–116.

48. B. Zhu and M.D. Swanson, and A.H. Tewfik, Image coding by folding, *Proceedings of IEEE International Conference on Image Processing, ICIP'97*, vol. 2, Santa Barbara CA, USA, October 1997, pp. 665–668.

49. B. Zhu and A.H. Tewfik, Media compression via data hiding, *Conference Record of the 31st Asilomar Conference on Signals, Systems & Computers*, vol. 1, Pacific Grove CA, USA, November 1997, pp. 647–651.

Part III

Domain-Specific Coding

11 Binary Image Compression

Charles Boncelet

CONTENTS

11.1   Introduction 285
11.2   Binary Images 285
11.3   Groups 3 and 4 Facsimile Algorithms 286
11.4   JBIG and JBIG2 287
       11.4.1   JBIG 287
       11.4.2   JBIG2 289
11.5   Context Weighting Applied to Binary Compression 292
       11.5.1   A Quick Introduction to Context Weighting 292
       11.5.2   Application to Binary Image Compression 294
       11.5.3   New Compression Algorithms 295
11.6   Conclusions 296
References 296

11.1 INTRODUCTION

This chapter discusses the problem of compressing binary images. A binary image is one where each pixel takes on one of two colors, conventionally black and white. Binary images are widely used in facsimile communication and in digital printing. This chapter reviews traditional compression methods, newer standards, and some of the recent research.

11.2 BINARY IMAGES

Binary images are a special case of digital images. A digital image is a rectangular array of pixels. Each picture element, or pixel, represents image intensity in one or more colors. For a color image, each pixel is a vector of numbers. Most color images have three colors, some have four, and some have many, even hundreds of colors. For gray-scale images, each pixel is a single color. Conventionally, the single color is taken to correspond to intensity. Values range from black to white with gray values in between. Typically, each number is quantized to b bits, where $b = 8$ is typical.

For a binary image, on the other hand, each pixel is quantized to a single bit. Each pixel is either black or white, but not any gray values in between.

Binary images arise in several applications, most notably in facsimile communication and in digital printing. Facsimile machines digitize the input documents, creating one or more binary images, and transmit the images over conventional phone lines. It is important that the binary images be compressed, to save telephone costs and time. However, it is notable that the compression algorithms used are quite inefficient by today's standards.

The other big application for binary images is in digital printing. Printers work by putting dots of ink on a page. Each dot is a spot of ink. The dot sizes are quantized in pixels just as digital images are. A printer can either put ink at a pixel or not; most printers have no ability to put partial dots on individual pixels. Thus, each color printed can be thought of as a binary image.

For example, most printers are CMYK printers. CMYK stands for cyan, magenta, yellow, and black. The printer can put a dot of each color at each pixel location or not. Thus, the printer prints a binary cyan image, a binary magenta image, a binary yellow image, and a binary black image.

The resolution of binary images is measured in pixels per centimeter or pixels per inch. Most often, the resolution in the horizontal and vertical dimensions are identical, but there are exceptions. Some printers have a greater resolution in one direction than the other.

Curiously, facsimile machines use nonsquare pixels. Each pixel is approximately 0.12 mm × 0.13 mm. A standard letter sized or A4 document has about 2 million pixels.

11.3   GROUPS 3 AND 4 FACSIMILE ALGORITHMS

The first facsimile machine was invented in 1843 (by Alexander Bain) and was demonstrated in 1851 (by Frederick Blakewell). It improved slowly over the next hundred years. The modern telephone-based facsimile machine was developed in the 1970s and became immensely popular in the 1980s. Somewhere along the way, the shortened form "fax" entered the lexicon.

The CCITT, now ITU-T, published the Group 3 fax standard in 1980. The standard is currently known as ITU-T Recommendation T.4 [9]. The Group 4 fax standard is now known as Recommendation T.6 [7]. A simple description of the Group 3 fax standard can be found in [6].

The Group 3 fax standard defines two modes, MH and MR. Both modes are relatively simple by today's standards. They were designed at a time when computer equipment was much more primitive and expensive. Nevertheless, the Group 3 algorithms are still widely used. They strike an acceptable balance between compression performance and susceptibility to transmission errors.

The modified Huffman (MH) code uses runlength coding to compress the bitstream. The pixels are scanned in raster order from top left to bottom right. Each row is compressed separately from every other row.

Starting at the left edge, the number of consecutive 0's (typically white pixels) are counted, then the number of 1's (black pixels), then the number of 0's, etc. These numbers, τ_0, τ_1, τ_2, ..., are encoded with a variable length code (e.g., a Huffman-like code). If a row starts with a 1, then the initial run of 0's has length 0.

Typically, there are longer runs of 0's than there are of 1's (most business documents are more white than black). The standard exploits this by using two different codes, one for white runs and one for black runs. Selected MH codes are given in Table 11.1. Note that runs longer than 63 require the use of "make-up" codes. The make-up code specifies how many times 64 goes into the runlength. Then the remainder is specified with a code from Table 11.1. For example, a runlength of $131 = 128 + 3$ would be represented by a make-up code for 128 and the code for 3. (Runlengths of 0 are relatively unimportant. They occur when a runlength is exactly equal to a multiple of 64 or when the line begins with a black pixel. In this case, the initial run of 0's has length 0.)

One important thing to note is that white runs compress only if the run is longer than 4; any white runs shorter than 4 expand when encoded. Black runs longer than 2 compress, but black runs of length 1 expand.

For example, a hypothetical row might look like 0010001000111001000. The runs are 2, 1, 3, 1, 3, 3, 2, 1, and 3. The row would be encoded as 0111 · 010 · 1000 · 010 · 1000 · 10 · 0111 · 010 · 1000 (where the "dots" are not transmitted, but are here to help the reader to parse the code).

The MR, or "Modified READ (Relative Element Address Designate)" algorithm encodes differences between one row and the previous row. Since two-dimensional information is used, the compression ratio for MR is higher than that for MH. However, MR is more susceptible to channel errors. A channel error not only affects the current row, but also the rows following which depend on this row. Accordingly, the standard specifies that every K (typically $K = 2$ or 4) rows, an original row must be encoded and sent (e.g., by MH). This limits the damage caused by a channel error to K rows.

TABLE 11.1
Selected MH Codes

Run	White Code	Length	Black Code	Length
0	00110101	8	0000110111	10
1	000111	6	010	3
2	0111	4	11	2
3	1000	4	10	2
4	1011	4	011	3
5	1100	4	0011	4
6	1110	4	0010	4
7	1111	4	00011	5
8	10011	5	000101	6
16	101010	6	0000010111	10
32	00011011	8	000001101010	12
63	00110100	8	000001100111	12

Neither MH nor MR compresses halftoned images well. Halftoned images have too many short runs (the halftoning process deliberately creates short runs to fool the eye into seeing shades of gray). In fact, halftoned images often expand with Group 3 fax.

The Group 4 fax algorithm achieves greater compression than MR or MH, especially on halftoned images. The Group 4 fax algorithm, however, is rarely used for telephone facsimile transmission. It was designed for noise-free transmission and breaks down when channel noise is present.

The Group 4 algorithm is known as Modified Modified READ (MMR). It is like MR, except that original lines are not encoded every K lines. MMR achieves better compression than MR, but is much more sensitive to channel noise.

In summary, MH and MR are widely used in telephone facsimile machines. While neither is state of the art in binary image compression, they are good enough and are reasonably resistant to channel errors. On typical business documents, MH achieves a compression ratio of about 7–10, MR of about 15, and MMR about 15–20.

11.4 JBIG AND JBIG2

In the 1990s, two new binary image compression standards were developed: JBIG (Joint Bi-level Image Group) and JBIG2. JBIG is a lossless compression standard and achieves much better compression than Group 3, especially on halftoned images [8]. JBIG2 is both lossless and lossy [10].

In this section, we outline both new standards: how they work and discuss their applications — or lack thereof.

11.4.1 JBIG

JBIG (ITU-T Recommendation T.82 ISO/IEC 11544) is a lossless compression algorithm, like MH and MR, but it works on a different principle than either of those and achieves much better compression ratios. The central idea of JBIG is that patterns of pixels tend to repeat. For each pattern, JBIG keeps track of how often a target pixel is a 1 vs. how often it is a 0. From these counts, probabilities of the pixel being a 1 can be estimated, and these probabilities can drive an arithmetic coder.

Arithmetic coders are entropy coders that do not parse the input into separate substrings. Rather, arithmetic coders encode whole strings at once. There are many variations of arithmetic coders. They go by curious acronyms including the Q coder, the QM coder, and the MQ coder. The QM coder is used in JBIG. We will not discuss the operation of these coders here, but the MQ coder (only slightly different from the QM coder) is discussed in Chapter 3. Various arithmetic coders are discussed in [1,5,11,12,14,15,17].

JBIG uses contexts, collections of previously encoded pixels to predict the next pixel to be encoded.

In both contexts, the "A" pixel is adaptive and can be moved to different locations, depending on the type of document being compressed. The 2-line context, Figure 11.1, can be slightly faster than the 3-line context, Figure 11.2, but the 3-line context achieves better compression.

Probabilities in JBIG are estimated using a complicated finite-state model. Rather than describe it (and the accompanying arithmetic coder), we will describe a simple table-based method. The table method is easier to understand (and achieves slightly better compression, but uses more memory) [2,13].

The user picks a context, either the 2- or 3-line context and a location for the "A" pixel. Each combination of the ten pixels defines an index into a table. The table needs 1024 rows. In each row, keep track of the number of times previously the unknown pixel is a 1, n_1, and the number of times it is a 0, n_0. The total number of times this context has been seen is $n = n_0 + n_1$. Note that these numbers vary with each 10-bit combination since some combinations may be more frequent than others.

Denote the unknown pixel as Z and the context as C. Probabilities can be estimated as follows:

$$\hat{P}(Z = 1|C) = \frac{n_1 + \Delta}{n + 2\Delta} \qquad (11.1)$$

where Δ is a small constant, typically 0.5. The idea is that Δ allows the process to start, i.e., one obtains a reasonable probability estimate even when $n_0 = n_1 = 0$. As n gets large, the bias caused by Δ becomes negligible. This probability is used in the arithmetic coder to encode Z. Then one of n_0 or n_1 is incremented (i.e., $n_Z := n_Z + 1$), so that the next time this context is used the probability estimate reflects this bit.

One simple change to this model usually results in a compression improvement: exponential weighting. Give increased weight to the more recent observations. Let α be a weighting constant.

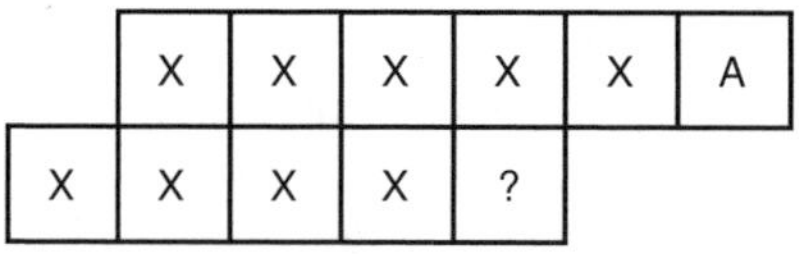

FIGURE 11.1 JBIG 2-line context. The "X" and "A" pixels are used to estimate the "?" pixel.

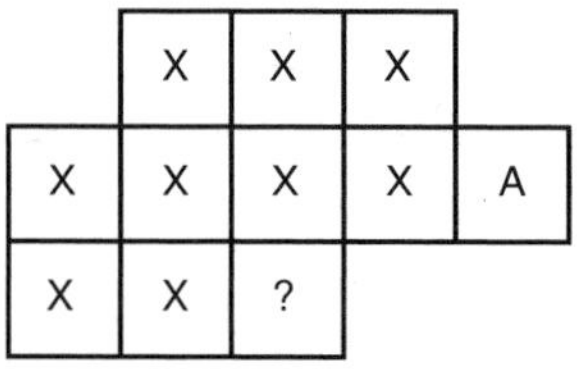

FIGURE 11.2 JBIG 3-line context. The "X" and "A" pixels are used to predict the unknown pixel, "?".

Typical values are in the range $0.95 \leq \alpha \leq 0.99$. Then,

$$n_0 := (1 - Z) + \alpha n_0 \tag{11.2}$$

$$n_1 := Z + \alpha n_1 \tag{11.3}$$

$$n := n_0 + n_1 \tag{11.4}$$

$$:= 1 + \alpha n \tag{11.5}$$

where Z is 0 or 1. Exponential weighting has two main advantages. The probabilities can adapt faster in the beginning and the probabilities can track changes in the image, i.e., the estimates can change as the image changes.

When exponential weighting is employed, the role of Δ needs to be carefully considered. Since n is now upper bounded (by $1/(1 - \alpha)$), the bias caused by Δ does not approach 0. There are two solutions. The first is to initialize n_0 and n_1 to Δ and replace (11.1) by n_1/n. This causes the bias to approach 0. The second solution is to reduce Δ, make it closer to 0, but keep (11.1) as is. This reduces, but does not eliminate the bias. However, our experiments indicate that a small bias is advantageous and results in better compression [13].

Overall, the JBIG model is much more sophisticated than either the MH or MR models. JBIG keeps track of patterns, not runlengths or simple changes from one line to the next. Accordingly, JBIG achieves better compression, typically achieving compression ratios of about 20 on business documents.

JBIG has three serious drawbacks, however. The first is that MH and MR were widely deployed by the time JBIG was developed. It is hard to displace an existing standard that is "good enough."

The second drawback is that JBIG is very sensitive to channel errors. The arithmetic coder can suffer catastrophic failure in the presence of channel errors (all decoded bits after the error can be erroneous). Also, the whole notion of context model and adaptive probabilities are sensitive to errors. When an error occurs, subsequent bits will use incorrect context values and incorrect counts of 0's and 1's. Both lead to further errors.

The third drawback is that JBIG is covered by approximately a dozen patents, owned by several different companies. This patent situation scared away many potential early adopters.

In summary, JBIG uses a context model, a finite-state machine for probability estimation, and an arithmetic coder. It achieves excellent compression, but is little used in practice.

11.4.2   JBIG2

In the late 1990s, the JBIG group developed a new standard for binary image compression. This standard is known as JBIG2 (ISO/IEC/ITU-T 14492 FCD). JBIG2 has been described as being both lossy and lossless. This is somewhat misleading. Like other multimedia compression standards, JBIG2 specifies the format of the bitstream for a compressed binary image. It defines the individual fields in the bitstream, how they are parsed, and how a JBIG2 compliant decoder will interpret them.

JBIG2 does not, however, define how the encoder creates the bitstream. The encoder has considerable flexibility in this regard. The encoder can create a bitstream that, when decoded, recreates the original image exactly. Or, the encoder can create a bitstream that results in an approximation to the original image. In this way, JBIG2 is both lossless and lossy. Both modes can be used in the same image: some parts encoded losslessly and other lossy.

In this section, we describe JBIG2 mostly from the point of view of the encoder. Much of the discussion is taken from the standard [10] and an excellent review article [4].

The first thing a JBIG2 encoder does is segment the image into three types of regions: text, halftones, and generic. The idea is that each region has different statistics and compresses best with an encoder specifically designed for that region.

Text regions are just that: regions containing text. JBIG2 does not specify or even care what language in which the text is written, or even whether the text is written left to right or right to left. The JBIG2 does, however, assume the text can be segmented further into individual letters. A letter is simply a bitmap of black and white dots (we assume black letters against a white background).

A halftone region is a representation of a gray-scale image. The gray-scale image is printed with a high-resolution pattern of black and white dots. The idea is that high-frequency dot patterns (alternating quickly between black and white) fool the eye into seeing gray levels.

A generic region is anything else. Note that a JBIG2 encoder may choose to label halftones as generic rather than as halftone. The reason for this is that in some cases the generic coder (described below) is more appropriate for halftones than is the halftone encoder.

Once the image is segmented in regions, the encoder must decide which of four encoders to use. These are the text encoder, the halftone encoder, the generic encoder, and the generic region refinement encoder.

The text encoder works by first segmenting the text region into individual "letters." Each letter is a pixel bitmap of black and white dots. (The encoder makes no effort to associate semantic information to the letters; the letters are simply bitmaps of 1's and 0's.) The encoder encodes each unique letter using Huffman or arithmetic coding and transmits it. The decoder builds a dictionary of these letters. For instance, the word "abracadabra" contains five different letters, "a," "b," "c," "d," and "r." Each of these bitmaps is encoded and transmitted. See Figure 11.3 for an example.

The encoder then pattern matches the letters in the document against those in the dictionary. The index of the best match and the coordinates of the letter on the page are transmitted. For instance, if the letter at (x, y) matches the ith letter in the dictionary, the triple (x, y, i) is encoded and transmitted. The decoder simply copies the ith letter to location (x, y).

This "build a dictionary and copy" is reminiscent of the way Lempel–Ziv (e.g., gzip) works on ordinary text [19]. The encoder and decoder keep track of previous symbols. The encoder tells the decoder which previously encoded symbols to copy into the current location. Since JBIG2 works on bitmaps and the locations are not necessarily regular, its compression is not nearly as good as gzip on ASCII text, but the idea is similar.

The JBIG2 text encoder can perform lossy compression by allowing imperfect matches. For instance, in "abracadabra," there are five "a"s. If the document is still in electronic form (e.g., as part of the output routine of a word processor), then the bitmaps for each "a" will be identical. The matches against the dictionary "a" will be perfect. However, if the document is a scanned from a paper copy, it is unlikely the matches will be perfect. The encoder then has three choices: (1) A partial match may be deemed sufficiently close and one dictionary "a" may be used for all of the "a"s in the document. (2) Multiple "a"s may be sent to the dictionary and used for matching. This allows perfect matches, but increases the bitrate two ways: more bitmaps are transmitted and the dictionary indices must be larger. (3) The encoder can improve image quality by using the refinement encoder (discussed below).

There is nothing in the standard that specifies how much loss (bitmap mismatch) is tolerable. This is entirely up to the encoder. More loss means more compression; less loss means less compression.

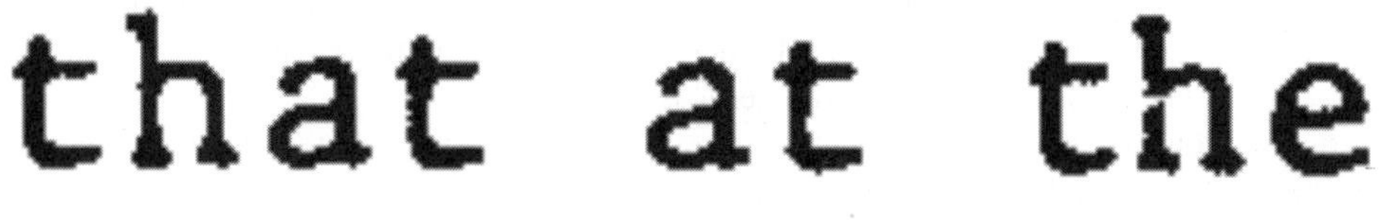

that at the

FIGURE 11.3 Excerpt from CCITT1, the first image in the CCITT test suite, showing the variation in bitmaps for the same letters. To a person, the letters are the same, but an encoder may have difficulty seeing that due to the noise in the scanning process.

Halftones are encoded similarly to text. The halftoned image is segmented into blocks. The blocks are designed to match the dither cell size in a dithered image. If the image was halftoned by a more complicated method, such as error diffusion or blue noise dithering, the blocks should match the scale of the original image pixels. For instance, if a 640×480 image is rendered to a size of 2560×1920 pixels (binary dots), each block would be 4×4. For each block, the encoder determines (guesses, if necessary) the gray level that generates that block. One block for each gray level is encoded and transmitted. The decoder builds a dictionary of these blocks.

The encoder interprets the gray levels as a gray-scale image and losslessly compresses and transmits the image. The decoder decodes the image and, for each pixel in the image, copies the corresponding halftone block from the dictionary to that location on the page.

The halftone encoder works best for ordered dithered images, especially if the encoder knows the original dither cell. However, modern halftoning algorithms often use large cells and more complicated algorithms. It becomes much more difficult for the encoder to properly segment the halftoned image and invert the halftoning process to determine the original gray-scale image. In these cases, it may be better to encode the halftoned image with the generic coder.

The generic coder works similarly to JBIG. It uses a context to estimate the probability that the current pixel is a 1 and uses this probability to drive an arithmetic coder. Some of its details are slightly different from JBIG, but the overall scheme is very similar. The principal improvement is that the JBIG2 generic coder can use a larger context, one with 16 pixels. See Figure 11.4.

The refinement coder is used to improve the performance of the lossy text coder. It combines the text and generic coders. The encoder sends a bitmap that may be only a partial match. This bitmap is used as part of the context in a generic-like coder. Since the bitmap is close to the actual pattern, the context does a good job of predicting the pixels in the binary image.

In summary, JBIG2 breaks new ground in binary image compression. It is the first standard that allows — and even encourages — lossy compression. The text coder can be very efficient, especially on a clean, noise-free, image. The combination of dictionary encoding followed by a generic refinement is novel. In lossless compression, JBIG2 is slightly better than JBIG2. In lossy compression, however, JBIG2 can be excellent, achieving compression ratios much greater than that can be achieved by lossless compression.

JBIG2 suffers much the same criticisms as JBIG, though. It is overly sensitive to channel errors, it is covered by numerous patents, and it has not been able to displace existing technology. In this case, the existing technology is not binary compression at all, but other document formats, such as Adobe's pdf. Documents are routinely transmitted in pdf format and printed at the destination. This is the market for which JBIG2 was developed, but pdf is cheaper (free), flexible, and faster (a pdf is smaller than a JBIG2 compression of the rendered document). Pdfs are also well suited for the Internet.

The last criticism of JBIG2 is that it is great technology, but that lossy compression of binary images is not much in demand. Most users want high-quality reproductions and are willing to tradeoff communications resources to achieve that.

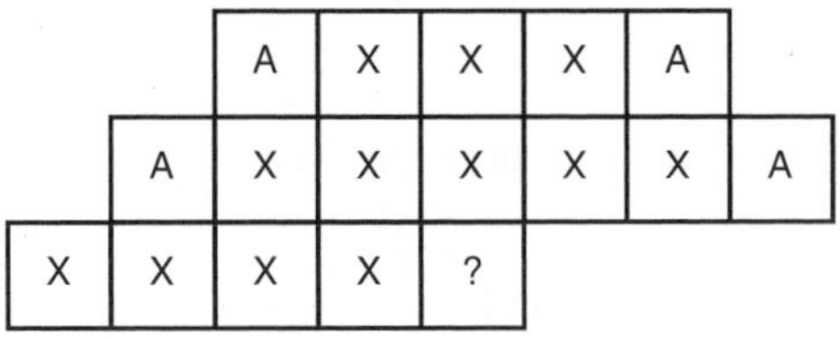

FIGURE 11.4 JBIG2 3-line context. The "X" and "A" pixels are used to predict the unknown pixel, "?."

11.5   CONTEXT WEIGHTING APPLIED TO BINARY COMPRESSION

In this section, we describe recent research in binary image compression. The research uses *context weighting*, described first in Section 11.5.1. The application to binary image compression is described next in Section 11.5.2.

11.5.1   A QUICK INTRODUCTION TO CONTEXT WEIGHTING

As shown in Section 11.4, modern binary image compression methods use context models. Unfortunately, the user must decide which context model to use. The choice, unfortunately, is not trivial. Different context models do well on different types of images. Also, small images do better with small contexts; larger images tend to do better with bigger contexts.

One solution to this dilemma is to weight two or more contexts and try to get the best performance features of each. The original context weighting idea considered the universal compression of one-dimensional strings (e.g., text) [16]. However, the idea can be applied to two-dimensional compression as well.

We will explain the context weighting idea, then describe its extension to binary image compression. Let $X_1^n = X_1, X_2, \ldots, X_n$ be a one-dimensional sequence of random variables. Let $P(X_1^n)$ be the probability of the sequence. Then, then number of bits necessary to encode the sequence, $L(X_1^n)$, obeys

$$L(X_1^n) \geq -\log P(X_1^n) \tag{11.6}$$

$$= -\sum_{i=2}^{n} \log P(X_i|X_1^{i-1})P(X_1) \tag{11.7}$$

On average, the number of bits required is determined by the entropy

$$EL(X_1^n) \geq \int P(x_1^n)(-\log P(x_1^n))\,\mathrm{d}x_1^n \tag{11.8}$$

$$= H(X_1^n) \tag{11.9}$$

Any of several various arithmetic coding algorithms can encode the data at rates very close to the entropy. Arithmetic coders can work on-line; all they need are the conditional probabilities,

$$P(X_n|X_1^{n-1}) = \frac{P(X_1^n)}{P(X_1^{n-1})} \tag{11.10}$$

See also the discussion in Chapter 3.

Now, consider two different probability models, $P_1(\cdot)$ and $P_2(\cdot)$, and let $P_1(X_1^n)$ and $P_2(X_1^n)$ be the sequence probabilities under each model. Let

$$P_{\mathrm{av}}(X_1^n) = \frac{P_1(X_1^n) + P_2(X_1^n)}{2} \tag{11.11}$$

be the averaged probability. Then the performance of the *average* model obeys the following:

$$-\log P_{\mathrm{av}}(X_1^n) = -\log \left(\frac{P_1(X_1^n) + P_2(X_1^n)}{2} \right) \tag{11.12}$$

$$\leq \min\left(-\log P_1(X_1^n), -\log P_2(X_1^n)\right) + 1 \tag{11.13}$$

Thus, we obtain the fundamental result of context weighting: the performance of the average model is within 1 bit of that of the better of the two individual models.

This result holds even though one may not know in advance which of the two models is going to outperform the other. What makes this result especially important is that the computation can be performed *on-line*. It does not require waiting for all the data before computing the probabilities and doing the compression. To see this, consider the following.

First, define the *conditional average probability*,

$$P_{\text{av}}(X_n|X_1^{n-1}) = \frac{P_{\text{av}}(X_1^n)}{P_{\text{av}}(X_1^{n-1})} \tag{11.14}$$

$$= \frac{P_1(X_1^n) + P_2(X_1^n)}{P_1(X_1^{n-1}) + P_2(X_1^{n-1})} \tag{11.15}$$

Then,

$$P_{\text{av}}(X_1^n) = \prod_{i=2}^{n} P_{\text{av}}(X_i|X_1^{i-1})P_{\text{av}}(X_1) \tag{11.16}$$

$$-\log P_{\text{av}}(X_1^n) = -\sum_{i=2}^{n} \log P_{\text{av}}(X_i|X_1^{i-1}) - \log P_{\text{av}}(X_1) \tag{11.17}$$

This last equation shows that the overall number of bits required can be built up as a sum of the conditional bits required for each symbol given in the previous ones. Note that it must be emphasized that the averaging is done on sequence probabilities, i.e., $P(X_1^n)$, not on conditional probabilities, i.e., $P(X_n|X_1^{n-1})$. The universality result above does not hold when the conditional probabilities are averaged.

Context weighting can be applied recursively, yielding what is known as *context tree weighting*. Each probability model can be a weighted average of other probability models.

For an example of how context weighting can be applied, consider a simple one-dimensional model. Let P_1 be a model where $P(X_n = k)$ is estimated by the number of times k is seen previously divided by the number of previous symbols, i.e., the empirical probability of k. Let $P_2(X_n = k) = P(X_n = k|X_{n-1} = j)$, where the conditional probability is the empirical probability of k given the previous symbol is j. So, the first probability model assumes that symbols are independent and the second assumes that symbols obey a simple Markov model. The user can average these two models and, without knowing in advance which one is better, can achieve a performance almost equal to the better of the two.

This idea can be applied recursively. Now, let P_2^* be a weighted average of two models: a first-order Markov model and a second-order Markov model. Then, P_2^* can be averaged with P_1 to yield an overall performance within at most 2 bits of the best of the three models. This recursive continuation of context weighting is known as *context tree weighting*.

Note that the averaging in the above example can be done in two ways: all three models can be averaged at once, yielding an overall performance within $\log_2 3 = 1.58$ bits of the best of the three. Or, the two Markov models can be averaged, then that average averaged with the independent model. This latter implementation is within 2 bits of the best, but can be simpler to implement. In practice, if the number of symbols is large, the difference in performance between the two implementations is negligible.

In many applications, the sequence is Markov or can be modeled as Markov and the conditional probabilities can be simplified:

$$P(X_i|X_1^{i-1}) \approx P(X_i|X_{i-l}^{i-1}) \tag{11.18}$$

This assumption helps simplify implementations, since only a finite amount of memory is needed.

In summary, in this section we showed how averaging over multiple probability models can result in coding lengths equal to the best model plus a small (fixed) amount. In actual practice, the flexibility offered by weighting often leads to better performance than any individual model. We show this in the application to binary image compression discussed next.

11.5.2 APPLICATION TO BINARY IMAGE COMPRESSION

In Section 11.4, we described the JBIG compression algorithms and showed how JBIG uses context models to achieve excellent compression. Let $P_1(\cdot)$ denote the probability estimates under one context model and $P_2(\cdot)$ be the probability under another context model. Then, by (11.13), the "average" model can do almost as well (within 1 bit) as the better of the two context models.

One implementation of this idea is described in [18] and can be described as follows: Context 1, denoted C_1, is a 3-line context model and Context 2, C_2, is a 5-line context. C_1 works very well on digitized text and error diffused images; C_2 works well on many kinds of dithered images. Figure 11.5 and Figure 11.6 show both contexts. Note both contexts use 12 previously encoded pixels.

Rather than simply averaging the two contexts, one obtains somewhat better performance by a slightly more complicated scheme. Consider the six common pixels. They can take on one of $2^6 = 64$ values and serve to identify which of 64 sets of tables to use. Within each table, three probabilities are estimated. The first is simply the probability of the unknown pixel given these previous six, the second is the probability of the unknown pixel given the 12 pixels in context 1, and the third is the probability of the unknown pixel given the 12 pixels in context 2. For each combination of the 6 common pixels, these three sequence probabilities are averaged.

This hybrid scheme has two main advantages. First, any unusual patterns, i.e., those patterns that occur too few times for a 12-bit context to perform well, are estimated with a 6-bit context. Thus, the probabilities are reasonably accurate. Second, using the 6-bit index first allows certain patterns

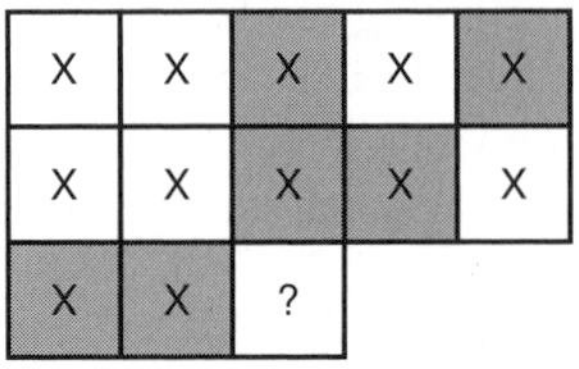

FIGURE 11.5 Context 1. The X's denote previously encoded pixels used to estimate the current pixel, denoted "?." The six shaded pixels are identical in Context 2.

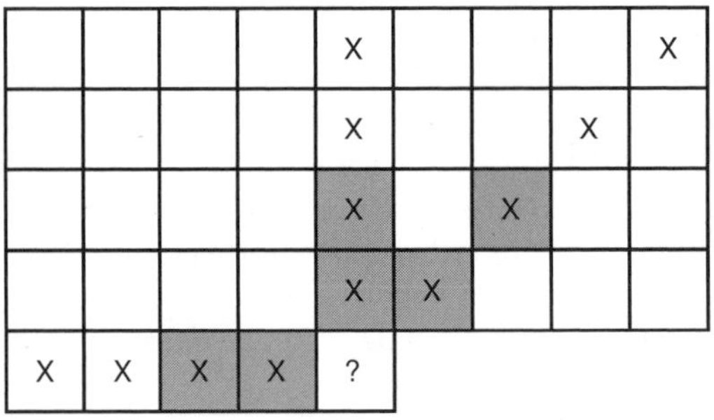

FIGURE 11.6 Context 2. The X's denote previously encoded pixels use to estimate the current pixel, denoted "?." The six shaded pixels are identical in Context 1.

TABLE 11.2
Compression Ratios for Various Documents (Higher Numbers Are Better)

Image Type	JBIG	6 Bits	3 Line	5 Line	Context Weight
Error diffusion	1.817	1.465	1.878	1.564	1.887
Ordered dither	3.666	2.335	2.919	4.435	4.535
CCITT test	19.64	17.47	18.99	18.15	19.90
Compound (1)	8.705	6.794	8.999	9.522	10.38
Compound (2)	7.712	6.691	7.844	8.082	8.875

Note: The "6 bits" numbers are for the 6 bit context alone. The "3 line" and "5 line" columns refer to contexts, C_1, and C_2, respectively. Finally, the "Context weight" column refers to the hybrid scheme that weights all three context models.

to be modeled by C_1 and others to be modeled by C_2. For instance, a composite document with both text and images might combine the best of both contexts: the text is modeled by C_1 and the images by C_2. The hybrid scheme can achieve good performance without having to segment the image into text and image regions.

For example, the compression ratios for various documents are presented in Table 11.2. The numbers are taken from Xiao and Boncelet [18]. "Error Diffusion" and "Ordered Dither" refer to documents created by halftoning images with error diffusion and clustered dot ordered dither, respectively. The "CCITT Test" images are the eight original documents used in testing the original CCITT fax algorithms. The two rows labeled "Compound" refer to two different sets of compound documents, containing both text and images.

As can be seen in the table, the context weighting scheme outperforms the others. The "6 bit" context is simply not detailed enough to compete with the larger, 12 bit, contexts on these documents (there are lots of pixels). However, averaging it in results in a negligible reduction in compression ratio.

Individually, the various context models perform differently on the various document types. As mentioned earlier, C_1 performs better on text and error-diffused images, while C_2 does better on halftoned images. In all cases, context weighting gives the best overall performance.

Also, context weighting frees the user from having to determine which context to use and largely eliminates any reason to segment the document.

The hybrid scheme discussed in this section works by averaging over two large contexts. Additional flexibility is exploited by indexing off the six common pixels. This allows some patterns to be modeled by C_1 and some by C_2, resulting in an overall performance superior to either context alone. It is possible that further research will improve performance by adding additional contexts or by adding additional pixels to the existing contexts.

11.5.3 NEW COMPRESSION ALGORITHMS

In this section, we briefly discuss some of the remaining research directions in binary image compression.

All lossless image compression, binary, gray-scale, and color, will likely benefit from the development of two-dimensional versions of Lempel–Ziv coding (e.g., gzip) or Burrows–Wheeler coding (e.g., bzip2) [3]. These algorithms give outstanding compression on one-dimensional data streams, such as text, but truly two-dimensional versions are lacking.

Related to the above comment, new algorithms for converting a two-dimensional object to a one-dimensional stream would be handy. Usually, a simple raster scan is used. However, more complicated scans, such as the Peano–Hilbert scan or a quadtree scan, might be superior at keeping pixels that are closely together in two dimensions close together in the scanned stream. Some work in this area has been done, but possibly new scans can still be developed, perhaps by introducing a controlled amount of redundancy.

Lossless binary image compression is directly related to lossless image compression. Since a binary image is a special case of a gray-scale image, any gray-scale compressor can be applied to a binary image. Conversely, a binary image compressor can compress gray-scale images 1 bit plane at a time. (Sometimes the bitplanes are Gray encoded first.) The standard advice is that lossless binary image compressors are competitive with the best gray-scale image compressors up to about 6 bits/pixel. This advice, though, needs to be better tested. Perhaps recent advances in binary image compression might lead to better lossless image compressors.

11.6   CONCLUSIONS

This chapter surveyed the important binary image compression standards, Group 3, Group 4, JBIG, and JBIG2. The main ideas behind each were briefly described. Over time, the algorithms have become more complex. The first algorithm, MH, uses runlength coding. The second and third, MR and MMR, use a simple two-dimensional model, encoding differences in one row from the previous row. The next standard, JBIG, uses complex context based probability models driving an arithmetic coder. The last standard, JBIG2, introduces the concept of lossy coding of binary images. It does this by segmenting the image into regions and encoding each region with specific algorithms designed for that region type. We then discussed one recent research direction, that of using context weighting to improve compression performance and eliminate a weak spot of JBIG, which the user must specify the proper context to use. Lastly, we discussed some potential research directions.

With each increase in complexity has come an increase in performance, with each new algorithm outperforming the previous ones. However, this increasing performance has come at a price: the new algorithms are more sensitive to channel errors. It is somewhat disappointing that the most important binary image application, telephone-based facsimile transmission, still uses technology developed 25 years ago. Perhaps some future research will produce an algorithm with both high compression and resistance to channel errors.

REFERENCES

1. C.G. Boncelet, Jr., Block arithmetic coding for source compression, *IEEE Transactions on Information Theory*, September 1993.
2. C.G. Boncelet, Jr. and J.R. Cobbs, Compression of halftone images with arithmetic coding, *42nd Annual Conference of the SPSE*, pp. 331–334, Boston, MA, May 1989.
3. M. Burrows and D.J. Wheeler. A Block-Sorting Lossless Data Compression Algorithm, Technical Report 124, Digital Systems Research Center, May 1994.
4. P. Howard, F. Kossentini, B. Martins, S. Forchhammer, and W. Rucklidge, The emerging JBIG2 standard, *IEEE Trans. Circuits Syst. Video Technol.*, 8, 838–848, 1998.
5. P.G. Howard and J.S. Vitter, Analysis of arithmetic coding for data compression, in J.A. Storer and J.H. Reif, Eds., *Proceedings of the Data Compression Conference*, Snowbird, UT, 1991, pp. 3–12.
6. R. Hunter and A.H. Robinson, International digital facsimile coding standards, *Proc. IEEE*, 68, 854–867, 1980.
7. ITU-T. ITU-T recommendation T.6, facsimile coding schemes and coding control functions for group 4 facsimile apparatus, November 1988, Group 4 fax.
8. ITU-T. ITU-T recommendation T.82, information technology — coded representation of picture and audio information — progressive bi-level image compression, March 1995. JBIG.

9. ITU-T. ITU-T recommendation T.4, standardization of group 3 facsimile apparatus for document transmission, October 1997, Group 3 fax.

10. ITU-T. ISO 14492 fcd, lossy/lossless coding of bi-level images. Available from http://www.jbig.org, July 1999. JBIG2.

11. G.G. Langdon, Jr., An introduction to arithmetic coding, *IBM J. Res. Dev.*, 28, 135–149, 1984.

12. W.B. Pennebaker, J.L. Mitchell, G.G. Langdon, Jr., and R.B. Arps, An overview of the basic principles of the Q-coder adaptive binary arithmetic coder, *IBM J. Res. Dev.*, 32, 717–726, 1988.

13. M. Reavy and C.G. Boncelet, Jr., An algorithm for compression of bi-level images, *IEEE Transactions on Image Processing*, 2001, pp. 669–676.

14. J. Rissanen, Generalized Kraft inequality and arithmetic coding, *IBM J. Res. Dev.*, Vol. 20, 198–203, 1976.

15. J. Rissanen and G.G. Langdon, Jr., Arithmetic coding, *IBM J. Res. Dev.*, 23, 149–162, 1979.

16. F. Willems, Y. Shtarkov, and T. Tjalkens, The context-tree weighting method: basic properties, *IEEE Trans. Inform. Theory*, 41(3), 653–664, 1995.

17. I.H. Witten, R.M. Neal, and J.G. Cleary, Arithmetic coding for data compression, *Comm. ACM*, 30, 520–540, 1987.

18. S. Xiao and C.G. Boncelet, Jr., On the use of context-weighting in lossless bilevel image compression, *IEEE Trans. Image Proc.*, 2004. Accepted for publication, 2006.

19. J. Ziv and A. Lempel, A universal algorithm for sequential data compression, *IEEE Trans. Inform. Theory*, It-23, 337–343, 1977.

12 Two-Dimensional Shape Coding

Joern Ostermann and Anthony Vetro

CONTENTS

12.1 Introduction 299
    12.1.1 Shape Coding Overview 301
    12.1.2 Related Work 301
        12.1.2.1 Implicit Shape Coding 301
        12.1.2.2 Bitmap-Based Shape Coding 302
        12.1.2.3 Contour-Based Shape Coding 302
        12.1.2.4 MPEG-4 Shape Coding 303
    12.1.3 Chapter Organization 303
12.2 MPEG-4 Shape Coding Tools 303
    12.2.1 Shape Representation 304
    12.2.2 Binary Shape Coding 304
        12.2.2.1 Intra-Mode 305
        12.2.2.2 Inter-Mode 306
        12.2.2.3 Evaluation Criteria for Coding Efficiency 307
    12.2.3 Gray-Scale Shape Coding 307
        12.2.3.1 Objects with Constant Transparency 307
        12.2.3.2 Objects with Arbitrary Transparency 307
    12.2.4 Texture Coding of Boundary Blocks 307
    12.2.5 Video Coder Architecture 308
12.3 Codec Optimization 308
    12.3.1 Preprocessing 309
    12.3.2 Rate–Distortion Models 310
    12.3.3 Rate Control 313
        12.3.3.1 Buffering Policy 313
        12.3.3.2 Bit Allocation 314
    12.3.4 Error Control 314
    12.3.5 Post-Processing 315
        12.3.5.1 Composition and Alpha Blending 315
        12.3.5.2 Error Concealment 315
12.4 Applications 316
    12.4.1 Surveillance 316
    12.4.2 Interactive TV 318
12.5 Concluding Remarks 318
References 319

12.1 INTRODUCTION

With video being a ubiquitous part of modern multimedia communications, new functionalities in addition to the compression as provided by conventional video coding standards like H.261, MPEG-1, H.262, MPEG-2, H.263, and H.264 are required for new applications. Applications like content-based storage and retrieval have to allow access to video data based on object descriptions, where objects are described by texture, shape, and motion. Studio and television postproduction applications require editing of video content with objects represented by texture and shape. For collaborative scene visualization like augmented reality, we need to place video objects into the scene. Mobile multimedia applications require content-based interactivity and content-based scalability in order to allocate limited bit rate or limited terminal resources to fit the individual needs. Security applications benefit from content-based scalability as well. All these applications share one common requirement: video content has to be easily accessible on an object basis. MPEG-4 Visual enables this functionality. The main part of this chapter describes MPEG-4 shape coding, the content-based interactivity enabling tool.

Given the application requirements, video objects have to be described not only by texture, but also by shape. The importance of shape for video objects has been realized early on by the broadcasting and movie industries employing the so-called chroma-keying technique, which uses a predefined color in the video signal to define the background. Coding algorithms like object-based analysis-synthesis coding (OBASC) [30] use shape as a parameter in addition to texture and motion for describing moving video objects. Second-generation image coding segments an image into regions and describes each region by texture and shape [28]. The purpose of using shape was to achieve better subjective picture quality, increased coding efficiency as well as an object-based video representation.

MPEG-4 Visual is the first international standard allowing the transmission of arbitrarily shaped video objects (VO) [21]. Each frame of a VO is called video object plane (VOP) consisting of shape and texture information as well as optional motion information. Following an object-based approach, MPEG-4 Visual transmits texture, motion, and shape information of one VO within one bitstream. The bitstreams of several VOs and accompanying composition information can be multiplexed such that the decoder receives all the information to decode the VOs and arrange them into a video scene; the composition of multiple video objects is illustrated in Figure 12.1. Alternatively, objects may be transmitted in different streams according to a scene description [11,44]. This results in a new dimension of interactivity and flexibility for standardized video and multimedia applications.

Two types of VOs are distinguished. For opaque objects, binary shape information is transmitted. Transparent objects are described by gray-scale α-maps defining the outline as well as the transparency variation of an object.

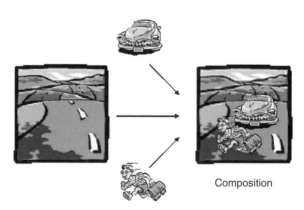

Composition

FIGURE 12.1 Content-based scalability requires individual objects to be transmitted and composited at the decoder. Depending on resources, only some of the objects might be composited to the scene and presented at the terminal.

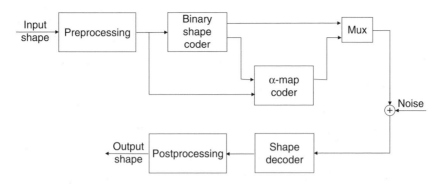

FIGURE 12.2 Processing steps for shape coding considering binary and gray-scale α-maps.

12.1.1 SHAPE CODING OVERVIEW

Figure 12.2 shows the processing steps related to shape coding. They apply to object-based coding systems that transmit shape information only, as well as to systems that transmit texture for the objects. The optional shape preprocessing may remove noise from the shape signal and simplify the shapes such that it can be coded more efficiently. Preprocessing usually depends on the shape coding algorithm employed.

For transparent objects, the preprocessed shape information is separated into a binary shape defining the pels belonging to the object and a gray-scale information defining the transparency of each pel of the object. For binary shapes and gray-scale shape information the binary shape coder codes the shape using lossless or lossy shape coding algorithms. In the case of transparent objects, an α-map coder codes the transparency information for the coded binary shape. The bitstreams get multiplexed, transmitted, and decoded at the decoder. The optional postprocessing algorithm provides error concealment and boundary smoothing.

The receiver decodes the VOs and composes them into a scene as defined by the composition information [11,44]. Typically, several VOs are overlayed on a background. For some applications, a complete background image does not exist. Foreground VOs are used to cover these holes. Often they exactly fit into holes of the background. In case of lossily coded shape, a pixel originally defined as opaque may be changed to transparent, thereby resulting in undefined pixels in the scene. Therefore, lossy shape coding of the background needs to be coordinated with lossy shape coding of the foreground VOs. If objects in a scene are not coded at the same temporal rate and a full rectangular background does not exist, then it is very likely that undefined pixels in the scene will occur. Postprocessing in the decoder may extend objects to avoid these holes.

12.1.2 RELATED WORK

There are three classes of binary shape coders. A *bitmap*-based coder encodes for each pel whether it belongs to the object or not. A *contour*-based coder encodes the outline of the object. In order to retrieve the bitmap of the object shape, the contour is filled with the object label. In the case where there is also texture transmitted with the shape information, an *implicit* shape coder, often referred to as chroma keying [7], can be used, where the shape information is derived from the texture using a predefined color for defining the outside of an object. Similar to texture coding, binary shapes can be coded in a lossless or lossy fashion.

12.1.2.1 Implicit Shape Coding

This class of coding defines a specific pixel value or a range of pixel values as the background of the image. The remaining pels are part of the object. An implicit shape coder is also specified in

GIF89a [9]. For each image, one number can be used to define the value of the transparent pels. All pels of this value are not displayed. Today, GIF89a is used in web applications to describe arbitrarily shaped image and video objects.

12.1.2.2   Bitmap-Based Shape Coding

Such coding schemes are used in the fax standards G4 [4] and JBIG [19]. The modified read (MR) code used in the fax G4 standard scans each line of the document and encodes the location of *changing pels* where the scanline changes its color. In this line-by-line scheme, the position of each changing pel on the current line is coded with respect to either the position of a corresponding changing pel in the reference line, which lies immediately above the present line, or with respect to the preceding changing pel in the current line [31].

12.1.2.3   Contour-Based Shape Coding

Algorithms for contour-based coding and the related contour representations have been published extensively. Different applications nurtured this research: for lossless and lossy encoding of object boundaries, chain coders [12,14] and polygon approximations [15,18,34,42] were developed. For recognition purposes, shape representations like Fourier descriptors were developed to allow translation, rotation, and scale-invariant shape representations [55].

A chain code follows the contour of an object and encodes the direction in which the next boundary pel is located. Algorithms differ by whether they consider a pel having four or eight neighbors for rectangular grids or six neighbors for hexagonal grids. Some algorithms define the object boundary between pels [41]. Freeman [14] originally proposed the use of chain coding for boundary quantization and encoding, which has attracted considerable attention over the last 30 years [23,27,32,38,39]. The curve is quantized using the grid intersection scheme [14] and the quantized curve is represented using a string of increments. Since the planar curve is assumed to be continuous, the increments between grid points are limited to the eight grid neighbors, and hence an increment can be represented by 3 bits. For lossless encoding of boundary shapes, an average 1.2 to 1.4 bits/boundary pel are required [12]. There have been many extensions to this basic scheme such as the generalized chain codes [39], where the coding efficiency has been improved by using links of different length and different angular resolution. In [23], a scheme is presented which utilizes patterns in a chain code string to increase the coding efficiency and in [38], differential chain codes are presented, which employ the statistical dependency between successive links. There has also been interest in the theoretical performance of chain codes. In [27], the performance of different quantization schemes is compared, whereas in [32], the rate–distortion characteristics of certain chain codes are studied. Some chain codes also include simplifications of the contour in order to increase coding efficiency [33,37]. This is similar to filtering the object shape with morphological filters and then coding with a chain code [35]. The entropy coder may code a combination of several directions with just one code word.

A polygon-based shape representation was developed for OBASC [17,18]. As a quality measure, the maximum Euclidean distance d_{max} between the original and the approximated contour is used. During subjective evaluations of CIF (352×288 pels) video sequences, it was found that allowing a peak distance of $d^*_{max} = 1.4$ pel is sufficient to allow proper representations of objects in low bit-rate applications. Hence the lossy polygon approximation was developed. Vertices of the spline approximation do not need to be located on the object boundary [24,26].

This polygon representation can be also used for coding shapes in inter mode. For temporal prediction, the texture motion vectors are applied to the vertices of the previously coded shape defining the predicted shape for the current shape. Then, all vertices within the allowable approximation error d^*_{max} define the new polygon approximation. It is refined as described above such that the entire polygon is within the allowable error d^*_{max}.

In [22], B-spline curves are used to approximate a boundary. An optimization procedure is formulated for finding the optimal locations of the control points by minimizing the mean-squared error between the boundary and the approximation. In [24], a polygon/spline representation is described which provides optimality in the operational rate distortion sense.

Fourier descriptors describing the contour of an object were developed for applications in recognition, where shape is an important key. Fourier descriptors allow translation, rotation, and scale-invariant representation [36]. In the first step, the coordinates of the contour are sampled clockwise in the xy-plane. This list of 2D coordinates (x_i, y_i) is then transformed into an ordered list $(i, (y_{i+1} - y_i / x_{i+1} - x_i))$ with $0 \leq i \leq i + 1$ being the contour point number and $(y_{i+1} - y_i / x_{i+1} - x_i)$ being the change in direction of the contour. Since the samples are periodic over the object boundary perimeter, they can be expanded into a Fourier series. In order to preserve the main characteristics of a shape, only the large Fourier coefficients have to be maintained. Fourier descriptors are not very efficient in reconstructing polygon-like shapes with only a few coefficients. This is one of the reasons, why they never became very competitive in coding efficiency.

12.1.2.4 MPEG-4 Shape Coding

The MPEG committee investigated implicit, bitmap-based, and contour-based shape coding techniques. Implicit shape coding requires high-quality texture coding in order to derive a high-quality shape. Since humans tend to be more sensitive to shape distortions than to texture distortions, this coupling of shape and texture distortion is not acceptable for many applications. Therefore, while many different approaches for shape coding were evaluated during the development of MPEG-4, proposals for polygon-based contour coding and binary context-adaptive arithmetic coding were the lead contenders. Ultimately, MPEG-4 adopted the binary context-adaptive arithmetic coding, which is elaborated further in the next section.

The publication of the MPEG-4 standard and its work on shape coding [3,26,33] inspired the invention of many shape coding algorithms that are able to outperform MPEG-4 shape coding in terms of coding efficiency while being more computationally demanding. Mainly due to rate distortion optimization, the vertex-based method described in [24] provides an average rate reduction of 7.8% with respect to the content-based arithmetic coder in MPEG-4. The skeleton-based method proposed in [54] gives bit rates 8 to 18% smaller than the MPEG-4 shape coder. In [1], digital straight lines are identified on the contour. These lines are then used to align a template for defining the context of an adaptive arithmetic encoder with the line. Although this algorithm only codes shapes in intra-mode, it requires 33% less bits than the MPEG-4 shape coder in inter-mode as described in Section 12.2.

12.1.3 Chapter Organization

This chapter provides an overview of the algorithms for shape coding in MPEG-4. In Section 12.2, binary and gray-scale shape coding techniques are first described. Then, in Section 12.3, a variety of issues related to the encoding, modeling, and postprocessing of shape are discussed. Section 12.4 presents a few promising application that rely on shape coding, and finally, Section 12.5 presents concluding remarks.

12.2 MPEG-4 SHAPE CODING TOOLS

This section will discuss the coding techniques for 2D shapes using MPEG-4 tools. First, we briefly introduce the general representation for the 2D shape of an object, including binary and gray-scale shape. Then, we describe techniques for coding each of these representations.

FIGURE 12.3 Comparison of scene composed with background image and foreground object with constant and gray-scale transparency. Top-left: background image; top-right: foreground object; bottom-left: scene composed with constant transparency; bottom-right: scene composed with gray-scale transparency, where the α-map of the foreground object is a horizontal ramp starting with $m_k(0, y) = 0$ on the left side and ending at $m_k(320, y) = 255$ on the right side.

12.2.1 SHAPE REPRESENTATION

The 2D shape of an object is defined by means of an α-map M_k of size XY pels:

$$M_k = \{m_k(x, y) \mid 0 \le x \le X, 0 \le y \le Y\}, \quad 0 \le m_k \le 255 \tag{12.1}$$

The shape M_k defines for each pel $\mathbf{x} = (x, y)^T$ whether it belongs to the object ($m_k(\mathbf{x}) > 0$) or not ($m_k(\mathbf{x}) = 0$). For an opaque object, the corresponding α-values are 255, and for transparent objects they range from 1 to 255. Usually, the α-map has the same spatial and temporal resolution as the luminance signal of the video sequence. In video-editing applications, the α-map is used to describe object shape and object transparency. Let us assume that we have a background image $s_b(\mathbf{x})$, the object represented by image $s_o(\mathbf{x})$, and the α-map $M_o(\mathbf{x})$. Overlaying the object on the background is done according to

$$s(\mathbf{x}) = \left(1 - \frac{M_o(\mathbf{x})}{255}\right) s_b(\mathbf{x}) + \frac{M_o(\mathbf{x})}{255} s_o(\mathbf{x}) \tag{12.2}$$

As shown in Figure 12.3, the amplitude of the α-map determines how visible the object becomes. We will describe the coding of binary object shapes, i.e., $m_k(\mathbf{x}) \in \{0, 255\}$ in the next subsection, followed by description of the coding of gray-scale shape.

12.2.2 BINARY SHAPE CODING

The MPEG-4 shape coder is known as the context-based arithmetic encoder (CAE) [2,3,24]. It works on macroblocks of size 16×16 pels that the MPEG-4 video encoder defines for a video object. In the following paragraphs, shape encoding in intra-mode is described. Then, this technique is extended to include an inter-mode. The evaluation criteria for lossy binary shape coding are presented last.

12.2.2.1 Intra-Mode

The CAE codes pelwise information only for boundary blocks exploiting the spatial redundancy of the binary shape information to be coded. Pels are coded in scan-line order and row by row. In intra-mode, three different types of macroblocks are distinguished: transparent and opaque blocks are signaled as macroblock type. The macroblocks on the object boundary containing transparent as well as opaque pels belong to the third type. For these boundary macroblocks, a template of 10 pels is used to define the causal context for predicting the shape value of the current pel as shown in Figure 12.4a. For encoding the state transition, a context-based arithmetic encoder is used. The probability table of the arithmetic encoder for the 1024 contexts was derived from several sequences. With 2 bytes allocated to describe the symbol probability for each context, the table size is 2048 bytes.

The template extends up to 2 pels to the left, to the right, and to the top of the pel to be coded. Hence, for encoding the pels in the two top and left rows of a macroblock, parts of the template are defined by the shape information of the already transmitted macroblocks on the top and on the left side of the current macroblock. For the two rightmost columns, each undefined pel of the context is set to the value of its closest neighbor inside the macroblock.

In order to increase coding efficiency as well as to allow lossy shape coding, a macroblock may be subsampled by a factor of 2 or 4 resulting in a subblock of size 8 × 8 pels or 4 × 4 pels, respectively. The subblock is encoded using the encoder as described above. The encoder transmits to the decoder the subsampling factor such that the decoder decodes the shape data and then up-samples the decoded subblock to the original macroblock size. Obviously, encoding the shape using a high subsampling factor is more efficient but the decoded shape after up-sampling may or may not be the same as the original shape. Hence, this subsampling is mostly used for lossy shape coding.

Depending on the up-sampling filter, the decoded shape can look somewhat blocky. During the MPEG evaluation process, two filters have been found to perform very well: a simple pel replication filter combined with a 3 × 3 median filter and slightly better performing adaptive nonlinear up-sampling filter. MPEG-4 decided to adopt the adaptive filter with a context as shown in Figure 12.5 [21]. The value of an up-sampled pel is determined by thresholding a weighted sum of the pels of its context.

The efficiency of the shape coder differs depending on the orientation of the shape data. Therefore, the encoder can choose to code the block as described above or transpose the macroblock prior to arithmetic coding.

Within the MPEG-4 standardization process, the quality of shape coding was controlled by a threshold, AlphaTH, which defines the maximum number of erroneously coded pels in a macroblock without considering effects like change in connectivity, impact on overall object shape, temporal shape

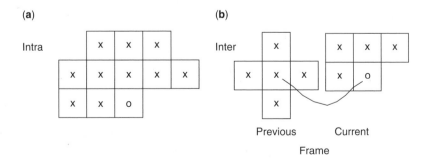

FIGURE 12.4 Templates for defining the context of the pel to be coded (o), where (a) defines the intra-mode context and (b) defines the inter-mode context. The alignment is done after motion compensating the previous VOP.

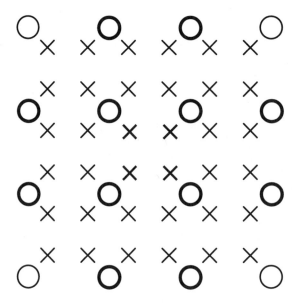

FIGURE 12.5 The up-sampled pels (×) lie between the locations of the transmitted pels (o) from the subsampled macroblock shape information. Neighboring pels (bold o) defining the values of the pels to be up-sampled (bold ×).

variation or the look of the object including its texture. Therefore, lossy shape coding achieves better results if these effects are considered.

12.2.2.2 Inter-Mode

In order to exploit temporal redundancy in the shape information, the coder described above is extended by an inter-mode requiring motion compensation and a different template for defining the context.

For motion compensation, a 2D integer pel motion vector is estimated using full search for each macroblock in order to minimize the prediction error between the previous coded VOP shape M'_{k-1} and the current shape M_k. The shape motion vectors are predictively encoded with respect to the shape motion vectors of neighboring macroblocks. If no shape motion vector is available, texture motion vectors are used as predictors. The shape motion vector of the current block is used to align a new template designed for coding shape in inter-mode as shown in Figure 12.4b.

The template defines a context of 9 pels resulting in 512 contexts. The probability for one symbol is described by 2 bytes giving a probability table size of 1024 bytes. Four pels of the context are neighbors of the pel to be coded, 5 pels are located at the motion-compensated location in the previous VOP. Assuming that the motion vector $(d_x, d_y)^T$ points from the current VOP_k to the previous coded VOP'_{k-1}, the part of the template located in the previously coded shape is centered at $m'_{k-1}(x - d_x, y - d_y)$ with $(x, y)^T$ being the location of the current pel to be coded.

In inter-mode, the same options as in intra-mode are available, such as subsampling and transposing. For lossy shape coding, the encoder may also decide that the shape representation achieved by simply carrying out motion compensation is sufficient thus saving bits by avoiding to code the prediction error. The encoder can select one of the seven modes for the shape information of each macroblock: transparent, opaque, intra, inter with or without shape motion vectors, and inter with or without shape motion vectors and prediction error coding. These different options with optional subsampling and transposition allow for encoder implementations of different coding efficiency and implementation complexity.

12.2.2.3   Evaluation Criteria for Coding Efficiency

In order to compare the performance of different shape coders, evaluation criteria have to be defined. Within MPEG-4, there are two quality measures for objectively assessing the quality of coded shape parameters. One is the maximum of the minimal Euclidean distance d_{max}^* (peak deviation) between each coded contour point and the closest contour point on the original contour. This measure allows for an easy interpretation of the shape quality. However, if lossy shape coding results in changing the topology of an object due to opening, closing, or connecting holes, the peak deviation d_{max}^* is not a useful measure. Therefore, a second measure d_n was used, which is the number of erroneously represented pels of the coded shape divided by the total number of pels belonging to the original shape. Since different objects can have very different ratios of contour pels to interior pels, a given value for d_n only allows to compare with other d_n of different approximations of the same video object. d_n by itself does not provide sufficient information about the shape quality. Hence, some evaluations are done just providing the number of erroneously coded pels.

It was found that the objective measures truthfully reflect subjective quality when comparing different bitmap-based shape coders or when comparing different contour-based shape coders. For lossy shape coding, the bitmap-based shape coders create blocky object shapes whereas contour-based shape coders create an object shape showing polygon edges. Since the two classes of shape coders give different results, a comparison between these two classes has to be done subjectively.

12.2.3   GRAY-SCALE SHAPE CODING

Gray-scale α-maps allow 8 bits for each luminance pel to define the transparency of that pel. As shown in Figure 12.3, transparency is an important tool for composing objects into scenes and special effects. Two types of transparencies are distinguished: binary α-maps with objects of constant transparency and arbitrary α-maps for objects with varying transparency.

12.2.3.1   Objects with Constant Transparency

For a transparent object that does not have a varying transparency, the shape is encoded using the binary shape coder and the 8-bit value of the α-map. In order to avoid aliasing, gray-scale α-maps usually have lower transparency values at the boundary. Blending the α-map near the object boundary can be supported by transmitting the coefficients of a 3×3 pel FIR filter that is applied to the α-map within a stripe on the inner object boundary. The stripe can be up to 3 pels wide.

12.2.3.2   Objects with Arbitrary Transparency

For arbitrary α-maps, shape coding is done in two steps [5,7]. In the first step, the outline of the object is encoded as a binary shape. In the second step, the actual α-map is treated like the luminance of an object with binary shape and coded using the MPEG-4 texture coding tools: motion compensation, DCT, and padding. The passing extrapolates the object texture for the background pels of a boundary block.

12.2.4   TEXTURE CODING OF BOUNDARY BLOCKS

For motion-compensated prediction of the texture of the current VOP, the reference VOP is motion-compensated using block motion compensation. In order to guarantee that every pel of the current

VOP has a value to be predicted from, some or all of the boundary and transparent blocks of the reference VOP have to be padded. Boundary blocks are padded using repetitive padding as follows. First, boundary pels are replicated in the horizontal direction, and then they are replicated in the vertical direction. If a value can be assigned to a pel by both padding directions, then an average value is assigned to that pel. Since this repetitive padding puts a significant computational burden on the decoder, a simpler mean padding is used in the second step. Transparent macroblocks bordering boundary blocks are assigned to an average value determined by the pels of its neighboring padded blocks.

In order to encode the texture of a boundary block, MPEG-4 treats the macroblock as a regular macroblock and encodes each block using an 8×8 DCT. The texture is decoded using conventional processing, then discards all information that falls outside of the decoded shape. In order to increase coding efficiency, the encoder can choose the texture of pels outside of the object such that the bit rate is minimized. This non-normative process is also called padding [25]. For intra-mode, a low-pass extrapolation filter was developed, while for inter-mode, setting these pels to 0 was found to perform well in terms of coding efficiency.

12.2.5 VIDEO CODER ARCHITECTURE

Figure 12.6a shows the block diagram of an object-based video coder [35]. In contrast to the block diagram shown in the MPEG-4 standard, this diagram focuses on the object-based mode in order to allow a better understanding of how shape coding influences the encoder and decoder. Image analysis creates the bounding box for the current VOP s_k and estimates texture and shape motion of the current VOP with respect to the reference VOP s_{k-1}. Shape motion vectors of transparent macroblocks are set to 0. Parameter coding encodes the parameters predictively. The parameters get transmitted and decoded, and the new reference VOP is stored in the VOP memory and also handed to the compositor of the decoder for display.

The increased complexity due to the coding of arbitrarily shaped video objects becomes evident in Figure 12.6b, which shows a detailed view of the parameter coding. The parameter coder first encodes the shape of the boundary blocks using shape and texture motion vectors for prediction. Then, shape motion vectors are coded. The shape motion coder knows which motion vectors to code by analyzing the possibly lossily encoded shape parameters. For texture prediction, the reference VOP is padded as described above. The prediction error is then padded using the original shape parameters to determine the area to be padded. Using the original shape as a reference for padding is again an encoder choice not implemented in the MPEG-4 Reference Software [20] (MoMuSys version). Finally, the texture of each macroblock is encoded using DCT.

12.3 CODEC OPTIMIZATION

The shape coding tools described in the previous subsection provide the basic techniques that are used to efficiently represent shape data. In order to use these tools most effectively, optimizations at various points in the encoder and decoder must be considered. Several key issues are outlined below and discussed further in this section.

Prior to encoding, preprocessing of shape data may be helpful to simplify the information to be coded, while still providing an accurate representation of the object. During the encoding, rate control is needed at the encoder to allocate bits to achieve a maximum quality subject to rate and buffer constraints. Rate–distortion (R–D) models of the shape (and texture) data may be used to alleviate some of the computations involved in making optimal coding decisions. Furthermore, error control may be used to minimize data loss incurred during transmission and error propagation in the reconstructed data.

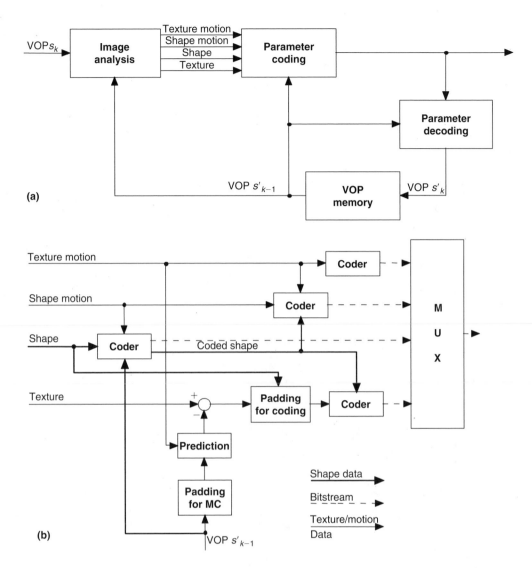

FIGURE 12.6 Block diagram of the video encoder (a) and the parameter coder (b) for coding of arbitrarily shaped video objects.

In addition to the the above encoder issues, postprocessing techniques after decoding also play an important role in reconstructing the object and scene data, and the scene. For one, if errors during transmission have corrupted the reconstructed data, then some form of error concealment should be applied. Also, given multiple objects in a scene, composition of these objects is also required.

12.3.1 PREPROCESSING

The decoder has no possibility to find out whether the encoder uses lossless or lossy shape coding and what shape coding strategy the encoder uses or padding algorithm for coding is used. In its reference implementation of a video coder, MPEG-4 chose to control lossy shape coding by using an AlphaTH threshold. This threshold defines the maximum number of incorrectly coded pels within a boundary block and allows the topology of the shape to change. Often, isolated pels at the object boundary are coded as part of the object. Using morphological filters to smooth object boundaries provide a much more predictable quality of a lossily encoded shape [35].

12.3.2 RATE–DISTORTION MODELS

For texture coding, a variety of models have been developed that provide a relation between the rate and distortion, e.g., [8,16]. These models are most useful for rate control of texture information. Given some bit budget for an object or frame, one can determine a quantizer value that meets a specified constraint on the rate. Additionally, such models can be used to analyze the source or sources to be encoded in an effort to optimize the coding efficiency in a computationally efficient way.

In the case of shape coding, analogous models have been developed, mainly for use in the analysis stage of an object-based encoder. The primary motivation to develop such models is to avoid performing actual coding operations to obtain the *R–D* characteristics of the binary shape. In the case of MPEG-4 shape coding, this involves down-sampling and arithmetic coding operations to obtain the rate, and up-sampling and difference calculations to compute the distortion at various scales. Figure 12.7 shows sample images for binary shapes at different resolutions and after up-sampling. Accurate estimates of the *R–D* characteristics can play a key role in optimizing the bit

FIGURE 12.7 Sample test shapes from the *Dancer*, *Coastguard*, and *Foreman* sequences. The first row shows the original shapes at full resolution. The second and third rows show the shapes down-sampled by factors 2 and 4 according to the MPEG-4 standard. The bottom row shows the up-sampled reconstructed shapes from quarter-scale images (*Dancer* and *Coastguard*) and half-scale image (*Foreman*) according to the MPEG-4 standard.

allocation for the binary shape among blocks and between shape and texture coding with much lower complexity. Also, knowing the expected rate for shape can stabilize the buffer occupancy of a video coder, especially for low bit-rate video coding [50].

In the following, we review several approaches that have been proposed for modeling the R–D characteristics of binary shape and discuss the performance of each. Throughout this section, we aim to model the characteristics of the MPEG-4 CAE-based shape coder described in Section 12.2. Also, we assume shapes are coded in intra-mode only, unless otherwise specified.

For the purpose of this subsection, the modeling problem is formally stated as follows. Let $(R, D)_k$ denote the R–D values for a binary shape that is coded at resolution k. Using input parameters, θ_i, which represent features extracted from the shape data, and modeling function, $f(\cdot)$, we consider the approximation,

$$(R, D)_k \approx f(\theta_i) \tag{12.3}$$

The first attempt to solve this problem attempted to categorize all possible binary patterns over a 2×2 neighborhood into N states [52]. This approach was referred to as state partitioning. As expected, the rate is predicted well by this approach, but the distortion suffers from large prediction error. The reason for this is because the rate could be accurately modeled from a fixed 2×2 neighborhood, such that the 10-bit states used by CAE can be correctly collapsed into one of the available states. The distortion, on the other hand, is not modeled so accurately because the actual up-sampling process uses a 12-pel neighborhood and estimating distortion based on the 2×2 pixels is not sufficient.

An improved probabilistic approach for modeling the R–D characteristics of binary shape was proposed in [53]. In this work, the shape model is based on the statistics (or moments) that one can extract from the data. With this approach, the aim was to have a distribution whose samples resemble the type of data that we are trying to code. At the same time, this model should be able to make distinctions between different shapes at the same scale, and also between the same shape at different scales.

In [10], a Markov random field (MRF) model that is able to represent the fine structures of an image is presented. The model relies on three parameters: edge, line, and noise. It has been shown in [53] that the moments of this model, which are considered sufficient statistics of the distribution, exhibit favorable properties for modeling as outlined above.

To estimate the rate and distortion at various scales, a simple multilayer feed-forward network has been proposed in [53], where the input to this network are the statistical moments of the MRF model that are calculated from the original shape image, and the output is the estimated rate and distortion of the shape data at different scales.

The plots in Figure 12.8 provide a comparison of the actual rates and output rates of the neural network for each of the three scales. In these plots, the first 100 frames correspond to the training set and the next 75 frames correspond to the testing set. We can see that that the neural network does indeed provide close estimates to the actual rate and follows the trends quite well for both the training and testing data. Similarly, the plots in Figure 12.9 provide a comparison of the actual distortions and output distortions of the neural network for the reconstructed binary maps from half and quarter scales. It should be noted that the absolute errors plotted in Figure 12.9 vary significantly in some cases due to size or complexity of the shape boundary. If these errors were normalized to the object size, the relative errors would be much closer.

While the results shown here demonstrate that MRF parameters can be used to model multiscale rate–distortion characteristics of binary shape, and do so with information provided at the full resolution only, the above method has only been applied to intra-coded shapes. In [6], a linear R–D model has been proposed based on parameters derived from the boundary blocks and a block-based shape complexity for the video object. Besides being much simpler than a neural network-based prediction, the linear model is applied to both intra- and inter-coded shape.

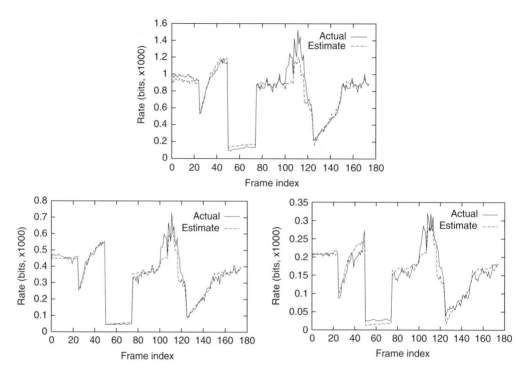

FIGURE 12.8 Comparison of actual rates and output rates of neural network from [53] at full-scale (top), half-scale (bottom-left), and quarter-scale (bottom-right). The first four clips (corresponding to the first 100 frames) are part of the training set, while the following three clips (corresponding to the remaining 75 frames) are part of the test set.

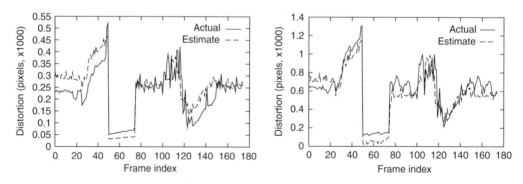

FIGURE 12.9 Comparison of actual distortions and output distortions of neural network from [53]. Distortion is measured between original and reconstructed binary map from half-scale (left) and quarter-scale (right). The first four clips (corresponding to the first 100 frames) are part of the training set, while the following three clips (corresponding to the remaining 75 frames) are part of the test set.

As stated in [6], the rate model is given by

$$\tilde{R}_i = a_i n + b_i \tag{12.4}$$

where n is the number of boundary blocks, a_i and b_i are model parameters, and i denotes the resolution scale. Linear regression is used to estimate the model parameters based on past data points.

Similar to the rate model, the distortion model is given by

$$\tilde{D}_i = c_i n + d_i \tag{12.5}$$

where c_i and d_i are model parameters, which are also calculated using linear regression on past data points. However, it has been observed that more complex boundaries will always produce more distortion, therefore a shape complexity measure based on a normalized perimeter has been introduced:

$$\kappa = \frac{\sum_{l=1}^{n} p_l}{S} \tag{12.6}$$

where p_l denotes the perimeter of each boundary block and S the number of nontransparent pixels in the shape. This complexity measure is factored into the distortion model in a multiplicative way such that the new distortion model becomes

$$\tilde{D}_i = c_i \kappa n + d_i \tag{12.7}$$

Simulation results in [6] confirm the accuracy of the above approach for both intra- and inter-coded shapes at various levels of resolution.

12.3.3 RATE CONTROL

In object-based coding, the problem of coding additional shape information becomes critical at lower bit rates. At higher bit rates, the shape bits occupy a much smaller percentage of the overall rate. In the following, we first describe a buffering policy to account for the additional shape information in object-based coding. With object-based coding, there is also some flexibility in allocating rate among texture and shape information, so we also discuss bit allocation among texture and shape. In both cases, the bits used for shape are estimated from the R–D models described in Section 12.3.2. Finally, we discuss how rate control decisions for different objects may impact the overall composition of a scene.

12.3.3.1 Buffering Policy

In frame-based rate control, the bits used to encode a frame, T_c, are added to the buffer. Let the buffer occupancy at time t be denoted by $B(t)$. To help ensure that the buffer will not overflow, a frameskip parameter, N, is set to zero and incremented until the following buffer condition is satisfied:

$$B(t) < \gamma B_s \tag{12.8}$$

where B_s is the size of the buffer, and the value of γ denotes a buffer margin having a typical value of 0.8. In the above, the updated buffer level is given by

$$B(t) = B(t - \tau) + T_c - T_d(N + 1) \tag{12.9}$$

where $B(t - \tau)$ denotes the buffer level of the previously coded frame and T_d is the amount of bits drained from the buffer at each coding time instant.

In object-based coding, the buffering policy must account for the high percentage of shape bits at low bit rates. Let the estimated number of bits to code the shape for all objects in the next frame be denoted by T_{shape}. Then, to ensure that there are a sufficient number of bits to code the texture in the next frame, the buffer condition given by Eq. (12.8) is slightly modified as

$$B(t) + T_{shape} < \gamma B_s \tag{12.10}$$

Another way to view the modified buffer condition is that the buffer margin is adaptively lowered according to the estimated bits for shape. In this way, if a relatively large number of bits are estimated for the shape, the rate control can account for these additional bits by skipping frames and allowing the buffer to drain. This will allow a sufficient number of bits to code all the information in the next frame, without compromising the spatial quality more than necessary.

12.3.3.2   Bit Allocation

Generally speaking, bit allocation aims to minimize distortion subject to rate and buffer constraints. In an object-based coding framework, allocation of bits used for both shape and texture data could be performed in a joint manner. However, since the distortion metrics for shape and texture are different, it is not straightforward to derive a single cost function that accounts for both metrics. It should be noted that binary shape distortion is usually defined as the ratio of error pixels to the total number of nontransparent pixels, while texture distortion is defined according to mean-squared error between original and reconstructed pixel values.

Let the total rate budget for an object be given by $R = R_t + R_s$, where R_t are the bits allocated for texture (including motion) and R_s are the bits allocated for shape. Given this allocation, the optimal texture and shape coding modes for each block in the object could be determined separately using, for example, conventional Lagrangian multiplier techniques. If R–D models are used to estimate the operating points for texture and shape at various quantizer values and conversion ratios, respectively, then some of the computational burden could be alleviated. The problem is then reduced to one of allocating the total bits for an object among R_t and R_s. This problem may also be posed at the frame level in which the optimization is conducted for all objects in a scene.

The main difficulty with the above approach is allocating bits between shape and texture. To overcome this problem, a practical approach has been proposed in [50], where rather than allocating a specific number of bits to shape, the maximum distortion for shape is controlled through the AlphaTH threshold instead. In this way, the shape will consume some portion of the total rate budget for a frame subject to a maximum distortion constraint and the remaining bits would be allocated to the texture. In [50], a method to dynamically adapt the AlphaTH threshold according to the buffer fullness and quality of the texture encoding has been proposed.

12.3.4   Error Control

It is largely recognized that intra-refresh can play a major role in improving error resilience in video coding systems. This technique effectively minimizes error propagation by removing the temporal relation between frames. Although the intra-refresh process decreases the coding efficiency, it will significantly improve error resilience at the decoder, which increases the overall subjective impact in the presence of errors in the transmission.

A key aspect to consider in applying intra-refresh schemes is to determine which components of the video to refresh and when. For object-based video, both shape and texture data must be considered. In [45], shape refreshment need (SRN) and texture refreshment need (TRN) metrics have been proposed. The SRN is defined as a product of the shape error vulnerability and the shape concealment difficulty, while the macroblock-based TRN is similarly defined as a product of the texture error vulnerability and the texture concealment difficulty. In both cases, the error vulnerability measures the statistical exposure of the shape/texture data to channel errors and the concealment difficulty expresses how difficult the corrupted shape/texture data is to recover when both spatial and temporal error concealment techniques are considered.

Based on the above refreshment need metrics, novel shape and texture intra-refreshment schemes have been proposed in [46]. These schemes allow an encoder to adaptively determine when the shape and texture of the various video objects in a scene should be refreshed in order to maximize the decoded video quality for a certain total bit rate.

12.3.5 POST-PROCESSING

Two specific needs for postprocessing shape data are considered in this subsection. First, given that the shape data defines objects in a scene, composition and alpha-blending techniques are required to generate a complete scene consisting of multiple objects. Second, considering that errors may occur during transmission, methods for error concealment become necessary. Both of these operations are performed at the receiver and will be discussed further below.

12.3.5.1 Composition and Alpha Blending

As mentioned earlier, composition issues may arise due to lossy coded shape or the coding of multiple objects at different temporal rates. In [50], undefined pixels due to lossy shape coding were simply assigned a gray value. Since maximum distortion in a block was controlled using AlphaTH, which only took values in the range [0, 64], the impact on visual quality was not noticeable. With higher levels of shape distortion, more sophisticated methods would be required to recover the pixel values for the texture. In [29], object coding with different temporal rates was considered. In this work, undefined pixels were assigned values from the reconstructed object with minimum distance to the undefined pixel coordinate.

12.3.5.2 Error Concealment

Concealment techniques could be divided into two categories: temporal and spatial. Temporal concealment techniques rely on shape information from previous time instants to do the concealment, while spatial concealment techniques use information only from the current time instant. Several techniques for spatial and temporal concealment are outlined below.

The earliest known attempt to address this concealment problem for shape was proposed in [13], where the idea of extending conventional motion-compensated concealment techniques for texture to shape was explored. In this way, when a given block of shape data is corrupted, the decoder conceals it by copying a block of shape data from the previous time instant. This could be achieved by simply copying the co-located block of shape data from the previous time instant. Alternatively, a motion vector estimate may be obtained such that the corrupted block in the current time instant is replaced by a displaced shape block from the previous time instant. Temporal concealment was further investigated in [40] considering that the boundary of the shape data of a given video object does not change significantly in time and thus these changes can be described by a global motion model. Based on this assumption, the global motion parameters are estimated at the encoder and sent along with the encoded bitstream to the decoder. With this method, the decoder would apply global motion compensation using parameters derived at the encoder to restore the corrupted contour, then fill in the concealed contour to recover the entire shape. An extension of this approach was proposed in [48], which eliminates the need for an encoder to send additional information by performing the estimation of global motion parameters at the decoder using available shape data. Additionally, this approach improves performance by adding a motion refinement step to better deal with shapes that have some local motion.

The above temporal concealment techniques are advantageous since there is access to past information that can significantly enhance the recovery of shape data, especially in cases where the shape does not change much. However, when the shape changes significantly over time or concealment is applied to still or intra-coded shapes, spatial concealment techniques should be used instead.

Spatial shape concealment was first addressed in [43]. In this work, the shape was modeled with a binary MRF and concealed based on maximum *a posteriori* (MAP) estimation. According to this algorithm, each missing shape element in a missing block is estimated as a weighted median of the neighboring shape elements that have been correctly decoded or concealed, where the weights are assigned adaptively based on the likelihood of an edge in that direction. For each missing shape

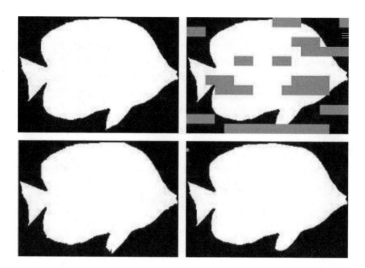

FIGURE 12.10 Performance comparison of error concealment techniques. Top-left: original shape image; top-right: corrupted shape image with 25% of total blocks lost; bottom-left: shape image recovered using method of Shirani et al. [43]; bottom-left: shape image recovered using method of Soares and Pereira [47].

block, this procedure is iteratively repeated until the algorithm converges. Additionally, if several consecutive missing blocks have to be concealed, the procedure is recursively applied to all the missing shape blocks. This method obviously has some computational burdens, but may also suffer in concealing isolated blocks since only local statistics of the data are considered. In an attempt to overcome these drawbacks, a method that interpolates the missing contours based on the available surrounding contours using Bezier curves has been proposed in [47]. These cubic parametric curves are fully determined by four points, which include two end points at the boundaries of a lost block and two additional control points within the region of the lost block. A method for determining the control points that ensure certain smoothness and continuity constraints has been presented. Figure 12.10 compares the performance between the concealment method of Shirani et al. [43] and that of Soares and Pereira [47]. Besides providing a lower complexity solution, the method by Soares and Pereira provides an improved recovery of the shape data that is characterized by a more accurate representation of the original shape and fewer artifacts.

12.4 APPLICATIONS

This section describes how shape coding and object-based video coding, in general, could be applied for surveillance and interactive TV applications.

12.4.1 Surveillance

Here we describe a surveillance application system that utilizes object-based coding techniques to achieve efficient storage of video content [49]. In this system, certain inaccuracies in the reconstructed scene can be tolerated; however, subjective quality and semantics of the scene must be strictly maintained. As shown in Figure 12.11, the target of this system is the long-term archiving of surveillance video, where several months of video content from multiple cameras would need to be stored.

One of the major advantages of object-based coding is that each object can vary in its temporal quality. However, in terms of rate–distortion metrics, only minimal gains in coding efficiency could be achieved [51] since the composition problem typically limits the amount of temporal frame skipping

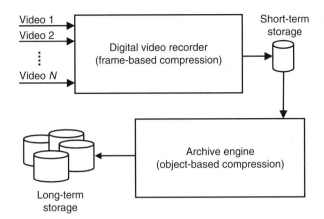

FIGURE 12.11 Application system for surveillance employing object-based coding for long-term archive of video content.

FIGURE 12.12 Sample reconstructed frame of *festA* sequence. Left: frame-based reconstruction; right: object-based reconstruction.

of the background object. Fortunately, in the surveillance system being discussed here, obtaining the full background without any foreground objects is not a problem. Also, subjective video quality is the main consideration.

In this system, a single background image is compressed using frame-based coding, and the sequence of segmented foreground objects are compressed using object-based coding; the background image is simply repeated for each reconstructed frame. Performing the object-based compression in this way gives rise to differences in the background pixels, especially for outdoor scenes that, for example, have swaying trees and objects due to wind conditions.

To demonstrate the effectiveness of this approach, the above object-based coding methodology is applied to several surveillance test sequences. Sample reconstructed frames of one sequence using frame- and object-based coding are shown in Figure 12.12. In this test sequence, wind is blowing the striped curtain, tree branches, and hanging ornaments; all of which are coded and accurately represented by the frame-based coding result. However, for the object-based coding result, these moving background elements are not recorded and only the moving foreground object is coded at each time instant. Semantically, however, these sample frames are equivalent. Table 12.1 summarizes the

TABLE 12.1
Comparison of Storage Requirements (in KB)
for Frame-Based and Object-Based Coding

Sequence	festA	festB	festC	festD	rain	tree
Frame-based	178	173	35	116	78	43
Object-based	17	18	13	18	22	10
% Savings	90.5	89.6	62.9	84.5	71.8	76.7

Note: Background image for object-based coding results are
included (fest: 9KB; rain: 4KB; tree: 4KB).

storage requirements over a broader range of sequences; it is clear that object-based coding provides favorable savings in the bits required to store the compressed video sequences.

12.4.2 INTERACTIVE TV

With the increasing demand for access to information for handicapped people, TV has to become accessible for the hearing impaired. Overlaying a person signing the spoken words over the video will enable this functionality. Transmitting this person using object-based coding and overlaying this object over the regular TV program enables the user to select whether this optional video object should be presented in order to not diminish the presentation for the nonimpaired viewers.

MPEG-4 shape coding can also be used for transmitting shape information independent of the video signal. This allows for transmitting a map of labels that can be used for interactive TV. If the user moves a cursor with the remote control over the video and clicks, the action to be performed is determined by the label of the map corresponding to the position of the cursor. This action can select items for teleshopping or request background information on the object or person identified.

Object-based video can also be used for overlaying news reporters over live footage at the TV receiver instead of at the broadcast studio. This is of special interest for Internet TV, where the same news story might require reporters talking in different languages. In this scenario, the background video is sent as an rectangular-shaped video object on one channel and the news reporter with the desired language represented as an arbitrarily shaped video object is selected on the second channel. The receiver composes these objects into one video.

Multipoint video conference systems would like to present the remote participants in an homogeneous environment to the local participant. This can be easily achieved when each participant is represented as an arbitrarily shaped video object. Scene composition can arrange the video objects on a common background and present the result to the local participant.

12.5 CONCLUDING REMARKS

This chapter has described the shape coding tools that have been adopted in the MPEG-4 coding standard. The context-based arithmetic encoder uses a template to define the context for coding the current pel of a binary shape bitmap. If objects have arbitrary transparency, then these values are coded using MPEG-4 tools like DCT and motion compensation. The integration of a shape coder with texture coding for object-based video coding requires use of texture extrapolation also known as padding. Lossy shape coding may create pels with undefined texture requiring the use of similar extrapolation techniques as well as concealment techniques. For video coding, binary shape coding is enabled in the MPEG-4 core, main and enhanced coding efficiency profiles, while gray-scale shape coding is only enabled in the main and enhanced coding efficiency profiles.

Beyond the fundamental shape coding techniques, we have discussed several issues related to encoder optimization, including R–D modeling of binary shape, rate control, and error control. A linear R–D model has shown to be effective in predicting rate and distortion of shape at various resolutions. Based on such models, we have reviewed buffering policies for maintaining a stable buffer as well as bit allocation techniques to distribute the rate among shape and texture. Furthermore, error control techniques that attempt to minimize error propagation in the decoder based on refreshment need metrics have been described.

Postprocessing of shape data has also been covered in this chapter. In particular, an overview of various error-concealment techniques for the recovery of lost data during transmission has been described. Temporal concealment techniques are advantageous since there is access to past information to improve recovery results; however, spatial concealment is still needed in cases where the shape changes significantly over time or for concealment of still or intra-coded images.

Finally, two promising applications of shape coding have been presented, including the use of object-based coding for long-term archiving of surveillance video resulting in bit savings between 60 and 90%, and interactive television.

REFERENCES

1. S.M. Aghito and S. Forchhammer, Context based coding of binary shapes by object boundary straightness analysis, *IEEE Data Compression Conference*, UT, USA, March 2004.
2. N. Brady, MPEG-4 standardized methods for the compression of arbitrarily shaped objects, *IEEE Trans. Circuits Syst. Video Technol.*, 9 (1999) 1170–1189.
3. N. Brady and F. Bossen, Shape compression of moving objects using context-based arithmetic coding, *Signal Process. Image Commun.*, 15 (2000) 601–618.
4. CCITT, Facsimile Coding Schemes and Coding Functions for Group 4 Facsimile Apparatus, CCITT Recommendation T.6, 1994.
5. W. Chen and M. Lee, Alpha-channel compression in video coding, *ICIP 97*, Special session on shape coding, Santa Barbara, 1997.
6. Z. Chen and K. Ngan, Linear rate-distortion models for MPEG-4 shape coding, *IEEE Trans. Circuits Syst. Video Technol.*, 14 (2004) 869–873.
7. T. Chen, C.T. Swain, and B.G. Haskell, Coding of subregions for content-based scalable video, *IEEE Trans. Circuits Syst. Video Technol.*, 7 (1997) 256–260.
8. T. Chiang and Y.-Q. Zhang, A new rate control scheme using quadratic rate-distortion modeling, *IEEE Trans. Circuits Syst. Video Technol.*, 7 (1997) 246–250.
9. Compuserve, *Graphics Interchange Format (sm), Version 89a*, CompuServe Incorporated, Columbus, OH, July 1990.
10. X. Descombes, R.D. Morris, J. Zerubia, and M. Berthod, Estimation of Markov random field parameters using Markov chain Monte Carlo maximum likelihood, *IEEE Trans. Image Process.*, 8 (1999) 954–962.
11. J.-C. Duford, BIFS: scene description, in *The MPEG-4 Book*, T. Ebrahimi and F. Pereira, Eds., IMSC Press Multimedia Series, Prentice-Hall, Upper Saddle River, 2002, pp. 103–147.
12. M. Eden and M. Kocher, On the performance of a contour coding algorithm in the context of image coding. Part I: contour segment coding, *Signal Process.*, 8 (1985) 381–386.
13. M.R. Frater, W.S. Lee, M. Pickering, and J.F. Arnold, Error concealment of arbitrarily shaped video objects, *Proceedings of IEEE International Conference on Image Processing*, vol. 3, pp. 507–511, Chicago, USA, Oct. 1998.
14. H. Freeman, On the encoding of arbitrary geometric configurations, *IRE Trans. Electron. Comput.*, EC-10 (1961) 260–268.
15. P. Gerken, Object-based analysis-synthesis coding of image sequences at very low bit rates, *IEEE Trans. Circuits Syst. Video Technol.*, 4 (1994) 228–235.
16. H.M. Hang and J.J Chen, Source model for transform video coder and its application – Part I: fundamental theory, *IEEE Trans. Circuits Syst. Video Technol.*, 7 (1997) 287–298.

17. M. Hoetter, Optimization and efficiency of an object-oriented analysis-synthesis coder, *IEEE Trans. Circuits Syst. Video Technol.*, 4 (1994) 181–194.

18. M. Hötter, Object-oriented analysis-synthesis coding based on two dimensional objects, *Signal Process.: Image Commun.*, 2 (1990) 409–428.

19. ISO/IEC JTC1/SC29/WG1, IS11544 — Coded Representation of Picture and Audio Information Progressive Bi-Level Image Compression, ISO, 1992.

20. ISO/IEC, JTC1/SC29/WG11, IS14496-5 — Information Technology — Coding of Audio-Visual Objects — Part 5: Reference Software, ISO, 1999.

21. ISO/IEC, JTC1/SC29/WG11, IS14496-2 — Information Technology — Coding of Audio-Visual Objects — Part 2: Visual, 3rd ed., ISO, 2004.

22. A.K. Jain, *Fundamentals of Digital Image Processing, Prentice-Hall*, Englewood Cliffs, NJ, 1989.

23. T. Kaneko and M. Okudaira, Encoding of arbitrary curves based on the chain code representation, *IEEE Trans. Commun.*, 33 (1985) 697–707.

24. A. Katsaggelos, L.P. Kondi, F.W. Meier, J. Ostermann, and G.M. Schuster, MPEG-4 and rate-distortion-based shape-coding techniques, *Proc. IEEE*, 86 (1998) 1126–1154.

25. A. Kaup, Object-based texture coding of moving video in MPEG-4, *IEEE Trans. Circuits Syst. Video Technol.*, 9 (1999) 5–15.

26. J.I. Kim, A.C. Bovik, and B.L. Evans, Generalized predictive binary shape coding using polygon approximation, *Signal Process.: Image Commun.*, 15 (2000) 643–663.

27. J. Koplowitz, On the performance of chain codes for quantization of line drawings, *IEEE Trans. Pattern Anal. Machine Intell.*, 3 (1981) 180–185.

28. M. Kunt, A. Ikonomopoulos, and M. Kocher, Second-generation image-coding techniques, *Proc. IEEE*, 73 (1985) 549–574.

29. J.-W. Lee, A. Vetro, Y. Wang, and Y.-S. Ho, Bit allocation for MPEG-4 video coding with spatio-temporal trade-offs, *IEEE Trans. Circuits Syst. Video Technol.*, 13 (2003) 488–502.

30. H.G. Musmann, M. Hötter, and J. Ostermann, Object-oriented analysis-synthesis coding of moving images, *Signal Process.: Image Commun.*, 1 (1989) 117–138.

31. A.N. Netravali and B.G. Haskell, *Digital Pictures — Representation and Compression*, Plenum Press, New York, 1988.

32. D.L. Neuhoff and K.G. Castor, A rate and distortion analysis of chain codes for line drawings, *IEEE Trans. Inf. Theory*, 31 (1985) 53–67.

33. P. Nunes, F. Marques, F. Pereira, and A. Gasull, A contour-based approach to binary shape coding using multiple grid chain code, *Signal Process.: Image Commun.*, 15 (2000) 585–600.

34. K.J. O'Connell, Object-adaptive vertex-based shape coding method, *IEEE Trans. Circuits Syst.*, 7 (1997) 251–255.

35. J. Ostermann, Efficient encoding of binary shapes using MPEG-4, *Proceedings of IEEE International Conference on Image Processing, ICIP'98*, Chicago, USA, October 1998.

36. P. van Otterloo, *A Contour-Oriented Approach for Shape Analysis*, Prentice-Hall, Hertfordshire, UK, 1991.

37. T. Ozcelik and A.K. Katsaggelos, Very low bit rate video coding based on statistical spatio-temporal prediction of motion, segmentation and intensity fields, in *Video Data Compression for Multimedia Computing*, H.H. Li, S. Sun, and H. Derin, Eds., Kluwer, Norwell, MA, 1997, pp. 313–353.

38. R. Prasad, J.W. Vieveen, J.H. Bons, and J.C. Arnbak, Relative vector probabilities in differential chain coded linedrawings, *Proceedings of IEEE Pacific Rim Conference on Communication, Computers and Signal Processing*, Victoria, Canada, June 1989, pp. 138–142.

39. J.A. Saghri and H. Freeman, Analysis of the precision of generalized chain codes for the representation of planar curves, *IEEE Trans. Pattern Anal. Machine Intell.*, 3 (1981) 533–539.

40. P. Salama and C. Huang, Error concealment for shape coding, *Proceedings of IEEE International Conference on Image Processing*, vol. 2, pp. 701–704, Rochester, USA, Sept. 2002.

41. P. Salembier, F. Marques, and A. Gasull, Coding of partition sequences, in *Video Coding*, L. Torres and M. Kunt, Eds. Kluwer Academic Publishers, Englewood Cliffs, pp. 125–170, 1996.

42. G.M. Schuster and A.G. Katsaggelos, An optimal segmentation encoding scheme in the rate-distortion sense, *Proceedings of ISCAS 96*, Atlanta, USA, vol. 2, pp. 640–643.

43. S. Shirani, B. Erol, and F. Kossentini, A concealment method for shape information in MPEG-4 coded video sequences, *IEEE Trans. Multimedia*, 2 (2000) 185–190.

44. J. Signes, Binary format for scenes (BIFS): combining MPEG-4 media to build rich multimedia services, *Proceedings of the Visual Communications and Image Processing*, San Jose, CA, Jan. 1999.

45. L.D. Soares and F. Pereira, Refreshment need metrics for improved shape and texture object-based resilient video coding, *IEEE Trans. Image Process.*, 12 (2003) 328–340.

46. L.D. Soares and F. Pereira, Spatial shape error concealment for object-based image and video coding, *IEEE Trans. Image Process.*, 13 (2004) 586–599.

47. L.D. Soares and F. Pereira, Adaptive shape and texture intra refreshment schemes for improved error resilience in object-based video coding, *IEEE Trans. Image Process.*, 13 (2004) 662–676.

48. L.D. Soares and F. Pereira, Temporal shape error concealment by global motion compensation with local refinement, *IEEE Trans. Image Process.*, Vol. 15, No. 6, June 2006.

49. A. Vetro, T. Haga, K. Sumi, and H. Sun, Object-based coding for long-term archive of surveillance video, *Proceedings of IEEE International Conference on Multimedia and Expo, ICME'03*, Baltimore, USA, July 2003.

50. A. Vetro, H. Sun, and Y. Wang, MPEG-4 rate control for coding multiple video objects, *IEEE Trans. Circuits Syst. Video Technol.*, 9 (1999) 186–199.

51. A. Vetro, H. Sun, and Y. Wang, Object-based transcoding for adaptable video content delivery, *IEEE Trans. Circuits Syst. Video Technol.*, 11 (2001) 387–401.

52. A. Vetro, H. Sun, Y. Wang, and O. Gulyeruz, Rate-distortion modeling of binary shape using state partitioning, *Proceedings of the IEEE International Conference on Image Processing, ICIP'99*, Kobe, Japan, October 1999.

53. A. Vetro, Y. Wang, and H. Sun, Rate-distortion modeling for multiscale binary shape coding based on Markov Random Fields, *IEEE Trans. Image Process.*, 12 (2003) 356–364.

54. H. Wang, G.M. Schuster, A.K. Katsaggelos, and T.N. Pappas, An efficient rate-distortion optimal shape coding approach using a skeleton-based decomposition, *IEEE Trans. Image Process.*, 12 (2003) 1181–1193.

55. C.T. Zahn and R.Z. Roskies, Fourier descriptors for plane closed curves, *IEEE Trans. Comput.*, 21 (1972) 269–281.

13 Compressing Compound Documents

Ricardo L. de Queiroz

CONTENTS

13.1   Introduction 323
13.2   Raster Imaging Models 324
      13.2.1   Mixed Raster Content 324
      13.2.2   Region Classification 325
      13.2.3   Other Imaging Models 326
            13.2.3.1   DjVu 326
            13.2.3.2   Soft Masks for Blending 327
            13.2.3.3   Residual Additive Planes 327
13.3   MRC for Compression 327
      13.3.1   Object Segmentation versus Region Classification 329
      13.3.2   Redundant Data and Segmentation Analysis 330
      13.3.3   Plane Filling 333
13.4   A Simple MRC: JPEG+MMR+JPEG 336
      13.4.1   Computing Rate and Distortion per Block 337
      13.4.2   Optimized Thresholding as Segmentation 338
      13.4.3   Fast Thresholding 340
      13.4.4   Performance 341
13.5   MRC within JPEG 2000 344
      13.5.1   JP2-Based JPM 347
Conclusions 348
References 349

13.1   INTRODUCTION

Electronic documents are commonplace. From PDF files [27] to fax transmissions, including internal raster representations, a number of electronic formats are used to convey text and pictorial information. Documents are present in a wide spectrum of printing systems and are basically represented in vectorial or raster forms. A document in a vector form is composed of a series of primitives, or instructions such as "render this text at that position." It often comes as a pair of object and instruction, e.g., the text is the object and the instruction was to render it at a particular position. The same happens to graphics. The document is imaged by "drawing" all "vectors" and applying instructions to image a number of objects. Examples are PostScript [29] and PDF files [27], printer languages such as PCL [13], graphics primitives in operating systems, and even ancient presentation protocols such as NAPLPS [2]. In contrast, raster documents are composed of images of the information to be rendered, e.g., a compressed image or a scan of a magazine page. The advantage of raster images is that they are ready to image, as opposed to vector formats that need to be rasterized first.

It is not much of a challenge to compress vectorized documents since each object can be compressed individually and the whole file can be further compressed losslessly. The real challenge is to compress rasterized compound documents. Compound documents are assumed here as images which contain a mix of textual, graphical, or pictorial contents. A single compression algorithm that simultaneously meets the requirements for both text and image compression has been elusive. As a rule, compression algorithms are developed with a particular image type, characteristic, and application in mind and no single algorithm is best across all image types or applications. When compressing text, it is important to preserve the edges and shapes of characters accurately to facilitate reading. The human visual system, however, works differently for typical continuous-tone images, better masking high-frequency errors. Roughly speaking, text requires few bits per pixel but many pixels per inch, while pictures require many bits per pixels but fewer pixels per inch. Document compression is frequently linked to facsimile systems, in which large document bitmaps are compressed before transmission over telephone lines. There is now a focus on new standards to provide color facsimile services over the telephone network and the Internet [5].

Compound raster documents have always been compressed as a single image, either converting the image to binary data, thus pretending the image is a black and white text, like in a regular fax, or by applying regular image compression to the whole scanned color document, as in color fax. When it comes to compound documents, different compression algorithms may be applied to each of the regions of the document. This can be accomplished by segmenting the regions or by generating multiple image layers. Apart from the multilayer or multiregion methods described in the next section, there are no viable and popular alternatives for compressing compound images, other than the inefficient single-coder approaches.

13.2   RASTER IMAGING MODELS

A raster bitmap is commonly generated immediately prior to imaging. Hybrid raster images are intermediate formats which can be easily rendered but allow for improved compression and editing. The main problem is that a raster, rather than a vector form, does not distinguish the different objects in a document. If one could separate the objects into different raster images, which somehow could be combined later on, it would facilitate processing the image. There are several imaging models to accomplish the separation of multiple raster by defining a method to recombine (render) multiple raster images into a single bitmap.

13.2.1   MIXED RASTER CONTENT

The mixed raster content (MRC) imaging model [16–19] enables a multilayer multiresolution representation of a compound document as illustrated in Figure 13.1. The basic three-layer MRC model represents a color image as two color-image layers (foreground [FG] and background [BG]) and a binary layer (Mask). The Mask layer describes how to reconstruct the final image from the FG/BG layers, i.e., to use the corresponding pixel from the FG or BG layers when the mask pixel is 1 or 0, respectively, in that position. Thus, the FG layer is essentially poured through the Mask plane onto the BG layer as depicted in Figure 13.1a. In reality, the MRC imaging model employs a sequence of image pairs: foreground layers and corresponding mask as illustrated in Figure 13.1b. With these, starting from a basic background plane, one can pour onto it a plurality of foreground and mask pairs as shown in Figure 13.1c. Actually, the MRC imaging model allows for one, two, three, or more layers. For example, a page consisting of a picture could use the background layer only. A page containing black-and-white text could use the Mask layer, with the FG and BG layers defaulted to black and to white, respectively. Layers may contain different dimensions and have offsets associated with them. If a plane contains only a small object, the effective plane can be made of a bounding box around the object. The reduced image plane is then imaged onto the larger reference plane, starting

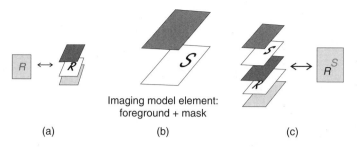

Imaging model element:
foreground + mask

(a)   (b)   (c)

FIGURE 13.1 (See color insert following page 336) Illustration of MRC imaging model. (a) The basic three-layer model where the foreground color is poured through the mask onto the background layer. In using a sequence of mask + FG pairs; (b) the three-layer model can be extended; (c) Less than three layers can be used by using default colors for mask and FG layers.

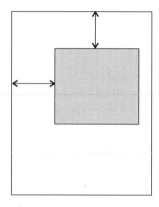

FIGURE 13.2 A layer in the MRC model can use only part of the imaging area, i.e., a layer can be smaller but properly positioned by indicating horizontal and vertical offset, along with the layer width and height. The remaining parts of the image plane will be assigned a default color.

from the given offset (top, left) with given size (width, height) as illustrated in Figure 13.2. This avoids representing large blank areas and improves compression. The portions of the imaging area of a layer which are unimaged are assigned a default color.

In effect, there are N layer pairs of foreground images f_k and masks m_k, $1 \leq k \leq N$, along with the background image f_0. If all layers and images are properly scaled to have the same dimension in pixels, then we start with filling the initial canvas pixels $x_0(i, j)$ with the background image $f_0(i, j)$. In the MRC imaging method, the layer pairs are sequentially imaged as

$$x_k(i,j) = m_k(i,j) f_k(i,j) + (1 - m_k(i,j)) x_{k-1}(i,j) \tag{13.1}$$

for $1 \leq k \leq N$, where the mask pixels $m_k(i,j)$ are binary, i.e., 1 or 0. We will describe later the continuous mask model. The final image is then $x_N(i,j)$.

The rationale for MRC is that once the original single-resolution image is decomposed into layers, each layer can be processed and compressed using different algorithms. The compression algorithm and processing used for a given layer would be matched to the layer's content.

13.2.2 REGION CLASSIFICATION

MRC implies a superposition of frames wherein a selector or mask plane decides from where to render a particular pixel, either from FG or BG planes. However, all the FG and BG layers must be

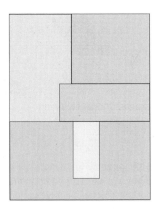

FIGURE 13.3 (**See color insert following page 336**) An image is divided into regions, each region being classified as belonging to one class, e.g., text, picture, background, etc. The final image is rendered by looking up this map and retrieving the pixel (or block) data from the assigned layer.

encoded somehow and there is redundant data available to the decoder. Alternatively, one can select which regions to assign to BG or FG, but, unlike MRC, not storing the image information that is not to be imaged. In this case, only one plane of data is made available to the decoder. In effect, the encoder prepares a map like the one depicted in Figure 13.3, which divides the image into several regions, each region labeled as pertaining to a given layer. Typically, the image is divided into blocks and each block is classified as, for example, text, picture, graphics, or background. Every class is generally assigned to an image compression method. When rendering the final raster, for a given block, one looks at the map as given in Figure 13.3 to find which coder (layer) contains the block information. The block image information is decoded, the block is rendered, and we move on to the next block.

Many image compression methods that deal with compound images use this model [7,11,30]. Note that a method is not MRC-compatible just because it employs multiple layers. The important distinction is data redundancy. The entire plane is represented in MRC. One can render it and print it if desired. When there is pure region classification, every image region goes in only one layer, so that layers are not entire images but data streams. For example, the renderer to form an image block will look at the map and retrieve compressed data from either the JBIG or JPEG stream. Region classification avoids redundancy and is more natural for efficient compression. However, it requires modifying compressors to deal with segmented image data. MRC, on the other hand, employs stock compression only.

13.2.3 OTHER IMAGING MODELS

13.2.3.1 DjVu

DjVu (pronounced as *déjà vu*) [3,4] is a software and technology for exchanging documents over the web. For that the target is very high compression ratios while preserving comfortable readability. In the DjVu model, there are a number of objects to be imaged onto the background plane (Figure 13.4). An object is a contiguous association of "on" pixels, like letters in a text. Each object is associated with a color and position in the page. In other words, one first images the background and then pastes color objects onto it. Note that it departs from the typical MRC model in which the "object" is within the selector plane and colors are conveyed as bitmaps in the foreground plane. In the DjVu model, the object itself is associated to the color. Note, also, that the DjVu package provides modes for compatibility with MRC. Nevertheless, this imaging model is an interesting solution for a mixed representation.

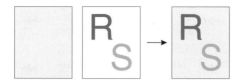

FIGURE 13.4 (See color insert following page 336) In the DjVu rendering model, objects are rendered, with their assigned colors, onto a background plane. There is neither an explicit mask plane nor FG bitmaps.

13.2.3.2 Soft Masks for Blending

In the typical three-layer MRC model, the output pixel $x(i,j)$ at position (i,j) will be imaged from either the FG, i.e., $x(i,j)=f(i,j)$, or the BG, i.e., $x(i,j)=g(i,j)$. The mask layer pixel $m(i,j)$ conveys the information of which plane to use for rendering at that point. This is a hard decision between planes. However, it is also possible to use continuous mask and blend the FG and BG planes, e.g.,

$$x(i,j) = m(i,j)f(i,j) + (1 - m(i,j))g(i,j) \tag{13.2}$$

for $0 \leq m(i,j) \leq 1$, i.e., the mask is continuous from 0 to 1 that controls the blending of the FG and BG planes. In the traditional hard decision case, there is no blending and $m(i,j)$ is either 0 or 1 but nothing in between. The continuous mask option allows for much more flexibility.

For multiple layers the initial canvas pixels $x_0(i,j)$ are the background pixels $f_0(i,j)$ and

$$x_k(i,j) = m_k(i,j)f_k(i,j) + (1 - m_k(i,j))x_{k-1}(i,j) \tag{13.3}$$

for $1 \leq k \leq N$, where the mask pixels $m_k(i,j)$ are continuous from 0 to 1. The final image is $x_N(i,j)$.

13.2.3.3 Residual Additive Planes

Additional to selecting or blending planes, one can also use additive planes to correct for errors in a linear way. For example, in the check compression standard [14], a MRC model plus an additive plane are provided, i.e.,

$$x(i,j) = m(i,j)f(i,j) + (1 - m(i,j))g(i,j) + r(i,j) \tag{13.4}$$

where $r(i,j)$ is the residual plane and $m(i,j)$ can be either binary (as is MRC and the check compression case) or continuous.

13.3 MRC FOR COMPRESSION

MRC is the international standard for compound image compression. It was originally approved for use in group 3 color fax and is described in ITU-T Recommendation T.44 [19]. For the storage, archiving, and general interchange of MRC-encoded image data, the TIFF-FX file format has been proposed [16]. TIFF-FX (TIFF for Fax eXtended) represents the coded data generated by the suite of ITU recommendations for facsimile, including single-compression methods MH, MR, MMR [20,21], JBIG [22], JBIG2 [23] and JPEG [28], JPEG 2000 [17,31], as well as MRC. As IETF RFC 2301, TIFF-FX [16] is an Internet standard and it is also the document compression framework for JPEG 2000 Part 6 [5,18,24]. MRC has been used in products such as Digipaper [15], DjVu [4], and LuraDocument [1].

As we have discussed, the MRC model can potentially use a number of mask plus foreground image pairs. Nevertheless, unless otherwise noted, we assume a basic three-layer MRC model, which the reader can easily extend to encompass multiple layers. Once the original single-resolution image is decomposed into layers, each layer can be processed and compressed using different algorithms

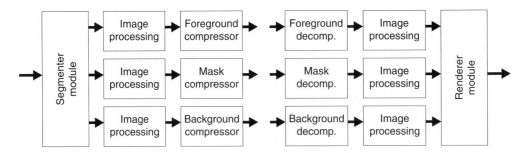

FIGURE 13.5 Block diagram of plane decomposition, compression, and rendering processes for a basic three-layer MRC representation.

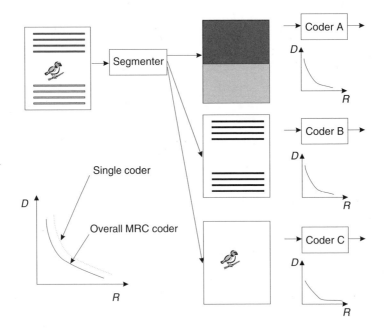

FIGURE 13.6 Interpretation of the multiplane approach as a means to modify the RD characteristics of coding mixed images. A mixed image is decomposed and each plane undergoes a better tailored coder, thus achieving better RD curves. The goal is to achieve a better overall RD curve than the one obtained by compressing the image itself with a single coder.

as shown in Figure 13.5. The image processing operations can include a resolution change or color mapping. The compression algorithm and resolution used for a given layer would be matched to the layer's content, allowing for improved compression while reducing distortion visibility. The compressed layers are then packaged in a format such as TIFF-FX [16] or as a JPM data stream [5,18] for delivery to the decoder. At the decoder, each plane is retrieved, decompressed, processed (which might include scaling), and the image is finally rendered using the MRC imaging model.

The reason why a model such as MRC works is because one can use processing and compressors tailored to each plane statistics. Thus, one can attain improved performance compressing each plane. Hopefully, these gains are enough to offset the expenses in representing redundant data (remember that every pixel position is represented in each of the FG–BG mask layers). The potential gain of the MRC model for compression can be analyzed under the light of its rate–distortion (RD) characteristics [9]. If the image in Figure 13.6 is compressed with a generic coder with fixed parameters except for a compression parameter, it will operate under a given RD curve. Another coder under the same circumstances is said to outperform the first coder if its RD curve is shifted to the left (down), i.e., has lower rate for a given distortion or less distortion for a given rate. The rationale for MRC is to split the image into multiple planes as shown in Figure 13.6 and to apply to each plane a coder (A, B, and C) whose RD curves are better than those of the single plane coder. In this case, there is a possibility that the equivalent coder will have better RD curves than the single plane coder, despite the overhead associated with a multiplane representation.

13.3.1 OBJECT SEGMENTATION VERSUS REGION CLASSIFICATION

The degrees of freedom in MRC-based compression are the layer decomposition process (the segmenter module in Figure 13.5) and the compressors with their associated parameters for each plane. The encoder and decoder would agree *a priori* on the compressors, which would be part of the standard employing MRC as an architectural framework. Decomposition affects the operation of the encoder, but not that of the decoder. There are two main approaches to decompose compound images which are illustrated in Figure 13.7. They are based on object segmentation or on region classification. The image in Figure 13.7a is associated with a BG plane in Figure 13.7b, which may contain the paper background along with the continuous-tone picture. The object decomposition relies on identifying text and graphics objects. The concept is that the text or graphics ink is poured through the mask plane onto the BG plane. For this, the mask should have the contours of text elements, as illustrated in Figure 13.7c, while the FG plane in Figure 13.7d contains solid colors of the objects. Thus, the

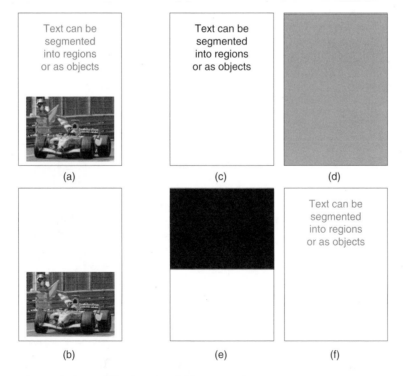

FIGURE 13.7 (**See color insert following page 336**) Two example segmentation strategies that yield the same image. (a) Original image containing a picture and colored text; (b) BG plane with the canvas and the picture; (c) mask plane containing the text shapes; (d) text associated colors are present at the FG plane; (e) another mask that simply marks the text areas; and (f) the corresponding FG plane containing the colored text.

mask image layer would contain text characters, line art, and filled regions, while the foreground layer contains the colors of the shapes in the mask layer, i.e., the color of text letters and graphics.

In region classification, regions containing text and graphics are identified and represented in a separate (foreground) plane. The whole region is represented in the FG plane including the spaces in between letters. The mask is very uniform with large patches indicating the text and graphics regions, while the BG plane contains the remaining regions, i.e., the document background itself, complex graphics, and continuous-tone pictures. In Figure 13.7e, the mask is actually made of large areas indicating where the text lies. The text itself is contained within the FG plane in Figure 13.7f. In the MRC model, which contains redundant image representation, if compression ratio is the main motivation, it is often more useful to employ object segmentation.

13.3.2   REDUNDANT DATA AND SEGMENTATION ANALYSIS

In an MRC model, image data is redundant. Even if the mask indicates that a particular image position is to be represented, for example, using the FG data, there is some image data in the BG plane for that position. That information is redundant and does not affect image reconstruction. We will later discuss efficient means to replace redundant data with something that is more easily compressible. This is referred here as plane filling, i.e., we plug holes into the data planes. Note that once the plane-filling algorithms are selected, the only degree of freedom in an MRC decomposition is the segmentation of the input compound image into a binary mask. As illustrated in Figure 13.8, the segmenter finds a suitable mask from the compound image data. Using the mask layer and the input image, the FG and BG layers are found in a deterministic way.

If the mask layer tells the decoder that some pixels are to be imaged from the BG plane, the corresponding pixels from the FG plane are redundant. In effect they are a waste of information which, nevertheless, has to be encoded. The opposite is true for the FG plane. Where the mask indicates the pixels to be imaged from the FG plane, the corresponding BG pixels are redundant too. Actually, any value we use for the redundant image region will be irrelevant for reconstruction. In

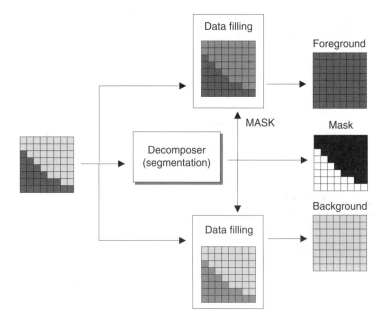

FIGURE 13.8 (See color insert following page 336) MRC plane decomposition diagram, consisting only of the segmenter and plane-filling modules.

Figure 13.8, the image block is analyzed by the segmenter which generates a mask with both colors, i.e., assigns pixels to both BG and FG planes. The data-filling algorithms will identify where the redundant regions are and replace their pixel values with some "smooth" data in a process that will be explained later. In each plane there are useful regions (labeled U) and redundant or "don't-care" regions (labeled X).

In light of the above discussion, we refer the reader to Figure 13.9 wherein a mask plane is illustrated on the left. For simplicity we used an example of a simple oval shape, but it could be anything, and is typically composed of a number of distinct objects. Let the "white" region be labeled A and the "black" region be labeled B. We can also define the transition region T, which encompasses a neighborhood near the border between A and B regions. This transition region crosses over the real border between regions and can be further subdivided into regions belonging to A (T_A) and B (T_B). In this case, the whole image I is made of subregions, i.e., $I = A \cup B \cup T_A \cup T_B$. If the A region means that the final pixels will be imaged from the FG plane, while B means one will use the BG plane, then the A region will mean a useful (U) region for the FG plane but a "don't-care" (X) region for the BG plane, and vice versa: region B means U for BG and X for FG. The scheme is illustrated in Figure 13.9, where either BG or FG is made of $U \cup X \cup T_U \cup T_X$.

Let us analyze the segmentation in a rate–distortion viewpoint [9]. If the original image is encoded using a single coder S, which does not use MRC, R_S bits will be spent yielding a reconstruction distortion D_S such that

$$R_S = R_S^A + R_S^B + R_S^{T_A} + R_S^{T_B} \quad \text{and} \quad D_S = D_S^A + D_S^B + D_S^{T_A} + D_S^{T_B} \tag{13.5}$$

where the distortion model was chosen to be linear, i.e., overall distortion is the sum of local distortions, while R^{A,B,T_A,T_B} and D^{A,B,T_A,T_B} are the rate and distortions for each of the plane regions (see Figure 13.9). If the image is split into the three planes (FG, BG, and mask) corresponding to the MRC model, then the overall rate and distortion are given by

$$R = R_M + \sum_{\Psi=FG,BG} \sum_{\Omega=A,B,T_A,T_B} R_\Psi^\Omega \quad \text{and} \quad D = \sum_{\Psi=FG,BG} \sum_{\Omega=A,B,T_A,T_B} D_\Psi^\Omega \tag{13.6}$$

Note that the mask is encoded without distortion and that X pixels, i.e., region B in the FG plane and region A in the BG plane, do not contribute to overall distortion. Thus

$$D = D_{BG}^B + D_{BG}^{T_B} + D_{FG}^A + D_{FG}^{T_A} \tag{13.7}$$

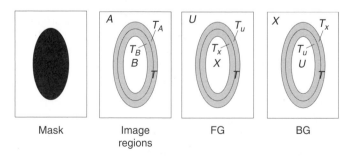

Mask   Image regions   FG   BG

FIGURE 13.9 An example mask dividing the image into regions A and B and the transition T, which can be further subdivided into its margins onto the AB regions as T_A and T_B. Each plane then has some regions that can be classified as useful (U), redundant (X), or transition (T). The layer transition can also be further subdivided (T_U and T_X).

If one requires the MRC scheme to outperform the single coder, it is necessary that either or both $R < R_S$ and $D < D_S$. It is sufficient to have

$$R < R_S \quad \text{and} \quad D < D_S \tag{13.8}$$

In a simple coding scenario where the coder for the FG and BG planes is the same as the single coder, we can make the following assumptions: $R_{BG}^B = R_S^B$, $R_{FG}^A = R_S^A$, $D_{BG}^B = D_S^B$, and $D_{FG}^A = D_S^A$. Furthermore, in general transform coding, the transform bases will likely extend across the region boundaries (that is why we made the transition regions in the first place) so that it is unlikely that one can separate the rate produced by either T_A or T_B. Hence we define $R_S^T = R_S^{T_A} + R_S^{T_B}$, $R_{FG}^T = R_{FG}^{T_A} + R_{FG}^{T_B}$, and $R_{BG}^T = R_{BG}^{T_A} + R_{BG}^{T_B}$, so that

$$D_S - D = D_S^{T_A} + D_S^{T_B} - D_{FG}^{T_A} + D_{BG}^{T_B} \tag{13.9}$$

$$R_S - R = R_S^T - R_M - R_{BG}^A - R_{FG}^B - R_{BG}^T - R_{FG}^T = R_S^T - R_o - R_{BG}^T - R_{FG}^T \tag{13.10}$$

where R_o is the overhead rate due to the mask and redundant data in the continuous planes. Reduction in rate and distortion are achieved iff

$$D_{FG}^{T_A} + D_{BG}^{T_B} < D_S^{T_A} + D_S^{T_B} \tag{13.11}$$

$$R_o + R_{BG}^T + R_{FG}^T < R_S^T \tag{13.12}$$

Therefore, following the analysis of this simple example, we see that transition regions are the main regions where compression can be improved by using MRC. In more detail, improvement comes when (1) distortion in the transition region is less than in the single coder and (2) the savings in encoding the transition regions (in both BG and FG) planes compared to the single coder are enough to offset the expenditure of bits to encode the overhead.

With text object segmentation, Equations (13.11) and (13.12) are usually satisfied. In general, $R_{BG}^T < R_S^T$ and $R_{FG}^T < R_S^T$. However, the decomposition has to be done in such a way that there will be enough transitions in the image to allow enough savings. Furthermore, the regions chosen to be transitions have to be such that they lead to large savings in bit rate in each plane in order to compensate for the redundant information and the overhead.

In an MRC approach, the plane-filling preprocessor (see Figure 13.8) can very well replace pixels in redundant regions with any computer-generated data which would reduce most of the distortion and bit rate, i.e., to ensure that Equations (13.11) and (13.12) are satisfied. With text segmentation, the transition in the mask occurs for edges in the original image. Hence, R_S^T and D_S^T are very high. If, for example, the transition region in each plane is made very smooth, not only will the distortion decrease but the bit rate can be kept very small. For smooth enough transitions and if we discard the terms R_{BG}^T and R_{FG}^T, then the trade-off of MRC can be summarized as

$$R_o < R_S^T \tag{13.13}$$

In other words, MRC is advantageous if the amount of bits saved by not encoding the transition regions is greater than the amount of overhead data (redundant and mask data). Of course, the main assumption is that the transition in both planes can be made "smooth" enough to significantly save in both R and D. Also, the input image has to contain a sufficient amount of those edges. An image with large text regions is a typical case. If there are only pictorial images, however, it is harder (but not impossible) to make a multiplane MRC outperform the single coder. In the limit, it may be advantageous to place the pictorial image in a single MRC layer, in which case the MRC behaves as a single coder.

In reality, a typical coding scenario is usually more favorable to MRC than the above example. This is because coders for FG and BG can be selected to outperform the single coder, while the mask plane often compresses very well. For example, if the text is placed into the mask, techniques such as JBIG, MMR, and JBIG-2 can compress text well. The FG would contain mainly text color and can be largely compressed. The BG plane would contain the pictorial images and the paper texture, which are features that do not contain high-resolution details. In that case, moderate subsampling can be carried before compression. The different nature of the data in each plane allows for very efficient compression with lower error visibility.

13.3.3 Plane Filling

In the previous section, we assumed that plane-filling algorithms can reasonably smooth transitions. Without loss of generality, we can address any plane (FG or BG) individually by referring to its X and U regions. We want to replace the data in the X region (and T_X in Figure 13.9) with any data that would improve compression. The overall goal is to reduce both rate and distortion, i.e., to minimize [8]

$$J = R + \lambda D \tag{13.14}$$

where λ controls the rate–distortion trade-off. Assuming rate and distortion are additive per regions (assuming they are independent), we have

$$J = (R^U + R^X + R^T) + \lambda(D^U + D^X + D^{T_U} + D^{T_X}) \tag{13.15}$$

where the superscript indicates the regions, and as we discussed $R^T = R^{T_U} + R^{T_X}$. Note that since a redundant (X) region is irrelevant for reconstruction, then $D^X = D^{T_X} = 0$. Also note that since the replacement of redundant data does not affect the U region, the minimization of the above cost function is equivalent to minimizing

$$J = R^X + R^T + \lambda D^{T_U} \tag{13.16}$$

i.e., it is equivalent to minimize rate in the X region and to make it RD-efficient at transitions.

True optimality is a very ambitious goal. The alternatives are too many to consider: layers can be resized or further processed and there are too many compression options, ranging from the transform type, through the wavelet type, to the choice of quantizers and entropy coders, etc. It is impractical to optimize the redundant data without fixing all these compression parameters; however there is a good compromise with a practical solution which aims to work well across a wide range of applications [8].

In order to minimize Equation (13.16), it seems to be reasonable to apply smooth (flat) data to the redundant region. That would definitely reduce R^X to its virtual minimum. The question is what to do with the transitions, i.e., how to minimize $R^T + \lambda D^T$. If we would do that blindly, the best intuitive solution is to make the transition as smooth as possible. Smooth patterns tend to produce less bits and cause less distortion in most popular image coders.

The problem is to generate smooth transitions. Figure 13.10 is an illustration of a 1D signal (a) which is masked using the mask in (b) yielding the signal in (c), where the redundant area was replaced by a constant value. If we simply filter the signal in Figure 13.10c, we obtain the signal in (d). After applying the mask, assuming we do not touch the useful region, the reconstructed signal is as shown in Figure 13.10e. Note that there still exist a discontinuity, which is caused by the symmetry of the filtering process around the edge. A good solution is to use a segmented filtering [8], where filter weights are exaggerated for pixels in the useful region of the layer. In 1D,

$$y(n) = \frac{\sum_{k=-L}^{L} h(k, n)\, x(n + k)}{\sum_{k=-L}^{L} h(k, n)} \tag{13.17}$$

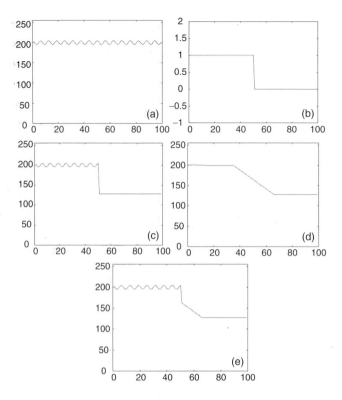

FIGURE 13.10 1D example: (a) Signal; (b) mask; (c) masked layer; (d) filtered version of (c); and (e) masked filtered layer.

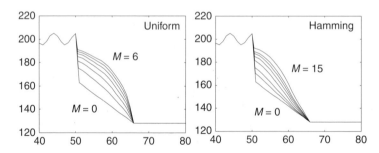

FIGURE 13.11 Segmented filtering for the 1D example for different M and for two windows.

where h is a time-varying filter of $2L + 1$ taps. Its weights are dependent on the mask values $m(n)$ as

$$h(k, n) = \begin{cases} 1 & \text{if } m(n + k) = 0 \\ Mf(k) + 1 & \text{if } m(n + k) = 1 \end{cases} \qquad (13.18)$$

where $f(k)$ is a filter window such as Hamming to de-emphasize distant pixels within the useful region. The result of applying the segmented filter is shown in Figure 13.11 for $L = 16$, by varying M and for a rectangular (uniform) and for a Hamming window. Note how the discontinuity is largely decreased. The case $M = 0$ is equivalent to a straight averaging filter without any emphasis. This filtering is very easy to implement. For uniform windows, like the averaging filter case, complexity is virtually independent of window size, and very large filters are easily implemented.

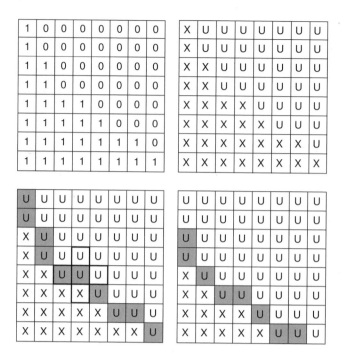

FIGURE 13.12 Top: an example block mask and respective labels for layer pixels. Bottom: two further stages of the iterative spatial domain block-filling algorithm. Shaded pixels are the X pixels that are replaced at each stage by the NSEW neighbor average, thus being relabeled as a U pixel.

This approach assumes a generic coder. If the coder is known, there are ways to further optimize the data-filling process. One example is the method used by DjVu in which the wavelet transform is known and a set of interactive projections are made in order to approach a stable and efficient solution slowly.

Other solutions exist for JPEG and operate over small 8×8 blocks. As in the general method, the preprocessor receives an input block and, by inspecting the binary mask, labels the input block pixels as useful (U) or "don't care" (X).

The spatial domain algorithm is relatively simple and inexpensive [8]. If there are 64 X-marked pixels, the block is unused and we output a flat block whose pixels are the average of the previous block (because of JPEG's DC DPCM). If there are no X-marked pixels, the input block is output-untouched. If there is a mix of U- and X-marked pixels, we follow a multipass algorithm. In each pass, X pixels that have at least one U pixel as a horizontal or vertical neighbor (i.e., in the N_4 or NSEW neighborhood as indicated in Figure 13.12) are replaced by the average of those neighbors as illustrated in Figure 13.12. In the next pass, those pixels that were replaced are marked U for the next pass. Figure 13.12 illustrates two steps, while the process is continued until there are no X pixels left in the block. The aim of the algorithm is to replace the unused parts of a block with data that will produce a smooth block based on the existing data in the U-marked pixels. Its disadvantage is that the X-marked pixels are just influenced by the bordering U pixels, i.e., an internal U pixel does not affect data filling. This is acceptable for most applications.

There is a more expensive data-filling alternative involving the computation of the DCT [8,9]. This method consists of the following steps: (1) initialize X pixels in any way, for example, as the average of the U pixels; (2) transform, quantize, and inverse transform the block, obtaining a new set of pixels in the block (call them X' and U' pixels); (3) replace X pixels by X' pixels in the original block; and (4) repeat the transformation and replacement process until convergence is reached. Convergence is

```
121  113  132  178  229  227  221  222
 73  101  118  102  198  227  221  221
 68   65   41   49  120  225  223  221
 38   43   41   32   54  220  222  224
 38   35   38   34   50  220  224  218
 96  105   64   35  210  224  223  217
124  129  144   68  229  223  222  218
 80   98  117  204  226  219  219  220

  1    1    1    1    0    0    0    0
  1    1    1    1    0    0    0    0
  1    1    1    1    1    0    0    0
  1    1    1    1    1    0    0    0
  1    1    1    1    1    0    0    0
  1    1    1    1    0    0    0    0
  1    1    1    1    0    0    0    0
  1    1    1    0    0    0    0    0
```

```
121  113  132  178   81   81   81   81      121  113  132  178  178  178  178  178
 73  101  118  102   81   81   81   81       73  101  118  102  111  116  118  119
 68   65   41   49  120   81   81   81       68   65   41   49  120  120  120  120
 38   43   41   32   54   81   81   81       38   43   41   32   54   54   54   54
 38   35   38   34   50   81   81   81       38   35   38   34   50   50   50   50
 96  105   64   35   81   81   81   81       96  105   64   35   43   46   48   49
124  129  144   68   81   81   81   81      124  129  144   68   68   68   68   68
 80   98  117   81   81   81   81   81       80   98  117   93   80   74   71   70
```

```
121  113  132  178   91   72   83   71
 73  101  118  102   91   82   86   81
 68   65   41   49  120   86   76   80
 38   43   41   32   54   83   69   80
 38   35   38   34   50   82   76   88
 96  105   64   35   66   84   83   81
124  129  144   68   71   82   88   67
 80   98  117   97   63   79   96   65
```

```
121  113  132  178  113   80   76   61
 73  101  118  102   97   83   85   86
 68   65   41   49  120   85   78   88
 38   43   41   32   54   84   67   77
 38   35   38   34   50   82   75   85
 96  105   64   35   52   79   88   89
124  129  144   68   56   78   92   72
 80   98  117   99   60   79   95   56
```

FIGURE 13.13 Iterative DCT-domain block filling. Top: Block and mask; left: three steps in the DCT domain algorithm; right: spatial domain method result.

achieved when X' and X pixels are identical or close within a certain prescribed tolerance. It usually happens after very few interactions. As an illustration, Figure 13.13 shows an example block, its respective mask, and the resulting block using DCT domain algorithm. For comparison, a block resulting from the spatial domain algorithm is also presented.

13.4 A SIMPLE MRC: JPEG+MMR+JPEG

A simple scheme that can yield good results utilizes a three-layer model and simple, standard coders. The most important step in an MRC representation is the segmentation which, along with plane filling, define the MRC plane decomposition. For this simple case, there are efficient segmentation algorithms and block-filling methods that enable efficient document compression. There are only the mask, BG, and FG planes. The FG plane should contain the text and graphics colors. The BG plane should contain the document background and pictorial data. The mask actually should contain the text and graphics shapes (Figure 13.14 and Figure 13.15).

FIGURE 7.3 Image IBB North: (a) original image, (b) wedge reconstruction $\lambda = 0.012$, and (c) with corresponding wedge grid superimposed.

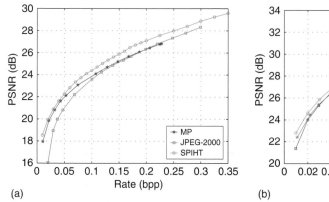

(a)

(b)

FIGURE 8.8 Distortion–rate performance for JPEG-2000, SPIHT, and the proposed MP coder for common test images. (a) *Cameraman* (256×256); (b) *Lena* (512×512).

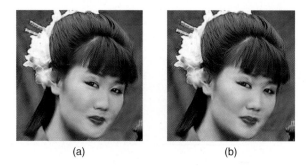

(a)　　　　　　　　　　　(b)

FIGURE 8.9 Japanese woman coded with 1500 MP atoms, using the most energetic channel search strategy in YUV color space (a) and in RGB color space (b).

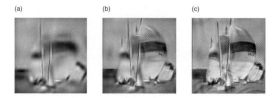

FIGURE 8.15 Matching pursuit bitstream of *sail* image decoded after 50, 150, and 500 coefficients. (a) 50 coefficients; (b) 150 coefficients; and (c) 500 coefficients.

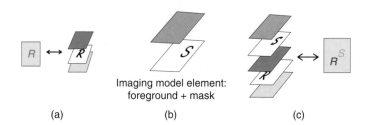

Imaging model element:
foreground + mask

(a)　　　　　　　　　　　(b)　　　　　　　　　　　(c)

FIGURE 13.1 Illustration of MRC imaging model. (a) The basic three-layer model where the foreground color is poured through the mask onto the background layer. In using a sequence of mask + FG pairs; (b) the three-layer model can be extended; (c) Less than three layers can be used by using default colors for mask and FG layers.

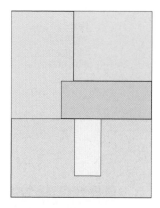

FIGURE 13.3 An image is divided into regions, each region being classified as belonging to one class, e.g., text, picture, background, etc. The final image is rendered by looking up this map and retrieving the pixel (or block) data from the assigned layer.

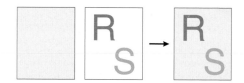

FIGURE 13.4 In the DjVu rendering model, objects are rendered, with their assigned colors, onto a background plane. There is neither explicit mask plane nor FG bitmaps.

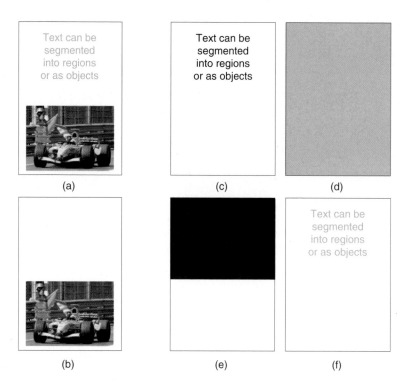

FIGURE 13.7 Two example segmentation strategies that yield the same image. (a) Original image containing a picture and colored text; (b) BG plane with the canvas and the picture; (c) mask plane containing the text shapes; (d) text associated colors are present at the FG plane; (e) another mask that simply marks the text areas; and (f) the corresponding FG plane containing the colored text.

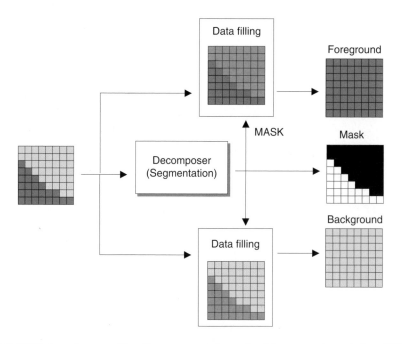

FIGURE 13.8 MRC plane decomposition diagram, consisting only of the segmenter and plane-filling modules.

January 31, 2001

Dear Mom and Dad,

How are both of you doing? I thought I would drop a line to say hi. Fanny, little Danny, and I are doing well. As you can see by the picture, little Danny isn't quite so little! Isn't this letter really great! I took a picture of Danny that was on a Kodak PhotoCD, and I merged it onto this letter using my computer. I then printed the letter using a color inkjet printer I just bought...

Danny's wearing the gorgeous BLUE sweater you gave him last time you were visiting. It just brings out the RED in his lips and cheeks. He definitely gets his good looks from his mother!

Take care of yourselves and write soon.

Love,

Michael

FIGURE 13.20 Another compound image for testing. Image size is 1400 × 1024 pixels.

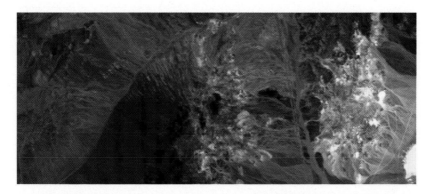

FIGURE 15.9 NASA/JPL AVIRIS *Cuprite Mine* image, 614 × 512 details shown as color compositions of red (band 30), green (band 16), and blue (band 4) radiances collected in 1997 with a wordlength of 16 bits.

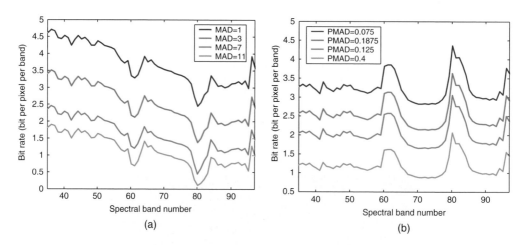

(a)

(b)

FIGURE 15.12 Bit rates produced by 3D RLPE on the data produced by the second spectrometer (NIR) of AVIRIS *Cuprite Mine '97*: (a) linear quantization to yield user-defined MAD values; (b) logarithmic quantization to yield user-defined PMAD values.

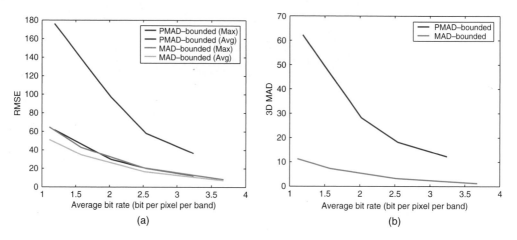

(a)

(b)

FIGURE 15.13 Radiometric distortions vs. bit rate for compressed AVIRIS *Cuprite Mine '97* data: (a) RMSE; (b) MAD.

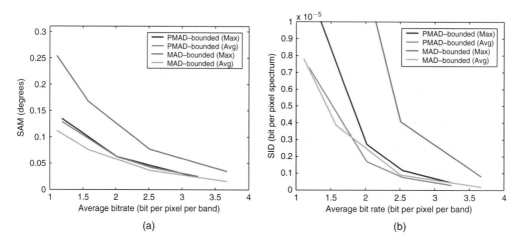

(a)

(b)

FIGURE 15.14 Spectral distortions vs. bit rate for compressed AVIRIS *Cuprite Mine '97* data: (a) spectral angle mapper (SAM); (b) spectral information divergence (SID).

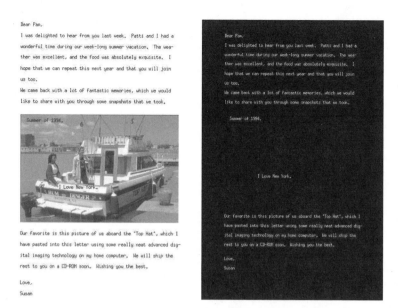

FIGURE 13.14 An example image and its ideal mask.

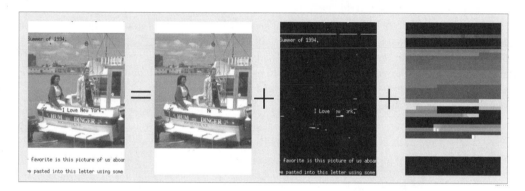

FIGURE 13.15 Zoomed portions of the MRC-decomposed layers. The fast variance-based block segmentation method was used along with the spatial block-based data-filling algorithm. It is shown the reconstructed image along with the BG–mask–FG layers.

The reason to use MRC for compression of compound images is because it is a standard. As a standard it is intended to employ standard compressors as well. JPEG is the most widely adopted compression method for color images and is well suited to compress both the FG and BG layers. For simplicity, one can use MMR to compress the binary plane, although any other binary coder may be applied. Because of the large number of variables involved, we need to keep the MRC model as simple as possible and it does not include resizing the layers before compression.

13.4.1 COMPUTING RATE AND DISTORTION PER BLOCK

JPEG is a block-based compression method. In the case of block-based compression we use block-based segmentation as well. The image is divided into blocks of 8×8 pixels. For the nth 8×8 input pixel block $\{x_n(i,j)\}$, the segmenter generates a binary mask block $\{m_n(i,j)\}$ with the same dimensions. The data-filling processor generates the layer blocks $\{L_n^{(FG)}(i,j)\}$ and $\{L_n^{(BG)}(i,j)\}$. The

FG/BG blocks are compressed and decompressed as $\{\hat{L}_n^{(FG)}(i,j)\}$ and $\{\hat{L}_n^{(BG)}(i,j)\}$, from which the block can be reconstructed as

$$\hat{x}_n(i,j) = m_n(i,j)\,\hat{L}_n^{(FG)}(i,j) + (1 - m_n(i,j))\,\hat{L}_n^{(BG)}(i,j) \tag{13.19}$$

The distortion for this block is

$$D_n = \sum_{ij} \left(x_n(i,j) - \hat{x}_n(i,j)\right)^2 \tag{13.20}$$

The (estimated) rate for a given block is given as

$$R_n = R_n^B + R_n^M + R_n^F \tag{13.21}$$

where R_n^B, R_n^M, and R_n^F are the estimated rates for compressing $\{L_n^{(BG)}(i,j)\}$, $\{m_n(i,j)\}$, and $\{L_n^{(FG)}(i,j)\}$, respectively. The reason for estimating as opposed to computing the actual rates is because of the interdependence among blocks. Even though the FG/BG layers are JPEG-compressed, the compression of the mask plane is not block-based. Binary coders generally rely on run-lengths, or line-by-line differential positions, or even object properties. Hence, for the mask layer block, it is very difficult (if not impossible) to accurately determine the amount of bits a single block will generate. Therefore, the contribution of a single block to the overall rate is not direct and one has to estimate the compressed rate for a given mask block. Furthermore, blocks in JPEG are not completely independent since there is a differential encoding of the DC coefficients. Fortunately, the number of bits saved by using block-to-block DC differential encoding in JPEG is not too significant compared to the overall bit rate per block. As for the mask, one solution is to estimate the mask rate by counting the number of horizontal transitions, to which one applies a fixed average penalty (e.g., 7 bits per transition). So, for a block with N_t transitions,

$$R_n^M = N_t * \text{penalty} \tag{13.22}$$

13.4.2 OPTIMIZED THRESHOLDING AS SEGMENTATION

As we discussed, the mask and the input block define the other layers (for fixed filling algorithms and without spatial scaling). Our goal is to find the best mapping from $\{x(i,j)\}$ to $\{m(i,j)\}$, which will optimize compression in a rate–distortion (RD) sense. We start by breaking the problem (image) into 8×8-pixel blocks $\{x_n(i,j)\}$ and finding the best mapping $\{x_n(i,j)\}$ to $\{m_n(i,j)\}$. Then, for each input block, there are 2^{64} possible mask blocks. In order to simplify the search algorithm we also impose restrictions on the quantizer table used to JPEG-compress each block. Not only do we use the same quantizer table $\{q(i,j)\}$ for both FG and BG planes, but we also use a scaled version of JPEG's *default* table $\{q_d(i,j)\}$ [28] as $q(i,j) = Qq_d(i,j)$. This simplification allows us to control rate and distortion as a function of a single variable Q as opposed to 128 of them. For a given image block $\{x_n(i,j)\}$, using a particular coding quantizer scale Q, the RD points for this block, i.e., R_n and D_n will depend on $\{x_n(i,j)\}$, Q and on the mask $\{m_n(i,j)\}$. Here, we want to optimize the segmentation in an RD sense, i.e., to minimize a block cost function

$$J_n = R_n + \lambda D_n \tag{13.23}$$

so that λ indicates an operating point to control the RD trade-off for the given block. Hence, $\{m_n(i,j)\}$ is a function of λ, Q, and of $\{x_n(i,j)\}$ of course.

We want to find each block mask $m_n(i,j)$. The simplest model for a compound image is based on the histogram of the block pixels. Pictorial, background, and text-edge areas should have histograms which are dense, flat, and bimodal, respectively. One simple approach is to find the bimodal blocks and to cluster the pixels around each of its modes. Irrespective of the method is used to perform clustering or test bimodality, the pixels will be divided by some sort of threshold. In block thresholding, the mask is found as

$$m_n(i,j) = u(t_n - x_n(i,j) - 1) \tag{13.24}$$

where t_n is the block's threshold and $u(k)$ the discrete step function ($=1$ for $k \geq 0$ and 0 otherwise). In effect, pixels below the threshold are placed in the BG layer. Since there are 64 pixels in a block, there are at most 64 different meaningful threshold values, whereby setting t_n to be less than the darkest pixel forces the mask block to be uniform, i.e., all samples imaged from one of the layers. A 65th threshold value, t_n greater than the brightest pixel, can be ignored since it achieves the same objective as the first. It has been shown that thresholding yields masks whose performance is among the best possible among all 2^{64} (R_n, D_n) pairs [10]. In other words, thresholding is RD-efficient.

The quest is to find the best threshold value t_n in an RD sense, from which one finds the mask using Equation (13.24). In a block, there are 64 pixels and therefore only up to 64 threshold values need to be tested. If we sort the block pixels $x_n(i,j)$ into a sequence $p(k)$, for each $t_n = p(k)$, we evaluate

$$J_k = R_n(k) + \lambda D_n(k) \tag{13.25}$$

where the index k denotes measurements for the kth threshold tested. We recall that λ was defined as a control parameter and is used here as a Lagrange multiplier, and that the computation of R_n and D_n assumed a particular Q. Both λ and Q are fixed for all image blocks. This is so, because it is well known that, for optimality, blocks should operate at the same slope on their RD curves [12], and because baseline JPEG does not allow for changing quantizer tables within an image. We test all $p(k)$ in a block and select the index $k = k_0$ for the minimum J_k. Then, $m_n(i,j)$ is found using Equation (13.24) for $t_n = p(k_0)$.

So, we optimize the sequence $\{t_n\}$, block by block, for fixed external variables λ and Q. Note that the overall R and D are functions of both λ and Q, i.e., $R(\lambda, Q)$ and $D(\lambda, Q)$. Given a budget R_b (or D_b), the goal is to minimize

$$\min_{\lambda,Q} D(\lambda, Q)\bigg|_{R(\lambda,Q) \leq R_b} \quad \text{or} \quad \min_{\lambda,Q} R(\lambda, Q)\bigg|_{D(\lambda,Q) \leq D_b} \tag{13.26}$$

or, equivalently, we are interested in the lower convex hull (LCH) for a bounded RD region. The search of the 2D space can be very expensive. However, there are simplifying circumstances that may reduce the search. It can be shown that an algorithm that will fulfill Equation (13.26) is as follows [10]:

1. Select a quantizer scale Q_d.
2. For every block, input $x_n(i,j)$ and find t_n which minimizes $J = R(t_n, Q_d)$.
3. Obtain mask $m_n(i,j)$ for each block using Equation (13.24).
4. With resulting mask layer in hand, compress FG/BG layers using another scaling factor Q_c.
5. Verify overall rate R (or D).
6. If R (or D) is not within parameters, adjust Q_c and go to step 4.

Let C_J denote the complexity (in terms of operations) of JPEG compressing a block. It takes about $3C_J$ to test one threshold. If, on average, k_t thresholds need to be tested per block, the segmenter complexity per block for the algorithm is $C = 3k_t C_J$. For predominantly binary images k_t is very small (minimum 2) while for pictures it can be up to 64, i.e., $6C_J \leq C \leq 192C_J$.

13.4.3 FAST THRESHOLDING

There is, however, a faster and yet efficient technique to obtain t_n [10]. The largest complexity in the RD-optimized algorithm comes from simulating compression of the FG/BG layer blocks in order to compute an estimation of the overall RD point. We want to avoid those computations as well as the block-filling operation. As we discussed, segmentation of bimodal blocks is essentially a method of finding a suitable threshold that would divide the block into two nearly uniform regions. One can modify the cost function to reflect this property. We also want to add a penalty for mask transitions. For nonbimodal blocks, the cost should be lower for uniform masks than for nonuniform ones. Once the block is thresholded into two sets, measures of variance or entropy should be good estimators of how similar the set members are. We have chosen to use the variance measure not only because the variance is simpler to compute than the entropy, but it also serves as a good rate estimator. Intuitively, if a block has low variance, it should be compressed well with a small distortion.

For each block, we sort its pixels in $p(k)$ just like in the RD case. However, we seek to minimize the following cost function:

$$J = \alpha_1 V_{BG} + \alpha_2 V_{FG} + \alpha_3 N_t \tag{13.27}$$

where α_i are the weighting factors, and V_{BG} and V_{FG} the variances of pixels in the BG and FG layer blocks. N_t is the number of horizontal transitions of the mask block (the first column of the block uses as reference the last column of the previous mask block, just like in the RD-optimized search case). For a given threshold, a mask block $\{m_n(i,j)\}$ is obtained and we define two sets:

$$\begin{aligned}
X_f &\equiv \{x_n(i,j) \mid m_n(i,j) = 1\} \\
X_b &\equiv \{x_n(i,j) \mid m_n(i,j) = 0\}
\end{aligned} \tag{13.28}$$

We define n_f and n_b as the number of pixels in the set X_f and X_b, respectively, where obviously $n_f + n_b = 64$. Then, variances are computed as

$$\begin{aligned}
V_{BG} &= \frac{\sum_{X_b} x_n(i,j)^2}{n_b} - \left(\frac{\sum_{X_b} x_n(i,j)}{n_b}\right)^2 \\
V_{FG} &= \frac{\sum_{X_f} x_n(i,j)^2}{n_f} - \left(\frac{\sum_{X_f} x_n(i,j)}{n_f}\right)^2
\end{aligned} \tag{13.29}$$

which can be efficiently implemented. Since thresholds are sorted, as we increment k, we will be effectively moving pixels from the set X_b to the set X_f. Thus, part of the computation does not change in each step. First, we set the mask to be all zeros effectively placing the whole block in the background, which is equivalent to set t_n to be the smallest of $x_n(i,j)$. Then, we set $k = 0$ and initialize the following variables:

$$s_b = \sum_{ij} x_n(i,j), \qquad v_b = \sum_{ij} x_n^2(i,j), \qquad n_b = 64 \tag{13.30}$$

$$s_f = v_f = n_f = N_t = 0 \tag{13.31}$$

We then compute

$$V_{BG} = \frac{v_b}{n_b} - \left(\frac{s_b}{n_b}\right)^2, \qquad V_{FG} = \frac{v_f}{n_f} - \left(\frac{s_f}{n_f}\right)^2 \tag{13.32}$$

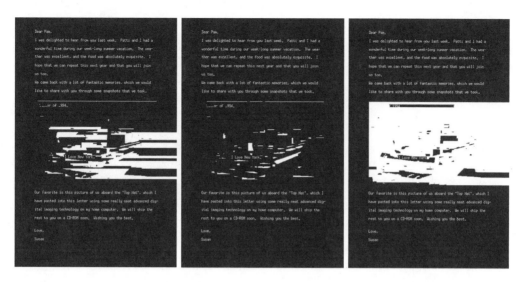

FIGURE 13.16 Mask examples for image "compound1" and $\alpha_3 = 200$. From left to right: $\alpha_1 = \alpha_2 = 1$; $\alpha_1 = 1$ and $\alpha_2 = 5$; $\alpha_1 = 5$; and $\alpha_2 = 1$

where, if n_f or n_b are 0, the corresponding variance is set to 0. Next, we compute Equation (13.27). As we increase the threshold to the next pixel in the sorted list, we increment k, and the mask changes so that we need to recompute N_t. Some n_p pixels that form a set X_p are then moved from X_b to X_f. Hence, we have to compute $s_p = \sum_{X_p} x_n(i,j)$ and $v_p = \sum_{X_p} x_n^2(i,j)$. Then, we update our variables as

$$s_f = s_f + s_p, \qquad v_f = v_f + v_p, \qquad n_f = n_f + n_p \qquad (13.33)$$

$$s_b = s_b - s_p, \qquad v_b = v_b - v_p, \qquad n_b = n_b - n_p \qquad (13.34)$$

and recompute V_{BG}, V_{FG}, and J_k. We repeat the process until X_b is empty. We test all 65 possibilities, i.e., from X_f empty to X_b empty in order to make the algorithm symmetric. We select $t_n = p(k)$ for $\min_k (J_k)$ and we compute the final mask using Equation (13.24).

The overall computation (in terms of operations) is much more reasonable than the RD-optimized counterpart. The overall processing has a computational complexity C not superior to simply compressing the block with JPEG once, i.e., $C \approx C_J$.

As for the weights, without loss of generality, we can normalize one of them (e.g., $\alpha_1 = 1$). The choice of weights is empirical; however there are some few guidelines that we can use for our advantage. We want to place the complex graphics and pictures in only one of the planes and that can be done with $\alpha_2 \neq \alpha_1$. Also, variances are much larger numbers than the number of transitions (given that pixels range from 0 to let us say 255). Thus, α_3 should be a very large number. In our experiments, the values of $\alpha_2 = 5$ and $\alpha_3 = 200$ were efficient for segmentation of images containing sharp contrasting black-on-white text. Some masks are shown in Figure 13.16 for the cases where $\alpha_1 = \alpha_2 = 1$, $\alpha_1 = 1$, and $\alpha_2 = 5$, and the case $\alpha_1 = 5$ and $\alpha_2 = 1$. Although all cases segment the text well, the placement of picture blocks differs. It might be advantageous for other reasons to use $\alpha_1 = 1$, $\alpha_2 = 5$ in order to place the picture along with the background. These values are recommended.

13.4.4 PERFORMANCE

We can compare the MRC performance using the images shown in Figure 13.17. The images range from purely graphics (graphics) to purely pictorial (baby), with two other mixed images with graphics

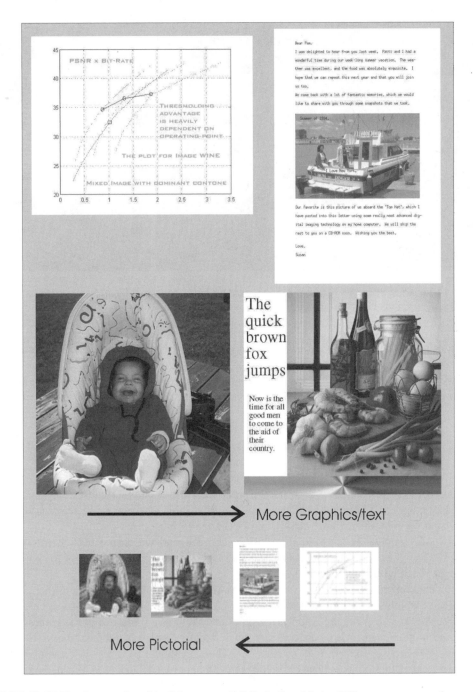

FIGURE 13.17 Test images: "graphics," "compound1," "baby," and "wine." The contents range from purely pictorial to purely graphics images.

(compound1) and pictorial (wine) dominance. Clearly, the more the graphics, the higher the advantage over a single coder such as JPEG. For an image such as "baby," our MRC approach has the disadvantage of encoding the overhead of two planes and is expected to be outperformed by JPEG. We compared the following compressors: (1) MRC with optimized thresholding segmentation and

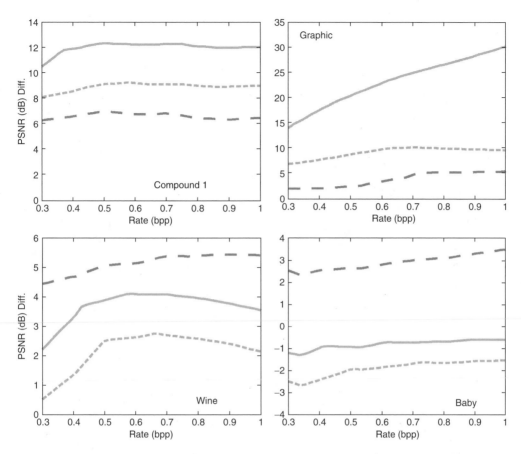

FIGURE 13.18 PSNR difference plots compared with the PSNR achieved by JPEG compression. The imaginary line at 0 is the JPEG reference performance. The solid line is for MRC compression using simple JPEG+MMR+JPEG and the optimized thresholding segmentation algorithm with $Q_d = 1$; The broken line is the same but with the fast variance-based thresholding algorithm. The dotted line is for compressing the whole image using single plane JPEG 2000. Image name is noted for each plot set.

$Q_d = 1$; (2) MRC with a variance-based thresholding and $Q_d = 1$; (3) single-plane JPEG 2000; and (4) single-plane JPEG. For the MRC approach, we computed RD curves by varying Q_c, which are shown in Figure 13.18. The distortion measure chosen was PSNR [12] and the plots present results in differential PSNR compared with JPEG, i.e., how many dB improvement would there be if we replace JPEG by the respective coder (MRC or JPEG 2000).

The PSNR difference against JPEG is extremely large for the graphics case since MRC quickly approaches the lossless state. The image compound1 is one of the best representatives of the target compound images. In this case, the PSNR difference is a staggering 12 dB over JPEG and several dB over JPEG 2000. The performance of the variance-based method is very close to that of the RD-based one, except for pictorial images. As the image becomes purely pictorial, the losses are about or below 1 dB for the RD-based segmentation compared with JPEG. This small loss is a very positive sign: even if by mistake a pictorial image is to be segmented, smart segmentation can minimize losses. Apart from MRC approaches, JPEG 2000 serves as an upper bound in single-layer performance. A comparison between JPEG 2000-based MRC and JPEG 2000 will be carried in the next section.

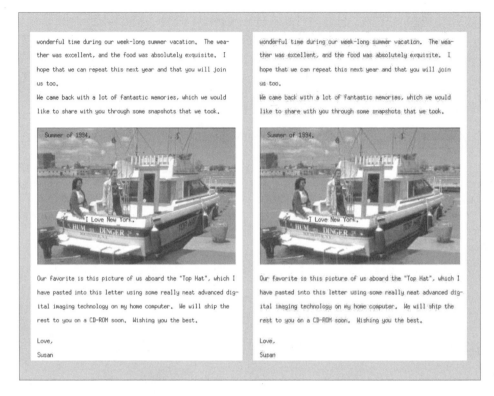

FIGURE 13.19 Reconstructed images after compression at 0.45 bpp. Left: MRC compressed, optimized thresholding segmentation method (PSNR = 37.49 dB). Right: JPEG (PSNR = 25.33 dB).

A sample of reconstructed images for comparison is shown in Figure 13.19. The superior quality of the MRC-coded image over the JPEG-coded one is easily noticeable in both pictorial and text areas. The images were compressed to 0.45 bpp before decompression, yielding a 12 dB difference in PSNR.

In another example, the image tested is shown in Figure 13.20. It is a typical compound color image, with graphics, text, and pictures. After compression at a 70:1 ratio, using JPEG, JPEG 2000, and MRC, the decompressed images are shown in Figure 13.21 and Figure 13.22. Figure 13.21 shows pieces of the text region compressed with different methods, while Figure 13.22 shows the same for a pictorial region. Note that MRC clearly yields better image quality at both regions for the given compression ratio.

13.5 MRC WITHIN JPEG 2000

JPEG 2000 is a newer compression standard designed to upgrade the original JPEG [31]. It is based on wavelet transforms and contextual coding of bit planes, achieving a near state-of-the-art performance by the time it was devised. The standard contains 12 parts, but only two of them are directly relevant to us here: parts 1 and 6. In JPEG 2000 part 1, it is defined as the core decoder, i.e., the basic compression algorithm that everyone needs to support. For example, part 2 contains extensions, while part 3 deals with motion images, but part 6 is the one directly related to document compression since it defines a compound image file format. In effect, JPEG 2000 part 6 defines an MRC format within JPEG 2000 [18–24]. While JPEG 2000 core files (implicitly part 1) are known as JP2, those for part 6 are known as JPM files (M for multilayer). We will refer to JPM as a short for JPEG 2000 part 6 and to JP2 as a short for the core coder in part 1.

January 31, 2001

Dear Mom and Dad,

*How are both of you doing? I thought I would drop a line
to say hi. Fanny, little Danny, and I are doing well. As
you can see by the picture, little Danny isn't quite so little!
Isn't this letter really great! I took a picture of Danny
that was on a Kodak PhotoCD, and I merged it onto this
letter using my computer. I then printed the letter using
a color inkjet printer I just bought...*

*Danny's wearing the gorgeous BLUE sweater you gave
him last time you were visiting. It just brings out the
RED in his lips and cheeks. He definitely gets his good
looks from his mother!*

Take care of yourselves and write soon.

Love,

Michael

FIGURE 13.20 (See color insert following page 336) Another compound image for testing. Image size is
1400 × 1024 pixels.

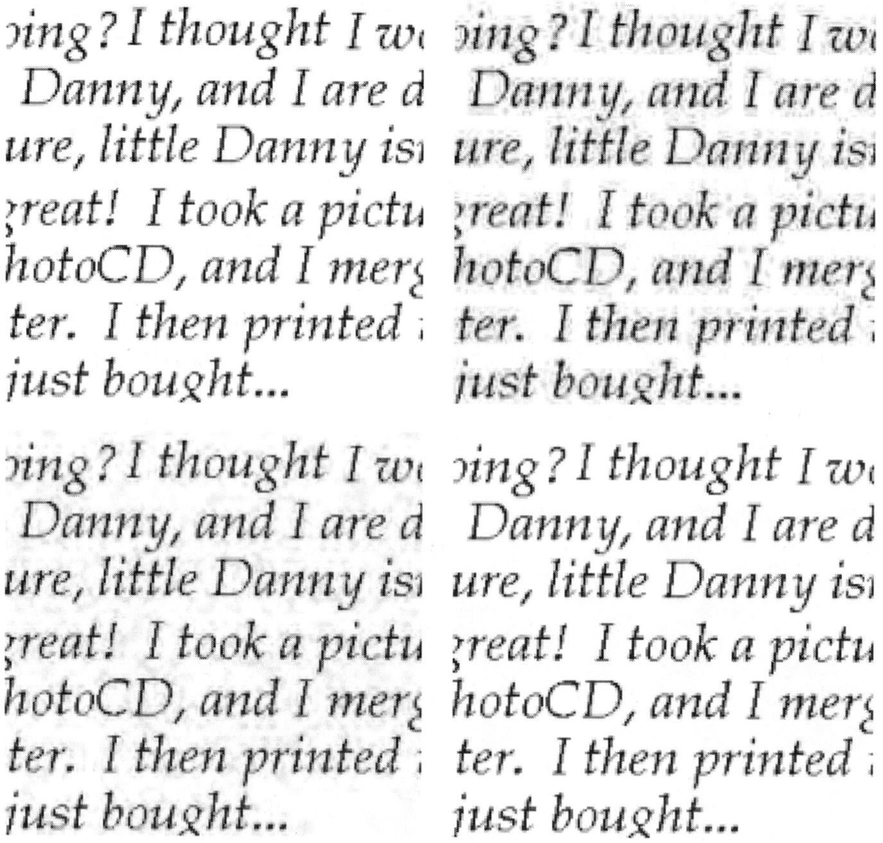

FIGURE 13.21 Enlarged portion of the text region of original and reconstructed images at a compression ratio of 70:1. Top left: original; top right: JPEG baseline; bottom left: JPEG 2000; and bottom-right: MRC using JPEG+MMR+JPEG. Portion size is 256 × 256 pixels.

JPM files allows for multipage documents, where each page can be made of a number of layout objects. Actually, JPM employs the concept of page collections wherein a number of individual pages are referenced together. A page collection is in effect a list of pointers to individual pages or to other page collections. An example is a book, where the main page collection points to the chapters' page collections which point to the sections' page collections which point to the individual pages. The ordering of pages in the document is defined by the topmost page collection and by those collections pointed by it.

Each page is imaged following the soft mask multilayer model in Section 13.2.3. It allows for planes with arbitrary size, resolution, and position, which are properly scaled and positioned before imaging. It contains a background and N mask+image pairs known as layout objects. Each layout object contains or assumes a mask and an image layer, e.g., if there is no image associated with the mask, a base color is specified and used. Each layer can be compressed using one of a number of coders including JP2, JPEG, JBIG-2, MMR, etc.

JPM inherits several benefits from the JPEG 2000 family such as the file format and the associated metadata support (e.g., XML). However, since JPM can contain a multipage document, metadata can be associated to a page or group of pages or even to the constituent MRC image layers. All references can be self contained within the same JPM file or can be a remote object. Each MRC layer would contain an URL, an offset, and a length. Note that pages or page collections can be referenced remotely.

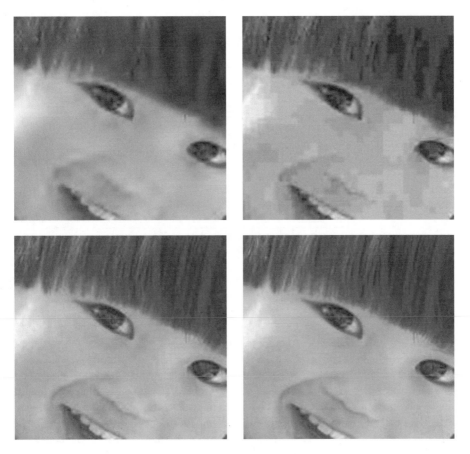

FIGURE 13.22 Enlarged portion of a pictorial region of original and reconstructed images at a compression ratio of 70:1. Top left: original; top right: JPEG baseline; bottom left: JPEG 2000; and bottom-right: MRC using JPEG+MMR+JPEG. Portion size is 256×256 pixels.

The description given here of JPEG 2000 part 6 multilayer support is very incomplete and can be found in much more detail elsewhere.

A successful MRC compression can only be achieved by an efficient segmentation strategy, as we have discussed. Since JPEG 2000 uses wavelet compression, which has no block structure, one shall use a general segmentation. Recently, an efficient segmentation for JP2-based MRC has been proposed [25], which is the follow-up of a block-based method [26]. DjVu [4] has a built-in segmenter which has been tested in products, typically for very high-compression scenarios. There is also multilayer products such as Digipaper [15] and LuraDocument [1] which employ general (non-block-based) segmentation. The goals of a segmentation algorithm can be either to select objects as to facilitate the image representation, or if compression is the main goal, then one may remove sharp edges from the FG–BG planes and move them into the masks. A possible strategy is to find background-contrasting, contiguous, uniform-color objects and move these object shapes onto layout objects.

13.5.1 JP2-Based JPM

Even though any of the many coders available can be used to compress each plane, we suggest using solely JP2 for compression. This is possible because JP2 allows for the compression of either

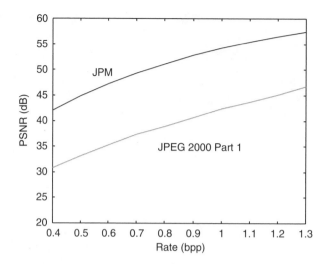

FIGURE 13.23 PSNR comparing JPEG 2000 and JPM for image compound1 with an ideal mask. For JPM, we used JPEG 2000 to compress all three planes.

gray-level or binary images. In order to compress binary images with JP2, one may set the number of wavelet levels to 0 and the bit depth to 1. This will essentially skip the wavelet transform and compress the binary data with the context-driven arithmetic coder which is typically used to compress the wavelet bit plane in JP2. Hence, the mask is lossless-compressed. One might see in the literature, comments to the effect that the JP2 binary coder has inferior performance than other binary coders such as MMR and JBIG-2. However, the difference is not constant and is very small. Furthermore, the mask plane does not spend too many bits in MRC anyway, so that JP2 is a very adequate substitute for any other binary coder in a JPM context.

We can use a three-plane decomposition: FG, BG, and the binary mask. The FG plane will have the text colors and can be highly compressed with JP2. The BG plane contains the paper background and pictures and is also compressed with JP2, but under moderate compression. The mask plane is compressed without loss with JP2, as described above. Thus, only the JP2 compression engine is needed to compress the JPM file. It is the same for decompression. As an example, if we use the compound image and mask shown in Figure 13.14, we compressed the image using JP2-based JPM, obtaining the RD curves shown in Figure 13.23. We also compared JPM with JP2 single-plane compression. Note the large disparity between JP2 and JP2–JPM: a staggering PSNR difference, beyond 10 dB. JPM is indeed an efficient compound image compression scheme.

13.6 CONCLUSIONS

In this chapter, we tried to provide an overview of predominant techniques for the compression of raster compound documents. We emphasize the term "raster" since we did not discuss any techniques for images in vector form. For raster images, the most important technique of all is the MRC model, mostly because it became an international standard. There are a number of variations to the MRC model as well as unrelated techniques, many of them attaining a very good performance. Nevertheless the MRC model is the predominant technique not only because it is already a standard, but also because it achieves very good RD performance.

The main backdraw of MRC (as well as any mixed mode compressor) is the need for reliable segmentation. We have shown how to overcome this problem and implement efficient segmentation for a simple JPEG–MMR–JPEG scheme. This method is block-based and would not work unless we use block-based compression for the FG–BG planes. There is ongoing work for MRC-driven document segmentation for the general case. For the general case, the JPM profile for JPEG 2000 provides an excellent vehicle for MRC compression (even allowing the JPEG–MMR–JPEG case). In JPM, documents can be efficiently compressed using only a JP2 engine for all planes.

Compound documents stress compression techniques. However, the tools we have at our disposal make it quite manageable to compress complex documents to very low bit rates with excellent quality. That enables a number of applications ranging from archival to web-based document retrieval.

We caution the reader for the fact that this chapter is merely introductory. We encourage the reader to seek the references for more detailed discussions and descriptions on MRC and on compound document compression.

REFERENCES

1. Algovision Luratech's Luradocument.jpm. See http://www.luratech.de
2. ANSI (American National Standard Institute) X3.110-1983 Videotex/Teletext Presentation Level Protocol Syntax (North American PLPS).
3. L. Bottou, P. Haffner, P. Howard, and Y. LeCun, Color documents on the Web with DjVu, *Proceedings of IEEE International Conference on Image Processing*, Kobe, Japan, Oct. 1999.
4. L. Bottou, P. Haffner, P. Howard, P. Simard, Y. Bengio, and Y. LeCun, High quality document image compression using DjVu, *J. Electron. Imaging*, 7, 410–425, 1998.
5. R.R. Buckley and J.W. Reid, A JPEG 2000 compound image fie reader/writer and interactive viewer. *SPIE Annual Meeting 2003*, San Diego, Proceedings of SPIE, Vol. 5203, no. 32, Aug. 2003.
6. R. Buckley, D. Venable, and L. McIntyre, New developments in color facsimile and internet fax, *Proceedings of IS&T's Fifth Color Imaging Conference*, pp. 296–300, Scottsdale, AZ, Nov. 1997.
7. H. Cheng and C. Bouman, Document compression based on multiscale segmentation, *Proceedings of IEEE International Conference on Image Processing*, 25PS1.8, Kobe, Japan, Oct. 1999.
8. R. De Queiroz, On data-filling algorithms for MRC layers, *Proceedings of IEEE International Conference on Image Processing, ICIP*, Vancouver, Canada, Vol. II, pp. 586–589, Sep. 2000.
9. R. De Queiroz, R. Buckley, and M. Xu, Mixed raster content (MRC) model for compound image compression, *Proceedings of EI'99, VCIP*, SPIE, Vol. 3653, pp. 1106–1117, Feb. 1999.
10. R. De Queiroz, Z. Fan, and T. Tran, Optimizing block-thresholding segmentation for multi-layer compression of compound images, *IEEE Trans. Image Process.*, 9, 1461–1471, 2000.
11. H.T. Fung and K.J. Parker, Segmentation of scanned documents for efficient compression, *Proceedings of SPIE: Visual Communication and Image Processing*, Orlando, FL, Vol. 2727, pp. 701–712, 1996.
12. A. Gersho and R.M. Gray, *Vector Quantization and Signal Compression*, Kluwer Academic, Hingham, MA, 1992.
13. HP PCL/PJL Reference (PCL 5 Color) — Technical Reference Manual, Aug. 1999.
14. J. Huang, Y. Wang, and E. Wong, Check image compression using a layered coding method, *J. Electron. Imaging*, 7, 426–442, 1998.
15. D. Huttenlocher and W. Rucklidge, DigiPaper: a versatile color document image representation, *Proceedings of IEEE International Conference on Image Processing*, 25PS1.3, Kobe, Japan, Oct. 1999.
16. IETF RFC 2301, File Format for Internet Fax, L. McIntyre, S. Zilles, R. Buckley, D. Venable, G. Parsons, and J. Rafferty, March 1998. ftp://ftp.isi.edu/in-notes/rfc2301.txt.
17. ISO/IEC JTC1/SC29 WG1, JPEG 2000 Committee, Final Draft International Standard, Sep. 25, 2000.
18. ISO-IEC 15444-6:2003, Information Technology — JPEG 2000 Image Coding System — Part 6: Compound Image File Format. http://www.jpeg.org/CDs15444.html.
19. ITU-T Study Group 8, Question 5, Draft Recommendation T.44, Mixed Raster Content (MRC), May 1997.

20. ITU-T Rec. T.4, Standardization of Group 3 Facsimile Apparatus for Document Transmission, July 1996.

21. ITU-T Rec. T.6, Facsimile Coding Schemes and Coding Control Functions for Group 4 Facsimile Apparatus, November 1988.

22. ITU-T Rec. T.82, Information Technology — Coded Representation of Picture and Audio Information — Progressive Bi-Level Image Compression, March 1995. Also ITU-T Rec. T.85, Application Profile for Recommendation T.82 — Progressive Bi-Level Image Compression (JBIG Coding Scheme) for Facsimile Apparatus, August 1995.

23. JBIG2 Working Draft WD14492, ISO/IECJTC1/SC29 JBIG Comm., 21 Aug. 1998, http://www.jpeg.org/public/jbigpt2.htm

24. K. Jung and R. Seiler, Segmentation and Compression of Documents with JPEG 2000, http://www.math.tu-berlin.de/ seiler/JPEG2000final1.1.pdf

25. D. Mukherjee, C. Crysafis, and A. Said, JPEG 2000-matched MRC compression of compound documents, *Proceedings of IEEE International Conference on Image Processing, ICIP*, Rochester, NY, ed. Sep. 2002.

26. D. Mukherjee, N. Memon, and A. Said, JPEG-matched MRC compression of compound Documents, *Proceedings of IEEE International Conference on Image Processing, ICIP*, Thesaloniki, Greece, Oct. 2001.

27. *PDF Reference, Version 1.4,* 3rd ed., Adobe Systems Inc., 2003.

28. W.P. Pennebaker and J.L. Mitchell, *JPEG: Still Image Compression Standard,* Van Nostrand-Reinhold, New York, 1993.

29. G. Reid, *Postscript Language: Program Design,* Addison-Wesley, Reading, MA, 1987.

30. A. Said and A. Drukarev, Simplified segmentation for compound image compression, *Proceedings of IEEE International Conference on Image Processing*, 25PS1.5, Kobe, Japan, Oct. 1999.

31. D.S. Taubman and M.W. Marcellin, *JPEG 2000, Image Compression Fundamentals, Standards and Practice*, Kluwer Academic, Norwell, MA, 2002.

14 Trends in Model-Based Coding of Multidimensional Medical Data

Gloria Menegaz

CONTENTS

14.1 Introduction 352
14.2 Requirements 353
14.3 State of the Art 354
    14.3.1 2-D Systems 354
        14.3.1.1 Lossless Techniques 355
        14.3.1.2 Lossy Techniques 356
        14.3.1.3 DCT-Based Techniques 356
        14.3.1.4 Wavelet-Based Techniques 357
    14.3.2 3-D Systems 360
        14.3.2.1 3-D Set Partitioning Hierarchical Trees 360
        14.3.2.2 Cube Splitting 361
        14.3.2.3 3-D Quadtree Limited 361
        14.3.2.4 CS-Embedded Block Coding (CS-EBCOT) 361
        14.3.2.5 3-D DCT 362
    14.3.3 3-D ROI-Based Coding 362
    14.3.4 Model-Based Coding 363
14.4 3-D/2-D ROI-Based MLZC: A 3-D Encoding/2-D Decoding Object-Based Architecture. 364
14.5 Object-Based Processing 365
    14.5.1 3-D Analysis vs. 2-D Reconstruction 368
14.6 Multidimensional Layered Zero Coding 368
    14.6.1 Layered Zero Coding 368
    14.6.2 MLZC Coding Principle 369
    14.6.3 Spatial Conditioning 369
    14.6.4 Interband Conditioning 371
    14.6.5 Bitstream Syntax 371
        14.6.5.1 Global Progressive (G-PROG) 371
        14.6.5.2 Layer per Layer Progressive (LPL-PROG) 371
        14.6.5.3 Layer per Layer (LPL) Mode 372
    14.6.6 3-D Object-Based Coding 372
        14.6.6.1 Embedded Zerotree Wavelet-Based Coding 373
        14.6.6.2 Multidimensional Layered Zero Coding 374
    14.6.7 3-D/2-D MLZC 374
    14.6.8 3-D/2-D Object-Based MLZC 374
14.7 Results and Discussion 375
    14.7.1 Datasets 375

14.7.2   3-D/2-D MLZC 377
14.7.3   3-D Object-Based MLZC 379
          14.7.3.1   System Characterization 379
          14.7.3.2   Object-Based Performance 381
14.7.4   3-D/2-D Object-Based MLZC 384
14.8   Conclusions 385
References 386

14.1   INTRODUCTION

It is a fact that medical images are increasingly acquired in digital format. The major imaging modalities include computed tomography (CT), magnetic resonance imaging (MRI), ultra sonography (US), positron emission tomography (PET), single photon emission computerized tomography (SPECT), nuclear medicine (NM), digital subtraction angiography (DSA), and digital flurography (DF). All these techniques have made the view of cross-sections of the human body available, permitted to navigate inside it and to design novel minimally invasive techniques to investigate many pathologies. The numeric representation enables new functionalities for both data archiving and processing that improve health care. The exploitation of image processing techniques in the field of medical imaging represented a breakthrough allowing to manipulate the visual diagnostic information in a novel perspective. Image analysis and rendering are among the many examples. Feature extraction and pattern recognition are the basis for the automatic identification and classification of different lesions (like melanomas, tumors, stenosis) in view of the definition of automatic systems supporting the formulation of the diagnosis by a human expert. Volume rendering and computer vision techniques permitted the development of computer-assisted surgery, stereotaxis, and many other image-based surgical and radiological operations.

The need to daily manipulate large volumes of data rose the issue of compression. The last two decades have seen an increasing interest for medical image coding. The objective is to reduce the amount of data to be stored and transmitted while preserving the features diagnosis is based on. The intrinsic difficulty of this issue when facing the problem with a wide perspective, namely without referring to a specific imaging modality and disease, led to the general agreement that only compression techniques allowing to recover the original data without loss (*lossless* compression) would be suitable. However, the increasing demand of storage space and bandwidth within the clinical environment has encouraged the development of *lossy* techniques providing a larger compression gain and thus a more efficient management of the data. Another push came from the picture archiving and communication systems (PACS) community, envisioning an *all digital* radiologic environment in hospitals including acquisition, storage, communication, and display. Image compression enables fast recovery and transmission over the PACS network. PACS aims at providing a system integration solution facilitating different activities besides computer-aided diagnosis, like teaching, reviewing of the patient's records, and the mining of the information of interest. It is worth mentioning that in these cases the preservation of the original information is less critical, and a loss due to compression could in principle be tolerated. Last but not the least, applications like teleradiology would be prevented without compression. The transmission over wide area networks enclosing low-bandwidth channels like telephone lines or Integrated Service Digital Networks (ISDN) could hinder the diffusion of this kind of applications.

As it was easy to expect, this rose a call for a regulation for setting the framework for the actual exploitation of such techniques in the clinical environment. The manipulation of medical information indeed implicates many complex legal and regulatory issues, at least as far as the processed data are supposed to be used for formulating a diagnosis.

Furthermore, as it is reasonable to expect in any multimedia multiuser framework, the need to exchange data and share resources, namely to *communicate*, calls for the definition of standards establishing a common language and protocol. In the digital signal processing framework, this led to the JPEG2000 [25] and MPEG4 [27] standards for still images and video, respectively, as well as MPEG7 [26] and MPEG21 [28] for more semantically related issues, like indexation via content-based description. Even though such standards address the problem for natural images, some attention has also been devoted to the particular case of medical images. JPEG2000 Part 10 addresses the issue of medical data compression, with focus on three-dimensional data distributions [56]. However, no agreement has yet been reached on the subject, which is still under investigation.

On top of this, with the introduction of digital diagnostic imaging and the increasing use of computers for clinical applications, the american college of radiologists (ACR) and the national manufacturers association (NEMA) already in the 1970s recognized the need for a standard method for transferring images and the associated information among devices manufactured by various vendors. The ACR and NEMA formed a joint committee in 1983 to face the problem. This led to the development of a standard, which is currently designated as the digital imaging and communications in medicine (DICOM) standard [2], meant to facilitate the interoperability of medical imaging equipments. More specifically, it sets forth the set of protocols to be followed by devices claiming conformance with the standard. This includes the syntax and semantics of the commands and the associated information that can be exchanged and, for media communication, a file format and a medical directory structure to facilitate the access to the images and the related information stored on the interchange media. Of particular interest here is that in its Part PS 3.5 it describes how DICOM standard is applied to construct and encode the data and the support of a number of standard image compression techniques. Among these are JPEG-LS and JPEG2000.

The field of medical data compression is challenging and requires the interest and efforts of both the signal processing and the medical communities.

This chapter is organized as follows. Section 14.2 summarizes the features and functionalities, which are required for a coding system suitable for the integration in a modern PACS system. Section 14.3 provides an overview on the state of the art; more specifically, Section 14.3.1 focuses on 2-D data (e.g., still images), and Section 14.3.2 is devoted to 3-D systems (e.g., volumetric data). Section 14.4 describes the 3-D encoding/2-D decoding object-based MLZC architecture. Section 14.5 describes the strategies employed for reaching independent object processing while avoiding border artifacts. Section 14.6 describes the coding techniques that have been used in the MLZC system as well as their generalization for region-based processing. The performance is discussed in Section 14.7, and Section 14.8 derives conclusions.

14.2 REQUIREMENTS

In the last decade, new requirements have emerged in the field of compression going beyond the maximization of the coding gain. Among the most important ones are progressively refinable up to lossless quality and region of interest (ROI)-based processing. Fast inspection of large volumes of data requires compression schemes be able to provide a swift access to a low-quality version of the images and to a given portion of them corresponding to the diagnostically relevant segment. Progressiveness (or *scalability*) allows to improve the quality of the recovered information by incrementally decoding the bitstream. In teleradiology this enables the medical staff to start the diagnosis at a very early stage of transmission, and to eventually delineate the region of interest to switch to a ROI-based mode. Two scalability options are possible: by quality and by resolution. In the scalability by quality mode, the resolution of the image does not change during the decoding process. This means that during the decoding process the image is initially recovered at full size but with low quality, e.g., different types of artifacts (depending on the coding algorithm) degrade its appearance. In the scalability by resolution mode, the encoded information is organized such that a reduced size version of the image is recovered at full quality just after the decoding starts; the resolution (size) then increases with the amount of the decoded information.

Another basic requirement concerns the rate–distortion performance. While both the lossless and lossy representation are needed to be available on the same bitstream, the system should be designed such that an *optimal* rate–distortion behavior is reached for any decoding rate. These features were not supported by the old JPEG standard, neither was ROI-based processing, making it obsolete. Besides these general requirements, other domain-specific constraints come into play when restraining to medical imaging. In this case, lossless capabilities become a must at least as far as the data are supposed to be used for diagnosis. Of particular interest in the medical imaging field are indeed those systems that are able to provide lossless performance and fast access to the information of interest, which translates in ROI-based capability and low computational complexity.

Historically, medical image compression has been investigated by researchers working in the wider field of image and video coding. As a natural consequence, the technological growth in this field is in some sense a byproduct of the progresses in the more general framework of natural image and video coding. As it is reasonable to expect, there is no golden rule: different coding algorithms are best fit to different types of images and scenarios. However, few global guidelines can be retained. One is the fact that exploitation of the full data correlation, in general, improves compression. Accordingly, three-dimensional coding systems are more suitable for the application to volumetric data, and the integration of some kind of "motion compensation" could lead to better results for time-varying data, including both image sequences (2-D + time) or volume sequences (3-D + time). On the other end, the same engines that proved to be the most effective for images and videos are also the most efficient when applied to medical images, in general. However, depending on the imaging modality and application, some are more suitable than others to fulfill a certain performance either in terms of compression factor or, generally, with respect to a desired functionality.

A particularly challenging requirement concerns the so-called *visually lossless* mode. The goal is to design a system that allows some form of data compaction without affecting the diagnostic accuracy. This implies the investigation of issues that go beyond the frontiers of classical signal processing, and involves different fields like vision sciences (to model the sensitivity of the visual system) and artificial intelligence (for the exploitation of the semantic a-priori knowledge on the image content). Even though the investigation of these issues is among the most promising paths of the current trends in image processing, it is beyond the scope of this chapter and it will not be discussed further.

14.3 STATE OF THE ART

In what follows, we review some of the most widespread classical coding methods that have been used to compress multidimensional medical data. Some of them respond better to the requirements summarized in the previous section while others fit best for some specific domains or applications. The choice of a radiologic compression scheme results from a complex trade-off between systemic and clinical requirements. Among the most important ones are image characteristic (resolution, signal-to-noise ratio, contrast, sharpness, entropy); image use (telemedicine, archiving, teaching, diagnosis); type of degradation introduced by lossy coding; practical issues like user-friendliness, real-time processing, and cost of implementation and maintenance.

14.3.1 2-D Systems

A first review of radiologic image compression appeared in 1995 [65]. In their paper, Wong et al. reviewed some of the techniques for lossless and lossy compression that have been applied to medical images so far. Among the lossless ones there were differential pulse code modulation (DPCM) [3], hierarchical interpolation (HINT) [49], bit-plane encoding (BPE) [34], multiplicative autoregression (MAR) [32], and difference pyramids (DP) [20]. However, the potential advantages of lossy techniques are clear, justifying the efforts of many researchers. Among those summarized in [65] were the 2-D discrete cosine transform (DCT), implemented either block-wise or on the entire image (e.g., *full-frame DCT*) [22,68], the lapped orthogonal transform (LOT) [23], and other classical methods like vector quantization [13] and adaptive predictive coding [33].

Hereafter, we mention some of the most interesting contributions.

14.3.1.1 Lossless Techniques

14.3.1.1.1 Differential Pulse Code Modulation
DPCM is a simple coding method based on prediction in the image domain. The value of the current pixel is approximated by the linear combination of some neighboring pixels according to some weighting factors. The prediction error is entropy-coded and transmitted. According to Wong [65], the compression factors that could be obtained with this technique were in the range 1.5–3, depending on the entropy coder. The main disadvantage of DPCM is that progressiveness is not allowed because the image is reconstructed pixel-wise.

14.3.1.1.2 Hierarchical Interpolation
Pyramids have extensively been exploited for data compression in different guises. Basically, subsampling is iterated on progressively lower-resolution versions of the original image up to a predefined level. The lowest resolution is encoded and transmitted. The lossless representation is obtained by successively encoding and transmitting the interpolation residuals between subsequent pyramid levels. This technique has been tested by different researchers. As an example, in the HINT implementation of Roos et al. [51] the compression ratios ranged from about 1.4 for 12-bit 512×512 MR images to 3.4 for 9-bit 512×512 angiographic images. This system was also generalized for 2-D image sequences to investigate the usefulness of exploitation of the temporal dimension. For interframe decorrelation, different approaches were considered, including extrapolation- and interpolation-based methods, methods based on local motion estimation, block motion estimation, and unregistered decorrelation. The test set consisted of sequences of coronary X-ray angiograms, ventricle angiograms, and liver scintigrams, as well as of a (nonmedical) videoconferencing image sequence. For the medical image sequences the authors concluded that the interpolation-based methods were superior to extrapolation-based methods and that the estimation of interframe motion, in general, was not advantageous [50].

14.3.1.1.3 Bit-Plane Encoding
BPE in the image domain can be seen as a successive approximation quantization that can lead to a lossless representation. When it is implemented in the image domain, the subsequent bit planes of the gray-level original image are successively entropy-coded and transmitted. Even though it does not outperform the other methods considered by Rabbani and Jones [46], the main interest of this technique is that it enables progressiveness by quality. This method has then been successfully applied for encoding the subband coefficients in wavelet-based coding, enabling scalability functionalities.

14.3.1.1.4 Multiplicative Autoregression
MAR is based on the assumption that images are locally stationary and as such they can be approximated by a 2-D linear stochastic model [15]. The basic blocks of a MAR encoder are a parameter estimator, a 2-D MAR predictor, and an entropy coder. A multi-resolution version (MMAR) has also been elaborated by Kuduvalli [32]. MAR and MMAR techniques have shown to outperform other methods on some datasets; the main disadvantage is implementation complexity.

More recently, different solutions have been proposed for lossless compression. Among these, the context-based adaptive lossless image codec (CALIC) [66] proposed by Wu has proved to be the most effective and, as such, it is often taken as the benchmark for lossless medical image compression.

14.3.1.1.5   CALIC

The basic principle of CALIC is to use a large number of modeling contexts for both (nonlinear) adaptive prediction and entropy coding. The problem of context dilution is avoided by decoupling the prediction and coding phases: CALIC only estimates the *expectation* of the prediction errors conditioned on a large number of compound contexts instead of estimating a large number of error conditional probabilities. Such expectation values are used to correct the prediction of the current pixel value as obtained from spatial prediction, and the resulting residual is entropy-encoded using only eight "primary" contexts. These contexts are formed by quantizing a local error energy estimator, which depends on the local gradients as well as on the noncorrected prediction error. CALIC was proposed to the ISO/JPEG as a candidate algorithm for the standard [14] and it was able to provide the lowest lossless rate on six out of seven image classes used as test set (medical, aerial, prepress, scanned, video, and compound documents). However, the low complexity context-based lossless image compression algorithm (LOCO-I) of Weinberger [63] was chosen owing to its lower computational cost and competitive performance. The superiority of CALIC with respect to the other state-of-the-art techniques was further proven in another set of tests described by Clunie [12]. In particular, it was run on a set of 3679 images including CT, MR, NM, US, IO, CR, and digitized x-rays (DX) and compared to different compression systems including JPEG-LS [1] and JPEG2000 in lossless mode. Again, results show that CALIC equipped with an arithmetic coder was able to provide the highest compression rate except for one modality for which JPEG-LS did better. These results are also consistent with those of Kivijarvi [31] and Deneker [17].

14.3.1.2   Lossy Techniques

Lossy methods implicitly rely on the assumption that a *visually lossless* regime can be reached which allows to obtain high coding gains while preserving those features which determine diagnostic accuracy. Though, this is still a quite ambitious challenge, which encloses many open issues going beyond the boundaries of the field of signal processing. The difficulty in identification of the features that are relevant for the formulation of the diagnosis of a given pathology based on a given imaging modality is what makes the human intervention unavoidable in the decision process. Systems for automatic diagnosis are still in their early stages, and are undergoing a vast investigation. If it is difficult to automatically extract the features, to *quantify* the amount of degradation, which is *acceptable* in order to preserve the diagnostically relevant information is still more ambitious. It is worth to outline that vision-related issues of both low level (related to stimulus encoding) and higher levels (concerning perception and even cognitive processes) come into play, which depend on vision-based mechanisms that are still far from being understood. Last but not the least, the validation of a lossy technique based on the subjective evaluation by medical experts is quite complex. The number of variables is large and probably they are not mutually independent. This makes it difficult to design an ad-hoc psychophysical experiment, where a large number of parameters must be controlled.

The most widespread approach to lossy compression is transform-based coding. Both the DCT and the discrete wavelet transform (DWT) have been exploited to this end.

14.3.1.3   DCT-Based Techniques

A revisitation of the DCT for compression of different types of medical images is due to Wu and Tai [68]. A block-based DCT was applied to the image (the block size is 8×8 pixels). Then the resulting subband samples were reordered by assigning all the samples at the same spatial frequency to disjoint sets. In this way, some kind of subband decomposition was obtained, where the blocks representing the high-frequency components had, in general, low energy, and were mostly set to zero after quantization. For the remaining bands, a frequency domain block-wise prediction was implemented. The residual error was further quantized and eventually entropy-coded. The same entropy coding method used by JPEG was adopted. The performance was evaluated on 30 medical images (US, angiograms, and x-ray) and compared to that of JPEG. The best results were obtained on the angiographic images, for which the proposed algorithm clearly outperformed JPEG, while the improvement was more modest for the others.

Another kind of subband decomposition was proposed by Chiu et al. [9] for the compression of ultrasound images of the liver. The rationale of this approach was in the fact that ultrasound images are characterized by a pronounced speckle pattern which holds a diagnostic relevance as representative of the tissue and as such should, in general, be preserved. The set partitioning hierarchical trees (SPIHT) algorithm proposed by Taubman [60] was chosen as the benchmark for performance.

The low-rate compression of this type of images by SPIHT, in general, produces artifacts that smooth the speckle pattern and are particularly visible in areas of low contrast. Instead, the authors proposed an image-adaptive scheme that chooses the best representation for each image region, according to a predefined cost function. Accordingly, the image would be split either in space (image domain) or in frequency, or both, such that the "best" basis would be selected for each partition. The space frequency segmentation (SFS) starts with a large set of basis and an associated set of quantizers, and then uses a fast tree-pruning algorithm to select the best combination according to a given rate–distortion criterion. In this work, however, only the lossy regime was allowed and tested. The proposed SFS scheme outperformed SPIHT over a wide range of bit rates and, as an important factor in this framework, this held true for both the objective (PSNR) and subjective quality assessment. The set of experiments performed by the authors for the investigation of the degradation of the diagnostically relevant information can only be considered as an indication. However, its importance is in the fact that it strengthened the point that the *amount* and *type* of distortion that could be tolerated depends hardly on the image features, and thus on both the imaging modality and the disease under investigation. This was also clearly pointed out by Cosman et al. in [13]. The amount of degradation that could be tolerated on medical images depends on the degree to which it would affect the subsequent formulation of a diagnosis. Noteworthy, in their paper the authors also outline that neither pixel-wise measures (like the PSNR) nor receiver operating curves (ROCs) would be suitable, the latter because the ROC analysis is only amenable for binary tasks, while a much more complex decisional process is generally involved in the examination of a medical image.

The key for combining high-coding efficiency with lossless performance is scalability. Allowing a progressively refinable up to lossless representation, systems featuring scalability functionalities are flexible tools that are able to adapt to the current user requests. The critical point is that for an embedded coding algorithm to be competitive with a nonembedded one in terms of rate and distortion, a quite effective coding strategy for the *significance map* must be devised. This has determined the success of the wavelet transform in the field of compression, since the resulting subband structure is such that both intraband and interband relationships among coefficients can be profitably exploited to the purpose. More generally, wavelets are particularly advantageous because they are able to respond to all the requirements summarized in Section 14.2.

14.3.1.4 Wavelet-Based Techniques

As mentioned in Section 14.2, the demand of the current multimedia clinical framework goes beyond the maximization of the compression ratio, calling for systems featuring specific functionalities. The most important ones are progressiveness and ROI-based processing. This has determined the success of the wavelet-based techniques, which can be implemented preserving the specificity of the input data (mapping integer to integer values) while being well suitable for region-based processing. Furthermore, the nonlinear approximation properties of wavelets inspired the design of an effective coding strategy, which has become the most widespread in the field of data compression: the Shapiro's embedded zerotree wavelet (EZW) based coding algorithm [53].

14.3.1.4.1   Embedded Zerotree Wavelet-Based Coding

The EZW [53] is the ancestor of a large number of algorithms that were successively devised with the purpose of improving its performance and adapting to particular data structures and applications. The basic idea is to exploit the correlation among the wavelet coefficients in different subbands with the same orientation through the definition of *parent–children* relationships. The core hypothesis (*zerotree hypothesis*) is to assume that if a wavelet coefficient w at a certain scale is below a given threshold T, then all its descendants (the coefficients in subbands at the analogous position, finer resolution, and same orientation) are also insignificant. Scalability is obtained by following a successive approximation quantization (SAQ) scheme, which translates into bit-plane encoding. The significance of the wavelet coefficients with respect to a monotonically decreasing set of thresholds is encoded into a corresponding set of *significance maps* in a two-steps process. The generated set of symbols is entropy coded by an arithmetic coder. The compression efficiency is due to the gathering of the (in)significance information of a set of wavelet coefficients forming a tree into a unique symbol afferent to the root (the ancestor).

Among the most relevant evolutions of the EZW is the SPIHT [52] algorithm introduced by Said and Pearlman.

14.3.1.4.2   SPIHT

Grounded on the same underlying stochastic model, SPIHT relies on a different policy for partitioning and sorting the trees for encoding. The basic steps of the SPIHT algorithm are partial ordering by magnitude of the wavelet coefficients; set partitioning into hierarchical trees (according to their significance); and ordered bit-plane encoding of the refinement bits. During the *sorting* pass, the data are organized into hierarchical sets based on the significance of the ancestor, the immediate offspring nodes, and the remaining nodes. Accordingly, three lists are defined and progressively updated during the "sorting pass": the list of insignificant pixels (LIP), the list of insignificant sets (LIS), and the list of significant pixels (LSP). During the "refinement pass" the value of the significant coefficients (i.e., those belonging to the LSP) is updated to the current quantization level. The SPIHT algorithm, in general, outperforms EZW at the expenses of an increased complexity. Both of them are often used as the benchmark for compression performance, in both the lossless and lossy modes.

The proven efficiency of these coding methods has pushed some researchers to use them in combination with subband decompositions resulting from the application of different families of linear or nonlinear filters, leading to a large set of algorithms.

It is worth pointing out here that the lack of a common database to be used for testing the performance of the different algorithms is one of the main bottlenecks for their comparison and classification. Some efforts in this direction are nowadays spontaneously done by the researchers in the data compression community, which is particularly important in view of the definition of a standard for multidimensional medical data compression. Here we summarize some of the more relevant contributions to draw a picture of the state-of-the-art scenario.

Gruter [21], applies the EZW coding principle to a subband structure issued from a so-called rank-order polynomial subband decomposition (ROPD) to a x-ray and a heart US images. Basically, the images are decomposed into a "lower" and a "higher" frequency subbands. The "approximation" subband is obtained by simply subsampling the image, while the "detail" subband represents the residual after rank-order polynomial prediction of the original samples. By an ad hoc definition of the prediction polynomials in vector form, both numerical and morphological (nonlinear) filters could be modeled. The ROPD algorithm was tested in both the lossless and lossy modes and compared to SPIHT, JPEG lossless, a previous version of the algorithm only using morphological filters [19] (morphological subband decomposition, MSD), and a codec based on wavelet/trellis-coded quantization (WTCQ) [36]. ROPD slightly outperformed SPIHT in terms of compression ratio, whereas a more sensible improvement was observed with respect to the other algorithms. This holds for all the images of the test set (two MRI images of a head scan and one x-ray of the pelvis). The interest of this work is mostly in the adaptation of the algorithm for ROI-based coding. In this case, a shape-adaptive nonlinear decomposition [19] was applied to the region of interest of a typical heart ultrasound image, which was then losslessly coded.

A solution for coding coronary angiographic images was proposed by Munteanu [42]. This wavelet-based algorithm reached both lossless and scalability by quality functionalities. The wavelet transform was implemented via the lifting steps scheme in the integer version [7] to enable the lossless performance. The strategy followed for coding only exploited intraband dependencies. The authors proved that if the zerotree hypothesis holds, the number of symbols used to code the zero regions with a fixed-size, block-based method is lower than the number of zerotree symbols that would be required following a EZW-like approach, for block sizes confined to some theoretical bounds. Basically, the wavelet image was partitioned into a lattice of squares of width v ($v = 4$ in their implementation). A starting threshold value was assigned to every square of every subband as the maximum power of two below the maximum absolute value of the coefficients in the square, and the maximum of the series of the starting threshold values T_{max} was retained. The subbands were then scanned in raster order, as well as the squares within the subbands, to record the significance of the coefficients according to a decreasing set of thresholds ranging from T_{max} to zero. The generated set of symbols was encoded using a high-order arithmetic coder. This coding strategy implies an ordering of the coefficients according to their magnitude, enabling scalability functionalities. Such an order can be modified for providing the algorithm region-based processing. After determining the correspondence between a spatial region and the corresponding set of squares, ROI-based coding could be reached by grouping such squares into a logical entity and attributing them the highest priority during the coding process. The algorithm was also tested in the case $v = 1$ for the sake of comparison. The performance was compared to that provided by a number of other methods including CALIC, lossless JPEG, EZW, and SPIHT. Both the wavelet-based coders were implemented on a subband structure issued by integer lifting. The results showed that the proposed algorithm outperformed the others with the exception of CALIC, which gave very close lossless rates.

14.3.1.4.3 ROI-Based Coding

The most effective wavelet-based coding methods have also been generalized for region of interest processing. Menegaz et al. [39–41], proposed an extension of the EZW algorithm for object processing. Among the main features of such an approach are the finely graded up-to-lossless representation of any object and the absence of discontinuities along the object's borders at any decoding quality.

Penedo et al. [44], provide a ROI-based version of SPIHT (OB-SPIHT) and of a similar algorithm, the set partitioning embedded block coding (SPECK) by Islam and Pearlman [24] (OB-SPECK).

The dataset consisted in this case of mammographies digitized at 12 bits/pixel over 4096×5120 pixels. The region of interest was first segmented and then transformed by a region-based DWT [4]. The shape of the region of interest was encoded by a two-link shape-coding method [35]. OB-SPIHT and OB-SPECK were obtained by simply pruning the tree branches falling outside of the objects. The performance was compared to that of SPIHT and JPEG2000 when applied to the entire image in order to quantify the rate saving provided by region-based processing. A pseudo-region-based mode was also obtained by running these two reference algorithms on the images with the background pixels set to zero. Results showed that OB-SPIHT and OB-SPECK perform quite similarly and that focusing on the region of interest improves the efficiency of the coding system, as it was reasonable to expect.

To conclude this section, we would like to summarize the main insights that can be drawn from the results of the research effort in this field. First, the lack of a common reference database of images representing different imaging modalities and pathologies impedes a clear classification of the many proposed algorithms. Second, only the algorithms featuring lossless functionalities are suitable in wide sense, allowing to recover the original information without loss. Lossy techniques could be adopted for specific applications, like education or postdiagnosis archiving. Whether lossy compression affects the diagnostically important image features is a difficult issue that still remains open.

14.3.2 3-D Systems

Most of the current medical imaging techniques produce three-dimensional data. Some of them are intrinsically volumetric, like MRI, CT, PET, and 3-D ultrasound, while others describe the temporal evolution of a dynamic phenomenon as a sequence of 2-D images or 3-D volumes. The huge amount of data generated every day in the clinical environment has triggered considerable research in the field of volumetric data compression for their efficient storage and transmission. The basic idea is to take advantage of the correlation among the data samples in the multidimensional space (3-D or 4-D) to improve compression efficiency. The most widespread approach combines a multidimensional decorrelating transform with some generalization of a coding algorithm that has proved to be effective in 2-D. Here we constrain to still images and volumes.

The design of a coding scheme results from the trade-off among the cost functions derived from a set of requirements. Among these are optimal rate–distortion performance over the entire set of bit rates as well as progressiveness capabilities, either by quality or by resolution. Besides these general requirements, which apply to any coding framework, there are some domain-specific constraint that must be fulfilled. In the case of medical imaging, lossless functionalities are a must. It is thus desirable that the type of chosen scalability will end up with a lossless representation.

Many solutions have been proposed so far. As was the case in 2-D, both the DCT and the DWT have been used for data decorrelation. The main problem with non-DWT-based schemes is that they hardly cope with the requirements mentioned above, which make them unsuitable despite in some cases they provide a good rate–distortion performance, eventually outperforming the DWT-based ones.

Among the wavelet-based methods, the most relevant ones all described hereafter.

14.3.2.1 3-D Set Partitioning Hierarchical Trees

Kim and Pearlman [29], applied the 3-D version of the SPIHT algorithm to volumetric medical data. The wavelet transform was implemented in its integer version and different filters were tested. The decomposition was performed first along the z-axis and then the x and y spatial dimensions were processed. The system was tested on five datasets including MR (chest, liver, and head) and CT (skull). The results were compared to those provided by 3-D EZW as well as to those of some two-dimensional techniques including CALIC for the lossless mode. The lossless rates provided by 3-D SPIHT have been improved up to about the 30–38% of those of the 2-D methods, and slightly outperformed 3-D EZW on almost all the test sets.

A similar approach was followed by Xiong et al. [69] where the problem of context modeling for efficient entropy coding was addressed. This algorithm provided a slightly higher coding gain than 3-D SPIHT on the MR chest set. Noteworthy, since it has often been used as the benchmark for performance evaluation of 3-D systems, the 3-D EZW algorithm has been tested by many researchers. Among the numerous contributions are those of Bilgin et al. [5] Menegaz et al. [39,41], Chan et al. [8], Tai et al. [58], and Wu et al. [67].

An extended study of the possible architectures is presented by Shelkens et al. [57], where the authors provide a comparative analysis of different 3-D wavelet coding systems. After a brief overview of the more interesting state-of-the-art techniques, they propose and compare different architectures. They developed new hybrid systems by combining the most effective methods, namely quadtrees, block-based entropy coding, layered zero coding (LZC), and context-adaptive arithmetic coding. The wavelet-based coding systems they proposed are the 3-D extensions of the square partitioning (SQP) [43], the quadtree limited (QT-L), and the embedded block coding (EBCOT) [59]. More specifically, the cube-splitting (CS) algorithm is based on quadtrees [54], the 3-D quadtree limited (3-D QT-L) combines the use of quadtrees with block-based coding of the significance map [42] and

3-D CS-EBCOT [55] integrates both the CS and the layered zero-coding strategies. All the wavelet-based systems share the same implementation of the DWT via integer lifting and different filters, and decomposition depths were allowed along the three axis. The benchmarks for performance were a JPEG-like 3-D DCT-based coding scheme (JPEG 3-D), 3-D SPIHT, and another 3-D subband-based set partitioning block coding (SB-SPECK) method proposed by Wheeler [64].

14.3.2.2 Cube Splitting

The Cube Splitting algorithm was derived as the generalization for 3-D data of the SQP method of Munteanu [43]. As it is the case for the nested quadtree splitting (NQS) [11] and the SB-SPECK, the same coding principle of EZW and SPIHT, consisting of a "primary" or "significance" pass and a "refinement" pass, is applied to a subband structure. Intraband instead of interband relationships are exploited for coding. Instead of using quadtrees, the subbands are split into squares. Following the SAQ policy, the significance of each square with respect to a set of decreasing thresholds is progressively established by an ad hoc operator. If the block is significant with respect to the current threshold, it is further split in four squares over which the test for significance is repeated. The procedure is iterated until the significance of the leaf nodes is isolated. Thus, the significance pass selects all the leafs that have become significant in the current pass. Then, the refinement pass for the significant leaf nodes is performed. Next, the significance pass is restarted to update the entire quadtree structure by identifying the new significant leaf nodes. In the SQP coder, the significance, refinement, and sign information were encoded by adaptive arithmetic coding. The generalization for 3-D is straightforward. The resulting CS coder aims at isolating small subcubes possibly containing significant wavelet coefficients. The same coding algorithm as for SQP is applied to oct-trees. A context-adaptive arithmetic coder is used for entropy coding employing four and two context models for the significance and refinement passes, respectively. Moreover, the 3-D data are organized into group of frames (GOFs) consisting of either 8, 16, and 32 slices to improve the accessibility.

14.3.2.3 3-D Quadtree Limited

This method is similar to the SQP and CS coders described in the previous section, the main difference being that in this case, the size of the blocks is upper-bounded. When the size of the corresponding area reaches a predefined minimum value, the partitioning is stopped and the subband samples are entropy-coded. A second difference is that the order of encoding of the coefficients that are nonsignificant in the current pass is partially altered by introducing another coding stage called "insignificance pass" [57]. The coefficients classified as nonsignificant during the significance pass are appended to a list named list of nonsignificant coefficients (LNC), and are coded at first at the beginning of the next significance step. The authors motivate this choice by considering that the coefficients in the LNC lie in the neighborhood of others that were already found to be significant during one of the previous passes (including the current one), and as such have an high probability to become significant in the next coding steps. The significance and refinement passes of the other coefficients follow. Finally, an extended set of contexts is used for both conditioning and entropy coding.

14.3.2.4 CS-Embedded Block Coding (CS-EBCOT)

The EBCOT coder proposed by Taubman [59] is a block-based coding strategy, which have become quite popular as the one chosen for the new JPEG2000 standard. Coding of the subband coefficients consists of two steps. During the first one, usually referred to as *Tier 1 (T1)*, the subbands are partitioned into blocks and each block is entropy-coded according to the LZC technique [60]. Thus each block is associated to an embedded bitstream featuring scalability by quality. The second coding step, *Tier 2 (T2)*, aims at identifying a set of truncation points of each block such that a given global rate–distortion function is optimized. The T1 pass is articulated in a "significance propagation," a "magnitude refinement," and a "normalization" pass. During the first one, the coefficients that have been classified as nonsignificant in all the previous passes and have at least one significant coefficient in a predefined preferred neighborhood are encoded. The second refines the quantization of the coefficients that were significant in one of the previous passes. Finally, the third pass processes all the coefficients that are significant regardless of their preferred neighborhood. On the other end, T2 performs the postcompression rate distortion (PCRD) optimization by searching the truncation point of every block-wise bitstream in order to reach the minimum distortion for the given rate. We refer to Taubman [59] for further details. The 3-D CS-EBCOT combines the principles utilized in the CS coder with a 3-D version of EBCOT.

14.3.2.5   3-D DCT

The 3-D DCT coding scheme represents in some sense the 3-D version of the JPEG standard. The three-dimensional DCT is performed on cubic blocks of $8 \times 8 \times 8$ voxels. The resulting coefficients are quantized by a uniform quantizer and scanned for coding along a space-filling 3-D curve. Entropy coding is implemented by a combination of run-length and arithmetic coding. The DCT-based coder normally does not provide lossless compression.

Lossy and lossless compression performance was evaluated over five datasets including CT (CT1 with $512 \times 512 \times 100 \times 12$ bpp and CT2 with $512 \times 512 \times 44 \times 12$ bpp), MRI ($256 \times 256 \times 200 \times 12$ bpp), PET ($128 \times 128 \times 39 \times 15$ bpp), and ultrasound images. The conclusions can be summarized as follows. First, the use of the wavelet transform boosts the performance: 3-D systems are advantageous when the interslice spacing is such that a significant correlation exists between adjacent slices. Second, for lossless coding, CS-EBCOT and 3-D QT-L provide the best results for all the datasets. Third, in lossy coding, 3-D QTL tends to deliver the best performance when using the (5/3) kernel [6]. At low rates, CS-EBCOT competes with JPEG2000-3-D. Conversely, the 3-D SPIHT would be the best choice in combination with the 9/7 filter, followed closely by CS and CS-EBCOT. Results show that the three proposed algorithms provide excellent performance in lossless mode, and rate–distortion results that are competitive with the reference techniques (3-D SB-SPECK and JPEG2000-3-D, namely the JPEG2000 encoder incorporating a 3-D wavelet transform).

In summary, the research in the field of medical image compression led to the common consensus that 3-D wavelet-based architectures are fruitful for compression as long as the correlation along the third dimension is sufficiently pronounced. On top of this, the availability of ad-hoc functionalities like fast access to the data and ROI-based capabilities are critical for the suitability of a given system for PACS. In the medical imaging field the users tend to give their preference to systems that are better tailored on their needs eventually sacrificing some gain in compression.

Probably, the major drawbacks of 3-D systems are the decoding delay and the computational complexity. A possible shortcut to the solution of this problem is ROI-based coding. This is particularly appealing for 3-D data due to the high demand in terms of transmission and storage resources.

14.3.3   3-D ROI-BASED CODING

Different solutions have been proposed so far for 3-D ROI-based coding. Of particular interest are those of Menegaz et al. [39,40], and Ueno and Pearlman [62]. Besides the strategy used for entropy coding, the main difference is in the way the region-based transform is implemented. The different solutions are application-driven and satisfy to different sets of constraints. The system presented in [62] combines a shape-adaptive DWT with scaling-based ROI coding and 3-D SPIHT for entropy coding. The dataset is partitioned into group of pictures (GOPs), which are encoded independently to save run-time memory allocation. The test set consists of a volumetric MR of the chest ($256 \times 256 \times 64 \times 8$ bpp). Performance is compared to that obtained with a conventional region-based DWT implementation. The advantage of this algorithm is that the number of coefficients to be encoded for each region is equal to the number of pixels corresponding to the object in the image domain.

A particularly interesting solution is provided by Menegaz et al. [39,40]. In this case, ROI-based functionalities are enabled on both 3-D objects and user-defined 2-D images, allowing a fast access to the information of interest with finely graded granularity. Furthermore, no border artifacts appear in the image at any decoding rate. The 3-D ROI-based multidimensional LZC (MLZC) technique [40] is an application-driven architecture. It was developed on the ground of the observation that despite the availability of advanced rendering techniques, it is still common practice for doctors to analyze 3-D data distributions one 2-D image at a time. Accordingly, in order to be suitable within PACS, a coding system must provide a fast access to the single 2-D images. On top of this, the availability of ROI-based functionalities enables to fasten the access to the portion of image that is crucial for diagnosis, permitting a prompt response by the experts. The 2-D decoding capabilities were accomplished by independently encoding each *subband image*, and making the corresponding information accessible through the introduction of some special characters (markers) into the bitstream. In this way, once the user had specified the position of the image of interest along the z-axis, the set of subband images that were needed for its reconstruction was automatically determined and the concerned information was decoded. The inverse DWT was performed locally. ROI-based functionality was integrated by assigning subsequent segments of the bitstream to the different objects, depending on their priority [38,39]. This led to a versatile and highly efficient coding engine allowing to swiftly recover any object of any 2-D image of the dataset at a finely graded up to lossless quality. Besides competitive compression rates and novel application-driven functionalities, the proposed system enables a *pseudo-lossless* mode, where the diagnostically relevant parts of the image are represented without loss, while a lower quality is assumed to be acceptable for the others. The potential of such an architecture is in the combination of a 3-D transform providing a concise description of the data, with the possibility to recover single 2-D images by decoding only the part of the bitstream holding the necessary information. Besides the improvement in the efficiency for accessing the information of interest, ROI-based processing enables the parallelization of the encoding/decoding of the different objects. Using the integer version of the wavelet transform, which is necessary for reaching the lossless mode, makes the algorithm particularly suitable for the implementation on a device. Last but not least, the analysis of the compression performance shows that the proposed system is competitive with the others state-of-the-art techniques. It is thus a good compromise between the gain in compression efficiency provided by 3-D systems and the fast access to the data of 2-D ones. The remaining part of this chapter is devoted to the description of such a system.

14.3.4   MODEL-BASED CODING

To conclude this section, we would like to briefly discuss the new trend for data compression: the model-based approach. In the perspective of model-based coding, high efficiency can be accomplished on the ground of a different notion of *redundancy*, grounded on semantic. This discloses a new perspective in the compression philosophy, based on a redefinition of the notion of *relevance*. If, in general, the goal of a coding system is to represent the information by reducing the mathematical or statistical redundancy, the availability of a priori information about the imaged data suggests a more general approach, where the ultimate goal is to eliminate all kinds of *redundancy*, either mathematical or semantical. Following a model-based approach means to focus on semantics instead of just considering the data as multidimensional matrices of numbers. This philosophy traces back to the so-called *second-generation coding techniques* pioneered by Kunt et al. in 1985 [33], and it has inspired many of the last-generation coding systems, like JPEG2000 [25] for still image coding, MPEG4 [27] for video-sequences, and MPEG7 [26] for content-based description of images and database retrieval.

Since semantic comes into play, such an approach must be preceded by an image analysis step. In the general case, the problem of object identification and categorization is ill-posed. However, in the particular case of medical imaging, the availability of a priori information about both the image content and the subsequent use of the images simplifies the task. To get a picture of this, just consider an MRI head scan performed in an oncology division. The object of interest will most probably be the brain, while the surrounding tissues and the skull could be considered as noninformative for the investigation under way. The features of the region of interest could then be progressively updated during the investigation and, for instance, focus on the tumor and the region of the brain that surrounds it, where the major changes due to the lesion are expected to be. In both cases, it is known a priori that the object of interest is the brain (or eventually a part of it), located at the center of the image and with certain shape and textural properties (the latter depending on the type of scan).

In most of the state-of-the-art implementations, the general idea of model-based coding specializes in region of interest coding. Such regions often correspond to physical objects, like human organs or specific types of lesions, and are extracted by an ad hoc segmentation of the raw data. Here we assume that the object of interest is given, and we refer to Duncan and Ayache [18] for a survey on medical image analysis. According to the guidelines described above, model-based coding consists in combining classical coding techniques on regions where the original data must be preserved (at least up to a given extent) with generative models reproducing the *visual appearance* on the regions where the lossless constraint can be relaxed.

14.4   3-D/2-D ROI-BASED MLZC: A 3-D ENCODING/2-D DECODING OBJECT-BASED ARCHITECTURE

The coding algorithm described in this section allows to combine the improvement in compression performance resulting from a fully 3-D architecture with a swift access to single imaged objects. In this framework, the qualification of *object* is used to identify the part of the data which is of interest for the user. Accordingly, it is used to indicate 3-D sets of voxels in the 3-D ROI-based working mode, a single 2-D image in the 3-D/2-D working mode and a region of a 2-D image in the ROI-based 3-D/2D working modality (as illustrated in Figure 14.1), respectively. The data are first decorrelated by a 3-D-DWT and subsequently encoded by an ad hoc coding strategy. The implementation via the lifting steps scheme proposed by Sweldens [16] is particularly advantageous in this framework.

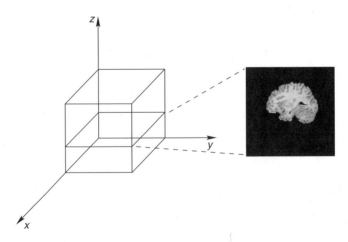

FIGURE 14.1 Volumetric data. We call z the third dimension and assume that the images are the intersections of the volume with a plan orthogonal to z-axis.

First, it provides a very simple way of constructing nonlinear wavelet transforms mapping integer-to-integer values [7]. Second, perfect reconstruction is guaranteed by construction for any type of signal extension along borders. This greatly simplifies the management of the boundary conditions underlying the independent object processing with respect to the classical filter-bank implementation. Third, it is computationally efficient. It can be shown that the lifting steps implementation asymptotically reduces the computational complexity by a factor of 4 with respect to the classical filter-bank implementation, as was proved by Reichel et al. [48]. Finally, the transformation can be implemented in-place, namely progressively updating the values of the original samples, without allocating auxiliary memory, which is quite important when dealing with large volumes.

The 3-D DWT is followed by SAQ, bit-plane encoding and context-adaptive arithmetic coding. Some markers are placed in the bitstream to enable random access. Tuning the coding parameters leads to different working modalities. In particular, 3-D encoding/2-D decoding capabilities are gained at the expense of a slight degradation in coding gain due to the extra information needed to enable random access to selected segments of the bitstream. The object-based functionality is reached by independently encoding the different objects, which can then be decoded at the desired bit rate. Finally, the working mode featuring both 3-D encoding/2-D decoding and object-based capabilities is obtained by concatenating one segment of bitstream built according to the rules enabling 2-D decoding for each object. The price to pay is an additional overhead, which lowers the compression factor. However, the possibility to focus the decoding process on a specific region of a certain 2-D image allows a very efficient access to the information of interest, which can be recovered at the desired up-to lossless quality. We believe this is important feature for a coding system meant to be used for medical applications, which largely compensates for the eventual loss in compression efficiency.

The next sections describe more in detail how the different functionalities are reached. In particular, Section 14.5 illustrates the object-based processing, Section 14.5.1 comments on the 3-D analysis/2-D reconstruction working modality and Section 14.6 illustrates the chosen coding strategy as well as its modifications for reaching to the different working modes.

14.5 OBJECT-BASED PROCESSING

In ROI-based compression systems, the management of objects involves both the transformation and the coding steps. In the perspective of transformation it brings up a boundary problem. As discrete signals are nothing but sets of samples, it is straightforward to associate the idea of *object* to a *subset* of samples. The issue of boundary conditions is greatly simplified when the DWT is implemented by the lifting steps scheme [16]. In this case, perfect reconstruction is ensured by construction, for any kind of signal extension at the borders. Nevertheless, perfect reconstruction is not the only issue when dealing with a complete coding system. Our goal was to avoid artifacts along borders in all working modalities in order to make object-based processing completely transparent with respect to the unconstrained general case where the signal is processed as a whole. Otherwise stated, we wanted the images decoded at a given graylevel resolution (i.e., quantization level) to be *exactly* the same in the following conditions: (a) the signal is encoded/decoded as a whole and (b) each object is independently encoded and decoded. The perfect reconstruction condition is not enough to ensure the absence of artifacts in terms of discontinuities at borders. Since quantized coefficients are approximations of the true values, any signal extension used to reconstruct two adjacent samples belonging to different objects (e.g., lying at the opposite sides of a boundary) would generate a discontinuity. To avoid this, the inverse transform must be performed *as if* the whole set of true coefficients were available. The use of the lifting scheme simplifies this task. The analysis and reconstruction filter-banks for the lifting steps implementation in 1-D case are shown in Figure 14.2. In the figure, s_m and t_m represent the mth *prediction* and *update* steps, respectively, and K a gain factor. At the output of the analysis chain (Figure 14.2a) the upper branch corresponds to the low-pass (LP) subband while the lower branch corresponds to the high-pass (HP) subband. The idea is to determine the spatial position of the wavelet coefficients in the considered subband at the *input* of the synthesis chain (Figure 14.2b) that are necessary to reconstruct a given wavelet coefficient at the *output*. The key point of the proposed solution is to start at the finest resolution ($l = 1$) and select the set of wavelet coefficients which are needed to reconstruct the object in the signal domain (full resolution, $l = 0$). At this point, the problem is solved for $l = 1$, or, equivalently, it has been projected to the next coarser level. Owing to the recursiveness of the implementation of the inverse transform, the approximation subband of level $l = 1$ becomes the reference set of samples that must be reconstructed without loss, and so on. By going through all the resolutions and successively iterating the procedure as described for $l = 1, \ldots, L - 1$, the appropriate set of wavelet coefficients is selected. Such set of coefficients is called *generalized projection* (GP) of the object in a given subband and is determined by an ad hoc GP operator [37]. The 3-D generalization is straightforward. Let *GP* be the generalized projection operator, and let GP_η be the set of samples obtained by applying *GP* in the direction $\eta = x, y, z$. The separability of the transform leads to the following composition rule:

$$GP_{zyx} = GP_z \{ GP_y \{ GP_x \{ \cdot \} \} \} \tag{14.1}$$

The set of wavelet coefficients to be encoded for each object are those belonging to its generalized projection.

As illustration, we consider a volumetric set consisting of $256 \times 256 \times 128$ images of a MR brain scan. It is assumed that the object of interest is the brain, while the surrounding part is the background (see Figure 14.3). Figure 14.4 illustrates the result of the application of the 3-D GP operator. For

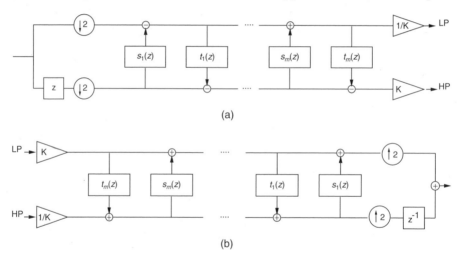

FIGURE 14.2 Filter banks for the lifting steps implementation. (a) Analysis; (b) reconstruction.

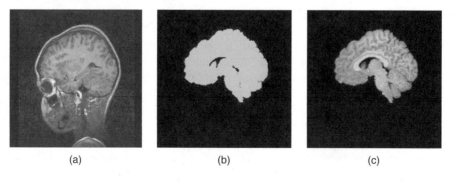

FIGURE 14.3 Saggital view of the MR of a brain: (a) original image, (b) object of interest, and (c) mask.

convenience, we define *border voxels BP(l, j)*, the samples belonging to the GP and *not* to the object projection (OP):

$$BP = GP \ominus (GP \cap OP) \tag{14.2}$$

For increasing levels of decomposition, the percentage of object voxels remains constant (being the outcome of the polyphase transform), while the percentage of border voxels increases. Figure 14.5 represents the percentage of object and border voxels as a function of a linear index obtained as $i = (l - 1) \times 7 + j$. The percentage of voxels corresponding to the $GP(l, j)$ in the different subbands increases with both the decomposition level and the length of the filter impulse response. It reaches 65% for the 9/7 and 25% for the 5/3 filter, on average, for $L = 4$. In the approximation subband the percentage of object voxels is 13.3%, while that of the border voxels amounts to the 19% and the 60.59% for the 5/3 and 9/7 filters, respectively.

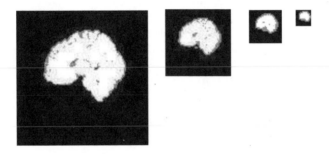

FIGURE 14.4 Three-dimensional GP of the brain for $l = 1, \ldots, 4$, $j = 1$ (LLH subband); $L = 4$ and the 5/3 filter was used. White voxels identify $O(l, j)$, while gray voxels represent the *border extension*.

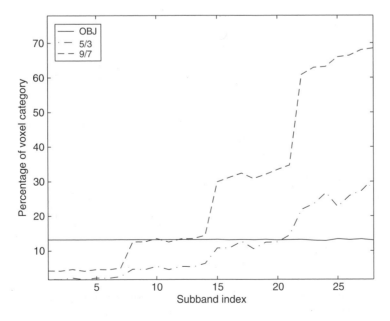

FIGURE 14.5 Percentage of object and border voxels across the subbands. Subband (l, j) corresponds to the index $i = (l - 1) \times 7 + j$. Continuous line: OP; dash-dot line: 5/3 filter; dashed line: 9/7 filter.

TABLE 14.1
Total Number of Subband Images to Decode for Reconstructing Image
k for $L = 3$

N_k/k	0	1	2	3	4	5	6	7
5/3	34	46	49	46	42	46	42	46
9/7	58	139	111	161	82	147	111	139

14.5.1 3-D ANALYSIS VS. 2-D RECONSTRUCTION

In the 3-D system, filtering is successively performed along the x, y, and z directions. We assume that the 2-D images are stacked along the z-axis. Then, the positions of the wavelet coefficients corresponding to $GP(l,j)$ in subband (l,j) in the 1-D case are mapped to the positions of the *subband images* along the z-axis in the 3-D case. More precisely, $GP(l,j)_k$ identifies the z-coordinates of all the images in subband (l,j) that are necessary to recover the image with index k. In this case, the index j selects either LP ($j = a$) or HP ($j = d$) filtering along z. The total number N_k of subband images needed for the reconstruction of the image of index k is given by [41]

$$N_k = 4[GP(L,a)_k + GP(L,d)_k]$$

$$+ \sum_{l=L-1}^{1} [3 \times GP(l,a)_k + 4 \times GP(l,d)_k] \qquad (14.3)$$

Table 14.1 shows the values for N_k for the 5/3 and the 9/7 filters. When using the 5/3 filter, the number of subband images needed is between 1/2 and 1/3 of those required when using the 9/7 filter, depending on k. Accordingly, the 5/3 filter allows a faster reconstruction of the image of interest, thus it is preferable when the fast access to the data is required.

14.6 MULTIDIMENSIONAL LAYERED ZERO CODING

MLZC is based on the LZC algorithm proposed by Taubman and Zakhor [60]. The main differences between LZC and the proposed MLZC algorithm concern the underlying subband structure and the definition of the *conditioning terms*. This section starts with an overview of the basic principles of the LZC method and then details the proposed system. Particularly, we first summarize the basics of LZC and then introduce the MLZC coding principle.

14.6.1 LAYERED ZERO CODING

In the LZC approach, each subband is quantized and encoded in a sequence of N quantization layers, following the SAQ policy, which provides scalability by quality features. The LZC method is based on the observation that the most frequent symbol produced by the quantizers is the zero symbol, and achieves high efficiency by splitting the encoding phase in two successive steps: *zero coding*, which encodes a symbol representing the *significance* of the considered coefficients with respect to the current quantizer and *magnitude refinement*, which generates and encodes a symbol defining the value of each nonzero symbol. Zero coding exploits some spatial or other dependencies among subband samples by providing such information to a *context-adaptive* arithmetic coder [45]. Different solutions are possible for the definition of the conditioning terms, accounting for both local and wide-scale neighborhoods. We refer to [60] for more details.

14.6.2 MLZC CODING PRINCIPLE

MLZC applies the same quantization and entropy coding policy as LZC to a 3-D subband structure. In order to illustrate how the spatial and interband relationships are exploited, we use the concepts of *generalized neighborhood* and *significance state* of a given coefficient. We define generalized neighborhood of a subband sample $c(l,j,\mathbf{k})$ in subband j of level l and position $\mathbf{k}$ the set $GN(l,j,\mathbf{k})$ consisting of both the coefficients in a given spatial neighborhood $N(l,j,\mathbf{k})$ and the parent coefficient $c(l+1,j,\mathbf{k}')$ in the same subband at the next coarser scale, where $\mathbf{k}' = \lfloor \mathbf{k}/2 \rfloor$:

$$GN(l,j,\mathbf{k}) = N(l,j,\mathbf{k}) \cup c(l+1,j,\mathbf{k}') \tag{14.4}$$

The MLZC scheme uses the significance state of the samples belonging to a *generalized* neighborhood of the coefficient to be coded for conditioning the arithmetic coding [61].

For each quantization level Q_i, the significance state of each coefficient is determined by scanning the subbands starting from the lowest resolution. For the resulting symbol, two coding modes are possible: *significance* and *refinement* mode. The significance mode is used for samples that were nonsignificant during all the previous scans, whether they are significant or not with respect to the current threshold. For the other coefficients, the refinement mode is used. The significance mode is used to encode the significance map. The underlying model consists in assuming that if a coefficient is lower than a certain threshold, it is reasonable to expect both its spatial neighbors and its descendants are below a corresponding threshold as well.

The significance map consists of the sequence of symbols

$$\sigma_i[\mathbf{k}] = \begin{cases} 1 & \text{if} \qquad \mathcal{Q}_i[x(\mathbf{k})] \neq 0 \\ 0 & \text{otherwise} \end{cases} \tag{14.5}$$

where $\mathbf{k} = n_x\mathbf{i} + n_y\mathbf{j} + n_z\mathbf{q}$ defines the position of the considered sample and the $\mathcal{Q}_i$ operator quantizes $x(\mathbf{k})$ with step Q_i. In what follows, we call *local scale* neighborhood of a coefficient $c(l,j,\mathbf{k})$ in subband (l,j) and position $\mathbf{k}$ the set of coefficients lying in the spatial neighborhood $N(l,j,\mathbf{k})$. Then, we will refer to the sequence of symbols resulting from the application of Q_i to the set $c(l,j,\mathbf{k})$ as to $\sigma_i[l,j,\mathbf{k}]$. The significance state of the samples in the generalized neighborhood of $c(l,j,\mathbf{k})$ is represented by some conditioning terms $\chi(\cdot)$. The *local-scale* conditioning terms $\chi^s(\cdot)$ concern spatial neighborhoods while *interband* terms $\chi_i^f(\cdot)$ account for interband dependencies:

$$\chi[l,j,\mathbf{k}] = \begin{cases} \chi^s[l,j,\mathbf{k}] + \chi^f[l,j,\mathbf{k}] & \text{for } l \neq L \\ \chi^s[l,j,\mathbf{k}] & \text{for } l = L \end{cases} \tag{14.6}$$

14.6.3 SPATIAL CONDITIONING

The $\chi^s[l,j,\mathbf{k}]$ are defined as linear combinations of functions representing the significance state of one or more samples in $N(l,j,\mathbf{k})$

$$\chi^s[l,j,\mathbf{k}] = \sum_{p=0}^{P-1} w_p \sigma[l,j,\mathbf{k}'] \quad \forall \mathbf{k}' \in N \tag{14.7}$$

where $p = p(\mathbf{k}')$. The weights $w_p = 2^p$ are such that each term of the summation contributes to the value of the pth bitplane of $\chi^s[l,j,\mathbf{k}]$, P is the bit depth of $\chi^s[l,j,\mathbf{k}]$, and σ is the distribution of the sequence of symbols $\sigma(l,j,\mathbf{k})$ generated by quantizer Q_i.

The set of local-scale 2-D contexts that were considered are illustrated in Figure 14.6. Contexts numbered from 1 to 5 only account for coefficients that have already been encoded in the current step, while those with label 6 to 8 also use samples that will be successively encoded in the current step, so that their significance state refers to the previous scan. The grouping of σ may become necessary for avoiding context dilution when dealing with local-space neighborhoods of wide support. This sets an upper limit on the number of possible contexts. The 3-D local-scale conditioning terms were obtained by extending the set of the M most effective 2-D contexts to the third dimension. The support $N(l, j, \mathbf{k})$ of each selected 2-D context was extended to the adjacent subband images as illustrated in Figure 14.7. According to our notations, the subband image with index $(\nu - 1)$ is scanned before that with index ν, making the significance state of the corresponding samples with respect to the current quantization level available for its encoding. Conversely, only the significance state relative to the previous scan is known for the subband image of index $(\nu + 1)$. Since we expect a more pronounced correlation among the significance states of adjacent samples within the same scan, we decided to give more degrees of freedom to the extension of the interscale conditioning term in the previous $(\nu - 1)$ than the next $(\nu + 1)$ subband images. Particularly, for $(\nu - 1)$, two possible configurations were tested, as illustrated in Figure 14.7. The name associated with each context is in the form $(n_1 n_p n_3)$, where the indexes identify the 2-D context in the previous $(\nu - 1)$, current (ν) and next $(\nu + 1)$ layer, respectively. The case $n_1, n_3 = 0$ reveals that no samples were considered in the corresponding layer. Results show that the spatial contexts leading to the best performance correspond to $n_p = 6, 7, 8$. Their 3-D extension leads to the triplets $(n_1 n_p n_3)$ with $n_1 = 1, 2, 3$ and $n_3 = 0, 1$. As an example, the

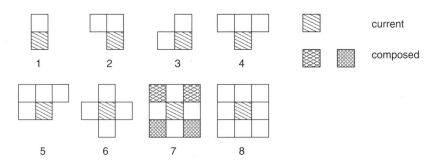

FIGURE 14.6 2-D contexts.

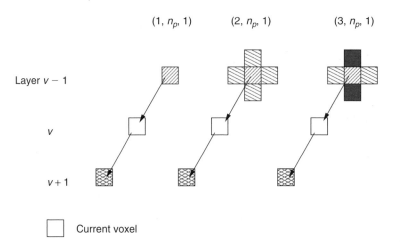

FIGURE 14.7 Extension of the spatial support in the previous $(\nu - 1)$ and the next $(\nu + 1)$ subband images. Squares with same pattern represent voxels whose significance states are combined in the definition of the corresponding $\chi^s[l, j, \mathbf{k}]$.

triplet (161) selects the 3-D context for which the 2-D spatial configuration in the current layer (v) is the one labeled as number 6 in Figure 14.6, and the samples with same coordinates are used in both the previous $(v+1)$ and subsequent $(v-1)$ layers (i.e., the leftmost configuration in Figure 14.7).

14.6.4 Interband Conditioning

The observed self-similarity among subbands within the subband tree makes the parent $c(l+1,j,\mathbf{k}')$ of the current coefficient $c(l,j,\mathbf{k})$ the most natural candidate for interband conditioning. Accordingly, the expression for the interband conditioning term is

$$\chi^f[l,j,\mathbf{k}] = w_{P_{\mathrm{MSB}}}\sigma[l+1,j,\mathbf{k}'] \tag{14.8}$$

where $w_{P_{\mathrm{MSB}}} = 2^{P_{\mathrm{MSB}}}$ is the weight needed to set the MSB of the *global* context

$$\chi[l,j,\mathbf{k}] = \chi^s[l,j,\mathbf{k}] + \chi^f[l,j,\mathbf{k}] \quad \forall l \neq L \tag{14.9}$$

This rule does not apply to the coarsest subbands for which no parents can be identified. In this case, only the local-space contribution is used.

14.6.5 Bitstream Syntax

The ability to access any 2-D image of the set constrains the bitstream structure. In all the modes (G-PROG, LPL-PROG, and LPL), the subbands are scanned starting from coarsest resolution. The signal approximation $LLL_{l=L}$ is encoded first, and all the subbands at level $(l+1)$ are processed before any subband at the next finer level l. What makes the difference among the considered working modalities are the order of encoding of the subband images and the placement of the markers. We describe them in what follows.

An illustration of the bitstream structure of a given 3-D subband in the different modes is given in Figure 14.8. In the figure, H is the bitstream header, L_i^v is the encoded information corresponding to the ith bitplane of layer v, I is the bit depth and N the number of layers. The placement of the markers within the bitstream and the ordering of the segments L_i^v determine the working mode, as described hereafter.

14.6.5.1 Global Progressive (G-PROG)

The scanning order follows the decomposition level: all subbands at level $l+1$ are scanned before passing to the next finer level l. During the ith quantization step, the quantizer Q_i is applied to each image of each subband. This enables the scalability by quality functionality on the volume, such that the decoding process can be stopped at any point into the bitstream. In this mode no markers are placed within the bitstream, the compression ratio is maximized, but the 3-D encoding/2-D decoding capability is disabled.

14.6.5.2 Layer per Layer Progressive (LPL-PROG)

This scheme is derived from the G-PROG mode by adding a marker into the bitstream after each segment representing one bitplane of one layer, namely between each couple (L_i^v, L_i^{v+1}), and between each couple of segments corresponding to changes in the bitplane (L_i^N, L_{i+1}^1). Since the quantizers are successively applied — as in the G-PROG mode — subband by subband and, within every subband, layer by layer, scalability by quality is allowed on both the whole volume and any 2-D image of the dataset, provided that 2-D local-scale conditioning is used. Owing to the large number of markers, the major drawback of this solution is the increase in the amount of the encoded information, and thus the reduction of the compression factor with respect to the G-PROG mode.

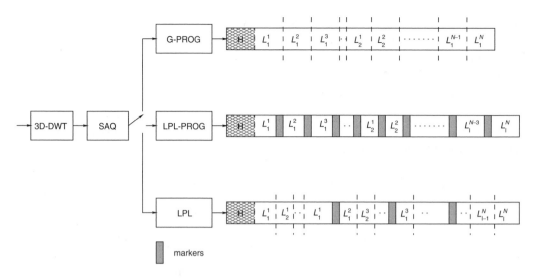

FIGURE 14.8 Structure of the bitstream for MLZC coding scheme in the different working modes. H is the bitstream header, L_i^v the encoded information corresponding to the ith bitplane of layer v, I the bit-depth and N the number of layers of the considered 3-D subband.

14.6.5.3 Layer per Layer (LPL) Mode

A possibility for reducing the overloading implied by the LPL-PROG mode is to reorder the segments and reduce the number of markers. In this case, segment units consist of the encoded information pertaining to all the bitplanes of a given layer, and markers are put in-between them, namely between couples of the type (L_I^v, L_1^{v+1}). The progressiveness by quality functionalities are suboptimal for both the single images and the whole volume. However, since this solution provides a compression factor in-between those corresponding to the G-PROG and LPL-PROG modes, it represents a good trade-off between compression efficiency and 2-D decoding time. Quality scalability could be improved by designing an ad-hoc procedure for rate allocation. We leave this subject for future investigation.

As previously mentioned, all these configurations were tested in conjunction with both the 2-D and 3-D contexts. However, the desired 3-D encoding/2-D decoding capabilities constrain the choice to two-dimensional contexts without interband conditioning.

14.6.6   3-D Object-Based Coding

We restrict our analysis to the case of two disjoint regions. For simplicity, we adopt the same terminology used in JPEG2000 and call ROI the object of interest and *background* the rest of the image or volume. In our implementation, the ROI is identified by a color code in a three-dimensional mask that we assume it is available at both the encoder and decoder sides. The problem of shape representation and coding is not addressed in this work.

Independent object coding has two major advantages. First, it is suitable for parallelization: different units can be devoted to the processing of the different objects simultaneously. Second, it is expected to improve coding efficiency when the objects correspond to statistically distinguishable sources. In what follows, the generalization of EZW-3-D and MLZC coding systems for region-based processing is detailed.

14.6.6.1   Embedded Zerotree Wavelet-Based Coding

The generalization of the classical EZW technique [53] for independent processing of 3-D objects is performed by applying the 3-D extension of the coding algorithm to the different objects independently. The definition of the parent–children relationship is slightly modified with respect to the general case where the entire volume is encoded, to emphasize the semantics of the voxels as belonging to a particular region. Accordingly, the set of descendants of a wavelet coefficient $c(l, j, \mathbf{k})$ at position $\mathbf{k}$ in subband (l,j) is identified by restricting the corresponding oct-tree to the domain of the generalized object projection $GP(l,j)$ in all the finer scales. More specifically, let T be a given oct-tree and let $T(l,j)$ identify the set of samples of the oct-tree in subband (l, j).

Definition 14.6.1

A semantic oct-tree *is the set of all subband samples ST:*

$$ST = \cup_{l,j} ST(l,j) \tag{14.10}$$

$$ST(l,j) = T(l,j) \cap GP(l,j) \tag{14.11}$$

Based on this, we derive a *semantically constrained* definition for a zerotree root.

Definition 14.6.2

A subband sample is a zerotree root *if all the coefficients, which belong to the afferent oct-tree, are nonsignificant with respect to the threshold.*

Figure 14.9 illustrates the semantically constrained oct-tree. Given a zerotree candidate point, the significance of all the descendents lying outside the generalized projection is assumed to be nonrelevant to the classification of the root as a zerotree. In consequence, we expect the number of

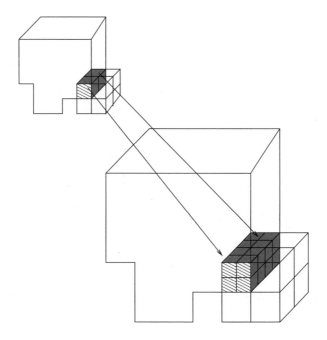

FIGURE 14.9 Semantic oct-tree.

zerotree roots to increase with respect to the general case where *all* the descendants within T are required to be nonsignificant. This potentially augments number of zerotree roots and thus the coding efficiency. The inherent embedding resulting from the quantization strategy allows PSNR scalability for any object. Accordingly, each object can be reconstructed with increasing quality by progressively decoding the concerned portion of the bitstream.

14.6.6.2 Multidimensional Layered Zero Coding

Very little modifications are needed to adapt the MLZC system for object-based processing. As for the EZW, each coefficient is encoded *if and only if* it belongs to the generalized projection of the considered object.

Equation (14.7) is generalized for this case by assuming that the significance state of any sample outside the generalized projection is zero:

$$\chi^s[\mathbf{k}, l, j] = \sum_{p=0}^{P-1} w_p \tilde{\sigma}[\mathbf{k}', l, j] \tag{14.12}$$

$$\tilde{\sigma}[\mathbf{k}', l, j] = \begin{cases} \sigma[\mathbf{k}', l, j] & \forall \mathbf{k}' \in GP(l, j) \\ 0 & \text{otherwise} \end{cases} \tag{14.13}$$

14.6.7   3-D/2-D MLZC

This section illustrates how the MLZC algorithm can be adapted for 3-D encoding/2-D decoding functionalities. In order to be able to access any 2-D image with scalable quality it is necessary to independently encode each bitplane of each subband layer (LPL-PROG mode). This implies the choice of an intraband two-dimensional context for spatial conditioning to avoid interlayer dependencies among the coefficients. This is a necessary condition for the independent decoding of the subband layers. Quality scalability on a given 2-D object in image k is obtained by successively decoding the bitplanes of the subband layers that are necessary for its reconstruction. In the LPL-PROG mode, the ith bitplane is encoded for each subband layer v before switching to the next $(i-1)$th. The markers separate the information related to the different bitplanes and subband layers.

Given the index of the image of interest, the concerned portions of the bitstream are automatically identified, accessed, and decoded. The required number of subband layers depends on the wavelet filter (see Table 14.1). As mentioned before, the 5/3 filter is preferable because it allows a significant saving of the decoding time compared with the 9/7. Moreover, while the two filters perform similarly in terms of lossless rate, the 5/3 minimizes the power of the round off noise implied by the integer lifting [48]. All this makes such a filter particularly suitable for this application.

A trade-off between overhead and scalability can be reached by removing the markers between the bitplanes and only keeping the random access to subband layers (LPL mode). In this case the coding order is layer by layer, each layer being represented with the entire set of bitplanes. However, this corresponds to a suboptimal embedding in the rate–distortion sense.

14.6.8   3-D/2-D Object-Based MLZC

The object-based 3-D/2-D MLZC system permits random access to any object of any 2-D image of the dataset with scalable up to lossless quality.

As was the case for the 3-D ROI-based system, in the encoding step, each object is assigned a portion of the bitstream, which can be independently accessed and decoded. Then, as explained in Section 14.6.7, to preserve quality scalability, the LPL-PROG mode must be chosen for both the ROI and the background in order to obtain the appropriate granularity for the encoded information. However, as already pointed out, this implies an overhead that degrades the compression performance. Quality scalability on a 2-D object of image k is obtained by successively decoding the bitplanes of its generalized projection in the subband layers that are required for the reconstruction of the corresponding image.

The bitstream structure shown in Figure 14.8 is applied to each subband of every object. The global bitstream consists of the concatenation of segments of this type, one for each object. Even though here we have considered the case of only two objects (the ROI and the background), this is not a constraint and the proposed system is able to handle any number of them. Given the index of the image of interest, the concerned portions of the bitstream are automatically identified, accessed, and decoded with the help of the masks. Besides providing competitive compression performance and allowing a fast access to the ROI at a progressively refinable quality, such a fine granularity also permits to easily reorganize the encoded information to get different functionalities. For instance, by changing the order of the coding units, or by randomly accessing the corresponding bitstream segments, progressiveness by resolution could be obtained.

14.7 RESULTS AND DISCUSSION

This section provides an overview of the results obtained using different system configurations and working modes. The datasets that were used are presented in Section 14.7.1. The performance of the MLZC system in the different working modalities (G-PROG, LPL-PROG, and LPL) are analyzed in Section 14.7.2 using four datasets with heterogeneous features. Section 14.7.3 analyzes the performance of the 3-D object-based system. It is assumed that the data consist of one ROI surrounded by the background. The system is first characterized by comparison with other 3-D and 2-D algorithms when applied to MR images of the brain. Then, the object-based performance is examined. At last, Section 14.7.4 is devoted to the 3-D/2-D object-based MLZC system, integrating both ROI-based functionalities and 2-D decoding capabilities. In this case, only 2-D spatial conditioning is allowed. Interband as well as 3-D spatial conditioning would indeed introduce the same type of dependencies among subband coefficients impeding the independent decoding of the subband layers.

14.7.1 DATASETS

The performance of the MLZC 3-D encoding/2-D decoding system was evaluated on the four datasets illustrated in Figure 14.10:

- *Dynamic spatial reconstructor (DSR)*. The complete DSR set consists of a 4D (3-D+time) sequence of 16 3-D cardiac CT data. The imaging device is a unique ultra-fast multislice scanning system built and managed by the Mayo Foundation. Each acquisition corresponds to one phase of the cardiac cycle of a canine heart and is composed of 107 images of size 128×128. A voxel represents approximately $(0.9 \text{ mm})^3$ of tissue.
- *MRI head scan*. This volume consists of 128 images of size 256×256 representing the saggital view of an human head.
- *MR-MRI head scan*. This volume has been obtained at the Mallinckrodt Institute of Radiology (Washington University) [5]. It consists of 58 images of a saggital view of the head of size 256×256. Since this dataset has also been used as a test set by other authors [5,30,69] it allows to compare the compression performance of the MLZC with that of other 3-D systems.
- *Opthalmologic angiographic sequence (ANGIO)*. The ANGIO set is a 3-D sequence (2-D+time) of angiography images of a human retina, consisting of 52 images of 256×256 pixels each.

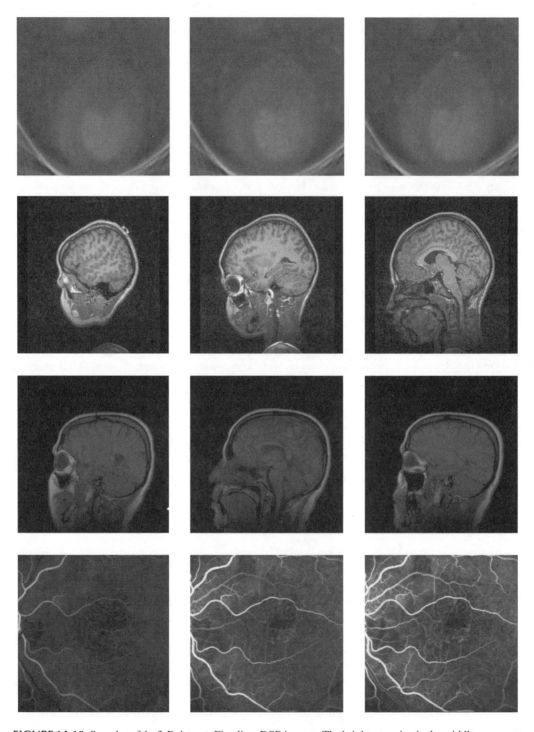

FIGURE 14.10 Samples of the 3-D dataset. First line: DSR images. The brightest region in the middle represents the left ventricle of a canine heart. Second line: human head MRI, saggital view. Third line: MR-MRI. Fourth line: opthalmologic angiography sequence (2-D + time). The brightness results from the diffusion of the contrast medium into the vessels.

The different features of the considered datasets make the resulting test set heterogeneous enough to be used for characterizing the system. The DSR volume is very smooth and high correlation is exhibited among adjacent voxels along all the three spatial dimensions. This makes it very easy to code and particularly suitable for the proposed coding system. It represents the "best case" test set, for which the coding gain of 3-D over 2-D systems is expected to be the highest.

Conversely, the ANGIO dataset can be considered as the "worst case" for a wavelet-based coding system. The images are highly contrasted: very sharp edges are juxtaposed to a smooth background. Wavelet-based coding techniques are not suitable for this kind of data. The edges spread out over the whole subband structure generating a distribution of nonzero coefficients whose spatial arrangement cannot be profitably exploited for coding. This is owing to the fact that wavelets are not suitable descriptors of images with sharp edges. The MR-MRI set has been included for sake of comparison with the results provided by other authors [5]. Nevertheless, we do not consider it as representative of a real situation because it went through some preprocessing. In particular, it has been interpolated, scaled to isotropic 8-bit resolution and thresholded. Finally, the characteristics of the MRI set lie in between. It is worthy to note that the structure and semantics of the MRI images make the volume suitable for an *object-based* approach to coding.

14.7.2 3-D/2-D MLZC

Different 2-D and 3-D coding algorithms were considered for the sake of comparison. The benchmark for the 3-D case was the 3-D EZW. The MLZC system was characterized by determining the lossless rates corresponding to the complete set of contexts, in each working mode. Referring to the notations introduced in Section 14.6, the 2-D context that has been chosen for the definition of the three-dimensional conditioning terms is the (060). Indeed, results show that it is among the three most effective 2-D contexts on all the datasets for both the LPL-PROG and LPL modes. As expected, the best performance in terms of lossless rate is obtained in the G-PROG mode. As it is the case for EZW-3-D, the G-PROG mode does not allow 2-D decoding. In the LPL and LPL-PROG modes such a functionality is enabled at the expenses of coding efficiency, which decreases because of the additional information to be encoded to enable random access.

One of the constraints posed by 2-D decoding is that no interband conditioning can be used. Even though the exploitation of the information about the significance of the parent within the subband hierarchy can be fruitful in some cases, results show that the compression performance is not much affected.

The observed dependency of the lossless rate on the design parameters of the conditioning terms (i.e., the spatial support and the use of interband conditioning) applies to both MLZC-2-D and MLZC. The efficiency of the entropy coding increases with the size of the spatial support up to a limit where the sparseness of the conditioning space does not allow an adequate representation of the statistics of the symbols to be encoded.

The benchmark for 2-D systems is the new coding standard for still images JPEG2000 [25,47]. Figure 14.11 compares the performance of the different 2-D algorithms for DSR. In this case, MLZC-2-D outperforms both JPEG2000 and JPEG-LS. For the old JPEG standard (JPEG-LS), all of the seven available prediction modes were tested and the one providing the best performance (corresponding to $K = 7$ for all the datasets) was retained.

Table 14.2 summarizes the performance of the different algorithms and working modes. The (060) and (160) contexts were chosen as references, and no interband conditioning was used. As was the case for JPEG2000, the data concerning the 2-D algorithms were obtained by running them on the whole set of 2-D images and taking the average of the resulting set of lossless rates. As expected, the coding gain provided by the 3-D over the 2-D systems depends on the amount of correlation and smoothness along the z-axis. Accordingly, it is quite pronounced for DSR (16.3%) and MR-MRI

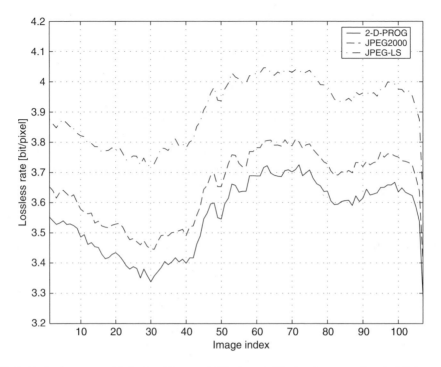

FIGURE 14.11 Performance of 2-D algorithms on DSR dataset. The lossless rate of each 2-D image is represented as a function of its position along the z-axis. Continuous line: MLZC-2-D; dashed line: JPEG2000; dash-dot line: JPEG-LS. The lossless rate provided by the MLZC algorithm in the LPL modality is 3.06 bit/voxel.

TABLE 14.2
Lossless Performances (Bit/Voxel) with 5/3 Filter

	G-PROG	LPL-PROG	LPL	EZW-3-D	MLZC-2-D	JPEG2K	JPEG-LS
DSR	2.99	3.11	3.03	2.88	3.56	3.62	3.90
	2.93	3.08	3.06				
MRI	4.58	4.63	4.55	4.46	4.62	4.65	5.10
	4.52	4.60	4.52				
MR-MRI	2.24	2.28	2.24	2.271	2.92	2.95	3.437
	2.19	2.23	2.22				
ANGIO	4.19	4.23	4.20	4.18	4.41	4.43	3.87
	4.16	4.22	4.21				

Note: The decomposition depth is $L = 4$ for DSR, MRI and MR-MRI, and $L = 3$ for ANGIO. The two values correspond to the contexts (060) and (160) for each dataset. No interband conditioning is used.

(33.06%), for which the LPL mode leads to a sensible rate saving over JPEG2000, while it is lower for both MRI (2.2%) and ANGIO (5.2%).

The best compression performance for ANGIO is obtained with JPEG-LS. As mentioned above, such a dataset is not suitable for wavelet-based coding, so that other algorithms can easily be more effective. Nevertheless, the LPL method provides an improvement of about 5% over JPEG2000. The 3-D encoding/2-D decoding approach can thus be considered as a good trade-off between compression

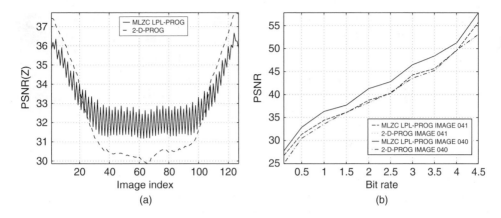

FIGURE 14.12 Performances in lossy regime of LPL-PROG and MLZC-2-D on MRI. (a) PSNR as a function of the image index (z coordinate); each image has been independently decoded at 0.5 bit/pixel. (b) PSNR as a function of the bit rate for images 40 and 41 of the dataset.

efficiency and the fast access to the data, while featuring many desirable functionalities which are not supported by JPEG-LS.

The evaluation of the performance in lossy regime is out of the scope of this paper. The observed oscillatory trend of the PSNR along the coordinate axis entails the analysis of both the rounding noise implied by integer lifting and the quantization noise [48]. Figure 14.12 gives an example. After encoding the volume in the LPL-PROG mode, every image of the dataset was independently decoded at 0.5 bit/pixel. Figure 14.12(a) compares the corresponding PSNR to that obtained by separately encoding and decoding each image with the MLZC-2-D at the same rate. It is important to notice that the control over the decoding bit rate on the single 2-D images is only possible when they are decoded one by one. On average, the 3-D method outperforms the 2-D counterpart on the central portion of the dataset (images 20 to 100), where the images are not dominated by the background. In this case, the oscillation has period one, namely every other image has higher PSNR. This makes the improvement provided by the 3-D system dependent on the position of the considered image. Figure 14.12(b) shows the PSNR for images of index 40 and 41 as a function of the decoding rate. The maximum and mean increase in the PSNR are about 4.7 and 2.7 dB for image 40, and about 0.8 and 2.7 for image 41, respectively. This issue is currently under investigation. The model proposed in [48] has been extended to the 3-D case [37]. The next step is the definition of a quantization policy ensuring a more uniform distribution of the residual error at a given rate.

14.7.3 3-D Object-Based MLZC

In medical images the background often encloses the majority of the voxels. For a typical MRI dataset, for instance, about 90% of the voxels belong to the background. A sensible rate saving can thus be achieved via ROI-based coding. In this experiment, the test set consisted of a MRI head scan of $256 \times 256 \times 128$ voxels with a graylevel resolution of 8 bpp. Again, it was assumed that the object of interest was the brain, and the rest was considered as the background. Figure 14.13 shows one sample image together with the corresponding masks selecting the ROI and the background, respectively. Some coding results are also provided for the 8-bit MR-MRI dataset.

14.7.3.1 System Characterization

The performance of the EZW-3-D and MLZC was analyzed by comparison with the 2-D counterparts, namely EZW-2-D and MLZC-2-D as well as with JPEG and JPEG2000. For 2-D encoding, the

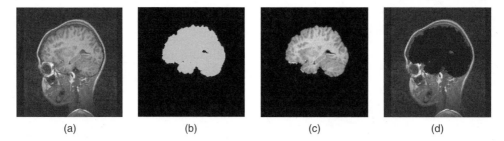

(a)   (b)   (c)   (d)

FIGURE 14.13 Saggital view of the MR of a brain: (a) original image; (b) mask; (c) object of interest; (d) background.

TABLE 14.3
Lossless Rates for MRI and MR-MRI Datasets

	EZW-2-D	MLZC-2-D	JPEG2K	JPEG-LS	EZW-3-D	MLZC
MRI	4.698	4.597	4.651	5.101	**4.456**	4.457
MR-MRI	2.878	2.848	2.954	3.437	2.271	**2.143**

Note: For the 2-D algorithms, the average lossless rate was retained for each mode. The filter is 5/3, $L = 4$, and global conditioning is used in the MLZC mode with the (070) and (271) contexts for 2-D and 3-D, respectively.

images were processed independently. For all the wavelet-based methods, namely EZW, MLZC, and JPEG2000, $L = 3, 4$ levels of decomposition and the 5/3 [6] filter were chosen. All of the seven prediction modalities of the lossless JPEG mode (JPEG-LS) were tested and the best one, corresponding to $k = 7$, was retained. Results show that the most effective context for these datasets using MLZC-2-D is the (070) with interband conditioning, so such a context was used to define the 3-D spatial conditioning terms. Accordingly, $\chi^s[l, j, \mathbf{k}]$ was constructed as illustrated in Figure 14.7, with (070) being the spatial support in layer v. The (271) context was used for the evaluation of the object-based performances. It consists of the corner pixels of the first-order neighborhood in the current layer (v), a cross-shaped support in the previous layer ($v - 1$), and the pixel in the same spatial position as the current one in the next ($v + 1$) layer (see Figure 14.6 and Figure 14.7).

For the head MRI dataset, performance tends to improve when extending the generalized neighborhood used for conditional arithmetic coding, in both the 2-D and 3-D cases.

The same set of experiments was run on the MR-MRI dataset. In general, the trend was the same as for the MRI set, namely the highest compression factors were obtained by using the context (070) and (271) with interband conditioning, for MLZC and MLZC-2-D, respectively. Table 14.3 compares the average lossless rates of each of the considered 2-D algorithms to those provided by MLZC and EZW-3-D, for both datasets. Among the 2-D algorithms, MLZC-2-D with context (070) outperforms the others. JPEG2000 results in a lossless rate slightly lower than EZW-2-D for MRI. All 2-D schemes provide a sensible improvement over JPEG-LS. For MRI, the lowest lossless rate corresponds to the EZW-3-D scheme, which in this case slightly outperforms MLZC. Nevertheless, the MLZC method is faster and less computationally demanding than EZW-3-D.

For MR-MRI, some results are available in the literature. We refer here to those presented by Bilgin et al. [5]. The first one was obtained for $L = 3$ and using the integer version of the $(2 + 2, 2)$ filter (as defined in [6]) on the whole volume. The second was based on a two-level integer transform with the $(1 + 1, 1)$ filter run on 16 slice coding units, and the compression efficiency data were averaged over the volume. The coding scheme — *3-D CB-EZW* — was a version of EZW-3-D exploiting context modeling. The corresponding lossless rates were 2.285 and 2.195 bits/voxel. The best MLZC mode results in 2.143 bits/voxel, slightly improving such a result. However, a wider set of measurements would be required for the comparative evaluation of the two competing systems.

14.7.3.2 Object-Based Performance

The results given in this section concern the MRI dataset. In the proposed system, the object of interest and the background are encoded *independently*. Each of them generates a self-contained segment of the bitstream. This implies that the *border information* is encoded twice: as side information for *both* the object *and* the background. In this way, each of them can be accessed and reconstructed *as if* the whole set of wavelet coefficients were available, avoiding artifacts along the contours for any quantization of the decoded coefficients. ROI-based EZW-2-D was taken as the benchmark for the object-based functionalities. Despite the availability of ROI-based functionalities, JPEG2000 was not suitable for the purpose. In JPEG2000, ROI-based coding is performed by the MAXSHIFT method [10]. Basically, the subband coefficients within the ROI mask are shifted up (or, equivalently, those outside the ROI are shifted down) so that the minimum value in the ROI is greater than the maximum value in the background. This splits the bitplanes, respectively, used for the ROI and the background in two disjoint sets. The rate allocation procedure assigns to each layer of each codeblock (in the different subbands) a coding priority (through the MAXSHIFT method), which depends on both the semantic classification and the gain in terms of rate–distortion ratio. This establishes the relative order of encoding of the ROI subband coefficients with respect to the background. Results showed that with the implementation of the procedure described by Taubman [59], for the head MRI dataset, high priority is assigned to the background layers in the codeblocks, moving the focus of the encoder out of the ROI. The ROI and background codeblocks are mixed up, compromising ROI-based functionalities. This can be easily verified by decoding the portion of the bitstream indicated by the encoder as representing the ROI. The resulting image is composed of both the ROI and the background. A possible solution would be to design an ad hoc rate allocation algorithm optimized for datasets having a background very easy to code, but this is out of the scope of our work. Instead, we independently compressed the ROI and the background with JPEG2000 and compared the respective bit rates to those provided by both our EZW-2-D object-based system and ROI-based JPEG2000. Such working conditions emphasize the *implicit* ROI mask encoding by JPEG2000. Even though the mask does not need to be separately coded, its encoding is implied by the exhaustive scanning of the subbands.

Results are given in Figure 14.14. The global lossless rate under different conditions is shown as a function of the image index. In particular, the dash-dot line represents ROI-based JPEG2000 and the continuous line is for EZW-2-D with independent object (IO) coding. The curve represents the sum of the lossless rates concerning the ROI and the background. Owing to the rate allocation policy, JPEG2000 outperforms EZW-2-D in compression efficiency. The drawback is that, as mentioned previously, the codeblocks of the ROI and the background are interlaced in such a way that the ROI-based functionalities are not always achieved. The dashed line represents the total rate needed for independently encoding the ROI and the background by JPEG2000. The gap between the corresponding curve and the one for EZW-2-D IO emphasizes the performance degradation due to the implict coding of the mask. Figure 14.14 points out that the EZW-2-D coding scheme represents a good compromise for the trade off between coding efficiency and random access to the objects. Figure 14.15 shows the lossless rates for the ROI (OBJ), the background (BGND), and the entire image (WHOLE) for EZW-2-D. The continuous and dashed lines correspond to $L = 3$ and $L = 4$, respectively. Here, the bit rates are calculated as the ratio between the size of the portion of the bitstream concerning the OBJ(BGND) and the size of the OBJ(BGND). While the curves for WHOLE

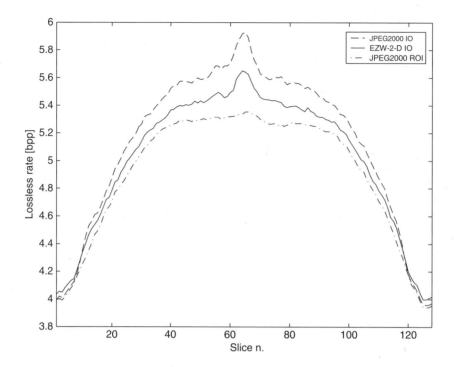

FIGURE 14.14 Lossless rates as a function of the position of the 2-D images along the z-axis. Continuous line: EZW-2-D; dashed line: JPEG2000 IO (independent object); dotted line: JPEG2000 ROI.

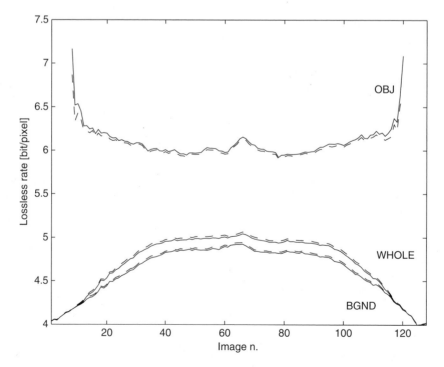

FIGURE 14.15 Lossless rates for the EZW-2-D algorithm as a function of the position of the 2-D images along the z-axis, for the 5/3 filter. Dashed line: $L = 3$; continuous line: $L = 4$.

TABLE 14.4
Lossless Rates (LR) for Head MRI

LR (bpp)	OBJ	BGND	WHOLE	OBJ + BGND	Δ%
EZW-3-D	0.9045	3.9012	4.4598	4.8057	+7.75
MLZC (271)	0.9188	3.8868	4.4566	4.8056	+7.83
EZW-2-D	0.9327	4.0835	4.6977	5.0162	+6.78
JPEG2K IO	1.0641	4.0656	4.6511	5.1297	+10.29
JPEG2K ROI	-	-	4.6511	4.9099	+5.56

Note: The filter is 5/3 $L = 4$. Global conditioning was used in the MLZC mode.

and BGND are close to each other, the one for OBJ is outdistanced. The volume and the background enclose a large number of black samples, which are simple to compress. Conversely, the region of interest is entirely structured, and necessitates more bit/pixel to be encoded. The steep slope at both ends of the curve corresponding to the object are due to the fact that the ROI takes only very few or no pixels, stretching the curve to infinity. This example points out the importance of the ROI-based approach. For this dataset, only the 19% — on average — of the bitstream corresponding to the entire volume is needed to represent the ROI. The random access to the objects allows fast access to the important information, with considerable improvement in compression *efficiency*.

Table 14.4 quantifies the degradation in compression efficiency due to independent object coding. The second and third columns (OBJ and BGND) show the lossless rates for the ROI and the background. The forth and fifth columns (WHOLE and OBJ+BGND) are the bitrates obtained when encoding the entire volume and the object and background, respectively. The last one shows the percentage increase of the lossless rate for independent encoding of the objects with respect to that corresponding to the entire volume. The increase of the lossless rate for independent object coding is measured by the difference between the required rate (OBJ+BGND) and the reference one (WHOLE). The differences between the compression ratios for the cases WHOLE and OBJ+BGND are due to two causes. First, the entropy coder performs differently in the two cases because of the different sources. Second, the total number of coefficients to be encoded is larger for OBJ+BGND because of the twofold encoding of the border voxels (BP) of the GP. The size of the bitstream increases by about 7% for $L = 4$ in case of separate object handling. According to Table 14.4, the gain in compression efficiency due to the exploitation of the full correlation among data is about 4–5%. The improvement in compression efficiency provided by MLZC over JPEG2000 depends on the working mode. Taking the OBJ+BGND as reference, the corresponding rate reduction is about 2.2% and 6.3%, respectively for JPEG2000 ROI and JPEG2000 IO.

The prioritization of the information inherent to separate object processing leads to a significant improvement in coding efficiency when it is possible to relax the lossless constraint in the background region. In this case, the BGND can be encoded/decoded at a lower resolution and combined with the object of interest — which was encoded/decoded without loss — in the final composed image. Figure 14.16 gives an example. Both the object and the background were compressed by the MLZC scheme, with context (271) and using interband conditioning. The OBJ was decoded at full quality (i.e., in lossless mode) while the BGND corresponds to a rate of 0.1 bit/voxels in Figure 14.16a and 0.5 bit/voxels in Figure 14.16b. The PSNR values for images of Figure 14.16a and Figure 14.16b are of 27.76 and 33.53 dB, respectively. Reconstructed images respecting the lossless constraint in the ROI and preserving a good visual appearance in the background can thus be obtained by decoding only the 20% of the information required for the lossless representation of the whole volume.

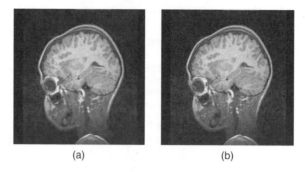

(a)   (b)

FIGURE 14.16 *Pseudo-lossless* regime for a sample MRI image. The OBJ was recovered without loss, while the BGND has been decoded at 0.1 bpv (a) and 0.5 bpv (b). The corresponding PSNR values are of 27.76 and 33.53 dB, respectively.

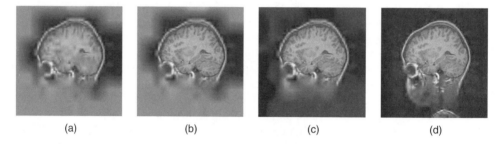

(a)   (b)   (c)   (d)

FIGURE 14.17 ROI-based JPEG2000. Images decoded at different rates: (a) 0.1 bpp; (b) 0.5 bpp; (c) 1.0 bpp; (d) reference rate for lossless ROI.

14.7.4 3-D/2-D Object-Based MLZC

The performances of the 3-D/2-D object-based MLZC coding system were tested in the LPL-PROG mode with the 2-D spatial context (030). This was chosen empirically as the one providing good compression performance while keeping the computational complexity low.

The JPEG2000 [25] standard in the ROI-based mode was chosen for comparison. Both the lossless and lossy regimes were tested. The test set consists of the same two MRI head scans used to characterize object-based functionalities as discussed in the previous section. Figure 14.17 shows the images for the decoding rates 0.1, 0.5, 1.0, and 1.6 bits/pixel. The last rate is the lossless rate of the ROI as provided by the JPEG2000 encoder. In order to make the comparison with our system, we independently compressed the ROI and the background with JPEG2000 and compared the respective bit rates to those provided by ROI-based JPEG2000 (see Figure 14.14). The average lossless rates are compared to those provided by MLZC in the LPL-PROG mode in Table 14.5. In particular, the OBJ column gives the lossless rate for the object of interest, when it is encoded independently by the different algorithms, while the WHOLE and OBJ+BGND columns provide the lossless rates obtained for the entire volume and the independent coding of the object and the background, respectively. The minimum lossless rate for OBJ is obtained by MLZC for both the datasets. For MRI, JPEG2000 ROI outperforms the proposed system for OBJ+BGND reducing the lossless rate of 2.7%. However, in this mode the ROI-based functionalities are not completely fulfilled. Conversely, MLZC OBJ+BGND outperforms JPEG2000 IO of about 1.6%, while preserving random access to every object.

TABLE 14.5
Lossless Rates (LR) for MRI and MR-MRI

Volume	LR (bpp)	OBJ	WHOLE	OBJ + BGND
MRI	MLZC (030)	0.959	4.666	5.048
	JPEG2K IO	1.064	4.651	5.130
	JPEG2K ROI	1.259	4.651	4.910
MR-MRI	MLZC (030)	0.788	2.310	2.933
	JPEG2K OI	1.062	2.950	3.300
	JPEG2K ROI	1.289	2.950	3.267

Note: The filter is 5/3, $L = 4$. Pure spatial conditioning was used for the
MLZC LPL-PROG mode.

Comparing these results with the lossless rates given in Table 14.4, one can observe that the performance of the MLZC system are slightly degraded. This is due to the different choice for the conditioning term, which makes the EZW-3-D outperform MLZC in terms of compression efficiency. However, the interest for the 3-D/2-D coding system lies in the functionalities that are not supported by the EZW-3-D. For the MR-MRI dataset, MLZC OBJ+BGND provides the best performances. The corresponding rate saving is about 11% and 12.5% over JPEG2000 ROI and IO, respectively.

14.8 CONCLUSIONS

Coding systems for medical imagery must focus on multidimensional data and cope with specific requirements. This constrains the design while facilitating the task through the availability of a-priori knowledge about the image content and its semantic. Besides good rate–distortion performance in any working mode, medical image coding systems must feature lossless capabilities as well as scalability (by quality or by resolution). Wavelets proved to be very suitable for this purpose leading to architectures that meet most of requirements. The semantic of the different objects present in the images pushes toward the definition of object-based techniques. Different solutions have been proposed by different researchers, each responding to some domain-specific requirements. On the same line, based on the fact that it is still common practice for medical doctors to analyze the volumes image by image, we proposed a coding system providing a fast access to any 2-D image without sacrificing compression performance. The integration of ROI-based functionalities led to a versatile and highly efficient engine allowing to swiftly recover any object of any 2-D image of the dataset at a finely graded up to lossless quality. Furthermore, independent object processing and the use of the integer version of the wavelet transform make the algorithm particularly suitable for the implementation on a device. We believe that the set of functionalities of 3-D/2-D object-based MLZC system makes it well suitable for the integration in PACS and largely compensates for the eventual loss in compression.

We would like to conclude by mentioning one of the most promising research directions. ROI-based capabilities are the first step along the path of *model-based* coding. The basic idea is to improve the compression performance by combining real and synthetic data. Real data would be classically encoded while and synthetic data would be generated according to an ad hoc recipe. We believe that such a philosophy opens the way to a novel approach to image representation leading to the next generation of *intelligent* coding systems.

REFERENCES

1. Lossless and Near-Lossless Coding of Continuous Tone Still Images-Baseline, Technical report, ISO/IEC 14495-1, 2000.
2. ACR-NEMA, *DICOM Digital Imaging and Communications in Medicine*, 2003.
3. J. Akune, H. Yonekawa, Y. Ishimitsu, H. Takeuchi, K. Doi, and H. MacMahon, Development of a data compression module for digital radiography, *Med. Biolog. Eng. Comp.*, 29(1991).
4. H. Barnard, Image and Video Coding Using a Wavelet Decomposition, Ph.D. thesis, Delf University, Delf, The Netherlands, 1994.
5. A. Bilgin, G. Zweig, and M. Marcellin, Three-dimensional image compression with integer wavelet transform, *Appl. Opt.*, 39(2000), 1799–1814.
6. A. Calderbank, I. Daubechies, W. Sweldens, and B. Yeo, Lossless image compression using integer to integer wavelet transforms, *Proceedings of the International Conference on Image Processing* (ICIP), 1997, pp. 596–599.
7. A. Calderbank, I. Daubechies, W. Sweldens, and B.-L. Yeo, Wavelet transforms that map integers to integers, *Appl. Comput. Harmon. Anal.*, 5(1998), 332–369.
8. K. Chan, C. Lau, K. Chuang, and C. Morioca, Visualization and volumetric compression, SPIE, vol. 1444, 1991, pp. 250–255.
9. E. Chiu, J. Vaisey, and M. Atkins, Wavelet-based space-frequency compression of ultrasound images, *IEEE Trans. Inform. Technol. Biomed.*, 5(2001), 300–310.
10. C. Christopoulos, A. Skodras, and T. Ebrahimi, The JPEG2000 still image coding system: an overview, *IEEE Trans. Consumer Electron.*, 46(2000), 1103–1127.
11. C. Chui and R. Yi, System and Method for Nested Split Coding of Sparse Datasets, U.S. Patent 005748 116a, Teralogic Inc., Menlo Park, CA, May 1998.
12. D. Clunie, Lossless compression of greyscale medical images — effectiveness of traditional and state of the art approaches, *Proceedings of SPIE Medical Imaging*, 2000.
13. P. Cosman, C. Tseng, R. Gray, R. Olshen, L. Moses, H. Devidson, C. Bergin, and E. Riskin, Tree-structured vector quantization of CT chest scans: image quality and diagnostic accuracy, *IEEE Trans. Medical Imaging*, 12(1993), 727–739.
14. K. S. D. Wu and N.D. Memon, A Context-Based, Adaptive, Lossless/Nearly Lossless Coding Scheme for Continuous Tone Images, ISO Working document n256, ISO/IEC/SC29/WG1, 1995.
15. M. Das and S. Burgett, Lossless compression of medical images using two-dimensional multiplicative autoregressive models, *IEEE Trans. Medical Imaging*, 12(1993), 721–726.
16. I. Daubechies and W. Sweldens, Factoring wavelet transform into lifting steps, *J. Fourier Anal. Appl.*, 41(1998), 247–269.
17. K. Deneker, J.V. Overloop, and I. Lemahieu, An experimental comparison of several lossless image coders for medical images, *Proceedings of the IEEE Data Compression Conference*, 1997.
18. J. Duncan and N. Ayache, Medical image analysis: progress over two decades and the challenges ahead, *IEEE Trans. on Pattern Anal. Machine Intelligence*, 22(2000), 85–106.
19. O. Egger, W. Li, and M. Kunt, High compression image coding using an adaptive morphological subband decomposition, *Proc. IEEE*, 83(1995), 272–287.
20. M. Goldberg and L. Wang, Comparative performance of pyramid data structures for progressive image transmission, *IEEE Trans. Commn.*, 39(1991), 540–548.
21. R. Gruter, O. Egger, J.-M. Vesin, and M. Kunt, Rank-order polynomial subband decomposition for medical image compression, *IEEE Trans. Medical Imaging*, 19(2000), 1044–1052.
22. B. Ho, J. Chao, C. Wu, and H. Huang, Full frame cosine transform image compression for medical and industrial applications, *Machine Vision Appl.*, 4(1991), 89–96.
23. C. Hwang, S. Wenaktraman, and K. Rao, Human visual system weighted progressive image transmission using lapped orthogonal transform/classified vector quantization, *Opt. Eng.*, 32 (1993), 1524–1531.
24. A. Islam and W. Pearlman, An embedded and efficient low-complexity hierarchical image coder, *Proc. SPIE*, 3653(1999), 294–305.
25. ISO/IEC JTC 1/SC 29/WG1, Information Technology — JPEG2000 Image Coding System, ISO/IEC International Standard 15444-1, ITU Recommendation T.800, 2000.

26. ISO/IEC JTC 1/SC 29/WG11, Working document N2861, July 1999.

27. ISO/IEC JTC 1/SC 29/WG11, Working document N3156, December 1999.

28. ISO/IEC JTC1/SC29/WG11/N5231, Working document N5231, October 2002.

29. Y. Kim and W. Pearlman, Lossless volumetric medical image compression, *Proceedings of SPIE Applications of Digital Image Processing XXII*, vol. 3808, 1999, pp. 305–312.

30. Y. Kim and W. Pearlman, Stripe-based sphit lossy compression of volumetric medical images for low-memory usage and uniform reconstruction quality, *Proceedings of the International Conference on Acoustics, Speech, and Signal Processing (ICASSP)*, vol. 4, 2000, pp. 2031–2034.

31. J. Kivijarvi, T. Ojala, T. Kaukoranta, A. Kuba, L. Nyul, and O. Nevalainen, A comparison of lossless compression methods for medical images, *Computerized Medical Imaging and Graphics*, 22(1998), 323–339.

32. G. Kuduvalli and R. Rangayyan, Performance analysis of reversible image compression techniques for high-resolution digital teleradiology, *IEEE Trans. Medical Imaging*, 11(1992), 430–445.

33. M. Kunt, A. Ikonomopoulos, and M. Kocher, Second-generation image coding techniques, *Proc. IEEE*, 73(1985), 549–574.

34. J. Liu and M. Engelhorn, Improvement in the locally optimum runlength compression of ct images, *J. Biomed. Eng.*, 12(1990), 108–110.

35. Z. Lu and W. Pearlman, Wavelet video coding of video objects by object-based speck algorithm, *Picture Coding Symposium*, April 2001, pp. 413–416.

36. M. Marcellin, Wavelet/TCQ Questionnaire, Technical report, ISO/IEC, JTC 1/SC29/WG1 X/275, 1997.

37. G. Menegaz, Model-based coding of multi-dimensional data with applications to medical imaging, Ph.D. thesis, Signal Processing Laboratory (LTS), Swiss Federal Institute of Technology (EPFL), May 2000.

38. G. Menegaz and L. Grewe, 3d/2d object-based coding of head mri data, *Proceedings of the International Conference on Image Processing (ICIP)*, vol. 1, Rochester, NY, 2002, pp. 181–184.

39. G. Menegaz and J.-P. Thiran, Lossy to lossless object-based coding of 3-d MRI data, *IEEE Trans. Image Process.*, 11(2002), 1053–1061.

40. G. Menegaz and J.-P. Thiran, 3-d encoding–2d decoding of medical data, *IEEE Trans. Medical Imaging*, 22(2003), 424–440.

41. G. Menegaz, V. Vaerman, and J.-P. Thiran, Object-based coding of volumetric medical data, *Proceedings of the International Conference on Image Processing (ICIP)*, vol. 3, 1999, pp. 920–924.

42. A. Munteanu, J. Cornelis, and P. Cristea, Wavelet-based lossless compression of coronary angiographic images, *IEEE Trans. Medical Imaging*, 18(1999), 272–281.

43. A. Munteanu, J. Cornelis, G.V. der Auwera, and P. Cristea, Wavelet-based lossless compression scheme with progressive transmission capability, *Int. J. Imaging, Sci. Tech.*, 10(1999), 76–85.

44. M. Penedo, W. Pearlman, P. Tahoces, M. Souto, and J. Vidal, Region-based wavelet coding methods for digital mammography, *IEEE Trans. Medical Imaging*, 22(2003), 1288–1296.

45. W. Pennebacker, J. Mitchell, G. Langdon, and R. Arps, An overview of the basic principles of the q-coder adaptive binary arithmetic coder, IBM *J. Res. Manage.*, 32(1988), 717–726.

46. M. Rabbani and P. Jones, Image compression techniques for medical diagnostic imaging systems, *J. Digital Imaging*, 4(1991).

47. M. Rabbani and R. Joshi, An overview of the JPEG2000 still image compression standard, *Signal Process.: Image Commn.*, 17(2002), 3–48.

48. J. Reichel, G. Menegaz, M. Nadenau, and M. Kunt, Integer wavelet transform for embedded lossy to lossless image compression, *IEEE Trans. Image Process.*, 10(2001), 383–392.

49. P. Roos and M. Viergever, Reversible intraframe compression based on hint (hierarchical interpolation) decorrelation and arithmetic coding, *Proceedings of SPIE, Medical Imaging V: Image Capture, Formatting, and Display*, vol. 1444, May 1991, pp. 283–290.

50. P. Roos and M. Viergever, Reversible 3-d decorrelation of medical images, *IEEE Trans. Medical Imaging*, 12(1993), 413–420.

51. P. Roos, M. Viergever, M.V. Dikje, and H. Peters, Reversible intraframe compression of medical images, *IEEE Trans. Medical Imaging*, 7(1988), 538–547.

52. A. Said and W. Pearlman, A new, fast, and efficient image coded based on set partitioning hierarchical trees, *IEEE Trans. Circuits Sys. Video Technol.*, 6(1996), 243–250.

53. J.M. Shapiro, Embedded image coding using zerotrees of wavelet coefficients, *IEEE Trans. Signal Process.*, 41(1993), 3445–3462.

54. P. Shelkens, J. Barbarien, and J. Cornelis, Compression of volumetric medical data based on cube splitting, *Proceedings of SPIE*, vol. 4115, July–August 2000, pp. 91–101.

55. P. Shelkens, X. Giro, J. Barbarian, and J. Cornelis, 3-d compression of medical data based on cube splitting and embedded block coding, *Proceedings of ProRISC/IEEE Workshop*, December 2000, pp. 495–506.

56. P. Shelkens and A. Munteanu, An overview of volumetric coding techniques, ISO/IEC JTC 1/SC29 WG1, Vrije Universiteit Brussel, July 2002.

57. P. Shelkens, A. Munteanu, J. Barbarien, M. Galca, X. Giro-Nieto, and J. Cornelis, Wavelet coding of volumetric medical datasets, *IEEE Trans. Medical Imaging*, 22(2003), 441–458.

58. S. Tai, Y. Wu, and C. Lin, An adaptive 3-d discrete cosine transform coder for medical image compression, *IEEE Trans. Inform. Tech. Biomed.*, 4(2000), 259–263.

59. D. Taubman, High performance scalable image compression with ebcot, *IEEE Trans. Image Process.*, 9(2000), 1158–1170.

60. D. Taubman and A. Zakhor, Multirate 3-d subband coding of video, *IEEE Trans. Image Process.*, 3(1994), 572–588.

61. G. Triantafyllidis and M. Strinzis, A context based adaptive arithmetic coding technique for lossless image compression, *IEEE Signal Process. Lett.*, 6(1999), 168–170.

62. I. Ueno and W. Pearlman, Region of interest coding in volumetric images with shape-adaptive wavelet transform, *Proc. SPIE Image Video Commn. Process.*, 5022(2003), 1048–1055.

63. M.J. Weinberger, G. Seroussi, and G. Sapiro, Loco-i: a low complexity, context-based, lossless image compression algorithm, *Proceedings of the Conference on Data Compression*, IEEE Computer Society, Washington, 1996, p. 140.

64. F. Wheeler, Trellis Source Coding and Memory Constrained Image Coding, Ph.D. dissertation, Renselaer Polytechnic Institute, Troy, New York, 2000.

65. S. Wong, L. Zaremba, D. Gooden, and H. Huang, Radiologic image compression: a review, *Proc. IEEE*, 83(1995), 194–219.

66. X. Wu and N. Memon, CALIC — a context based adaptive lossless image codec, *Proceedings of the IEEE International Conference on Acoustics, Speech, and Signal Processing (ICASSP)*, vol. 4, May 1996, pp. 7–10.

67. Y. Wu, Medical image compression by sampling dct coefficients, *IEEE Trans. on Inform. Tech. Biomed.*, 6(2002), 86–94.

68. Y. Wu and S.-C. Tai, Medical image compression by discrete cosine transform spectral similarity strategy, *IEEE Trans. Inform. Technol. Biomed.*, 5(2001), 236–243.

69. Z. Xiong, X. Wu, D. Yun, and W. Pearlman, Progressive coding of medical volumetric data using three-dimensional integer wavelet packet transform, *IEEE Second Workshop on Multimedia Signal Processing* (Cat.No.98EX175), Piscataway, NJ, USA, December 1998, pp. 553–558.

15 Remote-Sensing Image Coding

Bruno Aiazzi, Stefano Baronti, and Cinzia Lastri

CONTENTS

15.1 Introduction 389
15.2 Quality Issues in Remote-Sensing Data Compression 389
15.3 Distortion Measures 391
    15.3.1 Radiometric Distortion 391
    15.3.2 Spectral Distortion 392
15.4 Advanced Compression Algorithms for Remote-Sensing Images 393
    15.4.1 Context Modeling 394
15.5 Near-Lossless Compression through 3D Causal DPCM 395
15.6 Near-Lossless Image Compression through Noncausal DPCM 396
15.7 Experimental Results 399
    15.7.1 Multispectral Data 399
    15.7.2 Hyperspectral Data 401
    15.7.3 SAR Focused Data 406
    15.7.4 SAR Raw Data 407
15.8 Conclusions 409
References 409

15.1 INTRODUCTION

This chapter describes a category of data-compression algorithms capable of preserving the scientific quality of remote-sensing data, yet allowing a considerable reduction of the transmission bandwidth. *Lossless* compression applied to remote-sensing images guarantees only a moderate reduction of the data volume, because of the intrinsic noisiness of the data; on the other hand, conventional *lossy* techniques, in which the mean-squared error (MSE) of the decoded data is *globally* controlled by users, generally does not preserve the *scientific quality* of the images. The most suitable approach seems to be the use of near-lossless methods, which are capable of *locally* constraining the maximum error, either absolute or relative, based on the user's requirements. Advanced near-lossless methods may rely on differential pulse code modulation (DPCM) schemes, based on either *interpolation* or *prediction*. The former is recommended for lower quality compression, the latter for higher quality, which is the primary concern in remote-sensing applications. Experimental results of near-lossless compression of multispectral, hyperspectral, and microwave data from coherent imaging systems, like synthetic aperture radar (SAR), show the advantages of the proposed approach compared to standard lossy techniques.

15.2 QUALITY ISSUES IN REMOTE-SENSING DATA COMPRESSION

Data compression is gaining an ever increasing relevance for the remote-sensing community [41]. Since technological progresses allow observations of the Earth to be available at increasing spatial, spectral, radiometric, and temporal resolutions, the associated data volume is growing much faster than the transmission bandwidth does. The introduction of data compression can alleviate bandwidth requirements at the price of a computational effort for encoding and decoding as well as of a possible loss of quality.

Compression methods can be either reversible, i.e., *lossless* or irreversible (*lossy*). A variety of image compression methods exists for applications in which reconstruction errors are tolerated. In remote-sensing applications, data modeling, features extraction, and classifications are usually performed [20]. Hence, the original quality of the data must often be thoroughly preserved after compression/decompression. As a matter of fact, however, the intrinsic noisiness of sensors prevents strictly lossless techniques from being used to obtain a considerable bandwidth reduction. In fact, whenever reversibility is recommended, compression ratios higher than 2 can hardly be obtained, because the attainable bit rate is lower-bounded by the entropy of the sensor noise [10,37,38].

Noteworthy are those lossy methods that allow to settle *a priori* the maximum reconstruction error, not only globally but also locally. The maximum absolute error, also known as peak error (PE), or L_∞ distance between original and decoded image, is capable of guarantying a quality that is *uniform* throughout the image. If the L_∞ error is user-defined, besides being constrained to be small, the current definition of *near-lossless* compression [9] applies.

However, evaluation of the maximum-allowable distortion is an open problem; in other fields, objective measurements may be integrated with qualitative judgements of skilled experts, but in remote-sensing applications photoanalysis is not the only concern [34]. The data are often post-processed to extract information that may not be immediately available by user inspection. In this perspective, an attractive facility of near-lossless compression methods is that, if the L_∞ error is constrained to be near to one half of the standard deviation of the background noise, assumed to be additive and independent of the signal, the decoded image will be *virtually lossless* [6]. This term indicates not only that the decoded image is visually indistinguishable from the original, but also that possible outcomes of postprocessing are likely to be practically the same as if they were calculated from the original data. Thus, the price of compression will be a small and predictable increment in the equivalent sensor noisiness [9].

When multispectral or better hyperspectral data are being dealt with, *spectral* distortion becomes a primary concern, besides spatial and radiometric distortions. Spectral distortion is a measurement of how a pixel vector (i.e., a vector having as many components as spectral bands) changes because of an irreversible compression of its components.

Focusing on images generated by coherent systems, such as SAR systems, lossless compression methods are little effective for reducing the data volume, because of the intrinsic noisiness that weakens data correlation [12]. However, unpredictable local distortions introduced by lossy compression methods, though specifically tailored to SAR imagery [26,48], may be unacceptable in many applications. Furthermore, the signal-dependent nature of *speckle* [28] makes an L_∞ error control inadequate, since the errors to be bounded should be *relative*, i.e., measured as pixel ratios. In fact, larger errors should be encountered on homogeneous brighter areas than on darker ones. Hence, upper bounding of the PE no longer guarantees a quantifiable loss of quality. Therefore, near-lossless compression of SAR images should indicate that the pixel ratio of original to decoded image is strictly bounded within a prefixed interval [13]. If such an interval is comprised within the speckle distribution, then the decoded image will be *virtually lossless* as well [9].

The problem of data transmission to ground stations is crucial for remote-sensing imaging systems orbiting on satellite platforms. Recent developments of advanced sensors originate huge amounts of data; however, once these data were lossy compressed, they would not be available as they were acquired for the user community. The Consultative Committee for Space Data Systems (CCSDS) has issued a recommendation for the lossless compression of space data [23], which has been already adopted as an ISO standard [29]. Consequently, for concerns on-board data compression, only lossless methods are presently recommended, even if a final release of the CCSDS recommendation dealing with lossy compression is expected in the next future. On the other hand, to expedite dissemination and utilization of multispectral and especially hyperspectral images, near-lossless methods yielding constrained pixel error, either absolute or relative, are more suitable for obtaining a considerable bandwidth reduction and for preserving, at the same time, the spectral discrimination capability among pixel vectors, which is the principal source of spectral information.

15.3   DISTORTION MEASURES

15.3.1   RADIOMETRIC DISTORTION

Let $\{g(i,j)\}$, $0 \leq g(i,j) \leq g_{fs}$, denote an N-pixel digital image and σ its standard deviation, and let $\{\tilde{g}(i,j)\}$ be its possibly distorted version achieved by compressing $\{g(i,j)\}$ and decompressing the outcome bit stream.

Some distortion measurements used in this chapter are the following:
mean squared error (MSE), or L_2^2:

$$\text{MSE} = \frac{1}{N} \sum_i \sum_j [g(i,j) - \tilde{g}(i,j)]^2 \tag{15.1}$$

signal to noise ratio (SNR):

$$\text{SNR}_{\text{(dB)}} = 10 \log_{10} \frac{\sigma^2}{\text{MSE} + 1/12} \tag{15.2}$$

peak SNR (PSNR):

$$\text{PSNR}_{\text{(dB)}} = 10 \log_{10} \frac{g_{fs}^2}{\text{MSE} + 1/12} \tag{15.3}$$

maximum absolute distortion (MAD), or PE, or L_∞:

$$\text{MAD} = \max_{i,j} \{|g(i,j) - \tilde{g}(i,j)|\} \tag{15.4}$$

percentage maximum absolute distortion (PMAD):

$$\text{PMAD} = \max_{i,j} \left\{ \frac{|g(i,j) - \tilde{g}(i,j)|}{g(i,j)} \right\} \times 100 \tag{15.5}$$

Both in Equation (15.2) and in Equation (15.3), the MSE is incremented by the variance of the integer roundoff error, to handle the limit lossless case, when MSE $= 0$.

When multiband data are concerned, let $g_l(i,j)$, $l = 1, \ldots, L$, denote the lth component of the original multispectral pixel vector and $\tilde{g}_l(i,j)$, $l = 1, \ldots, L$, its distorted version. The radiometric distortion measurements may be extended to vector data as average RMSE (ARMSE), or $L_1(L_2)$ (the innermost norm refers to vector space (l), the outer one to pixel space (i,j)):

$$\text{ARMSE} = \frac{1}{N} \sum_{i,j} \sqrt{\sum_l [g_l(i,j) - \tilde{g}_l(i,j)]^2} \tag{15.6}$$

peak RMSE (PRMSE), or $L_\infty(L_2)$:

$$\text{PRMSE} = \max_{i,j} \sqrt{\sum_l [g_l(i,j) - \tilde{g}_l(i,j)]^2} \tag{15.7}$$

$$\text{SNR} = 10 \log_{10} \frac{\sum_{i,j,l} g_l^2(i,j)}{\sum_{i,j,l} [g_l(i,j) - \tilde{g}_l(i,j)]^2} \tag{15.8}$$

$$\text{PSNR} = 10 \log_{10} \frac{N \cdot L \cdot g_{fs}^2}{\sum_{i,j,l} [g_l(i,j) - \tilde{g}_l(i,j)]^2} \tag{15.9}$$

three-dimensional (3D) MAD, or $L_\infty(L_\infty)$:

$$\text{MAD} = \max_{i,j,l} \{|g_l(i,j) - \tilde{g}_l(i,j)|\} \tag{15.10}$$

The 3D PMAD:

$$\text{PMAD} = \max_{i,j,l} \left\{ \frac{|g_l(i,j) - \tilde{g}_l(i,j)|}{g_l(i,j)} \right\} \times 100 \tag{15.11}$$

In practice, ARMSE (15.6) and PRMSE (15.7) are, respectively, the average and maximum of the Euclidean norm of the distortion vector. SNR (15.8) is the extension of Equation (15.2) to the 3D data set. PSNR is the maximum SNR, given the full-scale of vector components. MAD (15.10) is the maximum over the set of pixel vectors of the maximum absolute component of the distortion vector. PMAD (15.11) is the maximum over the set of pixel vectors of the maximum percentage error over vector components.

15.3.2 SPECTRAL DISTORTION

Given two spectral vectors $\mathbf{v}$ and $\tilde{\mathbf{v}}$ both having L components, let $\mathbf{v} = \{v_1, v_2, \ldots, v_L\}$ be the original spectral pixel vector $v_l = g_l(i,j)$ and $\tilde{\mathbf{v}} = \{\tilde{v}_1, \tilde{v}_2, \ldots, \tilde{v}_L\}$ its distorted version obtained after lossy compression and decompression, i.e., $\tilde{v}_l = \tilde{g}_l(i,j)$. *Spectral* distortion measurement may be defined analogous to the *radiometric* distortion measurements.

The spectral angle mapper (SAM) denotes the absolute value of the spectral angle between the couple of vectors:

$$\text{SAM}(\mathbf{v}, \tilde{\mathbf{v}}) \triangleq \arccos \left(\frac{<\mathbf{v}, \tilde{\mathbf{v}}>}{||\mathbf{v}||_2 \cdot ||\tilde{\mathbf{v}}||_2} \right) \tag{15.12}$$

in which $<\cdot, \cdot>$ stands for scalar product. SAM can be measured in either degrees or radians.

Another measurement, especially suitable for hyperspectral data (i.e., for data with large number of components), is the spectral information divergence (SID) [22] derived from information-theoretic concepts:

$$\text{SID}(\mathbf{v}, \tilde{\mathbf{v}}) = D(\mathbf{v}||\tilde{\mathbf{v}}) + D(\tilde{\mathbf{v}}||\mathbf{v}) \tag{15.13}$$

with $D(\mathbf{v}||\tilde{\mathbf{v}})$ being the Kullback–Leibler distance (KLD), or entropic divergence, or *discrimination* [30], defined as

$$D(\mathbf{v}||\tilde{\mathbf{v}}) \triangleq \sum_{l=1}^{L} p_l \log \left(\frac{p_l}{q_l} \right) \tag{15.14}$$

in which

$$p_l \triangleq \frac{v_l}{||\mathbf{v}||_1} \quad \text{and} \quad q_l \triangleq \frac{\tilde{v}_l}{||\tilde{\mathbf{v}}||_1} \tag{15.15}$$

In practice, SID is equal to the symmetric KLD and can be compactly written as

$$\text{SID}(\mathbf{v}, \tilde{\mathbf{v}}) = \sum_{l=1}^{L} (p_l - q_l) \log \left(\frac{p_l}{q_l} \right) \tag{15.16}$$

which turns out to be symmetric, as one can easily verify. It can be proven as well that SID is always nonnegative, being zero iff $p_l \equiv q_l$, $\forall l$, i.e., if $\mathbf{v}$ is parallel to $\tilde{\mathbf{v}}$. The measure unit of SID depends on the base of logarithm: *nat/vector* with natural logarithms and *bit/vector* with logarithms in base 2.

Both SAM (15.12) and SID (15.16) may be either averaged on pixel vectors, or the maximum may be taken instead, as more representative of spectral quality. It is noteworthy that radiometric distortion does not necessarily imply spectral distortion. Conversely, spectral distortion is always accompanied by a radiometric distortion, that is minimal when the couple of vectors have either the same Euclidean length (L_2) for SAM, or the same city-block length (L_1) for SID.

15.4 ADVANCED COMPRESSION ALGORITHMS FOR REMOTE-SENSING IMAGES

Considerable research efforts have been recently spent in the development of lossless image compression techniques. The first specific standard has been the lossless version of JPEG [35], which may use either Huffman or arithmetic coding. More interestingly, a new standard, which provides also near-lossless compression, has been recently released under the name JPEG-LS [44]. It is based on an adaptive nonlinear prediction and exploits context modeling followed by Golomb–Rice entropy coding. A similar context-based algorithm, named CALIC, has also been proposed [45]. Eventually, it is worth mentioning that Part I of the JPEG2000 image coding standard [40] foresees a lossless mode, based on reversible integer wavelets, and is capable of providing a scalable bit stream that can be decoded from the lossy (not near-lossless) up to the lossless level [36]. However, image coding standards are not suitable for the compression of 3D data sets: in spite of their complexity, they are not capable of exploiting the 3D signal redundancy featured, e.g., by multispectral or hyperspectral imagery.

In the literature, there are only a few lossless compression algorithms tailored to multiband remote-sensed images, i.e., capable of a 3D decorrelation, and many of them are DPCM-based. A notable exception, especially for image archiving, is represented by the use of wavelet associated with inter-band prediction [19]. DPCM schemes, either *linear* or *nonlinear* [46], *causal* (prediction-based) or *noncausal*, i.e., interpolation-based or *hierarchical* [2], are indeed the only algorithms suitable for L_∞-constrained (near-lossless) compression of 2D or 3D data.

Concerning nonlinear prediction, a near-lossless 3D extension of the well-known CALIC algorithm has been recently proposed by Magli et al. [32], and seems quite promising for an on-board implementation. Unfortunately, the presence of empirical coefficients and thresholds, common to many nonlinear prediction schemes, may introduce instability when data change. Therefore, only linear prediction, yet adaptive, will be concerned in the following for a 3D extension suitable for multispectral and hyperspectral data.

Differential pulse code modulation basically consists of a decorrelation followed by entropy coding of the outcome prediction errors. The simplest way to design a predictor, once a *causal* neighborhood is set, is to take a linear combination of the values of such a neighborhood, with coefficients optimized to yield minimum mean squared error (MMSE) over the whole image. Such a prediction, however, is optimum only for stationary signals. To overcome this drawback, two variations have been proposed: adaptive DPCM (ADPCM) [35], in which the coefficients of predictors are continuously recalculated from the incoming new data, and classified DPCM [27], in which a preliminary training phase is aimed at recognizing some statistical classes of pixels and at calculating an optimized predictor for each class. Such predictors are then adaptively combined [5, 24] (as limit case the output is switched among one of them [3,11]) to attain the best space-varying prediction. This strategy will be referred to as adaptive combination/switching of adaptive predictors (ACAP/ASAP).

While details of up-to-date ASAP schemes, both 2D [17] and 3D [11] relaxation-labeled prediction encoder (RLPE), will be reviewed in Section 15.5, the ACAP paradigm underlies the development of a novel fuzzy logic-based prediction [16] (FMP), in which images are first partitioned into blocks and an MMSE linear predictor calculated for each block. From the large number of predictors obtained, a fuzzy-clustering algorithm produces an initial guess of a user-specified number of prototype predictors to be fed to an iterative procedure in which to each predictor pixels are given degrees of membership measuring the fitness of prediction on another causal neighborhood larger than the prediction support; then predictors are recalculated from pixels depending on their degrees of membership. The overall prediction will be fuzzy, being given by the sum of the outputs of each predictor weighted by the memberships of the current pixel to that predictor. The linearity of prediction makes it possible to formulate the above approach as a problem of approximating the optimum space-varying linear predictor at each pixel by projecting it onto a set of nonorthogonal prototype predictors capable of embodying the statistical properties of the image data.

The ACAP paradigm has been extended also to 3D data [12] by simply changing the 2D neighborhood into a 3D one spanning up to three previous bands. To enhance the entropy-coding performance, both schemes exploit context modeling (see Section 15.4.1) of prediction errors followed by arithmetic coding. Although RLPE is slightly less performing than FMP, its feature of real-time decoding is highly valuable in application contexts, since an image is usually encoded only once, but decoded many times.

15.4.1 CONTEXT MODELING

A notable feature of many advanced data-compression methods [16,17,24,39,44,45] is statistical context modeling for entropy coding. The underlying rationale is that prediction errors should be similar to stationary white noise as much as possible. As a matter of fact, they are still spatially correlated to a certain extent and especially are nonstationary, which means that they exhibit space-varying statistics. The better the prediction, however, the more noise-like prediction errors will be.

Following a trend established in the literature, first in the medical field, then for lossless coding in general [39,44,45], and recently for *near-lossless* coding [15,46], prediction errors are entropy-coded by means of a classified implementation of an entropy-coder, generally arithmetic or Golomb–Rice. For this purpose, they are arranged into a predefined number of statistically homogeneous classes based on their spatial *context*. If such classes are statistically discriminated, then the entropy of a *context-conditioned* model of prediction errors will be lower than that derived from a stationary memoryless model of the decorrelated source [43].

A context function may be defined and measured on prediction errors lying within a causal neighborhood, possibly larger than the prediction support, as the RMS value of prediction errors (RMSPE). The context function should capture the nonstationary of prediction errors, regardless of their spatial correlation. Again, causality of neighborhood is necessary to make the same information available both at the encoder and at the decoder. At the former, the probability density function (PDF) of RMSPE is calculated and partitioned into a number of intervals chosen as equally populated; thus, contexts are equiprobable as well. This choice is motivated by the use of adaptive arithmetic coding for encoding the errors belonging to each class. Adaptive entropy coding, in general, does not require previous knowledge of the statistics of the source, but benefits from a number of data large enough for training, which happens simultaneously with coding. The source given by each class is further split into sign bit and magnitude. The former is strictly random and is coded as it stands. The latter exhibits a reduced variance in each class, null if the context (RMSPE) of the current pixel is always equal its magnitude; thus, it may be coded with few bits. It is noteworthy that such a context-coding procedure is independent of the particular method used to decorrelate the data. Unlike other schemes, e.g., CALIC [45], in which context-coding is embedded in the decorrelation procedure, the above method [15] can be applied to any DPCM scheme, either lossless or near-lossless.

15.5   NEAR-LOSSLESS COMPRESSION THROUGH 3D CAUSAL DPCM

Whenever multispectral images are to be compressed, advantage may be taken from the spectral correlation of the data for designing a prediction that is both *spatial* and *spectral*, from a causal neighborhood of pixels [5,18,38,42,47]. This strategy is much more effective as the data are more spectrally correlated, as in the case of hyperspectral data [5]. If the *interband* correlation of the data is weak, as it usually occurs for data with few and sparse spectral bands, a 3D prediction may lead to negligible coding benefits. In this case, advantage may be taken from a *bidirectional* spectral prediction [7], in which once the $(k - 1)$th band is available, first the kth band is skipped and the $(k + 1)$th band is predicted from the $(k - 1)$th one; then, both these two bands are used to predict the kth band in a spatially causal but spectrally noncausal fashion.

The DPCM encoder utilized in this work is based on a classified linear-regression prediction according to the ASAP paradigm, followed by context-based arithmetic coding of the outcome residues. Image bands are partitioned into blocks, and an MMSE linear predictor is calculated for each block. Given a prefixed number of classes, a clustering algorithm produces an initial guess of as many classified predictors that are fed to an iterative labeling procedure, which classifies pixel blocks simultaneously refining the associated predictors.

To achieve reduction in bit rate within the constraint of a near-lossless compression [1], prediction errors are quantized with odd-valued step sizes, $\Delta = 2E + 1$, where E denotes the induced L_∞ error, with a quantization noise feedback loop embedded into the encoder, so that the current pixel prediction is formulated from the same "noisy" data that will be available at the decoder (see Figure 15.1a).

For the case of a relative-error-bounded compression, a rational version of prediction error must be envisaged. Let us define the *relative prediction error* (RPE) as ratio of original to predicted pixel value:

$$r(n) \triangleq \frac{g(n)}{\hat{g}(n)} \tag{15.17}$$

The *rational* nature of RPE, however, makes linear quantization unable to guarantee a strictly user-defined relative-error-bounded performance.

Given a step size $\Delta \in \mathbb{R}$, with $\Delta > 0$ and $\Delta \neq 1$, let us define as *logarithmic* quantization (Log-Q) of $t \in \mathbb{R}$, $t > 0$,

$$\mathcal{Q}_\Delta(t) \triangleq \text{round}[\log_\Delta(t)] = \text{round}[\log(t)/\log(\Delta)]$$
$$\mathcal{Q}_\Delta^{-1}(l) = \Delta^l \tag{15.18}$$

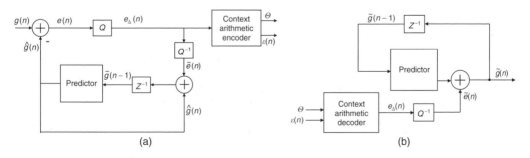

(a)   (b)

FIGURE 15.1 Flowchart of DPCM with quantization noise feedback loop at the encoder, suitable for error-bounded near-lossless compression: (a) encoder; (b) decoder.

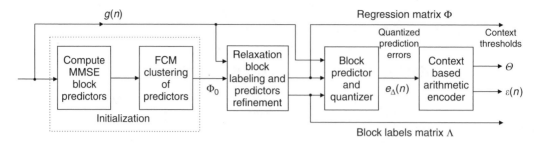

FIGURE 15.2 Flowchart of the causal DPCM encoder with context coding referred to as *relaxation-labeled prediction encoder* (RLPE).

Substituting Equation (15.18) into Equation (15.17) yields

$$\mathcal{Q}_\Delta[r(n)] = \text{round}\left[\frac{\log(g(n)) - \log(\hat{g}(n))}{\log \Delta}\right] \tag{15.19}$$

If a Log-Q with a step size Δ is utilized to encode pixel RPEs, it can be proven that the ratio of original to decoded pixel value is strictly bounded around 1

$$\min\left\{\sqrt{\Delta}, \frac{1}{\sqrt{\Delta}}\right\} \le \frac{g}{\tilde{g}} \le \max\left\{\sqrt{\Delta}, \frac{1}{\sqrt{\Delta}}\right\} \tag{15.20}$$

Now, let us introduce a peak measurement of rational error, namely the *peak rational error*, defined as

$$\text{PRE}_{(\text{dB})} \triangleq 20\log_{10}\left(\frac{\max\left[g(n)/\tilde{g}(n)\right]}{\min\left[g(n)/\tilde{g}(n)\right]}\right) \tag{15.21}$$

It is easily verified that if Log-Q is utilized, then

$$\text{PRE}_{(\text{dB})} = 20\log_{10}(\Delta) \tag{15.22}$$

Thus, the peak rational error may be easily user-defined.

Quantized prediction errors are then arranged into activity classes based on the spatial *context*, which are entropy-coded by means of arithmetic coding. Figure 15.2 shows a flowchart of the encoder. As it appears, the refined predictors are transmitted along with the label of each block and the set of thresholds defining the context classes for entropy coding.

15.6 NEAR-LOSSLESS IMAGE COMPRESSION THROUGH NONCAUSAL DPCM

Laplacian pyramids are multiresolution image representations obtained by means of recursive *reduction* (lowpass filtering followed by downsampling) and *expansion* (upsampling followed by lowpass filtering).

Start with $G_0(i,j) \equiv g(i,j)$; define the Gaussian pyramid (GP) as

$$G_{k+1} = \text{reduce}_2\{G_k\} \quad k = 0, 1, \ldots, K-1 \tag{15.23}$$

where 2^K is the highest power of 2 in which the image size can be factorized, and

$$\text{reduce}_2\{G_k\} \triangleq (G_k \otimes r_2) \downarrow 2 \tag{15.24}$$

in which the symbol $\otimes$ indicates linear convolution, $\downarrow 2$ decimation by 2, and r_2 the *anti-aliasing* (lowpass) filter.

Define the enhanced Laplacian pyramid (ELP) [2] as

$$L_k = G_k - \text{expand}_2\{G_{k+1}\}, \quad k = 0, 1, \ldots, K - 1 \tag{15.25}$$

with

$$\text{expand}_2\{G_{k+1}\} \triangleq (G_{k+1} \uparrow 2) \otimes e_2 \tag{15.26}$$

where $\uparrow 2$ stands for *upsampling* by 2, i.e., interleaving samples with 0s, and e_2 the interpolation (lowpass) filter, which can be the same as r_2, apart from a *dc* gain.

A spatial DPCM can also be *noncausal*, i.e., *interpolation-based*, or *hierarchical*: a coarse image version, i.e., the base band of the GP, G_K, is encoded followed by the ELP, L_k, $k = K - 1, \ldots, 0$. Quantization error feedback at each layer allows L_∞ error control via the quantization step at the finest resolution layer [2].

A *rational* ELP (RLP), matching the multiplicative nature of the speckle noise, was defined from the GP (15.23) and utilized for denoising [4]. The *ratio*, instead of the difference, between the kth level of the GP and the expanded version of the $(k + 1)$th level, yields a pyramid

$$L_k^*(i,j) \triangleq \frac{G_k(i,j)}{\tilde{G}_{k+1}(i,j)} \tag{15.27}$$

where $\tilde{G}_{k+1}$ is a shortcoming for $\text{expand}_2\{G_{k+1}\}$ and the domain of subscripts is the same as for Equation (15.25). Figure 15.3 shows GP and RLP of a test SAR image.

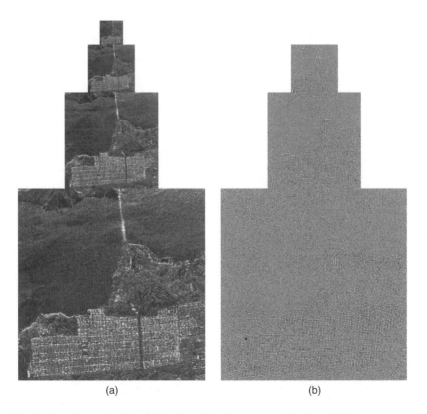

(a)   (b)

FIGURE 15.3 (a) Gaussian pyramid and (b) rational Laplacian pyramid of test SAR image.

The RLP approximates a *bandpass* image representation, thus retaining all the benefits of multi-resolution analysis, including those for data compression. The idea is to causally DPCM-encode the small baseband bitmap icon and to quantize and encode the RLP layers [8]. The quantizer step sizes except on the bottom layer may be chosen arbitrarily owing to the quantization noise feedback loop at the encoder, which consists of interpolating the same *noisy* data, i.e., affected by the distortion introduced during reconstruction, which will be available at the decoder.

Figure 15.4 shows that $\hat{G}_k$, the GP that will be reconstructed at the decoder for $k < K$, is recursively given by the product of the expanded $\hat{G}_{k+1}$ and of an approximate version of L_k^*, namely $\hat{L}_k^*$, due to quantization errors, in which $\hat{L}_k^* = G_k/\tilde{\hat{G}}_{k+1}$; $\hat{L}_K^* = \hat{G}_K$ for the pyramid top or *baseband*.

For the baseband and intermediate layers, linear quantizers are utilized. The step sizes of the linear quantizers are calculated for $k = K, K-1, \ldots, 1$, so as to minimize the bit rate for a given distortion, by exploiting the mechanism of quantization noise feedback. Following the procedure reported for the ELP [2], the entropy-minimizing step sizes are found out to be

$$\hat{\Delta}_k = \frac{2\bar{\sigma}_{k-1}^*}{P_E}, \quad k = K, K-1, \ldots, 1 \tag{15.28}$$

where $\bar{\sigma}_{k-1}^*$ is the average standard deviation of L_{k-1}^* and P_E the power gain of the 1D interpolation filter.

The last step size, Δ_0, as well as the type of quantizer is crucial when an error-bounded encoder is required, since it rules the PE: *absolute* for the ELP and *relative* for the RLP. In fact, G_0 cannot be exactly recovered from G_K and $\hat{L}_k^*$, $k = K-1, \ldots, 1, 0$, unless quantization on the last layer is extremely fine, which implies a large code rate. Thus, unlike the ELP [2], the RLP is unsuitable for a strictly lossless compression. Furthermore, the *rational* nature of RLP makes linear quantization unable to guarantee relative-error-bounded encoding.

Since the term $\tilde{\hat{G}}_0$ recursively accounts for previous quantization errors and $L_0^* = G_0/\tilde{\hat{G}}_0$, a logarithmic quantization (15.18) of L_0^* with a step size $\Delta_0 \neq 1$ implies that the pixel ratio of original to decoded image is strictly bounded through the step size Δ_0 of the last quantizer, as in Equation (15.20). Hence, a relationship identical to Equation (15.22) is found between the step size of the last quantizer, Δ_0, and the dynamic range of relative errors.

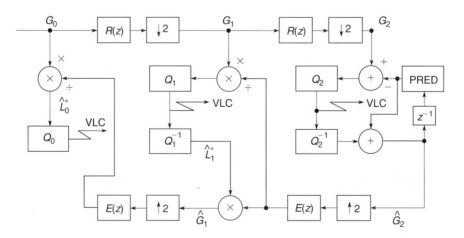

FIGURE 15.4 Flowchart of a hybrid encoder with quantization noise feedback loops on levels 2 (causal DPCM) and 1 (noncausal DPCM). VLC, variable length coding.

15.7   EXPERIMENTAL RESULTS

15.7.1   MULTISPECTRAL DATA

The optical data set comprises a Landsat Thematic Mapper (TM) image, with 8 bits/pel and 6 bands out of the 7 available: the 6th band (thermal infrared) was omitted mainly because of its poor resolution (120 m × 120 m) and scarce spectral correlation with the other bands. The test site is shown in Figure 15.5, which portrays the valleys of two rivers, Bradano and Basento, near Metaponto, in Southern Italy. Thematic mapper bands do not span the visible/infrared wavelength interval continuously. Apart from the visible spectrum, the infrared region is coarsely sampled. Thus, all the infrared bands are little correlated, both with the visible bands and with one another.

To achieve an optimal multispectral decorrelation, the different bands available should be arranged in a sequence that maximizes the average crosscorrelation between any couple of consecutive bands [42]. The optimum *causal* sequence was found to be $1 \rightarrow 2 \rightarrow 3 \rightarrow 7 \rightarrow 5 \rightarrow 4$. A bidirectional, i.e., spectrally noncausal, prediction yields bit rates that are slightly lower, on average [7]. The optimum bidirectional sequence was found to be $1 \rightarrow 3, 1 \rightarrow 2 \leftarrow 3 \rightarrow 7 \rightarrow 4 \rightarrow 5 \leftarrow 7$. The difference in rate between causal and noncausal prediction, however, is moderate: the latter provides an average gain of 400th of bit per pixel for the optical bands and of nearly 800th for the infrared channels.

Table 15.1 reports the estimated parameters for the six bands, the first of which is encoded in *intra* mode, i.e., without reference to any other previously encoded band. The noise variance is larger in the visible than in the infrared wavelengths also the signal variance follows such a trend; thus, the intrinsic SNRs are all comparable among the bands.

Bands 2 and 5 are bidirectionally predicted to a larger extent than the other bands of the visible and infrared group, respectively. The extremely fitting prediction is demonstrated by the associated prediction residues of Figure 15.7d, which practically comprise only the background noise in most

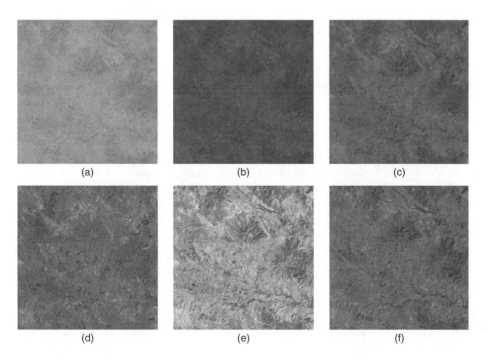

| (a) | (b) | (c) |
| (d) | (e) | (f) |

FIGURE 15.5 512 × 512 details from test TM image: (a) Band 1 (blue), (b) Band 2 (green), (c) Band 3 (red), (d) Band 4 (near-infrared), (e) Band 5 (short-wave infrared), (f) Band 7 (middle infrared).

TABLE 15.1
Average Variance ($\bar{\sigma}_g^2$), Estimated Noise Variance ($\hat{\sigma}_n^2$), SNR (dB), and Lossless Bit Rates (in Bit/Pel) of the Six 30 m Bands of the Test TM Image Achieved by RLPE

Band (Mode)	$\bar{\sigma}_g^2$	$\hat{\sigma}_n^2$	SNR (dB)	Bit Rate
TM-1 (I)	86.62	1.59	17.36	3.37
TM-2 (B)	48.99	0.32	21.85	1.86
TM-3 (P)	179.99	0.42	26.32	3.07
TM-4 (P)	124.40	4.41	14.50	3.85
TM-5 (B)	622.84	5.38	20.64	3.74
TM-7 (P)	245.41	1.85	21.23	3.59
Avg.	218.04	2.33	19.71	3.25

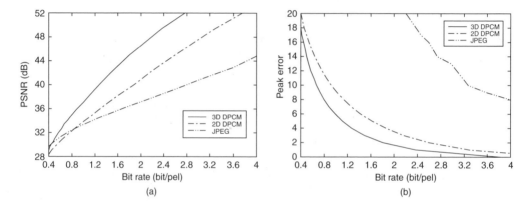

FIGURE 15.6 Band 5 of TM image (see Figure 15.7a) compressed by means of RLPE, in *intraband* mode (2D DPCM) and in bidirectional *interband* mode (3D DPCM) from bands 4 and 7, and JPEG (intraband): (a) PSNR vs. bit rate; (b) peak error vs. bit rate.

of the image. The work parameters of the algorithm are noncrucial [17] and have been chosen so as to balance coding performances with encoding time (decoding is always real time).

Rate distortion (RD) plots are reported in Figure 15.6a for the RLPE scheme, both in *intraband* (2D RLPE) and *interband* modes (3D RLPE), and for the DCT-based lossy JPEG. The test image is the 512×512 detail from band 5 (Figure 15.7a). In the *interband* mode (3D), the image is bidirectionally predicted from bands 4 and 7. 2D RLPE gains over JPEG for rates above 0.8 bit/pel; 3D RLPE crosses the RD plot of JPEG at 0.4 bit/pel. The knee for low rates is typical of all causal DPCM schemes and is an effect of quantization noise feedback in the prediction loop. From Figure 15.1a it appears that, since the "noisy" data reconstructed at the decoder are utilized for prediction, prediction becomes poorer and poorer as the bit rate, and hence the quality of dequantized samples, decreases. The near-lossless performance, shown in the PE vs. bit rate plots of Figure 15.6b, demonstrates that the two L_∞-bounded encoders are far superior to JPEG, which is L_2-bounded through the user-definable quality factor [35]. The standard deviation of the sensor's noise was found [14] to be approximately equal to 3 (2.36); hence the *virtually lossless* case, corresponding to a quantization step size $\Delta = 3$, with an induced MAD 1, is achieved at 3.45 bits/pel by 2D RLPE and at 2.43 bits/pel by 3D RLPE, with a PSNR gain over JPEG of 7.5 and 11 dB, respectively.

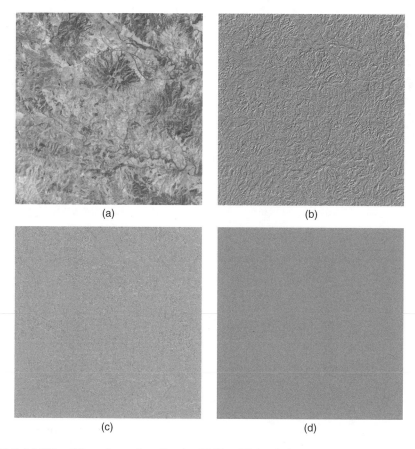

(a) (b)

(c) (d)

FIGURE 15.7 (a) 512×512 test image from Landsat TM band 5; prediction errors (stretched by 2 and biased by 128 for displaying convenience) produced by: (b) intraband lossless JPEG (predictor 7); (c) intraband predictor (2D RLPE); (d) interband predictor (3D RLPE) with bidirectional prediction from TM bands 4 and 7.

The meaning of the *virtually lossless* term is highlighted in Figure 15.8, reporting the bitmaps and the histograms of pixel differences between original and decoded images, for 3D RLPE and JPEG, respectively, both at 2.43 bits/pel. Besides the 11 dB PSNR gain of RLPE over JPEG, the error is practically uniformly distributed in $[-1, 1]$, as well as uniformly spread over the whole image. In the latter case the error is roughly Gaussian in $[-15, 17]$ and spatially heterogeneous, much larger around edges and in textured areas than on the background.

15.7.2 HYPERSPECTRAL DATA

The data set includes also a sequence of hyperspectral images collected in 1997 by the *Airborne visible infrared imaging spectrometer* (AVIRIS), operated by NASA/JPL, on the *Cuprite Mine* test site, in Nevada. The sequence is constituted of 224 bands recorded at different wavelengths in the range 380 to 2500 nm, with an average spectral separation between two bands of 10 nm. The image size is 614×2048 pixels. A 614×512 subimage was used in this experiment. The raw sequence was acquired by the 12-bit analog-to-digital converter (ADC) with which the sensor was equipped in 1995, in place of the former 10-bit ADC. The raw data from the digital counter have been radiometrically calibrated by multiplying by a gain and adding an offset (both varying with wavelengths), and are expressed as radiance values, rounded to integers, and packed in a 16-bit wordlength, including a

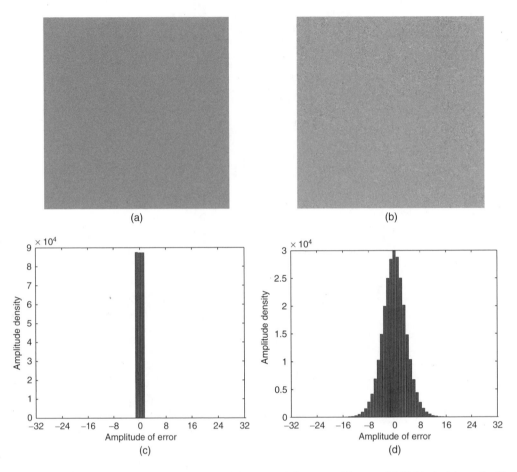

FIGURE 15.8 Error between original and decoded TM band 5, at same bit rate of 2.72 bits/pel (stretched by a factor 7 and biased by 128 for displaying convenience): (a) 3D RLPE; (b) JPEG; error distribution; (c) 3D RLPE; (d) JPEG.

sign bit. A detail of band 48 (808 nm) is shown in Figure 15.9. The second spectrometer, covering the near-infrared (NIR) spectrum, was analyzed in a recent work by Aiazzi et al. [10]. It was found that the noise affecting AVIRIS data is somewhat correlated spectrally and across track, and less along track, owing the "whisk-broom" scan mechanism as well as to postprocessing. First, the 3D RLPE was run on the test AVIRIS sequence in the reversible mode, i.e., with $\Delta = 1$. Each of the 224 bands was decorrelated and encoded with reference to its two previous bands. A larger number of bands for prediction is useless [5], besides being computationally more onerous. Figure 15.10 reports the bit rate produced by the encoder varying with the wavelength, for the proposed scheme and for the interband fuzzy-clustered DPCM encoder (3D FC-DPCM) [5]. The number of predictors $M = 4$ and the size of prediction support $S = 18$ are the same for both the encoders. As it appears, the former is slightly superior to the latter, which requires an encoding time more than ten times greater, and does not allow real-time decoding.

A distortion analysis varying with coding bit rate was carried out also on AVIRIS data. The test image is the band 48 (808 nm) of *Cuprite '97* portrayed in Figure 15.9. It is somewhat detailed and richly textured. Rate distortion plots are reported in Figure 15.11a for RLPE operating with $M = 4$ predictors, in both *intraband* (2D RLPE) mode ($S = 8$ coefficients per predictor), and *interband* mode (3D RLPE), with reference to either one or two previous bands, with $S = 12$ and $S = 16$, respectively.

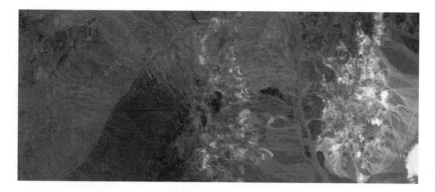

FIGURE 15.9 (See color insert following page 336) NASA/JPL AVIRIS *Cuprite Mine* image, 614×512 details shown as color compositions of red (band 30), green (band 16), and blue (band 4) radiances collected in 1997 with a wordlength of 16 bits.

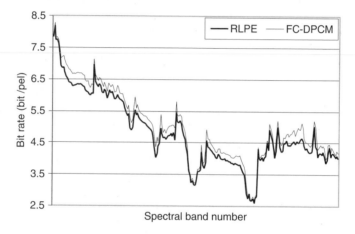

FIGURE 15.10 Bit rates (in bit/pel) produced by 3D RLPE and by the 3D *fuzzy-clustered* DPCM (FC-DPCM), for the *reversible* encoding of the 224 bands from the test AVIRIS sequences, varying with the wavelength. Each band is predicted both spatially and spectrally from the two previous bands.

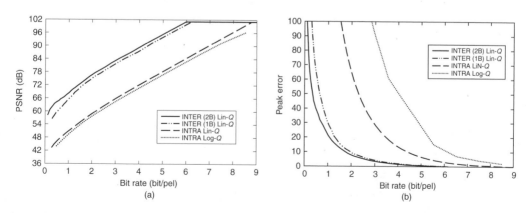

FIGURE 15.11 Band 48 of AVIRIS *Cuprite'97* compressed by means of the RLPE causal DPCM, in 2D, *intraband* mode (INTRA), and 3D, *interband* (INTER) from one (1B) and two (2B) previous bands, with either linear (Lin-Q), or logarithmic quantizer (Log-Q): (a) PSNR vs. bit rate; (b) peak error vs. bit rate.

Owing sign bit, the full scale g_{fs} in Equation (15.3) was set equal to $2^{15} - 1 = 32767$ instead of 65535 (negative values introduced by calibration never occur in the sample band). Hence, the PSNR attains a value of $10\log_{10}(12g_{fs}^2) \approx 102$ dB, due to integer roundoff noise only, when the reversibility is reached. The 3D RLPE gains 16 to 18 dB over 2D RLPE, corresponding to almost three code bit, depending on whether one or two previous bands are exploited for the 3D prediction. Two-band 3D prediction, instead of one, gains about 2 dB for medium to high bit rates, and up to 4 dB for low rates. Notice that according to RD theory [30], when a uniform quantizer is employed, all the SNR/PSNR bit rate plots are straight lines with slope ≈ 6 dB/bit, for rates larger than, say, 1 bit/pel. This does not happen for Log-Q which loses about 2 dB and drifts from the theoretical line as the lossless case is approached. The near-lossless performance is shown in the PE vs. bit rate plots of Figure 15.11b. Values of PE are far larger than those reported in Figure 15.6b, because the full scale is now 32,767 instead of 255. The trends are in accordance with those of PSNR, except for the Log-Q, which achieves a performance much poorer than that of Lin-Q for the *intra* experiment. The standard deviation of the noise was found [10] to be approximately 10; hence, the virtually lossless case is given by the 3D encoder at a bit rate around 3 bits/pel, yielding a compression ratio CR > 5.

Another experiment concerns assessments of PMAD-constrained coding performances. Bands 35 to 97, covering the NIR wavelengths, have been compressed in both MAD-constrained mode (linear quantization) and PMAD-constrained mode (logarithmic quantization). The work parameters of RLPE have still been chosen so as to balance performances with encoding time. The outcome bit rates varying with band number, together with the related distortion parameters, are shown in Figure 15.12. As it appears, the bit rate plots follow similar trends varying with the amount of distortion, but quite different trends for the two types of distortion (i.e., either MAD or PMAD). For example, around the water vapor absorption wavelengths ($\approx$ band 80) the MAD-bounded plots exhibit pronounced valleys that can be explained because the intrinsic SNR of the data becomes lower; thus the linear quantizer dramatically abates the *noisy* prediction errors. On the other hand, the PMAD-bounded encoder tends to quantize the noisy residues more finely when the signal is lower. Therefore, bit rate peaks are generated instead of valleys. More generally speaking, bit rate peaks from the PMAD-bounded encoder are associated with low responses from the spectrometer. This explains why the bit rate plots of Figure 15.12b never fall below 1 bit/pixel/band.

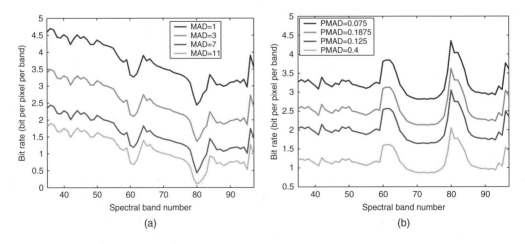

FIGURE 15.12 (**See color insert following page 336**) Bit rates produced by 3D RLPE on the data produced by the second spectrometer (NIR) of AVIRIS *Cuprite Mine '97*: (a) linear quantization to yield user-defined MAD values; (b) logarithmic quantization to yield user-defined PMAD values.

Eventually, some of the distortion measures defined in Section 15.3 have been calculated on the distorted hyperspectral pixel vectors achieved by decompressing the bit streams generated by the near-lossless encoder, both MAD- and PMAD-bounded. The RMSEs of the vector data, both *average* RMSE (15.6) and *peak* RMSE (15.7), are plotted in Figure 15.13a as a function of the bit rate from the encoder. The MAD-bounded encoder obviously minimizes both the radiometric distortions: average (ARMSE) and maximum (PRMSE) Euclidean norm of the pixel error vector.

A further advantage is that ARMSE and PRMSE are very close to each other for all bit rates. The PMAD-bounded encoder is somewhat poorer: ARMSE is comparable with that of the former, but PRMSE is far larger, owing the high-signal components that are coarsely quantized in order to minimize PMAD. Trivially, the MAD of the data cube (15.10) is exactly equal to the desired value (see Figure 15.12a), whereas the PMAD, being unconstrained, is higher. Symmetric results, not reported here, have been found by measuring PMAD on MAD- and PMAD-bounded decoded data.

As far as *radiometric* distortion is concerned, results are not surprising: radiometric distortions measured on vectors are straightforwardly derived from those measured on scalar pixel values. The introduction of such *spectral* measurements as SAM (15.12) and SID (15.16) may overcome the rationale of *distortion*, as established in the image processing community. Figure 15.14 shows spectral

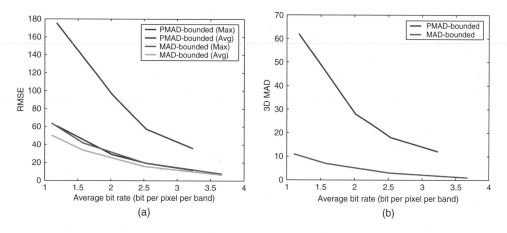

FIGURE 15.13 (**See color insert following page 336**) Radiometric distortions vs. bit rate for compressed AVIRIS *Cuprite Mine '97* data: (a) RMSE; (b) MAD.

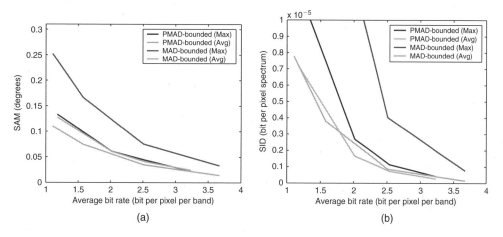

FIGURE 15.14 (**See color insert following page 336**) Spectral distortions vs. bit rate for compressed AVIRIS *Cuprite Mine '97* data: (a) spectral angle mapper (SAM); (b) spectral information divergence (SID).

distortions between original and decompressed hyperspectral pixel vectors. The PMAD-bounded algorithm yields plots (SAM, Figure 15.14a) lying in the middle between the corresponding ones produced by the MAD-bounded algorithm and that are very close to each other too.

Since the *maximum* SAM is a better clue of spectral quality of the decoded data than the *average* SAM may be, a likely conclusion would be that PMAD-bounded compression optimizes the *spectral* quality of the data, while MAD-bounded is superior in terms of *radiometric* quality. The considerations expressed for SAM are emphasized by the plots of Figure 15.14b reporting average and maximum SID. The latter is capable of discriminating spectral quality more finely than SAM does, as already noticed in the case of multiband classification [22].

15.7.3 SAR FOCUSED DATA

A thorough performance comparison aimed at highlighting the *rational* near-lossless approach was carried out among causal RLPE-DPCM (2D) with either linear or logarithmic quantization and noncausal DPCM achieved by RLP with linear quantization except on bottom layer where it is logarithmic. The standards JPEG [35] and JPEG 2000 [40] were also considered.

The well-known NASA/JPL AIRSAR image of San Francisco (4-look amplitude) remapped to 8 bits/pel was used for the coding experiment. The test SAR image is shown as the bottom layer of the GP in Figure 15.3a.

Figure 15.15a shows the PSNR vs. the bit rates produced by the five different encoders starting from the test SAR image. For the RLP and the RLPE-DPCM (Log-Q) the rightmost part of the plots (say, bit rate >2 bits/pel) correspond to a *virtually lossless* coding. As it appears, RLPE-DPCM gains more than 1.5 dB PSNR over RLP. The two plots are parallel for medium to high rates and cross each other at approximately 1 bit/pel. This is not surprising because both encoders utilize Log-Q, and it is widely known that noncausal DPCM is preferable to causal DPCM for low rates only [2]. Concerning the schemes utilizing linear quantizers, RLPE-DPCM and JPEG2K share the best RD performances: the former outperforms the latter for rates higher than 1.5 dB, and vice versa. However, it is evident that the logarithmic quantizer, introduced to allow relative-error-bounded coding, yields poorer RD performances than a linear quantizer does, especially for low rates. JPEG and JPEG2K having similar quantizers (psychovisual) follow similar RD trends. However, the two plots are closer for lower rates and farther apart as the rate increases, unlike what usually happens for "optical", i.e., noncoherent, images. The visual quality of the decoded images, however, is quite different. For lower rates, JPEG2K takes advantages from despeckling the image; instead JPEG introduces severe blocking impairments, especially annoying on the sea.

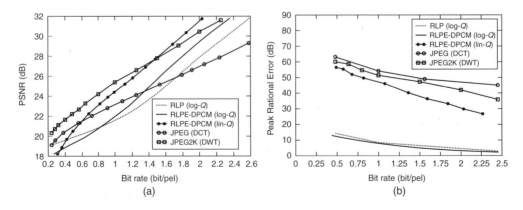

FIGURE 15.15 Comparisons on test SAR image (8 bits/pel): (a) PSNR vs. bit rate; (b) PRE vs. bit rate.

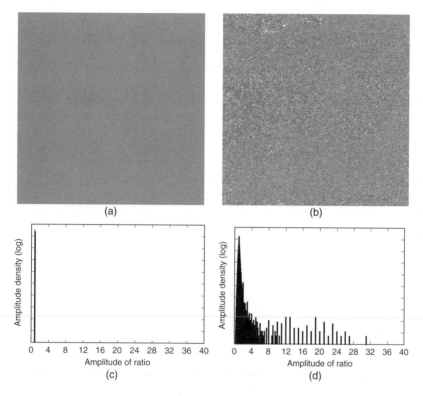

FIGURE 15.16 Ratio of original to reconstructed image at the same bit rate of 2.45 bits/pel: (a) RLP, (b) JPEG; distribution of original-to-decoded ratio: (c) RLP, (d) JPEG.

The relative-error-bounded *near-lossless* performance of the two encoder utilizing logarithmic quantization is highlighted in the plots of Figure 15.15b reporting PRE (15.21) vs. bit rate between original and decoded versions of the test SAR image. PRE is user defined and far lower than that of the other schemes, including RLPE-DPCM which exhibits the best results among the three encoders using linear quantization. Again RLP and RLPE-DPCM cross each other around 1 bit/pel.

The meaning of the *virtually lossless* term is demonstrated in Figure 15.16, reporting the bitmaps and the histograms of pixel ratio between original and decoded images (i.e., the image of noise), for RLP and JPEG, both at 2.45 bits/pel. Although the PSNR gain of RLP over JPEG is only 2 dB at 2.45 bits/pel, in the former case the relative error is small and uniformly spread; in the latter case it is heterogeneous, much larger around image edges. The variance of the ratio is less than one tenth of that of speckle (nominally 4-look amplitude) for RLP. Hence, the definition of *virtually lossless* applies [9].

15.7.4 SAR Raw Data

Many interesting papers [21,25,31,33] have reported previous work concerning raw data compression. All authors agree that the I and Q components of the complex SAR signal can be well modeled as zero-mean Gaussian-independent processes. Some nonstationarity is present causing a slow variation of the standard deviation of the data in range and azimuth. However, on reasonably small data blocks, the raw signal can be regarded as a stationary Gaussian process and it is quite natural to exploit the Gaussian signal statistics by employing a pdf-optimized nonuniform Lloyd–Max quantizer [30]. Given this premise it is apparent that the ELP [2] does not fit well the statistics of the raw data. This conclusion has been also strengthened by the analysis performed on the correlation of the data: some weak correlation exists in the range direction while data are extremely uncorrelated in the azimuth direction as shown in Figure 15.17a and Figure 15.17b where the autocorrelations of the real and imaginary parts of the raw SIR-C/XSAR data of Innsbruck, varying with lag, are reported, respectively. The interpolative 2D nature of ELP prevents this scheme from obtaining significant advantages on such data.

Relaxation-labeled prediction encoder seemed more promising for raw data and efforts have been thus concentrated on this scheme mainly. Preliminary results have shown that owing to the extremely scarce correlation in the azimuth direction, the causal neighborhood utilized for the prediction has to be 1D and oriented in the range direction. RLPE has been thus modified in this sense. According to [31], it has been found that when increasing the size of the causal neighborhood the performances of RLPE slightly increase. Such performances are better than those [31] reported up to now as shown in Figure 15.18a, where results of FBAQ and NBAQ are plotted for comparison.

Figure 15.18b shows instead the RD curves of ELP. Notwithstanding, the performances of RLPE are superior, once context-based arithmetic coding is adopted also ELP becomes better than NPAQ and FBAQ. On SAR raw data, a smart prediction can improve decorrelation performances up to 1 dB.

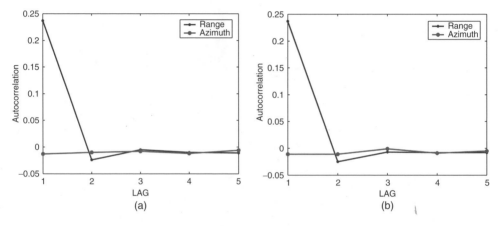

FIGURE 15.17 Autocorrelation of SIR-C/XSAR complex raw data of Innsbruck as a function of lag: (a) real part; (b) imaginary part.

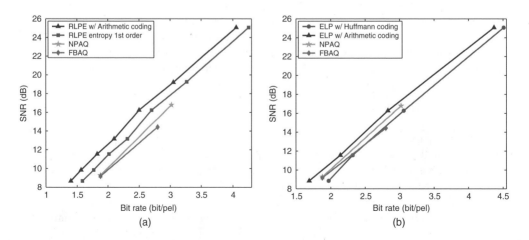

FIGURE 15.18 Compression performance on SIR-C/XSAR raw data of Innsbruck: (a) RLPE; (b) ELP.

Context modeling and context-based arithmetic coding can further increase performances of more than 1 dB. Overall, a gain of more than 2 dB can be obtained with respect to the most efficient schemes recently reported in the literature. The cascade of context modeling and context-based arithmetic coding is not advantageous: one of them is sufficient to guarantee the coding gain.

15.8 CONCLUSIONS

This work has demonstrated the potentialities of *near-lossless* compression, i.e., with bounded pixel error, either *absolute* or *relative*, when it is applied to remote-sensing data. Unlike lossless compression achieving typical CRs around 2, near-lossless compression can be adjusted to allow a *virtually lossless* compression with CRs higher than 3 for 8-bit multispectral data and larger than 5 for 16-bit hyperspectral data. The main result of this analysis is that, for a given CR, near-lossless methods, either MAD- or PMAD-constrained, are more suitable for preserving the spectral discrimination capability among pixel vectors, which is the principal outcome of spectral information. Therefore, whenever a lossless compression is not practicable, near-lossless compression is recommended in such applications where spectral quality is a crucial point. Furthermore, since the maximum reconstruction error is defined by the user before compression, whenever higher CRs are required, the loss of performance expected in application tasks can be accurately modeled and predicted. An original approach to *near-lossless* compression of detected SAR images is also reported, which is based on encoding the rational Laplacian pyramid of a speckled SAR image, after linearly quantizing its upper layers and logarithmically quantizing its bottom layer, to achieve near-lossless compression with constrained relative error. Besides virtually lossless compression, for which causal DPCM is recommended, noncausal pyramid-based DPCM outperforms causal DPCM when near-lossless compression at low to medium rates is desired. Eventually, preliminary results of RLPE on SAR raw data show the adaptation capabilities of this scheme, which results more efficient than the up-to-date algorithms reported in the literature.

REFERENCES

1. G.P. Abousleman, M.W. Marcellin, and B.R. Hunt, Hyperspectral image compression using entropy-constrained predictive trellis coded quantization, *IEEE Trans. Image Process.*, 6, 566–573, 1997.
2. B. Aiazzi, L. Alparone, S. Baronti, and F. Lotti, Lossless image compression by quantization feedback in a content-driven enhanced Laplacian pyramid, *IEEE Trans. Image Process.*, 6, 831–843, 1997.
3. B. Aiazzi, P.S. Alba, L. Alparone, and S. Baronti, Reversible compression of multispectral imagery based on an enhanced inter-band JPEG prediction, *Proceedings of IEEE International Geoscience and Remote Sensing Symposium*, Singapore, 1997, pp. 1990–1992.
4. B. Aiazzi, L. Alparone, and S. Baronti, Multiresolution local-statistics speckle filtering based on a ratio Laplacian pyramid, *IEEE Trans. Geosci. Remote Sensing*, 36, 1466–1476, 1998.
5. B. Aiazzi, P. Alba, L. Alparone, and S. Baronti, Lossless compression of multi/hyper-spectral imagery based on a 3-D fuzzy prediction, *IEEE Trans. Geosci. Remote Sensing*, 37, 2287–2294, 1999.
6. B. Aiazzi, L. Alparone, S. Baronti, and F. Lotti, Virtually lossless compression of medical images through classified prediction and context-based arithmetic coding, in *Visual Communications and Image Processing '99, Proc. SPIE*, Vol. 3653, K. Aizawa, R.L. Stevenson, and Y.-Q. Zhang, Eds., 1999, pp. 1033–1040.
7. B. Aiazzi, L. Alparone, and S. Baronti, Advantages of bidirectional spectral prediction for the reversible compression of multispectral data, *Proceedings of IEEE International Geoscience And Remote Sensing Symposium*, 1999, pp. 2043–2045.
8. B. Aiazzi, L. Alparone, and S. Baronti, Joint compression and despeckling of SAR images by thresholding and encoding the rational Laplacian pyramid, *Proceedings of EUSAR 2000*, Munich, Germany, 2000, pp. 2657–2659.
9. B. Aiazzi, L. Alparone, and S. Baronti, Information-preserving storage of Remote Sensing data: virtually lossless compression of optical and SAR images, *Proceedings of IEEE International Geoscience and Remote Sensing Symposium*, 2000, pp. 2657–2659.

10. B. Aiazzi, L. Alparone, A. Barducci, S. Baronti, and I. Pippi, Information-theoretic assessment of sampled hyperspectral imagers, *IEEE Trans. Geosci. Remote Sensing*, 39, 1447–1458, 2001.

11. B. Aiazzi, L. Alparone, and S. Baronti, Near-lossless compression of 3-D optical data, *IEEE Trans. Geosci. Remote Sensing*, 39, 2547–2557, 2001.

12. B. Aiazzi, L. Alparone, S. Baronti, and L. Santurri, Near-lossless compression of multi/hyperspectral images based on a fuzzy matching-pursuits interband prediction, in *Image and Signal Processing for Remote Sensing VII, Proc. SPIE*, Vol. 4541, S.B. Serpico, Ed., 2001, pp. 252–263.

13. B. Aiazzi, L. Alparone, and S. Baronti, Relative error-constrained compression for Synthetic Aperture Radar data, in *Mathematics of Data/Image Coding, Compression, and Encryption IV, with Applications, Proc. SPIE*, Vol. 4122, M.S. Schmalz, Ed., 2001, pp. 128–139.

14. B. Aiazzi, L. Alparone, A. Barducci, S. Baronti, and I. Pippi, Estimating noise and information of multispectral imagery, *J. Opt. Eng.*, 41, 656–668, 2002.

15. B. Aiazzi, L. Alparone, and S. Baronti, Context modeling for near-lossless image coding, *IEEE Signal Process. Lett.*, 9, 77–80, 2002.

16. B. Aiazzi, L. Alparone, and S. Baronti, Fuzzy logic-based matching pursuits for lossless predictive coding of still images, *IEEE Trans. Fuzzy Syst.*, 10, 473–483, 2002.

17. B. Aiazzi, L. Alparone, and S. Baronti, Near-lossless image compression by relaxation-labelled prediction, *Signal Process.*, 82, 1619–1631, 2002.

18. Z. Arnavut, Permutations and prediction for lossless compression of multispectral TM images, *IEEE Trans. Geosci. Remote Sensing*, 36, 999–1003, 1998.

19. A. Benazza-Benyahia, J.-C. Pesquet, and M. Hamdi, Vector lifting schemes for lossless coding and progressive archival of multispectral images, *IEEE Trans. Geosci. Remote Sensing*, 40, 2011–2024, 2002.

20. J.A. Benediktsson, J.R. Sveinsson, and K. Arnason, Classification and feature extraction of AVIRIS data, *IEEE Trans. Geosci. Remote Sensing*, 33, 1194–1205, 1995.

21. U. Benz, K. Strodl, and A. Moreira, A comparison of several algorithms for SAR raw data compression, *IEEE Trans. Geosci. Remote Sensing*, 33, 1266–1276, 1995.

22. C.-I. Chang, An information-theoretic approach to spectral variability, similarity, and discrimination for hyperspectral image analysis, *IEEE Trans. Inform. Theory*, 46, 1927–1932, 2000.

23. Consultative Committee for Space Data Systems, *Lossless Data Compression: Recommendation for Space Data Systems Standards (Blue Book)*, CCSDS, Washington, DC, May 1997.

24. G. Deng, H. Ye, and L.W. Cahill, Adaptive combination of linear predictors for lossless image compression, *IEE Proc.-Sci. Meas. Technol.*, 147, 414–419, 2000.

25. C. D'Elia, G. Poggi, and L. Verdoliva, Compression of SAR raw data through range focusing and variable-rate Trellis-coded quantization, *IEEE Trans. Image Process.*, 9, 1278–1297, 2001.

26. D. Gleich, P. Planinsic, B. Gergic, and Z. Cucej, Progressive space frequency quantization for SAR data compression, *IEEE Trans. Geosci. Remote Sensing*, 40, 3–10, 2002.

27. F. Golchin and K.K. Paliwal, Classified adaptive prediction and entropy coding for lossless coding of images, *Proceedings of IEEE International Conference on Image Processing*, Vols. III/III, 1997, pp. 110–113.

28. J.W. Goodman, Some fundamental properties of speckle, *J. Opt. Soc. Amer.*, 66, 1145–1150, 1976.

29. ISO 15887-2000: Space data and information transfer systems — Data systems — Lossless data compression, ISO TC 20/SC 13/ICS 49.140, 12-10-2000.

30. N.S. Jayant and P. Noll, *Digital Coding of Waveforms: Principles and Applications to Speech and Video*, Prentice-Hall, Englewood Cliffs, NJ, 1984.

31. E. Magli and G. Olmo, Lossy predictive coding of SAR raw data, *IEEE Trans. Geosci. Remote Sensing*, 41, 977–987, 2003.

32. E. Magli, G. Olmo, and E. Quacchio, Optimized onboard lossless and near-lossless compression of Hyperspectral data using CALIC, *IEEE Geosci. Remote Sensing Lett.*, 1, 21–25, 2004.

33. V. Pascazio and G. Schirinzi, SAR raw data compression by subband coding, *IEEE Trans. Geosci. Remote Sensing*, 41, 964–976, 2003.

34. S.D. Rane and G. Sapiro, Evaluation of JPEG-LS, the new lossless and controlled-lossy still image compression standard, for compression of high-resolution elevation data, *IEEE Trans. Geosci. Remote Sensing*, 39, 2298–2306, 2001.

35. K.K. Rao and J.J. Hwang, *Techniques and Standards for Image, Video, and Audio Coding*, Prentice-Hall, Engle Cliffs, NJ, 1996.

36. J. Reichel, G. Menegaz, M.J. Nadenau, and M. Kunt, Integer wavelet transform for embedded lossy to lossless image compression, *IEEE Trans. Image Processing*, 10, 383–392, 2001.

37. R.E. Roger and J.F. Arnold, Reversible image compression bounded by noise, *IEEE Trans. Geosci. Remote Sensing*, 32, 19–24, 1994.

38. R.E. Roger and M.C. Cavenor, Lossless compression of AVIRIS images, *IEEE Trans. Image Processing*, 5, 713–719, 1996.

39. A. Said and W.A. Pearlman, An image multiresolution representation for lossless and lossy compression, *IEEE Trans. Image Process.*, 5, 1303–1310, 1996.

40. D.S. Taubman and M.W. Marcellin, *JPEG2000: Image Compression Fundamentals, Standards and Practice*, Kluwer Academic Publishers, Dordrecht, The Netherlands, 2001.

41. V.D. Vaughn and T.S. Wilkinson, System considerations for multispectral image compression design, *IEEE Signal Process. Mag.*, 12, 19–31, 1995.

42. J. Wang, K. Zhang, and S. Tang, Spectral and spatial decorrelation of Landsat-TM data for lossless compression, *IEEE Trans. Geosci. Remote Sensing*, 33, 1277–1285, 1995.

43. M.J. Weinberger, J.J. Rissanen, and R.B. Arps, Applications of universal context modeling to lossless compression of gray-scale images, *IEEE Trans. Image Process.*, 5, 575–586, 1996.

44. M.J. Weinberger, G. Seroussi, and G. Sapiro, The LOCO-I lossless image compression algorithm: principles and standardization into JPEG-LS, *IEEE Trans. Image Process.*, 9, 1309–1324, 2000.

45. X. Wu and N. Memon, Context-based, adaptive, lossless image coding, *IEEE Trans. Commun.*, 45, 437–444, 1997.

46. X. Wu and P. Bao, L_∞ constrained high-fidelity image compression via adaptive context modeling, *IEEE Trans. Image Process.*, 9, 536–542, 2000.

47. X. Wu and N. Memon, Context-based lossless interband compression — Extending CALIC, *IEEE Trans. Image Process.*, 9, 994–1001, 2000.

48. Z. Zeng and I.G. Cumming, SAR image data compression using a tree-structure wavelet transform, *IEEE Trans. Geosci. Remote Sensing*, 39, 546–552, 2001.

16 Lossless Compression of VLSI Layout Image Data

Vito Dai and Avideh Zakhor

CONTENTS

16.1 Introduction 413
16.2 Overview of C4 414
16.3 Context-Based Prediction Model 415
16.4 Copy Regions and Segmentation 417
16.5 Hierarchical Combinatorial Coding 420
16.6 Extension to Gray Pixels 422
16.7 Compression Results 423
16.8 Summary 425
Acknowledgment 425
References 426

16.1 INTRODUCTION

For a next-generation 45-nm lithography system, using 25-nm, 5-bit gray pixels, a typical image of only one layer of a $2\,cm \times 1\,cm$ chip represents 1.6 terabits of data. A direct-write maskless lithography system with the same specifications requires data transfer rates of 10 terabits per second in order to meet the current industry production throughput of one wafer per layer per minute [7]. These enormous data sizes, and data transfer rates, motivate the application of lossless data compression to VLSI layout data.

VLSI designs produced by microchip designers consist of multiple layers of two-dimensional (2D) polygons stacked vertically, representing wires, transistors, etc. For pixel-based lithography writers, each layer is converted into a 2D image. Pixels may be binary or gray depending on the design of the writer. A sample of such an image is shown in Figure 16.1.

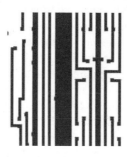

FIGURE 16.1 A sample of binary layout image data.

These lithography images differ from natural or even document images in several important ways. They are synthetically generated, highly structured, follow a rigid set of design rules, and contain highly repetitive regions cells of common structure.

Our previous experiments [4,5] have shown that Lempel–Ziv (LZ) style copying [13], used in ZIP, results in high compression ratios on dense, repetitive circuits, such as arrays of memory cells. However, where these repetitions do not exist, such as control logic circuits, LZ copying does not perform as well. In contrast, context-based prediction [10], used in JBIG [2], captures the local structure of lithography data, resulting in good compression ratios on nonrepetitive circuits, but fails to take advantage of repetitions, where they exist.

We have combined the advantages of LZ copying and JBIG context modeling into a new loss-less image compression technique called context copy combinatorial coding (C4). C4 is a single compression technique that performs well for all types of layout: repetitive, nonrepetitive, or a het-erogeneous mix of both. In addition, we have developed hierarchical combinatorial coding (HCC) as a low-complexity alternative entropy-coding technique to arithmetic coding [6] to be used within C4.

Section 16.2 describes the overall structure of C4. Section 16.3 describes the context-based prediction model used in C4. Section 16.4 describes LZ copying in two dimensions and how the C4 encoder segments the image into regions using LZ copying and context-based prediction. Section 16.5 describes HCC used to code prediction errors. Section 16.6 describes the extension of C4 to gray-pixel layout image data. Section 16.7 includes the compression results of C4 in comparison to other existing compression techniques for integrated circuit layout data.

16.2   OVERVIEW OF C4

The basic concept underlying C4 compression is to integrate the advantages of two disparate com-pression techniques: local context-based prediction and LZ-style copying. This is accomplished by automatic segmentation of the image into *copy regions* and *prediction regions*. Each pixel inside a copy region is copied from a pixel preceding it in raster-scan order. Each pixel inside a prediction region, i.e., not contained in any copy region, is predicted from its local context. However, neither predicted values nor copied values are 100% correct, so *error bits* are used to indicate the position of these prediction or copy errors. These error bits can be compressed using any binary entropy coder, but in C4, we apply our own HCC technique as a low-complexity alternative to arithmetic coding. Only the copy regions and compressed error bits are transmitted to the decoder.

In addition, for our application to direct-write maskless lithography, the C4 decoding algorithm must be implemented in hardware as a parallel array of thousands of C4 decoders fabricated on the same integrated-circuit chip as a massively parallel array of writers [4]. As such, the C4 decoder must have a low implementation complexity. In contrast, the C4 encoder is under no such complexity constraint. This basic asymmetry in the complexity requirement between encoding and decoding is central to the design of the C4 algorithm.

Figure 16.2 shows a high-level block diagram of the C4 encoder and decoder for binary layout images. First, a prediction error image is generated from the layout, using a simple 3-pixel context-based prediction model. Next, the resulting error image is used to determine the segmentation map between *copy regions* and the *prediction region*, i.e., the set of pixels not contained in any copy region. As specified by the segmentation map, the Predict/Copy block estimates each pixel value, either by copying or by prediction. The result is compared to the actual value in the layout image. Correctly predicted or copied pixels are indicated by a "0", and incorrectly predicted or copied pixels are indicated with a "1", equivalent to a Boolean XOR operation. These error bits are compressed without loss by the HCC encoder, which are transmitted to the decoder, along with the segmentation map.

The decoder mirrors the encoder, but skips the complex steps necessary to find the segmentation map, which are received from the encoder. Again as specified by the segmentation, the Predict/Copy block estimates each pixel value, either by copying or by prediction. The HCC decoder decompresses

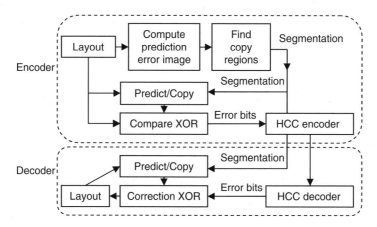

FIGURE 16.2 Block diagram of C4 encoder and decoder for binary images.

TABLE 16.1
The 3-Pixel Contexts, Prediction, and the Empirical Prediction Error Probability for a Sample Layout

Context	Prediction	Error	Error probability
			0.0055
			0.071
			0.039
			0
			0
			0.022
			0.037
			0.0031

the error bits from the encoder. If the error bit is "0" the prediction or copy is correct, and if the error bit is "1" the prediction or copy is incorrect and must be inverted, equivalent to a Boolean XOR operation. Since there is no data modeling performed in the C4 decoder, it is considerably simpler to implement than the encoder, satisfying one of the requirements of our application domain.

16.3   CONTEXT-BASED PREDICTION MODEL

For our application domain, i.e., integrated-circuit layout compression, we choose a simple 3-pixel binary context-based prediction model to use in C4, much simpler that the 10-pixel model used in JBIG. Nonetheless, it captures the essential "Manhattan" structure of layout data, as well as some design rules, as seen in Table 16.1.

The pixels used to predict the current-coded pixel are the ones above, left, and above-left of the current pixel. The first column shows the eight possible 3-pixel contexts, the second column the prediction, the third column what a prediction error represents, and the fourth column the empirical prediction error probability for an example layout. From these results, it is clear that the prediction mechanism works extremely well; visual inspection of the prediction error image reveals that prediction errors primarily occur at the corners in the layout. The two exceptional 0% error cases in rows 5 and 6 represent design rule violations. To generate the prediction error image, each correctly predicted pixel is marked with a "0", and each incorrectly predicted pixel is marked with a "1", creating a binary image which can be compressed with a standard binary entropy coder. The fewer the number of incorrect predictions, the higher the compression ratio achieved. An example of nonrepetitive layout for which prediction works well is shown in Figure 16.3a and its corresponding prediction error image is shown in Figure 16.3b.

In contrast to the nonrepetitive layout shown in Figure 16.3a, some layout image data contain regions that are visually "dense" and repetitive. An example of such a region is shown in Figure 16.4a. This visual "denseness" results in a dense, large number of prediction errors as seen clearly in the prediction error image in Figure 16.4b.

The high density of prediction errors translates into low compression ratios using prediction alone. In C4, areas of dense repetitive layout are covered by copy regions to reduce the number of errors, as described in Section 16.4.

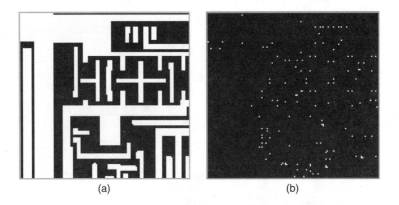

(a)   (b)

FIGURE 16.3 Nonrepetitive layout image data and its resulting prediction error image.

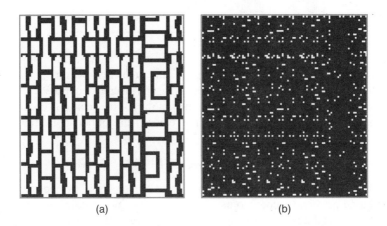

(a)   (b)

FIGURE 16.4 Dense repetitive layout image data and its resulting prediction error image.

16.4 COPY REGIONS AND SEGMENTATION

As seen in Figure 16.4a of the previous section, some layout images are highly repetitive. We can take advantage of this repetitiveness to achieve compression by specifying *copy regions*, i.e., a rectangular region that is copied from another rectangular region preceding it in raster-scan order. In the remainder of this section, we describe the procedure the C4 encoder uses to find these copy regions.

An example of a copy region is shown in the dashed rectangle in Figure 16.5. As seen, a copy region is specified with six copy parameters: position of the upper left corner x, y, width w, height h, distance to the left to copy from dx, and distance above to copy from dy. For the copy region in Figure 16.5, every pixel inside the region is copied from dx pixels to its left, and d$y = 0$. Although the entire region is copied, the copy itself need not be 100% correct. Similar to the prediction error map, there is a corresponding copy error map within the copy region. Each correctly copied pixel is indicated by a "0", and each incorrectly copied pixel is marked by a "1", creating a binary subimage, which can be compressed with a standard binary entropy coder.

As described in Section 16.2, the C4 encoder automatically segments the image into copy regions and the prediction region, i.e., all pixels not contained in any copy region. Each copy region has its own copy parameters and corresponding copy error map, and the background prediction region has a corresponding prediction error map. Together, the error maps merge to form a combined binary prediction/copy error map of the entire image, which is compressed using HCC as a binary entropy coder. The lower the number of the total sum of prediction and copy errors, the higher the compression ratio achieved. However, this improvement in compression by the introduction of copy regions is offset by the *cost* in bits to specify the copy parameters (x, y, w, h, dx, dy) of each copy region. Moreover, copy regions that overlap with each other are undesirable: each pixel should only be coded once, to save as many bits as possible.

Ideally, we would like the C4 encoder to find the set of nonoverlapping copy regions, which minimizes the sum of number of compressed prediction/copy error bits, and the number of bits necessary to specify the parameters of each copy region. An exhaustive search over this space would involve going over all possible nonoverlapping copy region sets, a combinatorial problem, generating the error bits for each set, and performing HCC compression on the error bits. This is clearly infeasible. To make the problem tractable, a number of simplifying assumptions and approximate metrics are adopted.

First we use entropy as a heuristic to estimate the number of bits generated by the HCC encoder to represent error pixels. If p denotes the percentage of prediction/copy error pixels over the entire image, then error pixels are assigned a per-pixel cost of $C = -\log_2(p)$ bits, and correctly predicted or copied pixels are assigned a per-pixel cost of $-\log_2(1 - p) \approx 0$. Of course, given a segmentation map, p can be easily calculated by counting the number of prediction/copy error bits; at the same

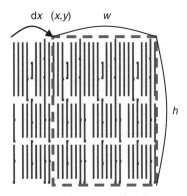

FIGURE 16.5 Illustration of a copy left region.

time, p affects how copy regions are generated in the first place, as discussed shortly. In C4, we solve this chicken and egg problem by first guessing a value of p, finding a segmentation map using this value, counting the percentage of prediction/copy error pixels, and using this percentage as a new value for p as input to the segmentation algorithm. This process can be iterated until the guess p matches the percentage of error pixels, but in practice we find that one iteration is sufficient if the starting guess is reasonable. Empirically, we have found a good starting guess to be the percentage of error pixels when no copy regions are used, then discounted by a constant factor, e.g., a factor of 4.

Next, for any given copy region, we compare the cost, in bits, of coding that region using copy vs. the cost of coding the region using prediction. If the cost of copying is lower, then the amount by which it is lower is the *benefit* of using this region. The cost of copying is defined as the sum of the cost of describing the copy parameters, and the cost of coding the copy error map. For our particular application domain, the description cost is 51 bits. Here we have restricted x, y, w, h to 10-bits each, which is reasonable for our 1024×1024 test images. In addition, we assume that copies are either from above, or to the left, so (dx, dy) is replaced by $(left/above, d)$ and represented with 11 bits, where d, represented by 10 bits, denotes the distance left or above to copy from, and $left/above$, represented by 1 bit, denotes the direction left or above to copy from. This assumption is in line with the Manhattan structure of layout data. The cost of coding the copy error map is estimated as CE_{copy}, where C denotes the estimated per-pixel cost of an error pixel, as discussed previously, and E_{copy} denotes the number of copy error pixels in the region. Correctly copied pixels are assumed to have 0 cost, as discussed previously. So the total cost of copying is $51 + CE_{copy}$.

The cost of coding the region using prediction is the cost of coding the prediction error map of that region. It is estimated as $CE_{context}$, where $E_{context}$ denotes the number of prediction error pixels in the region. Finally, the *benefit* of a region is the difference between these two costs, $C(E_{context} - E_{copy}) - 51$. Note that it is possible for a region to have negative benefit if $E_{context} - E_{copy} \leq (51/C)$. The threshold $T = (51/C)$ is used to quickly disqualify potential copy regions in the search algorithm presented below.

Using *benefit* as a metric, the optimization goal is to find the set of nonoverlapping copy regions, which maximizes the sum of *benefit* over all regions. This search space is combinatorial in size, so exhaustive search is prohibitively complex. Instead we adopt a greedy approach, similar to that used in the 2D-LZ algorithm described in [4]. The basic strategy used by the *find copy regions* algorithm in Figure 16.2 is as follows: start with an empty list of copy regions; and in raster-scan order, add copy regions of maximum benefit, which do not overlap with regions previously added to the list. The completed list of copy regions is the *segmentation* of the layout. A detailed flow diagram of the *find copy regions* algorithm is shown in Figure 16.6, and described in the remainder of this section.

In raster-scan order, we iterate through all possible (x, y). If (x, y) is inside any region in the *segmentation* list, we move on to the next (x, y); otherwise, we iterate through all possible $(left/above, d)$. Next for a given $(x, y, left/above, d)$, we maximize the size of the copy region (w, h) with the constraint that a *stop pixel* is not encountered; we define a *stop pixel* to be any pixel inside a region in the *segmentation* list, or any pixel with a copy error. These conditions prevent overlap of copy regions and prevent the occurrence of copy errors, respectively. Later, we describe how to relax this latter condition to allow for copy errors. The process of finding maximum size copy regions (w, h) is discussed in the next paragraph. Finally, we compute the benefit of all the maximum sized copy regions, and, if any region with positive benefit exists, we add the one with the highest positive benefit to the *segmentation* list.

We now describe the process of finding the maximum size copy region (w, h). For any given $(x, y, left/above, d)$, there is actually a set of maximum size copy regions bordered by stop pixels, because (w, h) is a 2D quantity. This is illustrated in the example in Figure 16.7. In the figure, the position of the stop pixels are marked with $\otimes$ and three overlapping maximum copy regions are shown as (x, y, w_1, h_1), (x, y, w_2, h_2), and (x, y, w_3, h_3). The values w_1, h_1, w_2, h_2, w_3, and h_3 are found using the following procedure: initialize $w = 1$, $h = 1$. Increment w until a stop pixel is encountered; at

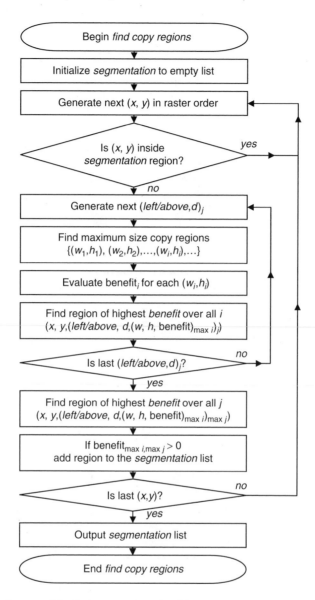

FIGURE 16.6 Flow diagram of the *find copy regions* algorithm.

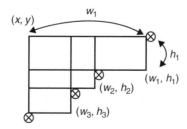

FIGURE 16.7 Illustration of three maximum copy regions bordered by four stop pixels.

this point $w = w_1$. Next increment h, and for each h increment w from 1 to w_1, until a stop pixel is encountered; at this point $h = h_1$ and $w = w_2$. Again increment h, and for each h increment w from 1 to w_2, until a stop pixel is encountered; at this point $h = h_2$ and $w = w_3$. Finally, increment h, and for each h increment w from 1 to w_3, until a stop pixel is encountered; at this point $h = h_3$ and $w = 1$. The maximum size algorithm is terminated when a stop pixel is encountered at $w = 1$.

As stated previously, any pixel inside a region in the *segmentation* list, and any pixel with a copy error, is a *stop pixel*. The latter condition prevents any copy errors inside a copy region. We relax this condition to merge smaller, error-free, copy regions into larger copy regions with a few number of copy errors. The basic premise is to tradeoff the 51 bits necessary to describe a new copy region against the introduction of bits needed to code copy errors, by excluding some copy error pixels from being *stop pixels*. For each copy error pixel, we look at a window of W pixels in a row, where the leftmost pixel is the copy error. If, in that window, the number of copy errors is less than the average number of errors expected, $E_{copy} < Wp$, and the number of copy errors is less than the number of prediction errors, $E_{copy} < E_{predict}$, then the pixel with the copy error is no longer considered to be a stop pixel. The size of the look-ahead window W is a user-defined input parameter to the C4 algorithm. Larger values of W correspond to fewer, larger copy regions, at the expense of increasing the number of copy errors.

16.5 HIERARCHICAL COMBINATORIAL CODING

We have proposed and developed combinatorial coding (CC) [6] as an alternative to arithmetic coding to encode the error bits in Figure 16.2. The basis for CC is universal enumerative coding [3], which works as follows. For any binary sequence of known length N, let k denote the number of ones in that sequence. k ranges from 0 to N, and can be encoded using a minimal binary code [12], i.e., a simple Huffman code for uniform distributions, using $\lceil \log_2(N + 1) \rceil$ bits. There are exactly $C(N, k) = N!/(N - k)!k!$ sequences of length N with k ones, which can be hypothetically listed. The index of our sequence in this list, known as the *ordinal* or *rank*, is an integer ranging from 1 to $C(N, k)$, which can again be encoded using a minimal binary code, using $\lceil \log_2 C(N, k) \rceil$ bits. Enumerative coding is theoretically shown to be optimal [3] if the bits to be compressed are independently and identically distributed (i.i.d.) as Bernoulli(θ), where θ denotes the unknown probability that "1" occurs, which in C4 corresponds to the percentage of error pixels in the prediction/copy error map. The drawback of computing an enumerative code directly is its complexity: the algorithm to find the rank corresponding to a particular binary sequence of length N, called *ranking* in the literature, is $O(N)$ in time, is $O(N^3)$ in memory, and requires $O(N)$ bit precision arithmetic [3].

In CC, this problem is addressed by first dividing the bit sequence into blocks of fixed size M. For today's 32-bit architecture computers, $M = 32$ is a convenient and efficient choice. Enumerative coding is then applied separately to each block, generating a (k, rank) pair for each block. Again, using the same assumption that input bits are i.i.d. as Bernoulli(θ), the number of k ones in a block of M bits are i.i.d. as Binomial(M, θ). Even though the parameter θ is unknown, as long as the Binomial distribution is not too skewed, e.g., $0.01 < \theta < 0.99$, a dynamic Huffman code efficiently compresses the k values with little overhead, because the range of k is small. Given there are k ones in a block of M bits, the rank remains uniformly distributed, as in enumerative coding. Therefore, *rank* values are efficiently coded using a minimum binary code.

The efficiency of CC, as described, is on par with arithmetic coding, except in cases of extremely skewed distributions, e.g., $\theta < 0.01$. In these cases, the probability that $k = 0$ approaches 1 for each block, causing the Huffman code to be inefficient. To address this issue, we have developed an extension to CC, called hierarchical combinatorial coding (HCC). It works by binarizing sequence of k values such that $k = 0$ is indicated with a "0" and $k = 1$ to 32 is indicated with a "1". CC is then applied to the binarized sequence of "0" and "1", and the value of k, ranging from 1 to 32 in the "1" case, is Huffman-coded. Clearly, this procedure of CC encoding, binarizing the k values, then CC

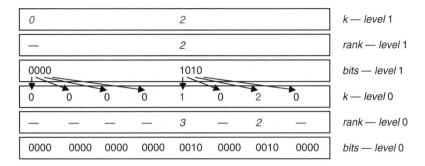

0				2				k — *level* 1

—				2				*rank* — *level* 1

0000				1010				*bits* — *level* 1

0	0	0	0	1	0	2	0	k — *level* 0

—	—	—	—	3	—	2	—	*rank* — *level* 0

0000	0000	0000	0000	0010	0000	0010	0000	*bits* — *level* 0

FIGURE 16.8 Two-level HCC with a block size $M = 4$ for each level.

encoding again can be recursively applied in a hierarchical fashion to take care of any inefficiencies in the Huffman code for k values as θ approaches 0.

Figure 16.8 is an example of HCC in action with two levels of hierarchy and block size $M = 4$. Only values in bold italics are coded and transmitted to the decoder. Looking at rows from bottom to top, the original data are in the lowest row labeled "bits — level 0". Applying CC with $M = 4$, the next two rows show the *rank* and k value for each block in level 0. Note that when $k = 0$ no *rank* value is needed as indicated by the hyphen. The high frequency of 0 in "k — level 0" makes it inefficient for coding directly using Huffman coding. Instead, we binarize "k — level 0", to form "bits — level 1", using the binarization procedure described in the previous paragraph. CC is recursively applied to "bits — level 1", to compute "*rank* — level 1" and "k — level 1". Finally, to code the data, "k — level 1" is coded using a Huffman code, "*rank* — level 1" is coded using a minimal binary code, *nonzero* values of "k — level 0" are coded using a Huffman code, and "*rank* — level 0" is coded using a minimal binary code.

The rationale for choosing Huffman coding and minimal binary coding is the same as CC. If the input is assumed to be i.i.d. as Bernoulli(θ), all level *rank* values are uniformly distributed, given the corresponding k-values in the same level. Furthermore, although the exact distribution of k values is unknown, a dynamic Huffman code can adapt to the distribution with little overhead, because the dynamic range of k is small. Finally, for highly skewed distributions of k, which hurts the compression efficiency of Huffman coding, the binarization process reduces the skew by removing the most probable symbol $k = 0$.

Studying the example in Figure 16.8, we can intuitively understand the efficiency of HCC: the single Huffman coded *0* in "k — level 1" decodes to M^2 zeros in "bits — level 0". In general, for L-level HCC, a single Huffman coded *0* in level $L - 1$ corresponds to M^L zeros in "bits — level 0". HCC's ability to effectively compress blocks of zeros is critical to achieving high compression ratios, when the percentage of the error pixels is low.

In addition to achieving efficient compression, HCC also has several properties favorable to our application domain. First, the decoder is extremely simple to implement: the Huffman code tables are small because the range of k values is small, unranking is accomplished with a simple table lookup, comparator, and adder, and minimal binary decoding is also accomplished by a simple table lookup and an adder. Second, the decoder is fast: blocks of $M^{(L+1)}$ zeros can be decoded instantly when a zero is encountered at level L. Third, HCC is easily parallelizable: block sizes are fixed and block boundaries are independent of the data, so the compressed bitstream can be easily partitioned and distributed to multiple parallel HCC decoders. This is in contrast to run-length coding schemes such as Golomb codes [8], which also code for runs of zeros, but have data-dependent block boundaries.

Independent of our development of HCC, a similar technique called hierarchical enumerative coding (HEC) has been developed in [9]. The main difference between HEC and HCC is the method of coding k values at each level. HCC uses binarization and simple Huffman coding, whereas HEC uses hierarchical integer enumerative coding, which is more complex [9]. In addition, HEC requires

TABLE 16.2
Result of 3-Pixel Context-Based Binary Image Compression on a 242 kb Layout Image for a P3 800 MHz Processor

Metric	Huf8	Arithmetic	Golomb	HEC	HCC
Compression ratio	7.1	47	49	48	49
Encoding time (sec)	0.99	7.46	0.52	2.43	0.54
Decoding time (sec)	0.75	10.19	0.60	2.11	0.56

more levels of hierarchy to achieve the same level of compression efficiency as HCC. Consequently, HCC is significantly less complex to compute than HEC.

To compare HCC with existing entropy coding techniques, we apply 3-pixel context-based prediction as described in Section 16.3 to a 242 kb layout image and generate eight binary streams. We then apply Huffman coding to blocks of 8-bits, arithmetic coding, Golomb run-length coding, HEC, and HCC to each binary stream, and report the compression ratio obtained by each algorithm. In addition, we report the encoding and decoding times as a measure for complexity of these algorithms. The results are shown in Table 16.2.

Among these techniques, HCC is one of the most efficient in terms of compression, and one of the fastest to encode and decode, justifying its use in C4. The only algorithm comparable in both efficiency and speed, among those tested, is Golomb run-length coding. However, as mentioned previously, HCC has fixed, data-independent block boundaries, which are advantageous for parallel hardware implementations, run-length coding does not. Run times are reported for 100 iterations on an 800 MHz Pentium III workstation. All algorithms are written in C# and optimized with the assistance of VTune to eliminate bottlenecks. The arithmetic coding algorithm is based on that described in [12].

16.6 EXTENSION TO GRAY PIXELS

So far, C4 as described is a binary image compression technique. To extend C4 to encode 5-bit gray-pixel layout image, slight modifications need to be made to the prediction mechanism, and the representation of the error. Specifically, the local 3-pixel context-based prediction, described in Section 16.3, is replaced by 3-pixel linear prediction with saturation, to be described later; furthermore, in places of prediction or copy error, where the error bit is "1", an *error value* indicates the correct value of that pixel. A block diagram of the C4 encoder and decoder for gray-pixel images is shown in Figure 16.9.

First, a prediction error image is generated from the layout, using a simple 3-pixel linear prediction model. The error image is a binary image, where "0" denotes a correctly predicted gray-pixel value and "1" denotes a prediction error. The copy regions are found as before in binary C4, with no change in the algorithm. As specified by the copy regions, the Predict/Copy generates pixel values either using copying or linear prediction. The result is compared to the actual value in the layout image. Correctly predicted or copied pixels are indicated by a "0", and incorrectly predicted or copied pixels are indicated by a "1" with an error value generated indicating the true value of the pixel. The error bits are compressed with a HCC encoder, and the actual error values are compressed with a Huffman encoder.

As in binary C4, the gray-pixel C4 decoder mirrors the encoder, but skips the complex steps necessary to find the copy regions. The Predict/Copy block generates pixel values either using copying or linear prediction according to the copy regions. The HCC decoder decodes the error bits, and the Huffman decoder decodes the error values. If the error bit is "0" the prediction or copy is correct, and if the error bit is "1" the prediction or copy is incorrect and the actual pixel value is the error value.

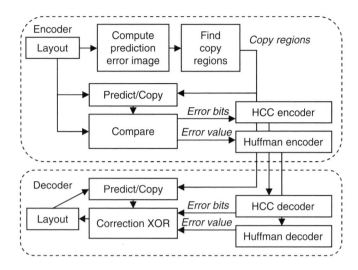

FIGURE 16.9 Block diagram of C4 encoder and decoder for gray-pixel images.

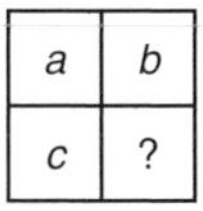

		$x = b - a + c$
a	b	if $(x < 0)$ then $? = 0$
c	$?$	if $(x > \text{max})$ then $? = \text{max}$
		otherwise $? = x$

FIGURE 16.10 Three-pixel linear prediction with saturation used in gray-pixel C4.

The linear prediction mechanism used in gray-pixel C4 is analogous to the context-based prediction used in binary C4. Each pixel is predicted from its 3-pixel neighborhood as shown in Figure 16.10. "?" is predicted as a linear combination of its local 3-pixel neighborhood "a," "b," and "c." If the prediction value is negative or exceeds the maximum allowed pixel value max, then the result is clipped to 0 or *max*, respectively. Interestingly, this linear predictor can also be applied to a binary image by setting *max* = 1, resulting in the same predicted values as binary context-based prediction described in Section 16.3. It is also similar to the median predictor used in JPEG-LS [11].

16.7 COMPRESSION RESULTS

We apply a suite of existing and general lossless compression techniques as well as C4 to binary layout image data. Compression results are listed in Table 16.3. The original data are 2048 × 2048 binary images with 300 nm pixels sampled from an industry microprocessor layout, which corresponds to a 0.61 mm × 0.61 mm section, covering about 0.1% of the chip area. Each entry in the table corresponds to the compression ratio for one such image.

The first column "type" indicates where the sample comes from: memory, control, or a mixture of the two. Memory circuits are typically extremely dense but highly repetitive. In contrast, control circuits are highly irregular, but typically much less dense. The second column "layer" indicates which layer of the chip the image comes from. Poly and Metal1 layers are typically the densest, and mostly correspond to wire routing and formation of transistors. The remaining columns from left to right are compression ratios achieved by JBIG, ZIP, 2D-LZ our 2D extension to the LZ77 copying [4], BZIP2 based on the Burrows–Wheeler transform [1], and C4. The bold numbers indicate the highest compression for each row.

TABLE 16.3

Compression Ratios of JBIG, ZIP, 2D-LZ, BZIP2, and C4 for 2048 × 2048 Binary Layout Image Data

Type	Layer	JBIG	ZIP	2D-LZ	BZIP2	C4
Memory	M2	59	88	233	260	**332**
cells	M1	10	48	79	56	**90**
	Poly	12	51	120	83	**141**
Control	M2	47	22	26	32	**50**
logic	M1	20	11	11	11	**22**
	Poly	42	19	20	23	**45**

TABLE 16.4

Compression Ratio of Run Length, Huffman, LZ77, ZIP, BZIP2, and C4 for 5-Bit Gray Layout Image Data

Layer	RLE	Huf	LZ77 256	LZ77 1024	ZIP	BZIP2	C4
M2	1.4	2.3	4.4	21	25	28	**35**
M1	1.0	1.7	2.9	5.0	7.8	11	**15**
Poly	1.1	1.6	3.3	4.6	6.6	10	**14**
Via	5.0	3.7	10	12	15	24	**32**
N	6.7	3.2	13	28	32	42	**52**
P	5.7	3.3	16	45	52	72	**80**

As seen, C4 outperforms all these algorithms for all types of layouts. This is significant, because most layouts contain a heterogeneous mix of memory and control circuits. ZIP, 2D-LZ, and BZIP2 take advantage of repetitions resulting in high compression ratios on memory cells. In contrast, where the layout becomes less regular, the context modeling of JBIG has an advantage over ZIP, 2D-LZ, and BZIP2.

Table 16.4 gives compression results for more modern layout image data with 65 nm pixels and 5-bit gray layout image data. For each layer, five blocks of 1024 × 1024 pixels are sampled from two different layouts, three from the first, and two from the second, and the *minimum* compression ratio achieved for each algorithm over all five samples is reported. The reason for using minimum rather than the average has to do with limited buffering in the actual hardware implementation of maskless lithography writers. Specifically, the compression ratio must be consistent across all portions of the layout as much as possible. From left to right, compression ratios are reported in columns for a simple run-length encoder, Huffman encoder, LZ77 with a history buffer length of 256, LZ77 with a history buffer length of 1024, ZIP, BZIP2, and C4. Clearly, C4 still has the highest compression ratio among all these techniques. Some notable lossless gray-pixel image compression techniques have been excluded from this table including SPIHT and JPEG-LS. Our previous experiments [5] have already shown that they do not perform well as simple ZIP compression on this class of data.

In Table 16.5, we show results for ten sample images from the data set used to obtain Table 16.4, where each row is information on one sample image. In the first column "type", we visually categorize each sample as repetitive, nonrepetitive, or containing a mix of repetitive and nonrepetitive regions. The second column is the chip layer from which the sample is drawn. The third column "LP" is the compression ratio achieved by linear prediction alone, equivalent to C4 compression with copy regions disabled. The fourth and fifth columns are the compression ratio achieved by ZIP and the

TABLE 16.5
Percent of Each Image Covered by Copy Regions (Copy%), and Its Relation to Compression Ratios for Linear Prediction (LP), ZIP, and C4 for 5-Bit Gray Layout Image Data

Type	Layer	LP	ZIP	C4	Copy%
Repetitive	M1	3.3	7.8	18	94
	Poly	2.1	6.6	18	99
Nonrepetitive	M1	14	12	16	18
	Poly	7.3	9.6	14	42
Mixed	M1	7.5	12	15	44
	Poly	4.1	10	14	62
	M2	15	26	35	33
	N	18	32	52	21
	P	29	52	80	33
	Via	7.1	15	32	54

full C4 compression, respectively. The last column "Copy%" is the percent of the total sample image area covered by copy regions, when C4 compression is applied. Any pixel of the image not covered by copy regions is, by default, linearly predicted from its neighbors.

Clearly, the Copy% varies dramatically from image to image ranging from 18% to 99% across the 10 samples, testifying to C4's ability to adapt to different types of layouts. In general, a high Copy% corresponds to repetitive layout, and low Copy% corresponds to nonrepetitive layout. Also, the higher the Copy%, the more favorably ZIP compares to LP compression. This agrees with the intuition that LZ-style techniques work well for repetitive layout, and prediction techniques work well for nonrepetitive layout. At one extreme, in the nonrepetitive-M1 row, where 18% of the image is copied in C4, LP's compression ratio exceeds ZIP. At the other extreme, in the repetitive-Poly row, where 99% of the image is copied, ZIP's compression ratio is more than three times that of LP. This trend breaks down when the compression becomes high for both LP and ZIP, e.g., the rows labeled Mixed-N and Mixed-P. These layouts contain large featureless areas, which are easily compressible by both copying and prediction. In these cases, C4 favors using prediction to avoid the overhead of specifying copy parameters.

16.8 SUMMARY

C4 is a novel compression algorithm, which successfully integrates the advantages of two very disparate compression techniques: context-based modeling and LZ-style copying. This is particularly important in the context of layout image data compression, which contains a heterogeneous mix of data: dense repetitive data, better suited to LZ-style coding, and less dense structured data, better suited to context-based encoding. In addition, C4 utilizes a novel binary entropy-coding technique called combinatorial coding, which is simultaneously as efficient as arithmetic coding and as fast as Huffman coding. Compression results show that C4 achieves superior compression results over JBIG, ZIP, BZIP2, and 2D-LZ for a wide variety of industry lithography image data.

ACKNOWLEDGMENT

This research is conducted under the Research Network for Advanced Lithography, supported jointly by SRC (01-MC-460) and DARPA (MDA972-01-1-0021).

REFERENCES

1. M. Burrows and D.J. Wheeler, A Block-Sorting Lossless Data Compression Algorithm, Technical report 124, Digital Equipment Corporation, Palo Alto CA, 1994.

2. CCITT, ITU-T Rec. T.82 & ISO/IEC 11544:1993, Information Technology — Coded Representation of Picture and Audio Information — Progressive Bi-Level Image Comp., 1993.

3. T.M. Cover, Enumerative source coding, *IEEE Trans. Inf. Theory*, IT-19, 73–77, 1973.

4. V. Dai and A. Zakhor, Lossless layout compression for maskless lithography systems, *Emerging Lithographic Technologies IV, Proceedings of the SPIE*, Vol. 3997, 2000, pp. 467–477.

5. V. Dai and A. Zakhor, Lossless compression techniques for maskless lithography data, *Emerging Lithographic Technologies VI, Proceedings of the SPIE*, Vol. 4688, 2002, pp. 583–594.

6. V. Dai and A. Zakhor, Binary combinatorial coding, *Proceedings of the Data Compression Conference*, Vol. 2003, 2003, p. 420.

7. V. Dai and A. Zakhor, Advanced low-complexity compression for maskless lithography data, *Emerging Lithographic Technologies VIII, Proceedings of the SPIE*, Vol. 5374, 2004, pp. 610–618.

8. S.W. Golomb, Run-length encodings, *IEEE Trans. Inf. Theory*, IT-12, 399–401, 1966.

9. L. Oktem and J. Astola, Hierarchical enumerative coding of locally stationary binary data, *Electron. Lett.*, 35, 1428–1429, 1999.

10. J. Rissanen and G.G. Langdon, Universal modeling and coding, *IEEE Trans. Inf. Theory*, IT-27, 12–23, 1981.

11. M.J. Weinberger, G. Seroussi, and G. Sapiro, The LOCO-I lossless image compression algorithm: principles and standardization into JPEG-LS, *IEEE Trans. Image Process.*, 9, 1309–1324, 2000.

12. I.H. Witten, A. Moffat, and T.C. Bell, *Managing Gigabytes*, 2nd ed., Academic Press, New York, 1999.

13. J. Ziv and A. Lempel, A universal algorithm for sequential data compression, *IEEE Trans. Inf. Theory*, IT-23, 337–343, 1977.

INDEX

2-D systems, in multidimensional medical data, 354–360
  DCT-based techniques, 356–357
  lossless techniques, 355–356
  lossy techniques, 356
  ROI-based coding, 359–360
  wavelet-based techniques, 357–360
3-D encoding/2-D decoding object-based architecture, 364–365
3-D object-based coding, 372–374
  EZW technique, 373
  MLZC coding, 374
3-D object-based MLZC, 379–384
  object-based performance, 381–384
  system characterization, 379–381
3-D systems, 360–362
  3-D DCT coding scheme, 362
  3-D quadtree limited, 361
  3-D set partitioning hierarchical trees, 360–361
  CS-embedded block coding, 361–362
  cube splitting algorithm, 361

Algorithm context, 131
Aliasing component, in image compression, 9
Alphabet of X, in image compression, 36
ANGIO (opthalmologic angiographic sequence), 375
Anti-collage theorem, 151–152
Application domains, of lossless image coding, 134–140
  band ordering, 135
  color and multiband, 134–137
  color-indexed images and graphics, 139–140
  error modeling and coding, 136
  hyperspectral images, 136–137
  interband prediction, 135–136
  predictive techniques, 134–135
  reversible transform-based techniques, 137
  video sequences, 137–139
Arithmetic coding and context modeling, in image compression, 40–52
  adaptive probability estimation, 48–49
  arithmetic coding variance, 50–51
  binary arithmetic coding, 50
  conditional coding and context modeling, 47–50
  Elias coding, 42–43
  from coding to intervals on (0, 1), 40–42
  in JBIG, 51–52
  practical arithmetic coding, 43–47
Auto-regressive (AR) model, 29
AVIRIS (airborne visible infrared imaging spectrometer), 401–406
AWGN (additive white Gaussian noise) attacks, 262

Bandelets, 210
Basis pursuit (BP) algorithm, 211
Binary image compression, 285–296
  binary images, 285–286, *see also* Binary images
  context weighing applied to, 292–296, *see also* Context weighing
  group 3 and 4 facsimile algorithms, 286–287
  JBIG and JBIG2, 287–291
  new compression algorithms, 295–296
Binary images, 285–286
  in digital printing, 285
  in facsimile communication, 285
Biorthogonal filter bank, 9
Bitmap-based shape coding, 301–302
Bit-plane coder, 55
Bit-stuffing technique, 51
Boundary filters, in image compression, 10
BPE (bit-plane encoding) technique, 355
Brushlets, 192
Burt's LP, modified version, *see* ELP

C4 (context copy combinatorial coding), 414
  overview, 414–415
CAE (context-based arithmetic encoder), 304–305
CALIC (context-based adaptive lossless image codec), 123–126, 355–356, 359
CELP (code excited linear prediction) voice compression method, 264
CHA (computational harmonic analysis), 208
Chroma-keying technique, 300
CMF (conjugate mirror filter) banks, 9
Codeblocks
  in EBCOT algorithm, 26
  in JPEG 2000 standard, 103
Codec optimization, 308–316
  bit allocation, 314
  buffering policy, 313–314
  composition and alpha blending, 315
  error concealment, 315–316
  error control, 314
  post-processing, 315
  preprocessing, 309–310
  rate control, 313–314
  rate–distortion models, 310–313
Coding passes, 104
Coding strategy
  definition, 244–246
  for sensor broadcast problem, 237
Coding task, in lossless image coding, 115
Collage theorem, 148
Color video coding, 71

Compressing compound documents, 323–349
  JPEG+MMR+JPEG, 336–344, *see also*
    JPEG+MMR+JPEG
  MRC for compression, 327–336
  raster imaging models, 324–327, *see also* Raster imaging
    models
Compression algorithms
  classical image compression scheme, 4
  nature, 3
Compression methods, types, 390
Conditional average probability, 293
Conditional coding primitives, 58
Conditional entropy, 39
Conditional probability, 39
Content-driven progressive transmission, 27–29
Context-based adaptive lossless image coding, 123–127
  coding context selection and quantization, 125–126
  context modeling of prediction errors and error feedback,
    126
  entropy coding of prediction errors, 126–127
  gradient-adjusted predictor, 124–125
'Context dilution', 96
Context-induced information sequencing, 57
Context modeling, in lossless image coding, 116–117
Context weighing
  binary image compression applied to, 292–296
  context tree weighting, 293
  description, 292–294
Continuous wavelet transform and frames, in image
  compression multiresolution analysis, 5
Contour-based shape coding, 301–303
Contraction, of fractal image model, 148
COVQ (channel optimized vector quantization), 264–266
CS-EBCOT (CS-embedded block coding), 361–362
Curvelet transform construction, 210
Curvelets constructions, 179, 187–191
CWT (continuous wavelet transform), 4, 82

DCT (discrete cosine transform), 4, 69, 158–159, 208,
  257–258
Deadzone quantizers, 55
Decoder complexity reduction, 162–164
  codebook update, 163–164
  fast decoding with orthogonalization, 162
  hierarchical decoding, 162–163
Decorrelator, in compression algorithms, 4
Delaunay triangulations, 156, 192
Descendants coefficients, 21
DH-EC (data-hiding-based error concealment) algorithm,
  273
DICOM (digital imaging and communications in medicine)
  standard, 353
Differential pulse code modulation, 393
Digital image compression, 208–209
Digital wedgelets, 193–196
  wedge domains, rapid summation, 194–195
'Directionlets', 192
Discrete green's theorem, 194–195
Distortion scalability, 54

Distributed compression, of sensor networks field
  snapshots, 235–251
  bandlimited images distributed decorrelation, transforms
    for, 243–248
  data compression structures, 238–239
  data compression using independent encoders, 239
  distributed computation of decorrelating transforms, 241
  distributed image coding and wireless sensor networks,
    236
  exploiting correlations in the source, 239
  images constrained by physical laws, 241–243
  information-theoretic bounds for bandlimited images,
    240–241
  linear signal expansions with bounded communication,
    244–248, *see also* Linear signal expansions with
    bounded communication
  of sensor measurements, 240–243
  physically constrained nonbandlimited images, 248–250,
    see also Physically constrained nonbandlimited images
  sensor broadcast problem, 236–238, *see also* Sensor
    broadcast problem
Divisive normalization model, 72
DjVu software/technology, 326–327, 335
Dominant list, in EZW algorithm, 21–22
Dominant pass, in EZW algorithm, 22
DPCM (differential pulse code modulation) technique, 4,
  355
'Drop-data' transform, 243–244
DSR (dynamic spatial reconstructor), 375
DWT (discrete wavelet transform), 4, 9, 12, 53, 65,
  82, 259

EBCOT (embedded block coding with optimal truncation)
  algorithm, 26–27
  embedded block coding primitives, 58–61
  fractional bit-plane scan, 61–62
  magnitude refinement coding, 61
  multiple quality layers, 63–64
  optimal truncation, 62–63
  overview, 58–64
  second tier coding strategy in, 64
  sign coding, 60
  significance coding, 59–60
Elias coding, 42–43
ELP (enhanced Laplacian pyramid), 16–20
  quantization, with noise feedback, 18–20
Embedded zero-tree wavelet coder, in image compression
  multiresolution analysis, 20–22
Encoder complexity reduction, 157–162
  classification by intensity and variance, 159–160
  classification schemes, 159–160
  clustering methods, in classification, 160
  fast search via fast convolution, 161–162
  feature vectors, 157–159
  fractal image compression without searching, 162
  Jacquin's approach, in classification, 159
  multiresolution approaches, in classification,
    160–161
  tree-structured methods, in classification, 160
Energy compaction, 54

Entropy coding, in image compression, 36–40
    information and entropy, 36
    joint and conditional entropy, 38–40
Entropy coding, in JPEG coding standards, 91–94
    arithmetic coding, 93–94
    Huffman coding, 91–93
Entropy coding, in lossless image coding, 117–121
    arithmetic coding, 119–121
    Huffman coding, 117–119
Entropy coding stage, in compression algorithms, 4
Entropy of X, in image compression, 36
Error modeling, in lossless image coding, 115
EZW (embedded zero-tree wavelet) algorithm, 4, 20–22,
    83, 357–358
    lists, 21–22
    passes, 22

Fat zero, 55
First-generation wavelets, 11
Fixed-length coding, 37
Fractal decoder, 148
Fractal encoder, 148
Fractal image compression, 145–171
    attractor coding, 164–165
    channel coding, in extensions, 167
    color images compression, in extensions, 168
    decoder complexity reduction, 162–164, *see also*
        Decoder complexity reduction
    encoder complexity reduction, 157–162, *see also*
        Encoder complexity reduction
    extensions, 167–168
    fractal image model, 148–154, *see also* Fractal image
        model
    hybrid methods, in extensions, 167
    image partitions, 154–157, *see also* Image partitions
    postprocessing, 167–168
    progressive coding, in extensions, 167
    rate-distortion coding, 166–167
    state of the art fractal image compression, 168–170
    video coding, in extensions, 168
Fractal image model, 148–154
    variants, 153–154
Fractal transform, 148
Frequency sensitivity, in perception, 69

GAP (gradient-adjusted predictor), 123–125
Golomb–Rice codes, 98
G-PROG (global progressive), 371
Gradient-adjusted predictor, in Lossless image coding,
    124–125
'Graduate student algorithm', 147
Group 3 and 4 facsimile algorithms, 286–287

HEC (hierarchical enumerative coding), 421
Hierarchical Interpolation technique, 355
'Holes in the code-space', 42
Horizontal–vertical (HV) partitioning, 155
HS (half-sample symmetric) extension, in image
    compression, 10
Huffman coding, 91–93

HVS (human visual system), and image coding, 70–72
    bleaching process, 70
    cones, 71
    rods, 71

IDWT (inverse discrete wavelet transform), 260
IFS (iterated function systems), 145–146
Image and video coding, data hiding for, 225–278
    data hiding for error concealment, 268–277
    data in image, 257–263
    data in video, 264–267
    error concealment, for resynchronization, 269–270
    image and video compression, data hiding, 256–267
    MV values recovery, error concealment for, 270–271
    pixel values recovery, error concealment for, 271–275,
        see also Pixel values recovery
Image coding, perceptual aspects, 69–84
    human visual system, 70–72, *see also* HVS
    perceptual distortion metrics, 76–78
    physiological models, 72–76
    vision signals treatment, 71
Image coding using redundant dictionaries, 207–231
    current methods, limitations, 209
    digital image compression, 208–209
    greedy algorithms, 212–214
    highly nonlinear approximations, 210–212
    nonlinear algorithms, 210–214
    reduntant expansions, 209–230, *see also* Redundant
        expansions
Image compression multiresolution analysis, 3–32
    and filter banks, 6–7
    coding schemes, 20–32
    content-driven ELP coder, 27–29
    continuous wavelet transform and frames, 5
    embedded block coding with optimized truncation, 26–27
    embedded zero-tree wavelet coder, 20–22
    lifting scheme, 11–14
    multiresolution spaces, 5–6
    orthogonal and biorthogonal filter banks, 8–9
    reconstruction at boundaries, 9
    synchronization tree, 29
    wavelet analysis and filter banks, 4–16
    wavelet decomposition of images, 14–15
Image compression, advanced modeling and coding
    techniques for, 35–66
    arithmetic coding and context modeling, 40–52, *see also*
        Arithmetic coding and context modeling
    EBCOT, overview, 58–64, *see also* EBCOT
    entropy and coding in, 36–40
    fixed- and variable-length codes, 37–38
    information sequencing and embedding, 52–58, *see also*
        Information sequencing and embedding
Image compression, application, 199–204
    coding schemes, 202–204
    experimental approximation properties, 199–202
Image multiresolution analysis, 14–15
Image partitions, in fractal image compression, 154–157
    hierarchical partitions, 155–156
    quadtrees, 155
    split–merge partitions, 156–157

Implicit shape coding, 301–302
Information sequencing and embedding, in image
     compression, 52–58
  coding vs. ordering, 57–58
  embedded quantization and bit-plane coding, 54–56
  fractional bit-plane coding, 56–57
  multiresolution compression with wavelets, 53–54
  quantization, 52–53
Integer-to-integer wavelets, 128
ISDN (integrated service digital networks), 352

Jacquin–Barnsley operator, 149–154, 162–166
JBIG (joint bi-level image group)
  arithmetic coding in, 51–52
  in binary image compression, 287–291
  JBIG2, 289–291
  probabilities in, 288
JND (just noticeable distortion) threshold, 69
  evaluation, 78–79
JPEG 2000 standard, 101–109
  advanced features, 106–108
  codestream syntax, progression orders and
     codestream generation, 105–106
  data organization, 103–104
  DC level shifting, 101–102
  entropy coding, 104–105
  error resilience, 107–108
  extensions, 108
  motion JPEG 2000, 108
  multicomponent transformation, 102
  quantization, 103
  region of interest coding, 106–107
  transform and quantization, 101–103
  wavelet transformation, 102–103
JPEG family of coding standards, 87–110
  advanced research related to, 109–110
  available software, 110
  DCT-based coding, 109
  history, 88
  JPEG 2000 standard, 101, *see also* JPEG 2000 standard
  JPEG standard, 89–96, *see also* JPEG standard
  JPEG-LS standard, 96–101, *see also* JPEG-LS standard
  wavelet-based coding and beyond, 109–110
JPEG standard, 89–96
  codestream syntax, 95–96
  entropy coding, 91–94
  Huffman coding, 91–93
  lossless mode, 94
  progressive and hierarchical encoding, 95
  quantization, 90–91
  transform, 89–90
JPEG+MMR+JPEG, 336–344
  computing rate and distortion per block,
     337–338
  fast thresholding, 340–341
  optimized thresholding as segmentation,
     338–339
  performance, 341–344
JPEG2000 lossless coding option, 128–129
JPEG2000 sign coding primitive, 60

JPEG-LS standard, 96–101
  codestream syntax, 100–101
  context-based prediction, 97–98
  entropy coding, 98–99
  Golomb coding, 98–99
  JPEG-LS part 2, 99–100
  near-lossless mode, 99
  run mode, 99

KLT (Karhunen–Loeve transform), 208

Learning penalty, 49
Lifting scheme, in wavelet transforms implementation,
     12–14
Linear predictive techniques, in lossless image coding,
     116
Linear signal expansions with bounded communication,
     244–248
  coding strategy, definition, 244–246
  communication within a coherence region, 246
  global communication, 246–247
  local and global communication requirements,
     differentiation, 246
  wavelets and sensor broadcast, 244
LIP (list of insignificant pixel), 24–26, 358
LIS (list of insignificant sets), 358
L-level wavelet transformation, 102–103
LNC (nonsignificant coefficients), 361
Local contrast sensitivity, in perception, 69
LOCO-I, 127
Lossless DPCM, *see* Linear predictive techniques
Lossless image coding, 113–140
  application domains, 134–140, *see also* Application
     domains
  context modeling, 116–117
  entropy coding, 117–121
  methods
  optimizations, 129–134, *see also* Optimizations
  prediction, 115–116
  principles, 114–121
Lossless image coding methods, 121–129
  and prediction, 123
  arithmetic coding procedures, 122–123
  context-based adaptive lossless image coding, 123–127,
     see also Context-based adaptive lossless image coding
  experimental results, 129
  gradient-adjusted predictor, 124–125
  Huffman coding procedures, 121–122
  JPEG lossless, 121–123
  JPEG-LS, 127–128
  reversible wavelets, 128–129
LPL (layer per layer) mode, 372
LPL-PROG (layer per layer progressive), 371–372
LSP (list of significant pixel), 24–26
LTI (linear time-invariant) filters, 13

Mallat's DWT, quantization, 15–16
  wavelet pyramid, 19
MAP (maximum a posteriori) estimation, 315
MAR (multiplicative autoregression) technique, 355

MAV (maximum absolute value), 28
MAXSHIFT method, 106–107
MCECQ (minimum conditional entropy context
    quantization), 131–132
McMillan condition, 37
MDI (multidirectional interpolation) method, 272
MED (median edge detection), 116
Minkowski metrics, 77
MLZC (multidimensional layered zero coding),
    368–375
    3-D object-based coding, 372–374, *see also* 3-D
        object-based coding
    3-D/2-D MLZC, 374
    3-D/2-D object-based MLZC, 374–375
    bitstream syntax, 371
    global progressive, 371
    interband conditioning, 371
    layer per layer mode, 372
    layer per layer progressive, 371–372
    layered zero coding approach, 368
    MLZC coding principle, 369
    spatial conditioning, 369–371
MMSE (minimum mean-squared error), 4
    MMSE decoder, 263
Modeling task, in lossless image coding, 115
Modified Huffman (MH) code, 286
MP (Matching pursuit) algorithm, 212–213
MPEG-4 shape coding tools, 303–308
    binary shape coding, 304–307
    boundary blocks texture coding, 307–308
    codec optimization, 308–316
    coding efficiency evaluation criteria, 307
    gray-scale shape coding, 307
    inter-mode, 306–307
    intra-mode, 305–306
    shape representation, 304
    video coder architecture, 308
MPS (more probable symbol), 120
MR (magnitude refinement) primitive, in EBCOT
    algorithm, 27
MR (modified READ (relative element address designate))
    algorithm, 286–287
MRC (mixed raster content) imaging model, 324–325
    for compression, 327–336
    JP2-based JPM, 347–348
    JPEG+MMR+JPEG, 336–344, *see also*
        JPEG+MMR+JPEG
    object segmentation versus region classification,
        329–330
    plane filling, 333–336
    redundant data and segmentation analysis, 330–333
    within JPEG 2000, 344–348
MRF (Markov random field) model, 311
MSB (most significant bits), in pixel values recovery, 273
MTF (modulation transfer function), 69
Multidimensional medical data, model-based coding trends,
    351–385
    2-D systems, 354–360, *see also* 2-D systems
    3-D analysis versus 2-D reconstruction, 368

3-D object-based MLZC, 379–384
3-D ROI-based coding, 362–363
3-D systems, 360–362, *see also* 3-D systems
3-D/2-D MLZC, 377–379
3-D/2-D object-based MLZC, 384–385
3-D/2-D ROI-based MLZC, 364–365
datasets, 375–377
MLZC, 368–375, *see also* MLZC
model-based coding, 363–364
object based processing, 365–368
requirements, 353–354
Multiresolution analysis, for image compression, 3–32, *see*
    also Image compression
Multiresolution spaces, in image compression
    multiresolution analysis, 5–6

NCD (normalized color distance), 261
Near-lossless compression, 390
New image representation paradigms, 179–205
    alternative approaches, 192–193
    curvelets, 187–191
    digital curvelets, 196–198
    digital wedgelets, 193–196, *see also* Digital wedgelets
    implementation, 195–196
    problems and solutions, 180–193
    wedgelets, 184–187
Noiseless source coding theorem, 36
Nonlinear approximation error, 180
Nonspecific suppression, 72
Normal mode, in significance coding, 60

OBASC (object based analysis-synthesis coding), 300
Optimal context quantization, 131–134
Optimizations, in lossless image coding, 129–134
    multiple prediction, 130–131
    optimal context quantization, 131–134
Orthogonal and biorthogonal filter banks, in image
    compression multiresolution analysis, 8–9

PACS (picture archiving and communication systems), 352
PCRD-opt (postcompression rate-distortion
    optimization), 62
Perception
    and image coding, 69–84, *see also* Image coding,
        perceptual aspects
    characterization mechanisms, 69
    effects, in DCT domain, 79–81
    perception metrics, in wavelet domain, 81–84
Perceptual coding, 256
Perceptual distortion metrics, 76–78
Perfect reconstruction (PR) system, in image
    compression, 9
Physically constrained nonbandlimited images, 248–250
    compression, 250
    sampling and interpolation, 249–250
    wave field models, 248–249
Pixel values recovery
    error concealment for, 271–275
    self-embedding methods, 275–277
Postcompression rate-distortion optimization, 109

Posteriori quantization, 218
Precincts, 104
PWC (piecewise-constant) codes, 139

QMF (quadrature mirror filter) banks, 9
Quad-tree hierarchy, 28
Quality progressive pack-stream, in EBCOT, 64
Quantizer, in compression algorithms, 4

RAPP (runs of adaptive pixelpatterns), 139
Raster imaging models, 324–327
    DjVu, 326–327
    mixed raster content, 324–325
    region classification, 325–326
    residual additive planes, 327
    soft masks for blending, 327
Reduntant expansions, 209–230
    anisotrophy, benefits, 220–222
    benefits, 209–210
    coding performance, 222–223
    experimental results, 220–223
    extensions to color images, 223–226
    high adaptivity, 227–230
    high adaptivity, importance, 227
    rate scalability, 229–230
    scalable image encoder, 214–220, *see also* Scalable
        image encoder
    spatial adaptivity, 227–229
Reed Solomon (RS) codes, 262
Relaxation-labeled prediction encoder, 408
Remote-sensing image coding, 389–409
    advanced compression algorithms for, 393–395
    context modeling, 394–395
    distortion measures, 391–393
    experimental results, 399–409
    hyperspectral data, 401–406
    multispectral data, 399–401
    near-lossless compression through 3D causal DPCM,
        395–396
    near-lossless image compression through noncausal
        DPCM, 396–398
    quality issues, 389–391
    radiometric distortion, 391–392
    SAR focused data, 406–407
    SAR raw data, 407–409
    spectral distortion, 392–393
Repeat-accumulate (RA) codes, 262
Resolution scalability, 26, 54
RLC (run-length coding) primitive, in EBCOT algorithm,
    27
ROB (region of background), 276
ROI (region of interest), in pixel values recovery, 276
ROPD (rank-order polynomial subband decomposition),
    358
Run mode, in significance coding, 60

Scalable image encoder, 214–220
    anisotrophy and orientation, 216–217
    coding stage, 218
    coefficient quantization, 218–219

dictionary functions, generation, 215–216
    dictionary, 217–218
    matching pursuit search, 215
    rate control, 220
Second-generation coding techniques, 363–364
Second-generation wavelets, 11
Sensor broadcast problem, 236–238
    problem formulation, 236–237
    problem relevance, 237–238
Sensor networks field snapshots, distributed compression,
    235–251, *see also* Distributed compression
SFS (space frequency segmentation), 357
Sign coding primitive, in EBCOT algorithm, 27
Significance coding, in EBCOT algorithm, 59–60
Significance propagation pass, 61
Skewed sources, 37
SNR scalability, 26
Spatial orientation tree, 23
SPECK (set partitioning embedded block coding), 359
SPIHT (set partitioning hierarchical trees) algorithm,
    22–26, 57, 357–359
SRN (shape refreshment need), 314
Subordinate list, in EZW algorithm, 21–22
Subordinate pass, in EZW algorithm, 22
Synchronization tree, in image compression, 29

TRN (texture refreshment need), 314
TRPP (template relative pixel patterns), 139
TSGD (two-sided geometric distribution) modeling, 127
Two-dimensional shape coding, 299–319
    applications, 316–318
    implicit shape coding, 301–302
    interactive TV, 318
    MPEG-4 shape coding, 303, *see also* MPEG-4 shape
        coding tools
    shape coding, 301

UCM (universal context modeling) scheme, 123

Variable-length code, 37
Virtually lossless compression, 390
Virtually lossless term, 407
Visually lossless mode, in multidimensional medical data,
    354
Visually weighted MSE, 62
VLSI layout image data, lossless compression,
    413–425
    and C4, *see also* C4 (context copy combinatorial coding)
    compression results, 423–425
    context-based prediction model, 415–416
    copy regions and segmentation, 417–420
    extension to gray pixels, 422–423
    hierarchical combinatorial coding, 420–422
    ranking in, 420
VOP (video object plane), 300

Wavelet analysis and filter banks, in image compression
    multiresolution analysis, 4–16
Wavelet footprints, 210
Wavelet orthonormal bases, shortcomings,
    179–180

Wavelet transforms implementation, via lifting, 12–14

Wavelets, new image representation paradigms, 179–205

Wedgelets constructions, 179, 184–187

Wedgeprints, 210

WS (whole-sample symmetric) extension, in image compression, 10

W-SFQ (wedgelet-space frequency quantization) compression, 203–204

WTCQ (wavelet/trellis-coded quantization), 358

YUV color space, 225

ZC (zero coding) primitive, in EBCOT algorithm, 27